Engineering Mechanics and Strength of Materials

Roger Kinsky

McGRAW-HILL BOOK COMPANY Sydney

New York St Louis San Francisco Auckland Bogotá
Caracas Hamburg Lisbon London Madrid Mexico Milan
Montreal New Delhi Oklahoma City Paris San Juan
São Paulo Singapore Tokyo Toronto

The McGraw·Hill Companies

First Published 1986
Reprinted 1989, 1990, 1992, 1993, 1994, 1995, 1998, 1999, 2003, 2004, 2006, 2007, 2008, 2009

National Library of Australia Cataloguing-in-Publication Data:

Kinsky, Roger.
Engineering mechanics and strength of materials.
Includes index.
ISBN 10: 0 07 452155 1
ISBN 13: 978 0 074521557

1. Mechanics, Applied. 2. Strength of materials.
I. Title

620.1

Published in Australia by
McGraw-Hill Australia Pty Limited
Level 33 World Square, 680 George Street Sydney, NSW 2000, Australia
Sponsoring Editor: Isabel Hogan
Production Editor: Daphne Rawling
Designer: George Sirett
Technical Illustrator: Colin Webster
Printed in Australia by SOS Print + Media

Contents

PART III STRENGTH OF MATERIALS

Preface

This book was written in response to the need for a new and modern textbook at technician level in the combined subject areas of applied mechanics and strength of materials. There are many good books available treating the subject matter at an introductory level, for example *Mechanical Engineering Science* by Val Ivanoff. Similarly there are many books at a very comprehensive and usually advanced level suitable for graduate engineering students. In between these two regions lay a void which needed filling and when McGraw-Hill threw out the challenge to me I succumbed once again despite the previous promise to myself "no more textbooks" after completion of *Applied Fluid Mechanics* and the second edition of *Applied Heat*.

In accordance with modern trends towards the use of numerical methods, these are stressed rather than the graphical approach which was popular before the widespread use of hand-held calculators and microcomputers. I was particularly attracted to the idea of using computer methods and this philosophy is also in agreement with that of the School of Mechanical Engineering in NSW Colleges of TAFE where more widespread use of computers has been mainly hindered by lack of suitable software. Consequently when writing the book I wrote computer programs wherever this seemed to offer advantages. Having used the programs myself to check the solution to many of the worked examples and to solve many of the problems, I am convinced that this is the way of the future. I do not claim that the programs I have written are necessarily the most elegant and no doubt a specialist computer programmer could improve on them, but the programs do work and are readily comprehensible and usable for those possessing a reasonable familiarity with computers and the BASIC language. Copies of all programs and sample solutions using them are included in Appendix 3. The programs may be stored on a single floppy disc as used with most microcomputers. I plan to make the disc available to students and teachers alike and I trust it will prove a useful learning and teaching aid.

Drawing on the experience of my other two books, I have tried to write this book with minimum theoretical dissertation and verbiage and a maximum measure of illustrations, worked examples, concise summary points and solved problems. However, in order to facilitate understanding, formulas have been derived from first principles wherever this could be done without introducing advanced or complex mathematical treatment. Analysis has been generally restricted to the two-dimensional level as most engineering problems may be reduced to this level and a three-dimensional treatment usually introduces unnecessary complexity.

Believing that a picture is worth a thousand words, the book contains a large number of illustrations. These will, I hope, be clear but they are generally not to scale, or at best to approximate scale, so that data should not be scaled off them. All dimensions shown in the illustrations are in millimetres unless otherwise stated. The book deals only with solids and any reference to the mechanics or properties of fluids has been deliberately avoided.

Although at the present time there is no standard set of symbols applicable to all branches of engineering, I have used those which appear to me to be the most logical and accepted symbol in each case. I have tried to avoid duplication but in some cases this was not feasible, for example the symbol k is used for radius of gyration as well as for spring constant. It is felt that in such cases the meaning of the symbol will be clear from the equation or context in which the symbol appears. Also the meaning of the symbols is explained in the text and a list of symbols is given in Appendix 1.

Calculations given in the worked examples and problems have been done with a calculator or computer and subtotals retained in memory for greatest accuracy. However answers have generally been rounded off to three significant figures.

It is generally accepted that mechanics and strength of materials lies midway between the fields of engineering science and mechanical design and I have kept this in mind at all times. Consequently the preliminary treatment of the subject matter is brief in order to avoid duplication of material already covered previously in engineering science or equivalent subjects. On the other side of the spectrum, there is no in-depth treatment of topics normally covered in engineering design subjects and hence comprehensive tables of material properties or section properties have not been included. Similarly I have avoided "cookbook" methods which follow procedures or practices prescribed by codes or standards and which do not aid the general understanding or application of the underlying principles. However, in the theory, worked examples and problems, I have tried to make the book relevant to engineering practice and to lead the student into a better appreciation of subsequent design procedures.

I would like to thank John Harris, Royal Melbourne Institute of Technology, and Clyde Paton, Regency Park College, for reviewing the manuscript and for their helpful suggestions.

I have made every effort to ensure that the book is as free from error as possible but despite careful checking and double checking there will no doubt still be some errors in the final publication. I would be grateful if these were brought to my attention as well as any suggestions for improving the book which could be incorporated in future print runs or editions.

ROGER KINSKY

Introduction

The mechanical engineer is vitally concerned with the determination of the forces acting upon members of machines and structures and the effect of these forces, in order to determine whether the member can withstand the stresses and strains imposed upon it. That is, the engineer seeks to determine the suitable dimensions and material of manufacture for the member so that it can safely and economically carry the loads to which it will be subjected.

Traditionally, there has been a division of the analysis into two separate areas:

1. the analysis of forces acting upon members considered to be rigid or non-deformable; this area is known as **applied mechanics**;
2. the analysis of the effect of the forces on members considered to be deformable, that is the analysis of the stresses and strains imposed upon members as a result of the forces; this area is known as **strength of materials**.

Applied mechanics is also usually subdivided further into areas known as statics and dynamics; **statics** considers members or force systems at rest or in uniform motion, and **dynamics** involves members or force systems where there is relative motion or more particularly accelerated masses.

It would seem logical that the study of applied mechanics should precede the study of strength of materials, that is the engineer should first calculate the forces and moments acting on a member before proceeding to determine the effect of these forces on the member. However, this approach is not always feasible and sometimes the two areas must be combined in order to provide a complete solution to a given problem.

For example, consider the column shown in Figure I.1 in which an inclined force F acts. This force may be broken into two components—a vertical one F_V and a horizontal one F_H. In Figure I.1(a), the member is considered rigid; in Figure I.1(b) it is considered to be deformable and deflects under the action of the load as shown.

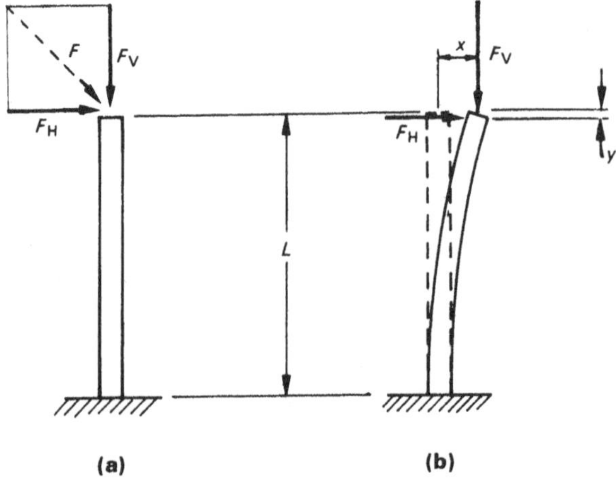

Fig. I.1 *(a) Rigid and (b) deformable members*

The reactive moment at the base may be determined by taking moments about the base:

(a) $M = F_H L$

(b) $M = F_H(L - y) + F_V x$

$= F_H L - F_H y + F_V x$

Since every real member is in fact deformable, an error has resulted in (a) by considering the member as rigid. The magnitude of the error is:

$F_V x - F_H y$

The significance of this error depends upon the magnitude of the vertical deflection y and the horizontal deflection x. In most cases in engineering, these deflections would be comparatively small (less than about 0.5%) so that the error would also be of this order and therefore justifiably neglected. However, in some cases where deflections are of significance (for example in springs or torsion bars) the deflection should be considered in conjunction with the loads in order to determine the true forces and moments in the member. Such a combined approach goes beyond the intended scope of this book and members are considered as non-deformable for the purpose of calculating forces and moments.

Another class of problems where a combined approach is required for solution is where the forces or moments depend upon the deflection of the member and cannot be determined by assuming non-deformability.

Consider the continuous beam resting on three supports as shown in Figure I.2. If the beam and the supports are considered as non-deformable, the equations of statics yield less equations than there are unknowns, so that a solution using these equations alone is not possible. Such problems are known as statically indeterminate and are beyond the scope of this text.

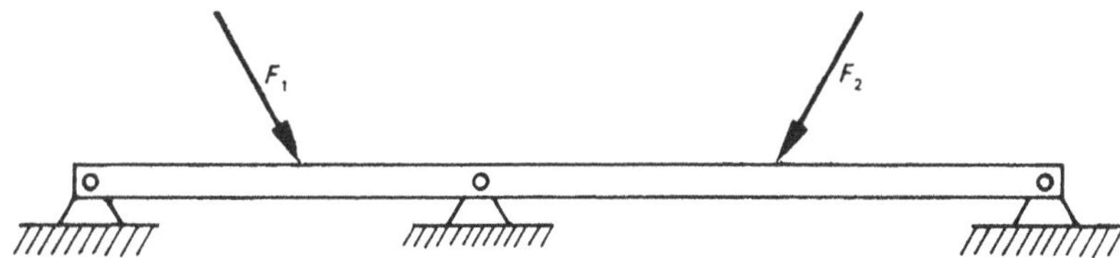

Fig. I.2 *Continuous beam*

PART I

STATICS

1

Resultant force and moment

This chapter considers the methods by which forces can be combined in such a manner that a single force has exactly the same effect as the original group of forces. This single force is known as the resultant force. For example, consider a number of men pulling on a rope anchored to a wall. Each of the men exerts an individual force on the rope. However, at the anchor point the individual forces have been combined to a single total force—the resultant.

Determination of the resultant force of a group of forces is often necessary in mechanics as it enables complex groups of forces to be considerably simplified. In order to determine the location of the resultant force it is often necessary to take moments of the forces to determine the resultant moment. Hence this chapter also considers the combination of moments to obtain the resultant moment.

Note that since force is a vector quantity, the direction of any unknown force or resultant must be specified in addition to the magnitude when the solution is stated. Also since moment is a vector quantity, the direction (sign) of the moment must similarly be stated.

1.1 Horizontal and vertical forces

Many forces encountered in mechanics are horizontal or vertical. In accordance with accepted mathematical convention, the sign that applies to positive and negative forces relative to the x, y axes is shown in Figure 1.1.

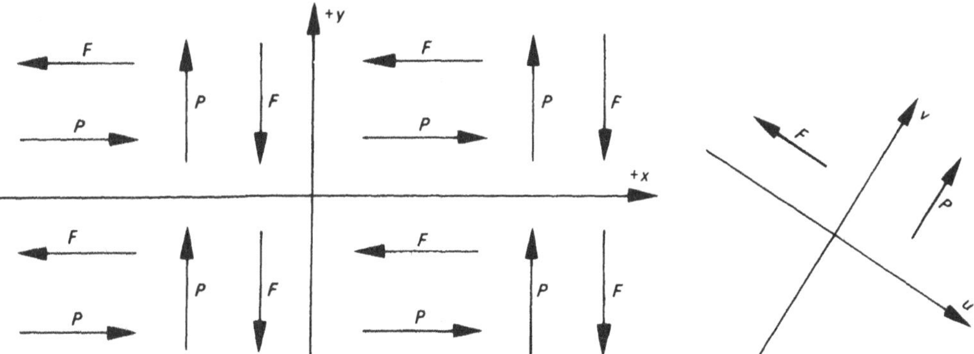

Fig. 1.1 *x, y axes (frame of reference)*

Fig. 1.2 *Inclined frame of reference*

Using this convention, forces shown P are positive and those shown F are negative. The x and y axes define a frame of reference; the same sign convention may be applied to any frame of reference even though it is not necessarily horizontal and vertical, as shown in Figure 1.2. Note that the position of the force in the frame of reference is immaterial in determining its sign—only the direction is of consequence.

1.2 Components of a force

The magnitude (or effect) of a force in any direction other than the original direction of the force is known as the component of the force in that direction. The component of a force in any direction is obtained by drawing a perpendicular from the force to the direction as shown in Figure 1.3.

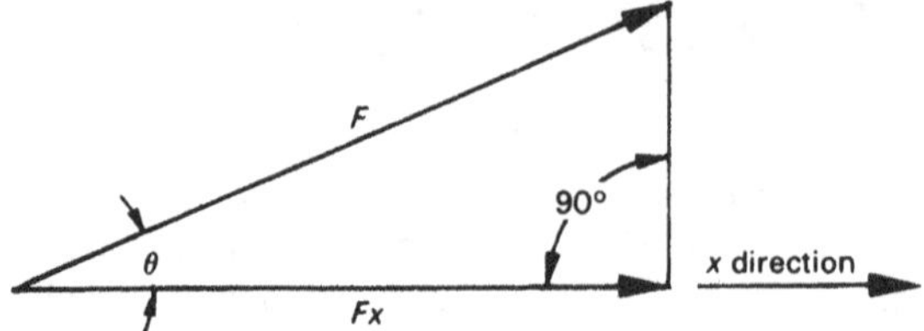

Fig. 1.3 *Component of a force (x direction)*

It follows that the component of force F in the x direction is:

$$F_x = F\cos\theta \tag{1.1}$$

Similarly for the y direction:

$$F_y = F\sin\theta \tag{1.2}$$

Several important observations follow:

1. A force has *no component* in a direction perpendicular to itself.
2. A single force cannot be completely replaced by one component. However two components may replace a single force; the force is then said to be resolved into its components.
3. The direction (sign) of the component may be obtained by inspection. For example in Figure 1.4 it is evident that force F has a positive y component and a negative x component, whereas force P has both x and y components negative.

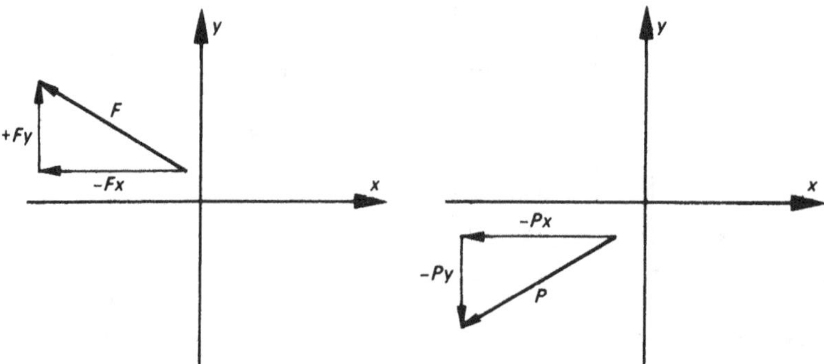

Fig. 1.4 *Positive and negative components*

4. The correct sign of the x and y components will also be obtained mathematically (without the necessity of inspection) provided the following rule is observed:

Forces pointing away from the origin are positive and angles measured anticlockwise from the x axis are positive. This rule is illustrated in Figure 1.5.
With modern calculators there is no difficulty in obtaining trigonometric functions for angles in the range 0–360° or for negative angles.

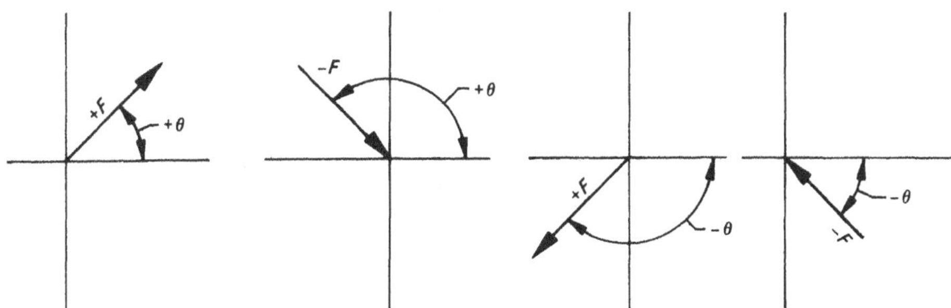

Fig. 1.5 *Sign convention for forces at an angle*

Example 1.1
A 5 kN force acts at 120° to the x axis as shown in Figure 1.6(a). Using the rule for force specification, this force is also specified in three other possible ways, Figure 1.6(b), (c) and (d).

Obtain the x and y components using equations 1.1 and 1.2 for each of the four cases and show that the correct magnitude and sign is obtained in each case.

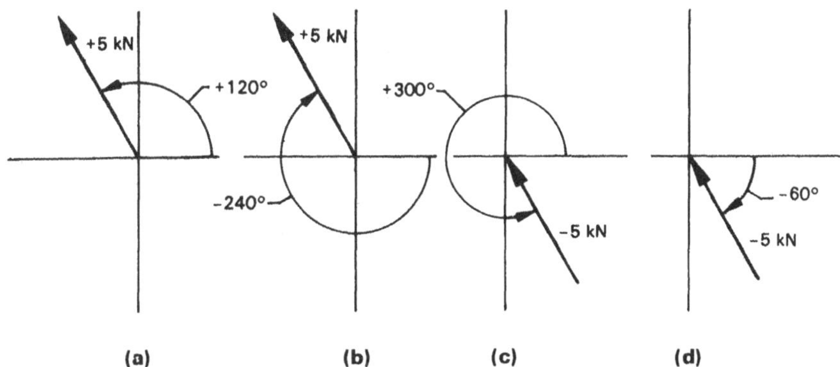

Fig. 1.6

Solution

Case	*Vertical component F_y (kN)*	*Horizontal component F_x (kN)*
(a)	5 sin 120° = 4.33	5 cos 120° = −2.5
(b)	5 sin −240° = 4.33	5 cos −240° = −2.5
(c)	−5 sin 300° = 4.33	−5 cos 300° = −2.5
(d)	−5 sin −60° = 4.33	−5 cos −60° = −2.5

Therefore complete consistency has been obtained as in each case there is the same magnitude and sign for the horizontal and vertical components of the force.

1.3 Resultant of concurrent forces: Analytical solution

When forces are combined so that a single force replaces two or more other forces, then that single force so found is called the resultant. The resultant of any number of forces acting at a point (concurrent forces) may readily be obtained analytically by the following method:

1. Determine the x and y components of each force.
2. Determine the sum of all the x and all the y components.
3. Determine the resultant from the combined total found in step 2.

This method is illustrated in example 1.2.

Example 1.2

Determine the resultant of the forces shown in Figure 1.7 (analytically).

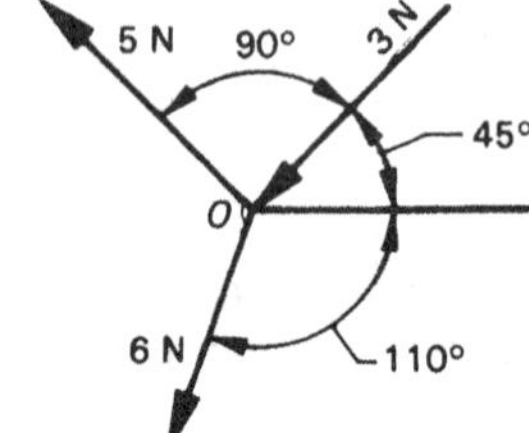

Fig. 1.7

Solution

Using the sign convention, the following table may be derived:

Force (N)	*Angle (°)*	*F_y (N)*		*F_x (N)*	
−3	45	−3 sin 45°	−2.121	−3 cos 45°	−2.121
5	135	5 sin 135°	3.536	5 cos 135°	−3.536
6	−110	6 sin −110°	−5.638	6 cos −110°	−2.052
		Total	−4.223		−7.709

The forces may now be combined as shown in Figure 1.8.

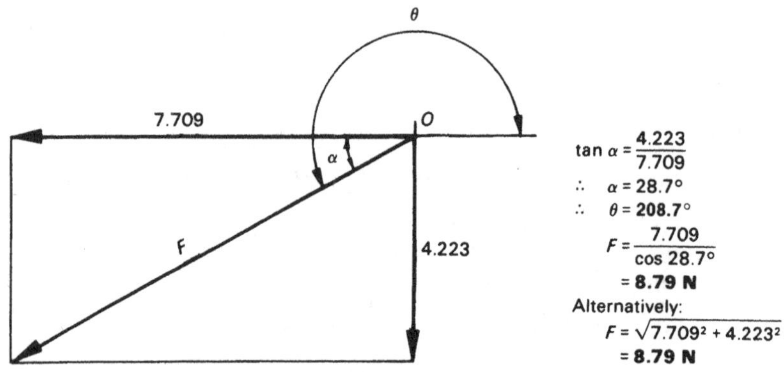

Fig. 1.8

Therefore the resultant force is **8.79 N** passing through O at an angle of **208.7°**.

1.4 Resultant of concurrent forces: Computer solution

A computer program named FORCE 1 written in BASIC language for determining the resultant of any number of concurrent forces is listed in Appendix 3.1.

The program inputs are:

1. number of forces
2. magnitude of each force
3. angle made by each force with the x axis (in degrees anticlockwise).

It is necessary to input 2. and 3. for each force in turn.

The program outputs are:

1. magnitude of the resultant force
2. angle made by the resultant force with the x axis (in degrees anticlockwise)
3. sum of the vertical components
4. sum of the horizontal components.

Notes

1. The convention for specifying a force as illustrated in Figure 1.5 is used for the input of forces and angles. This means that the same force could be specified by any one of four combinations:

   (a) $+F + \theta$

   (b) $-F + \theta$

   (c) $+F - \theta$

   (d) $-F - \theta$

   The program will accept any one of these four variations.
2. The sequence in which the forces are input is immaterial.
3. The output is given only as variation (a) namely $+F + \theta$.
4. Consistent units must be used for the magnitude of the forces throughout, e.g. either N or kN, but the units may not be mixed.
5. Output forces and angles are rounded off to two decimal places.

The solution to example 1.2 using the computer program is also given in Appendix 3.1 below the listing of the program and should be self-explanatory.

1.5 Resultant of concurrent forces: Graphical solution

Two methods are widely used, namely the parallelogram of forces method and the vector addition method.

Parallelogram of forces method

This method may be used when there are only two forces. The method is illustrated in example 1.3. Note that when this method is used *the forces must be drawn so that they are positive*, i.e. radiating outward from the point where their lines of action intersect.

The method may of course be extended to any number of forces by successively combining pairs of forces and resultants of pairs of forces.

Vector addition method (force polygon method)

This method may be used with any number of forces. It is illustrated in example 1.3 for two forces and in example 1.4 for more than two forces. Note that when using this method *the heads and tails of the forces must follow each other* and the resultant force direction is from the initial tail (start point) to the final head (finish point). The order or sequence in which the forces are drawn is immaterial.

Example 1.3

Determine the resultant of the forces shown in Figure 1.9, using

(a) parallelogram of forces method and (b) vector addition method.

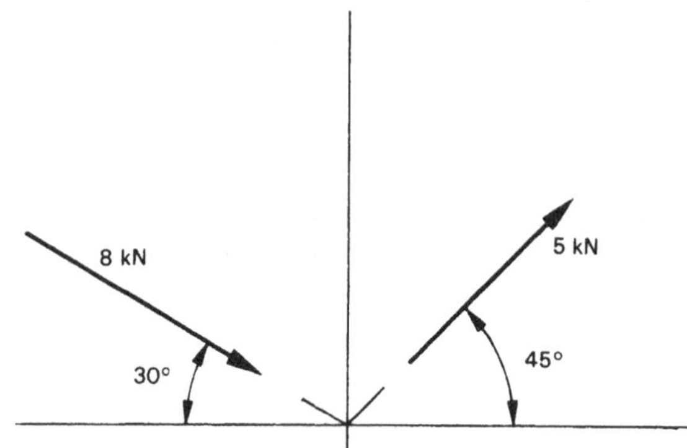

Fig. 1.9

Solution

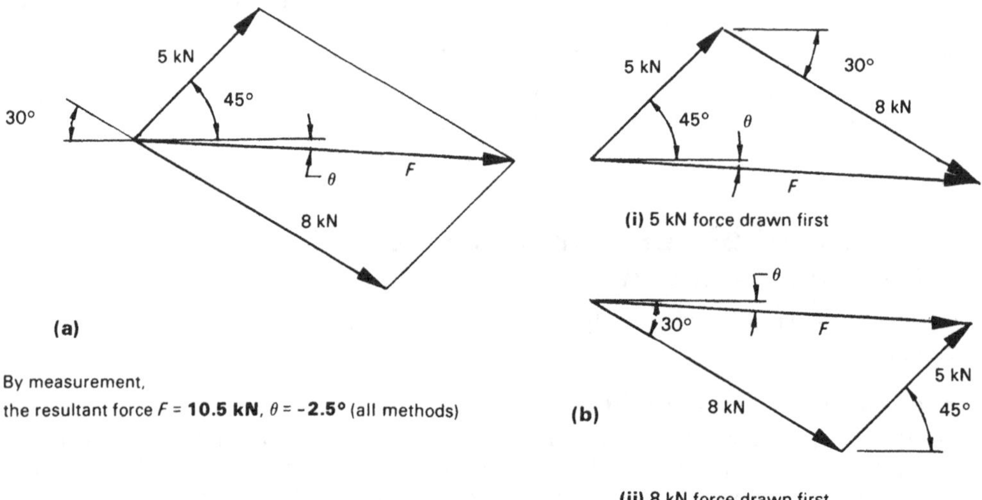

Fig. 1.10 *(a) Parallelogram of forces method. (b) Vector addition method*

Example 1.4

Determine the resultant of the forces shown in Figure 1.11 using the force polygon method.

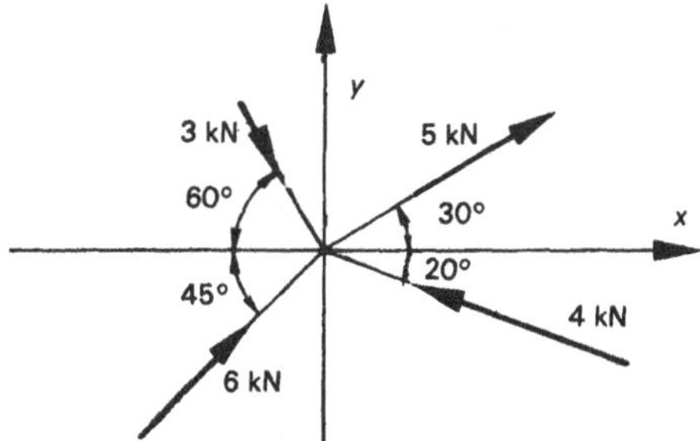

Fig. 1.11

Solution

The sequence in which the forces are drawn is immaterial. Two different sequences are shown in Figure 1.12 both giving the same resultant.

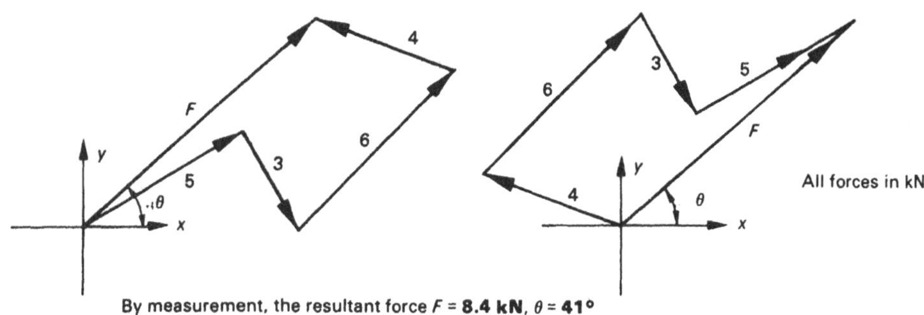

Fig. 1.12

1.6 Moment of a force (M)

The moment of a force is the turning effect of that force about a point. It is obtained by multiplicaton of the force and the perpendicular distance of the force from the point.

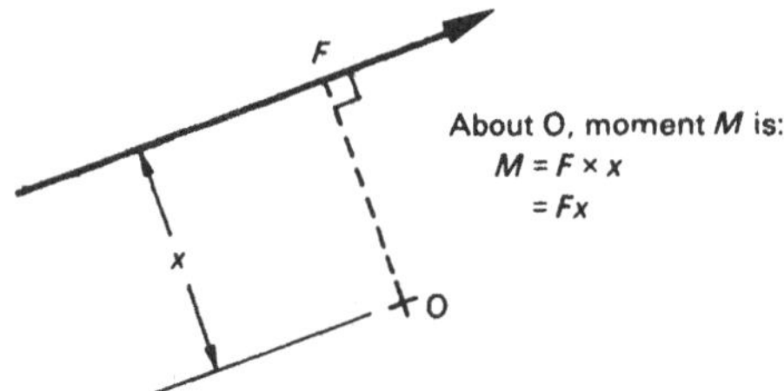

Fig. 1.13 *Moment of a force*

$$M = Fx \tag{1.3}$$

where M is the moment of the force (Nm)
F is the force (N)
x is the distance (m)

Notes

1. The following sign convention will be followed:
   (a) Clockwise moments are positive.
   (b) Anticlockwise moments are negative.

   This means that the direction of the force *itself* does not determine the sign of the moment and a positive force may give a negative moment and vice versa, depending upon the location of the line of action of the force relative to the point about which moments are taken.
2. The resultant moment of a number of forces about a common point is the sum of the moments of each force considered alone.
3. A force has *no moment* about a point which lies on its line of action.
4. The point about which moments are taken *should always be stated* since the moment is different about any other point except for a couple (pair of displaced equal and opposite forces) which produce the same moment about any point.
5. The resultant moment for a shaft turning about its centre is also known as *torque*.
6. The moment of a force may also be found by multiplication of the perpendicular component of the force and its distance from the point (see Fig. 1.14).

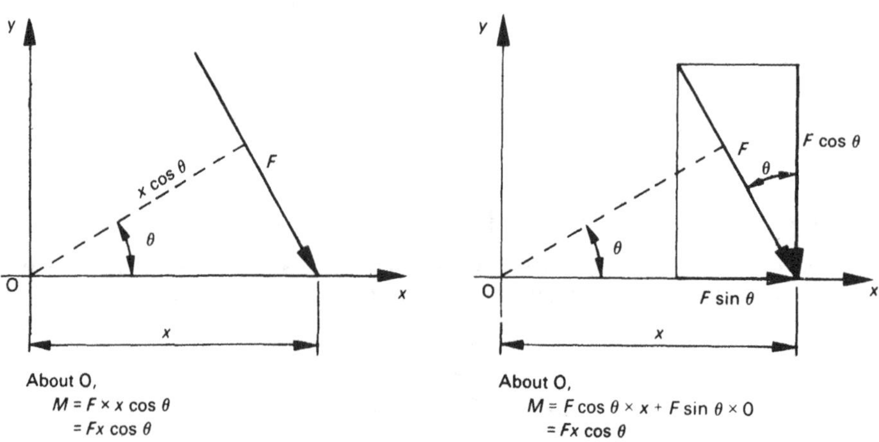

Fig. 1.14 *Two equivalent ways of determining the moment of a force*

Example 1.5
Determine the resultant moment about O of the forces shown in Figure 1.15.

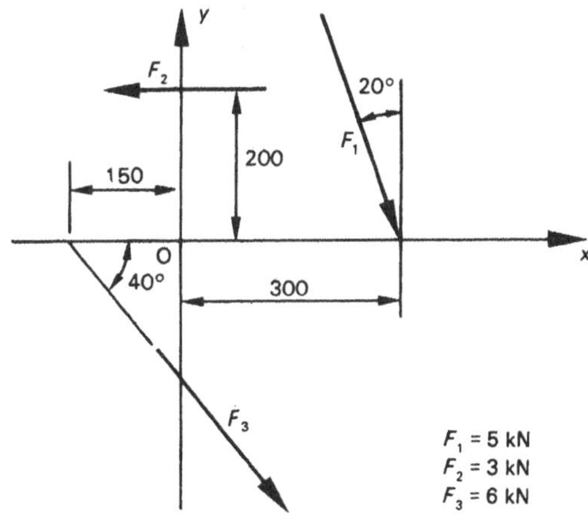

Fig. 1.15

Solution

Force (kN)	*Moment (kNm) about O*	
$F_1 = 5$	$5 \cos 20° \times 0.3$	+1.4095
$F_2 = 3$	-3×0.2	−0.6
$F_3 = 6$	$-6 \times 0.15 \sin 40°$	−0.5785
	Total	0.231

The resultant moment is therefore **0.231 kNm (231 Nm)** and is positive (clockwise).

1.7 Resultant of parallel forces: Analytical solution

When forces are parallel they may be combined by simple addition (bearing in mind the correct sign of the force). The resultant force so found will have its line of action parallel to the applied forces. The location of the line of action relative to the applied forces may be found by taking moments about any convenient point. The method is illustrated in example 1.6.

Example 1.6

Determine the resultant of the forces shown in Figure 1.16 using an analytical method.

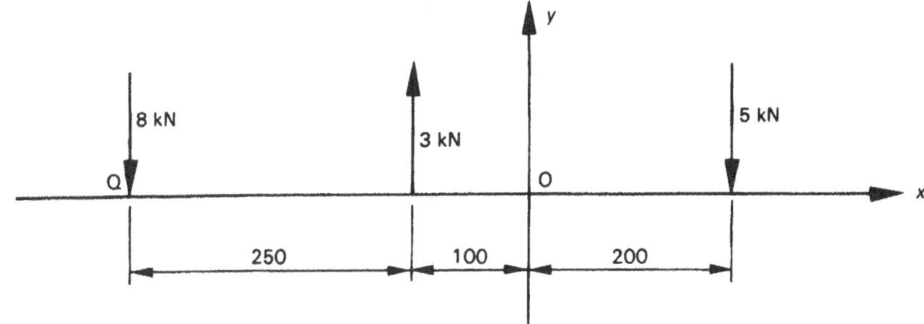

Fig. 1.16

Solution

The resultant force F is vertical and has a value:

$F = -5 + 3 - 8 = -10$ kN (the minus sign means that the force acts downward).

The location of the line of action of F may be determined by taking moments about a convenient point. Choosing the origin O and taking moments about O:

$$M = \underset{\text{(clockwise)}}{5 \times 0.2} + \underset{\text{(clockwise)}}{3 \times 0.1} - \underset{\text{(anticlockwise)}}{8 \times 0.35}$$

$= -1.5$ kNm (the minus sign means that the moment is anticlockwise).

Since $M = Fx$, $-1.5 = 10x$

$\therefore \quad x = -0.15$ m or -150 mm (to the left of the origin).

That is, the resultant force is **10 kN** acting downward at a distance of **150 mm** to the left of O.

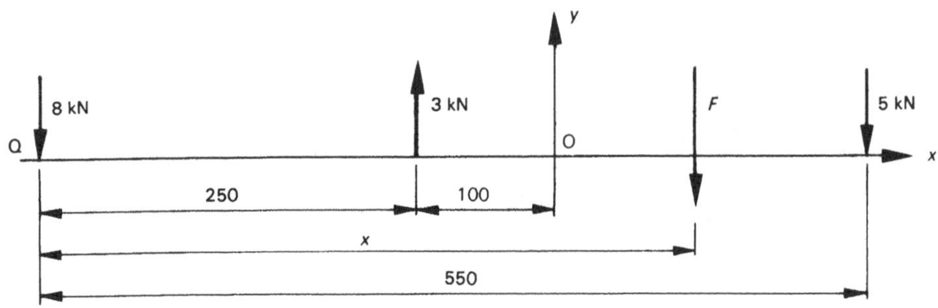

Fig. 1.17

Note that the location of the line of action may also be determined by taking moments about any other point, say where the 8 kN force cuts the x axis. Calling this point Q and taking moments about Q:

$M = 5 \times 0.55 - 3 \times 0.25 - 8 \times 0$

$= 2.0$ kNm

Since $M = Fx$, $2.0 = 10x$

$\therefore \quad x = 0.2$ m or 200 mm

That is, the resultant force acts 200 mm to the right of Q (i.e. 150 mm to the left of O).

1.8 Resultant of parallel forces: Graphical solution

When forces are parallel, the magnitude of the resultant may be determined by simple addition (or subtraction) so that a graphical solution has no advantage for this (although a graphical solution is still valid). However, it is often useful to determine the *location* of the resultant force by a graphical method. There are several methods available but the one with the widest application is the string (or funicular polygon) method. The string polygon is a vector force polygon where each force is replaced by two components (also called strings or rays). Referring to Figure 1.18, it is seen that the force F may be replaced by two components (labelled a and b). The point of intersection of these components is the pole point O which may be located at any convenient position on the page.

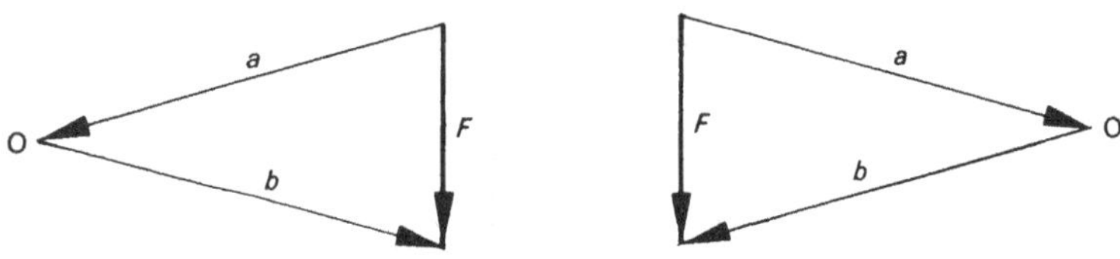

Fig. 1.18 *Force F may be replaced by two components a and b*

When drawing the string polygon for a number of forces, a common pole point is used so that there is a common component between each pair of adjacent forces which has the same magnitude and line of action but opposite sense (direction).

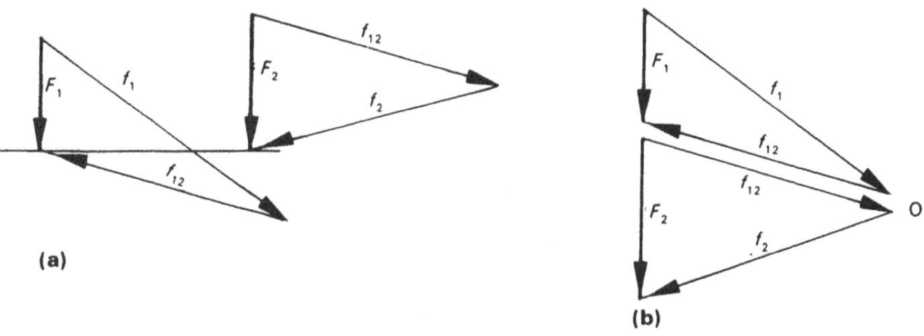

Fig. 1.19 *f_{12} is the common component between F_1 and F_2: (a) space diagram, (b) string polygon*

In order to determine the location of the resultant force on the space diagram, the common component for each pair of forces is drawn between these forces so that its effect is cancelled (being equal and opposite). The point of intersection of the non-common components f_1 and f_2 on the space diagram is a point through which the line of action of the resultant force passes. This is illustrated in Figure 1.20 which has been drawn for two parallel forces having the same direction.

The method may be used equally well when the parallel forces do not all have the same direction. This is illustrated in Figure 1.21 which has been drawn for two parallel forces having opposite directions.

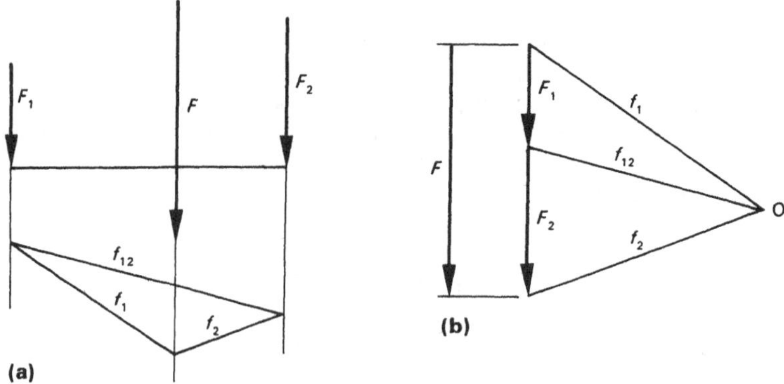

Fig. 1.20 *Location of the line of action of resultant force F: (a) space diagram, (b) string polygon*

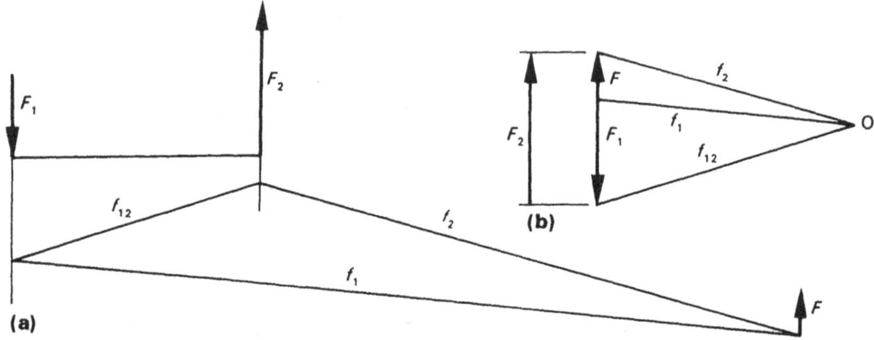

Fig. 1.21 *Line of action of resultant (forces in opposite directions): (a) space diagram, (b) string polygon*

Note that in this case, the common component f_{12} does not lie between f_1 and f_2 on the string polygon (although it still does so in the space diagram).

Example 1.7

Determine the magnitude and location of the resultant force in the system of forces shown in Figure 1.22 using the string polygon method.

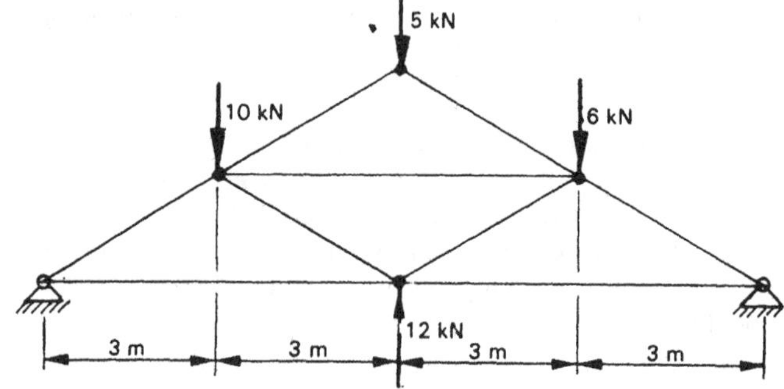

Fig. 1.22

Solution

Since the 5 kN force and the 12 kN force have the same line of action, they may be combined to a 7 kN force acting upward. The space diagram and string polygon are then as shown in Figure 1.23.

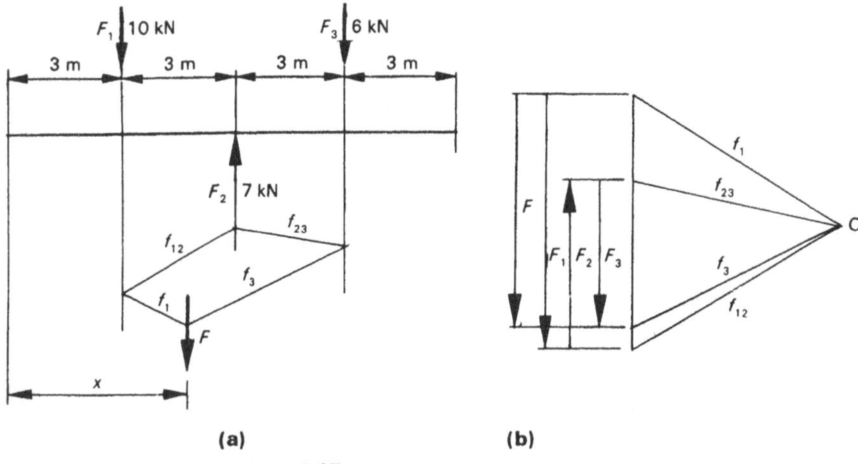

By measurement, **F** = **9 kN** and ***x*** = **4.67 m**

Fig. 1.23 *(a) Space diagram, (b) String polygon*

1.9 Resultant of forces (general case): Analytical solution

In the most general case when the forces are non-concurrent and non-parallel, the resultant may be obtained in magnitude and direction by determining the horizontal and vertical components of each of the forces and then combining them. The location of the resultant may be obtained by taking moments about any convenient point. The method is illustrated in example 1.8.

Example 1.8

Determine the magnitude, direction and location of the resultant (relative to O) of the forces acting on the bracket shown in Figure 1.24.

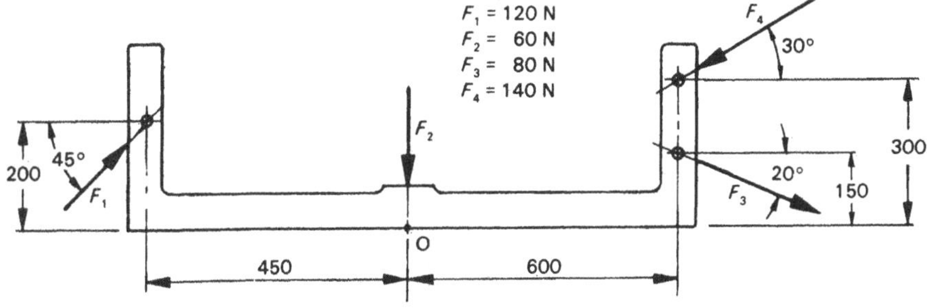

Fig. 1.24

Solution

Force (N)	F_y *(N)*		F_x *(N)*	
F_1	120 sin 45°	84.85	120 cos 45°	84.85
F_2		−60.0		0
F_3	80 sin − 20°	−27.36	80 cos − 20°	75.175
F_4	− 140 sin 30°	−70.0	− 140 cos 30°	−121.24
	Total	−72.51		38.785

combining:

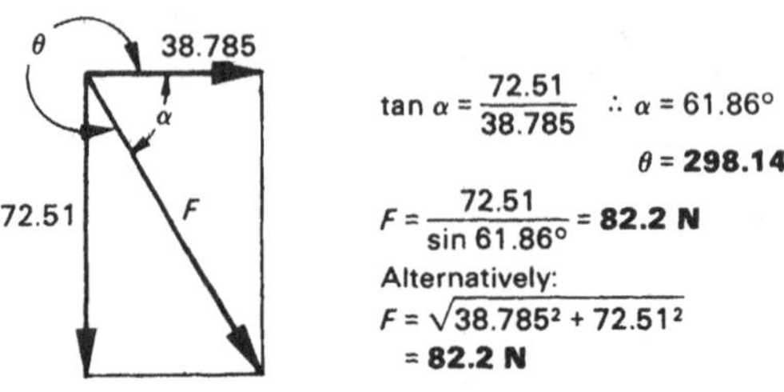

Fig. 1.25

However in order to determine the location of the resultant force relative to point O, take moments about O. Assume that the resultant force is located to the right of O as shown in Figure 1.26.

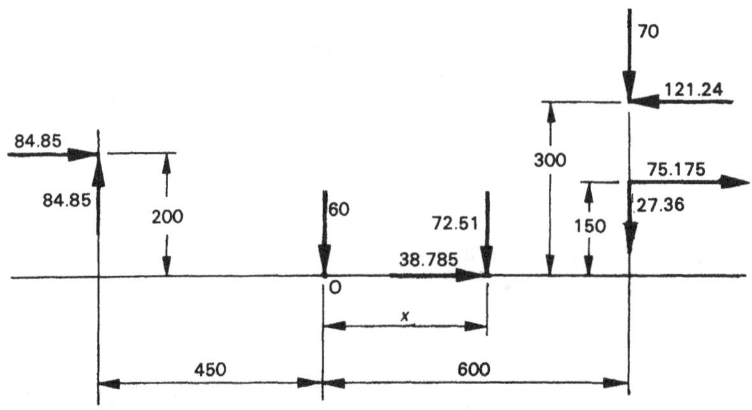

Fig. 1.26

Taking moments about O:

$$72.51 \times x = 84.85 \times 0.45 + 84.85 \times 0.2 + 27.36 \times 0.6 + 75.175 \times 0.15 + 70 \times 0.6 - 121.24 \times 0.3$$

$$\therefore \quad x = \mathbf{1.22\ m}$$

The line of action of the resultant force therefore crosses the base line of the bracket 1.22 m to the right of O.

Note that when taking moments in order to determine the location of the line of action of the resultant force on the x axis, only the vertical component of the resultant force is included since the horizontal component at the x axis has no moment about O.

1.10 Resultant of forces (general case): Computer solution

A computer program named FORCE 2 written in BASIC language for determining the magnitude, direction and location of the resultant of any number of non-concurrent forces is listed in Appendix 3.3.

The program inputs are:

1. the number of forces
2. the magnitude of each force
3. the angle made by each force with the x axis (in degrees anticlockwise)
4. the x and y coordinates of a point on the line of action of each force.

It is necessary to input 2, 3 and 4 for each force in turn.

The program outputs are:

1.–4. given for program FORCE 1
5. the resultant moment (about O)
6. the location of the line of action of the resultant force on the x axis. If the resultant force is parallel to the x axis, the location of the line of action of the resultant force on the y axis is output instead.

Notes

The same notes apply as given for FORCE 1 (section 1.4) with the addition of the following:

1. Consistent units must be used throughout for the x and y coordinates—namely either mm or m but not a mixture of both.
2. If a force is parallel to the y axis (vertical force), the y coordinate is immaterial and may be taken as zero or any other value. Similarly for horizontal forces, the x coordinate is immaterial.
3. If the frame of reference (x and y axis) is not given, the user may locate the frame of reference at any position or angle. Of course, the forces must be specified relative to the frame of reference being used and similarly the output will also be relative to this frame of reference.

The solution to example 1.8 using the computer program FORCE 2 is also given in Appendix 3.2 below the listing of the program and should be self-explanatory. The frame of reference used in example 1.8 is such that the four given forces are specified thus:

F_1 120, 45, −450, 200

F_2 −60, 90, 0, 0

F_3 80, −20, 600, 150

F_4 −140, 30, 600, 300

where the first figure is the magnitude of the force (N), the second is the angle made by the line of application of the force with the x axis (degrees anticlockwise) and the third and fourth figures are the x and y coordinates of a point on the line of action of the force (in this case in mm).

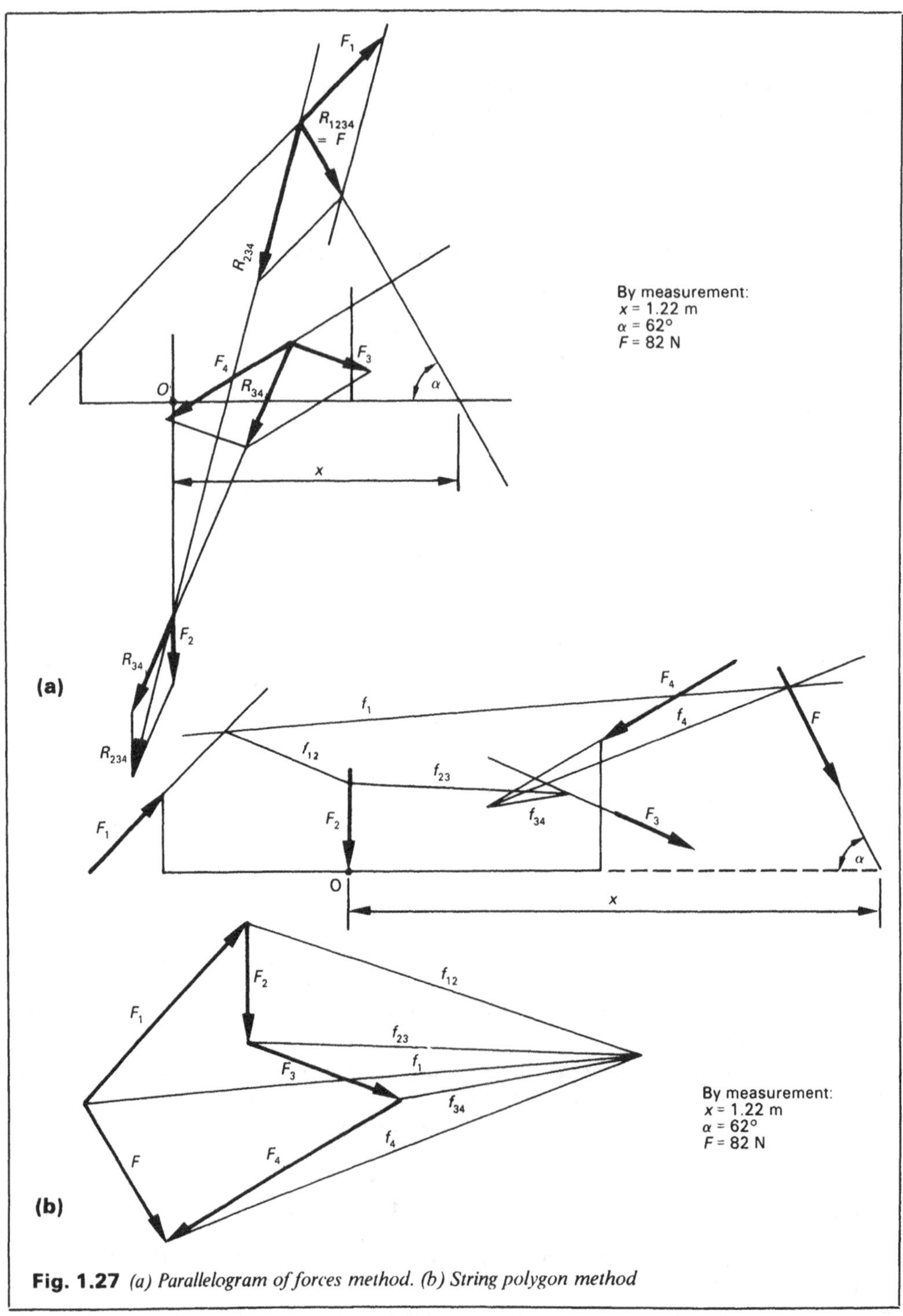

Fig. 1.27 *(a) Parallelogram of forces method. (b) String polygon method*

1.11 Resultant of forces (general case): Graphical solution

The magnitude, direction and location of the resultant of any number of non-concurrent forces may be determined graphically by successive use of the parallelogram of forces method or by the string polygon method. The use of both methods is illustrated in example 1.9.

Example 1.9

Solve example 1.8 using both graphical methods.

Solution

See Figure 1.27.

Problems

1.1 to 1.9 Determine the resultant of the force systems in Figures P1.1 to P1.9, using

(a) analytical method

(b) graphical method

(c) computer method.

All forces are in kN.

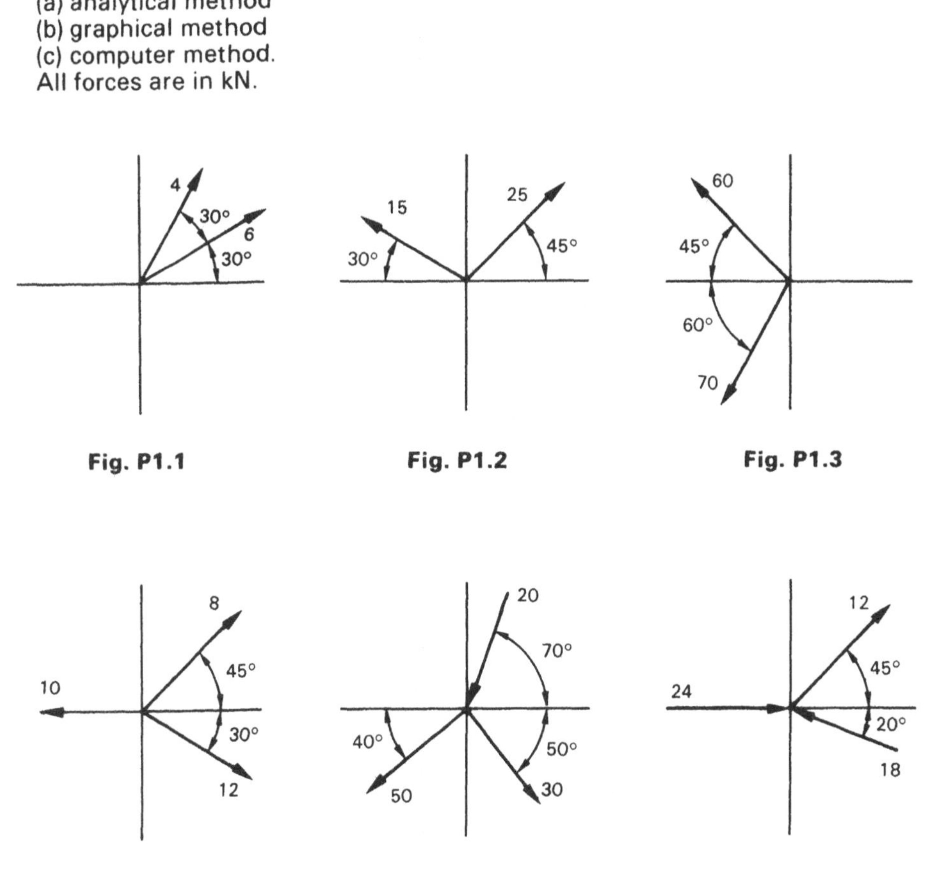

Fig. P1.1 **Fig. P1.2** **Fig. P1.3**

Fig. P1.4 **Fig. P1.5** **Fig. P1.6**

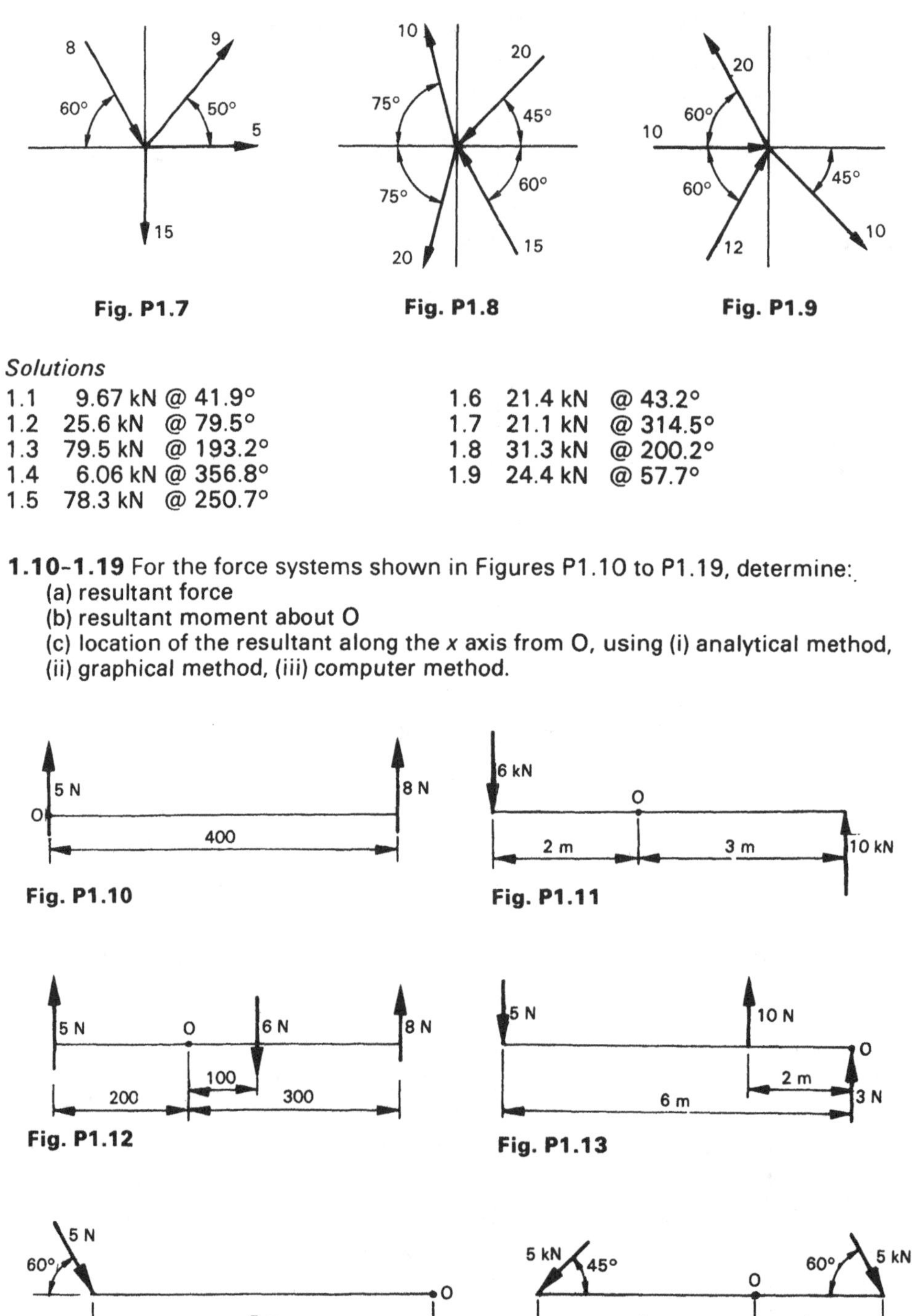

Fig. P1.7

Fig. P1.8

Fig. P1.9

Solutions

1.1	9.67 kN	@ 41.9°
1.2	25.6 kN	@ 79.5°
1.3	79.5 kN	@ 193.2°
1.4	6.06 kN	@ 356.8°
1.5	78.3 kN	@ 250.7°
1.6	21.4 kN	@ 43.2°
1.7	21.1 kN	@ 314.5°
1.8	31.3 kN	@ 200.2°
1.9	24.4 kN	@ 57.7°

1.10–1.19 For the force systems shown in Figures P1.10 to P1.19, determine:

(a) resultant force

(b) resultant moment about O

(c) location of the resultant along the *x* axis from O, using (i) analytical method, (ii) graphical method, (iii) computer method.

Fig. P1.10

Fig. P1.11

Fig. P1.12

Fig. P1.13

Fig. P1.14

Fig. P1.15

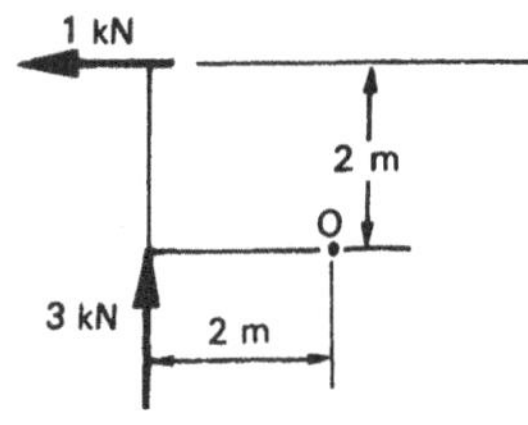

Fig. P1.16

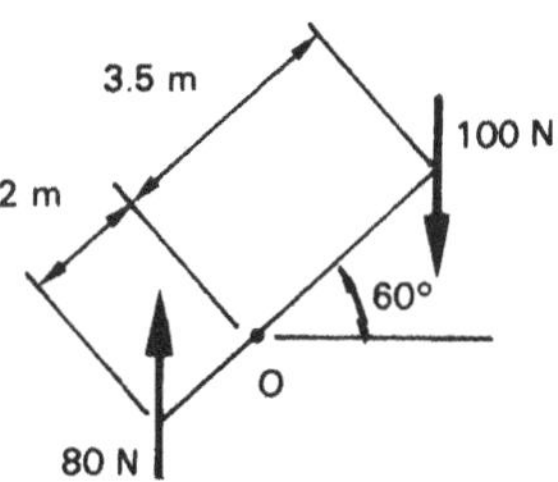

Fig. P1.17

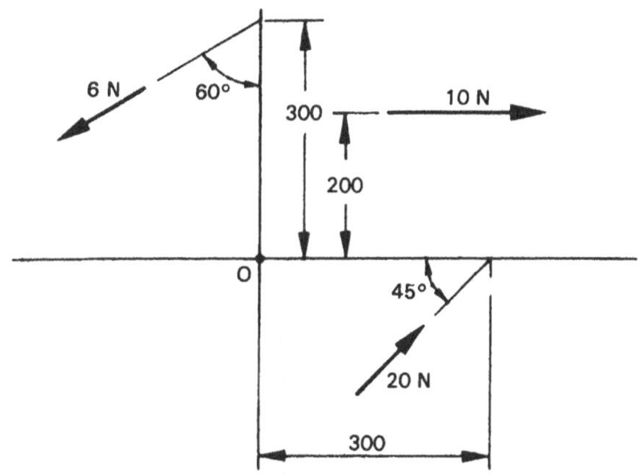

Fig. P1.18

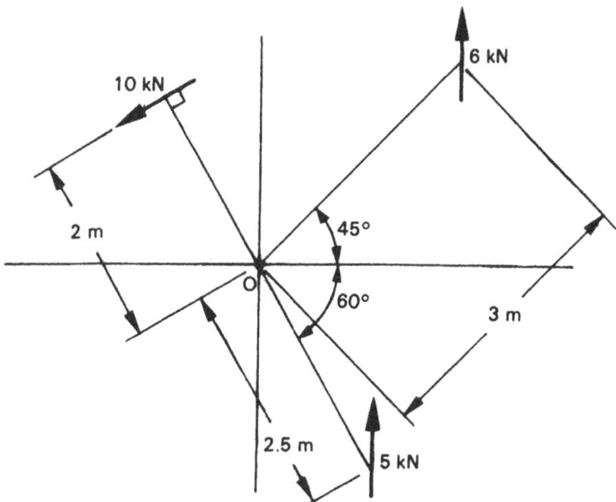

Fig. P1.19

Solutions

	(a)	(b)	(c)
1.10	13 N @ 90°	-3.2 Nm	246 mm
1.11	4 kN @ 90°	-42 kNm	10.5 m
1.12	7 N @ 90°	-0.8 Nm	114 mm
1.13	8 N @ 90°	-10 Nm	1.25 m
1.14	5 N @ 300°	-21.65 Nm	-5 m
1.15	7.93 kN @ 262.5°	-2.74 kNm	-0.35 m
1.16	3.16 kN @ 108.4°	4 kNm	-1.33 m
1.17	20 N @ 270°	255 Nm	12.75 m
1.18	22 N @ 30.5°	-3.8 Nm	341 mm
1.19	10.54 kN @ 145.3°	-39 kNm	6.5 m

2

Equilibrium of forces

In Chapter 1 it was seen that a number of forces may be combined into a single force known as the resultant force. Also each force exerts a turning moment about a given point so that there will also be a resultant turning moment about that point.

When the forces acting on a body are in balance so that there is no change in the motion of the body (either linear or rotational) then the forces and moments acting on the body must be balanced or in equilibrium. This follows from Newton's famous equation, $F = ma$.

If the acceleration is zero then the sum of the forces acting on the body must be zero. The rotational equivalent of this formula is $\tau = I\alpha$ and the same conclusion applies namely that if the rotational acceleration (α) is zero the sum of the moments (or torque) must also be zero.

Equilibrium of forces and moments therefore applies to:

1. all bodies at rest
2. all bodies moving with constant velocity (either linearly or rotationally).

Equilibrium therefore applies to a vast number of situations in practice and is a very important condition in mechanics.

2.1 Systems of forces

In mechanics the term "system" is often used in connection with the forces acting on a body. The word "system" also arises in the study of fluid mechanics and thermodynamics and the fundamental meaning is the same. In mechanics, it is necessary to isolate groups of forces from all other forces which act and react on one another, otherwise there would be an impossibly large number of forces to be considered in every problem. This is done by introduction of a system boundary, that is a dividing line between the forces to be considered and those which do not need to be considered. The forces acting inside the boundary then become the force system. That is, a force system may be defined as *a group of forces which mutually interact with one another*.

Note that the boundary of the force system does not need to be a real one, that is to correspond with any physical boundaries. Indeed in mechanics the boundary is most often an imaginary one. Nor does there need to be only one boundary for any given problem. For example in the consideration of the forces acting on a framed structure, the boundary may be taken around the entire structure in order to determine the reactions at the supports as in

Figure 2.1(a), around any of the joints as in Figure 2.1(b), or around any member as in Figure 2.1(c) in order to determine the internal forces in the joints or frame members.

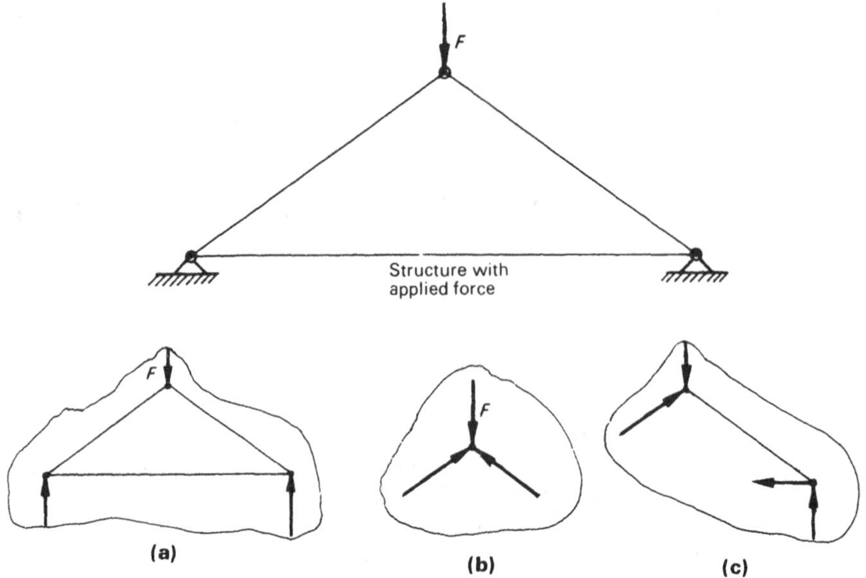

Fig. 2.1 *Some force systems for a framed structure*

Notes

1. It is often not necessary actually to draw the system boundary as it may be quite clear where the boundary is being taken from by the way the forces are shown.
2. The diagrams (a), (b) and (c) in Figure 2.1 are also called *free body diagrams* since they in effect show the force system on a body as if it were free in space (cut off from the supports).
3. The system boundary can be drawn anywhere but must form a closed loop.

2.2 Conditions for equilibrium

Consider the three force systems as shown in Figure 2.2. It is evident that none of these systems are in equilibrium because:

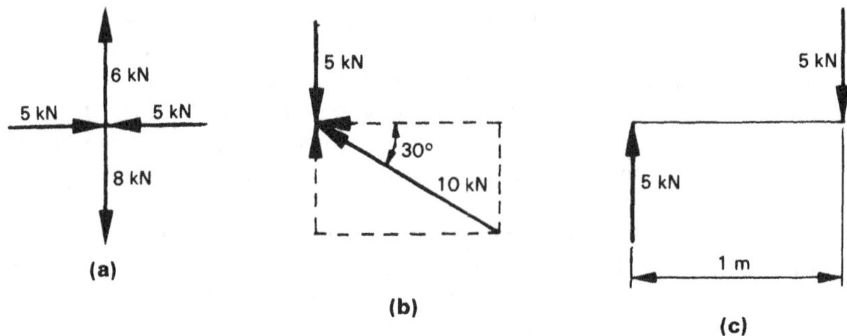

Fig. 2.2 *Systems of forces not in equilibrium*

system a: the horizontal forces are in equilibrium but the vertical forces are not;
system b: the vertical force of 5 kN is balanced by the vertical component of the 10 kN force (10 sin 30° = 5); however the 10 kN force also has a horizontal component of 8.66 kN (10 cos 30° = 8.66) which is unbalanced;
system c: there are no horizontal forces and the vertical forces are balanced; however there is an unbalanced moment of 5 kNm (in a clockwise direction).

Therefore for equilibrium of a force system three conditions must be satisfied:

1. The sum of the forces or components of the forces in any direction must be zero, that is their resultant in this direction must be zero.
2. The sum of the forces or components of the forces in another direction (other than that given in 1. must also be zero.
3. The sum of the moments about any point must be zero.

Notes

1. Each of these conditions is necessary but not sufficient on its own.
2. The direction specified for the forces or force components are usually mutually perpendicular and correspond with the x and y axes.
3. If the sum of the forces is zero then the polygon of forces will close (i.e. there will be no resultant force). This is illustrated in Figure 2.3.

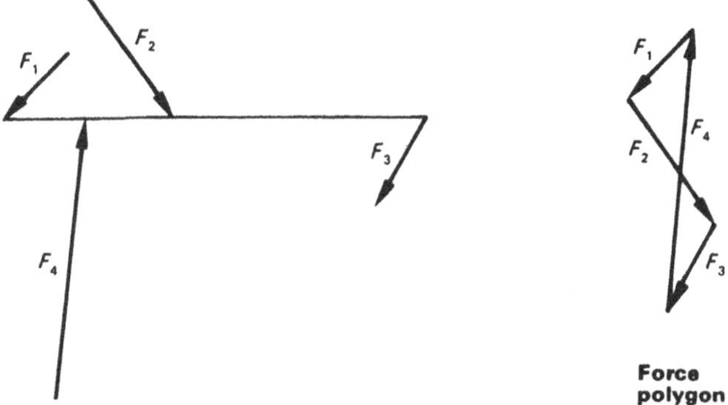

Fig. 2.3 *For a system of forces in equilibrium, the force polygon closes*

4. The fact that the force polygon closes means that the forces are in equilibrium but this fact on its own does not necessarily prove that the system is in equilibrium as the moments could still be unbalanced.
5. The moments may be taken about any point. If the sum of the moments about this point is zero, then the sum of the moments about any other point will also be zero.

The conditions for equilibrium of a force system may be expressed mathematically by the following equations:

$$\Sigma F_V = 0 \quad \textbf{(2.1)}$$

$$\Sigma F_H = 0 \quad \textbf{(2.2)}$$

$$\Sigma M = 0 \quad \textbf{(2.3)}$$

Example 2.1

Determine whether the system of forces shown in Figure 2.4 is in equilibrium.

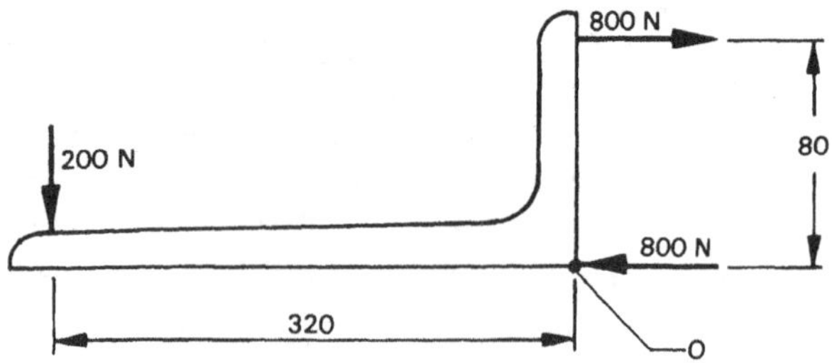

Fig. 2.4

Solution

Using the equations for equilibrium:

$\Sigma F_V = -200$ N

$\Sigma F_H = 800 - 800 = 0$

ΣM (about O) $= -200 \times 0.32 + 800 \times 0.08 = 0$

Therefore the horizontal forces and moments are balanced but the vertical forces are not in equilibrium as the sum of the vertical forces is not equal to zero. The system is therefore *not in equilibrium*.

2.3 Equilibrant force

When a system of forces is not in equilibrium, then the force necessary to restore equilibrium is known as the equilibrant force. It is evident that the equilibrant force must be equal in magnitude to the resultant and have the same line of action but must act in the opposite direction (sense).

Referring to Figure 2.5 it will be seen that the forces F_1, F_2 and F_3 are not in equilibrium and they have a resultant F.

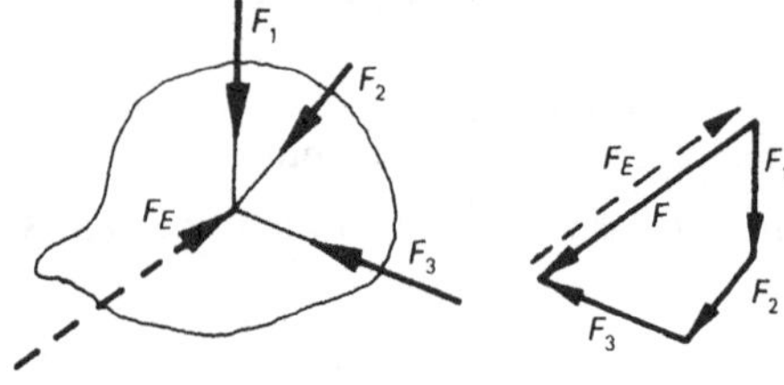

Fig. 2.5 *The equilibrant force F_E is equal and opposite to the resultant force F*

In order to restore equilibrium to the system, the force F_E equal and opposite to the resultant must be applied.

Example 2.2

Determine the magnitude and location of the equilibrant force for the system of forces shown in Figure 2.4 (example 2.1).

Solution

It was found in example 2.1 that the sum of the vertical forces was not zero but $\Sigma F_V = -200$ N. Therefore for equilibrium it is necessary to introduce an additional vertical force of +200 N into the system. It was also found that, for the system as given, ΣM (about O) = 0. Therefore the equilibrant force of 200 N must also act at O since this is the only place to introduce the force without unbalancing the moment.

If this is done, the system of forces will now be as shown in Figure 2.6. Note that the force polygon now closes.

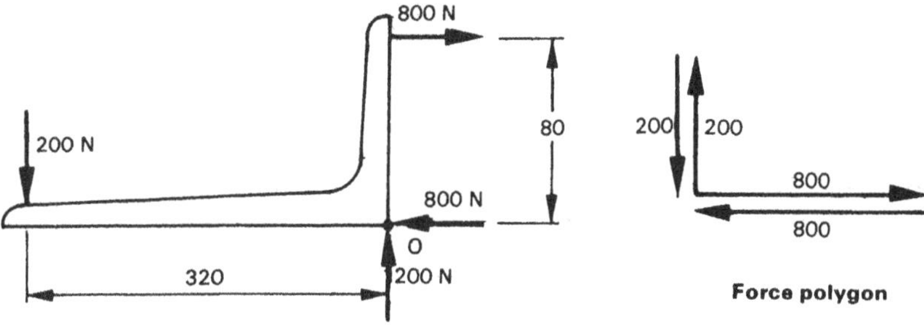

Fig. 2.6

Location of the equilibrant force

When the forces acting in a system are concurrent but not in equilibrium, as was the case for the force system shown in Figure 2.5, then it is evident that the equilibrant must act at the point of concurrency also, otherwise there would be an unbalanced moment. If the forces are non-concurrent and not in equilibrium, then the equilibrant force must act along the line of action of the resultant of the applied forces but have the opposite direction. This is clarified in example 2.3.

Example 2.3

Determine the magnitude, direction and location of the equilibrant force for the system of forces shown in Figure 2.7.

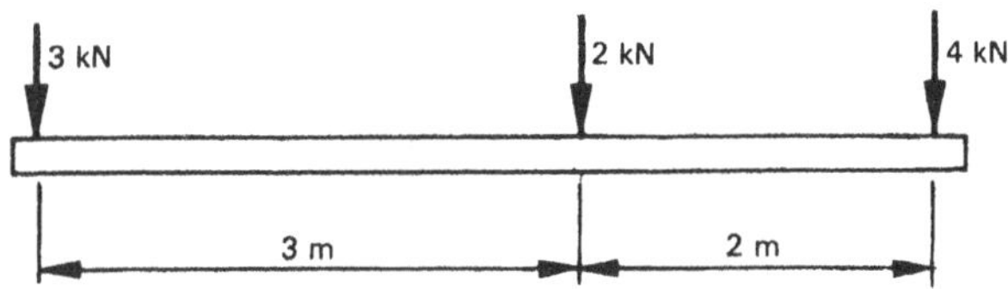

Fig. 2.7

Solution

Using the principles given in Chapter 1, first determine the magnitude and location of the resultant of these forces either graphically or analytically (or by computer).

Using the analytical method:

$$\Sigma F_V = -3 -2 -4 = -9 \text{ kN (minus sign, acting down)}$$

Since there are no horizontal forces, the resultant force is this vertical force. Its location may be determined by taking moments about any point. Taking moments about the line of action of the 3 kN force:

$$Fx = 2 \times 3 + 4 \times 5$$

$$\therefore \quad x = \frac{6 + 20}{9} = 2.89 \text{ m}$$

The resultant force therefore acts downward 2.89 m to the right of the 3 kN force. Hence the equilibrant force is **9 kN** acting **vertically upward** at a distance of **2.89 m** to the right of the 3 kN force.

2.4 Equilibrant moment

It is possible for a system of forces to be balanced with regard to horizontal and vertical components but to be unbalanced with regard to moment. For example, consider the force system shown in Figure 2.8.

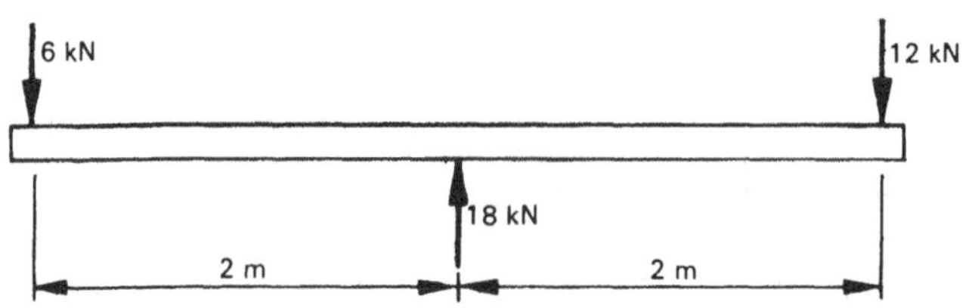

Fig. 2.8 *Force system with balanced forces but unbalanced moments*

The vertical components are balanced and the horizontal components are zero but the moments are not balanced. For example, taking moments about the 6 kN force:

$$\Sigma M = -18 \times 2 + 12 \times 4 = 12 \text{ kNm}$$

Therefore $\Sigma M \neq 0$, the system is not in equilibrium as there is an unbalanced moment of 12 kNm clockwise. Indeed the same unbalanced moment is obtained by taking moments about any other point (as may be readily verified).

Therefore in order to restore equilibrium to the system, an equilibrant mcment of −12 kNm (anticlockwise) is required. However this moment cannot be applied by application of a single force since this would unbalance the forces (which are in balance). Hence it is necessary to apply a couple to the system, i.e. a pair of equal and opposite forces. The couple produces pure moment with no additional resultant force. The moment of −12 kNm required may be produced by any combination of equal and opposite forces which are spaced so as to produce this moment. Two possibilities are shown in Figure 2.9.

Fig. 2.9 *Equivalent couples (– 12 kNm)*

Location of the equilibrant moment

Since a couple produces no resultant force, it may be located at any position in the force system with the same result.

Example 2.4

For the system of forces shown in Figure 2.10, determine how equilibrium may be obtained by addition of 2 kN forces to the system.

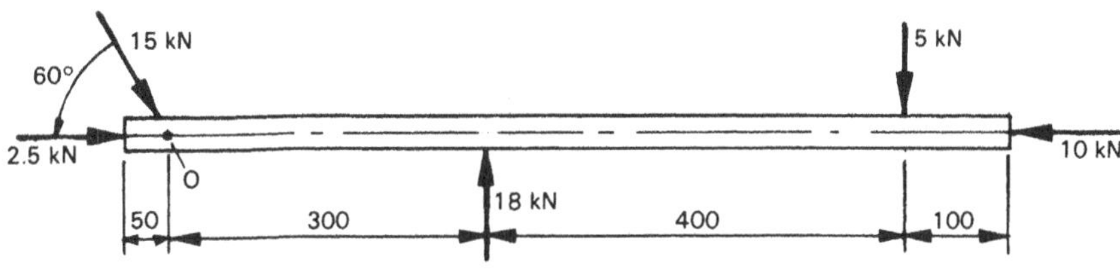

Fig. 2.10

Solution

$\Sigma F_V = -15 \sin 60° + 18 - 5 = 0$ (very closely)

$\Sigma F_H = 2.5 + 15 \cos 60° - 10 = 0$

ΣM (about O) $= -18 \times 0.3 + 5 \times 0.7 = -1.9$ kNm

Therefore there is an unbalanced moment (anticlockwise) of 1.9 kNm and equilibrium will be restored by addition of a couple of this magnitude to the system. Since 2 kN forces are to be used, they must be placed at a distance of **0.95 m apart** anywhere in the system provided that they produce a **clockwise** moment.

2.5 The three force system

It was seen that if any force system is in equilibrium there will be no resultant force (in any direction) and no resultant moment (about any point). This may be expressed in another way by regarding any one of the forces as an equilibrant force. Then the resultant of the other forces must be equal and opposite to this "equilibrant" force.

Consider the three parallel forces acting in the system shown in Figure 2.11.

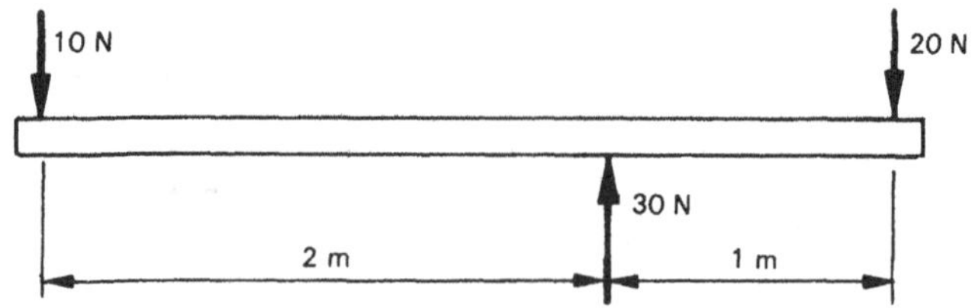

Fig. 2.11 *Force system in equilibrium with three parallel forces*

It is evident that this system is in equilibrium because

$\Sigma F_V = 0, \Sigma F_H = 0, \Sigma M = 0$ (about any point).

Now regard the 30 N force as an "equilibrant" force. Then the resultant of the 10 N and 20 N forces is 30 N, acting downward and 2 m to the right of the 10 N force. That is the resultant of the two remaining forces is equal and opposite to the initial force and has the same line of action. The same conclusion applies to a three force system where the forces are not all parallel and it leads to an important new principle: *for such a system to be in equilibrium the forces must all be concurrent*, that is their lines of action must all pass through a common point. This principle may readily be verified by reference to Figure 2.12.

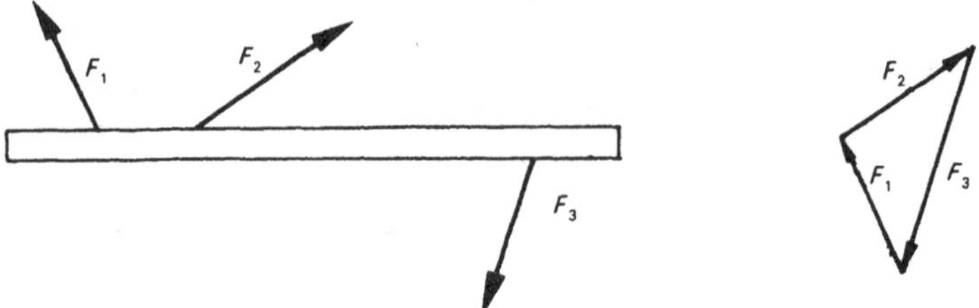

Fig. 2.12 *Force system with three non-parallel forces*

The forces are such that the force polygon closes; however the system cannot be in equilibrium because the forces are non-concurrent. For example, combining any two of the forces (say F_1 and F_2) their resultant F will be equal in magnitude and opposite in direction to F_3 (because the force polygon closes) but will not have the same line of action (see Fig. 2.13).

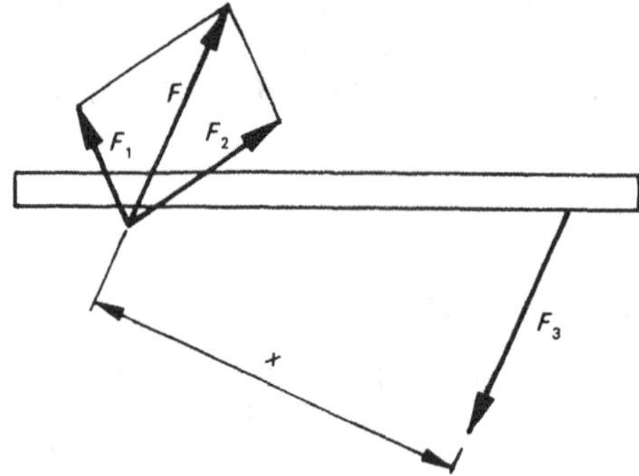

Fig. 2.13 *The resultant of F_1 and F_2 is F which is equal and opposite to F_3 but displaced a distance x from it*

It is evident that there is an unbalanced moment of magnitude F_3x (or Fx) acting clockwise so the system cannot be in equilibrium. This moment will only be zero when x is zero, that is the lines of action of the forces pass through a common point.

Notes

1. The principle of concurrency for equilibrium of a three force system also implies that for equilibrium *no two of the forces may be parallel.* Either all three forces must be parallel or all three forces must be concurrent.
2. The principle of concurrency does *not* apply to three force systems where an external moment acts on the system. By external moment is meant a moment not caused by one of the three forces.
3. The principle of concurrency may also be applied where more than three forces act in the system by combining forces until a three force system is obtained.

Example 2.5

Show that the principle of concurrency of three forces applies to the system of forces shown in Figure 2.6 by combining the two forces acting at O.

Solution

The system is reproduced in Figure 2.14.

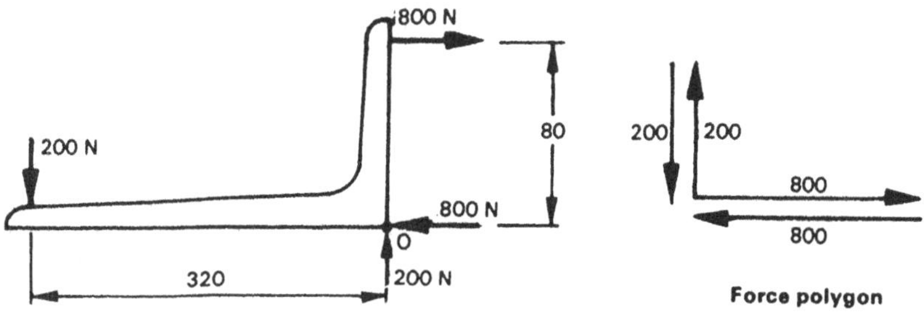

Fig. 2.14

Combining the 800 N and 200 N forces at O:

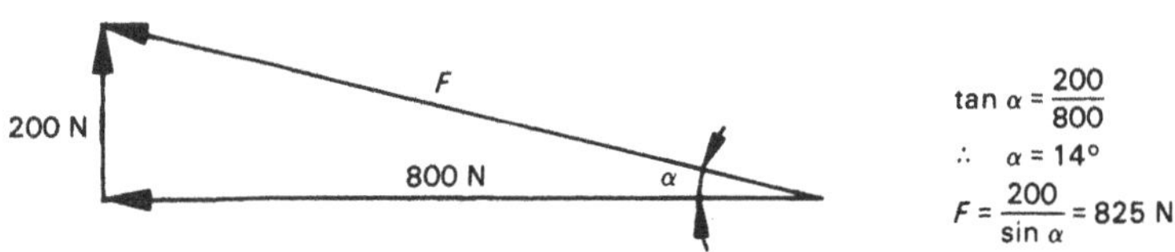

Fig. 2.15

Now by drawing to scale it is evident that the forces are concurrent, the point of concurrency being X (see Fig. 2.16).

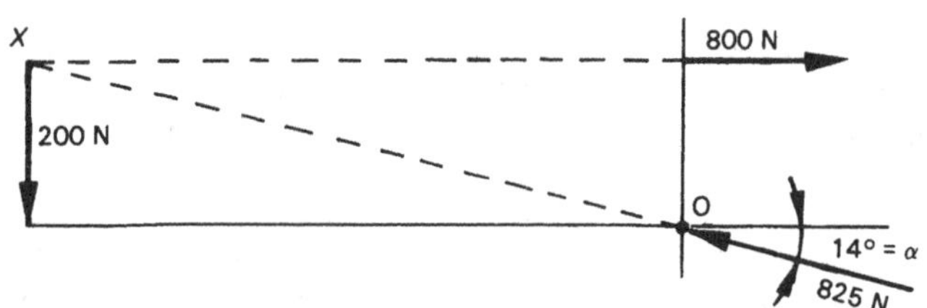

Fig. 2.16

Or by trigonometry:

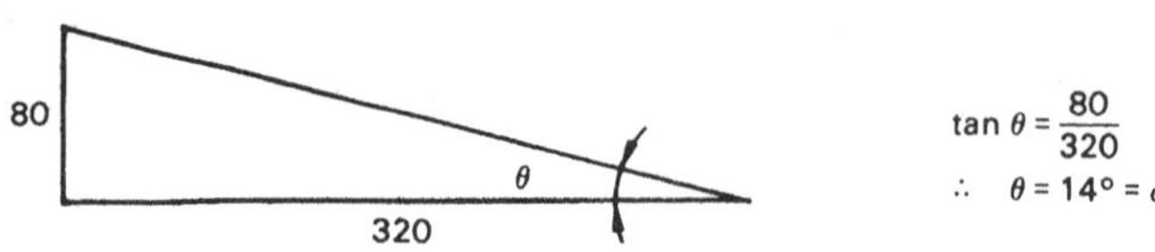

Fig. 2.17

Therefore the line of action of the 825 N force must also pass through the intersection point of the lines of action of the other two forces.

2.6 Members, joints and supports

Before using the theoretical principles of force equilibrium to solve engineering problems in statics, it may be necessary to make some simplifying assumptions regarding the members or supports used in the system. These are now outlined:

Joints or supports

(*a*) *Rigid*

The rigid joint or support may support both force or moment in any direction.

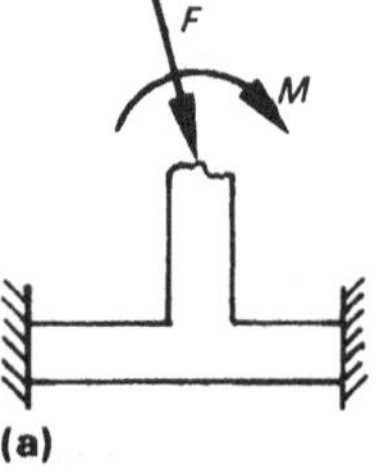

(a)

(*b*) *Sliding*

A sliding joint may have a resultant force *toward* the sliding faces but not away from them. The component of the force parallel to the face ($F \cos \theta$) is limited by the frictional force F_F.

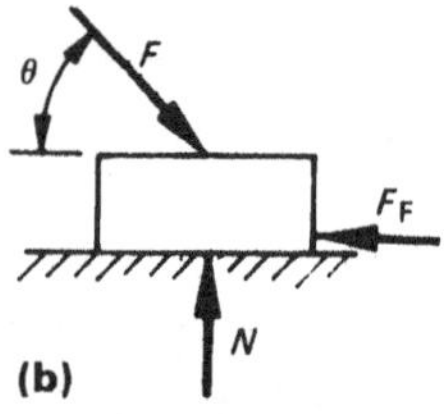

(b)

(*c*) *Frictionless*

Where there is negligible friction, $F_F = 0$. The sliding joint can then only have a resultant force in a perpendicular (or normal) direction to the surface.

(*d*) *Roller*

The roller has negligible friction and hence may be regarded as frictionless.

(*e*) *Bearing*

The friction in a bearing is usually negligible. Therefore the bearing may transmit force in every direction but no moment.

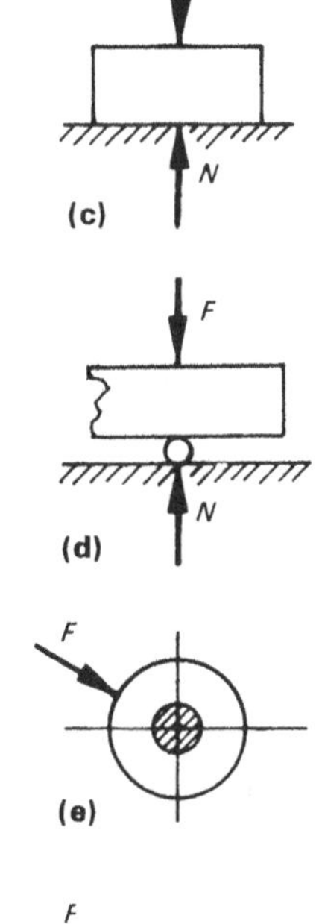

(*f*) *Pin or bolt*

In engineering, a single pinned or bolted joint is not designed to transmit moment loads since this load can only be transmitted by friction. Under these circumstances a single pinned or bolted joint may be considered as a bearing.

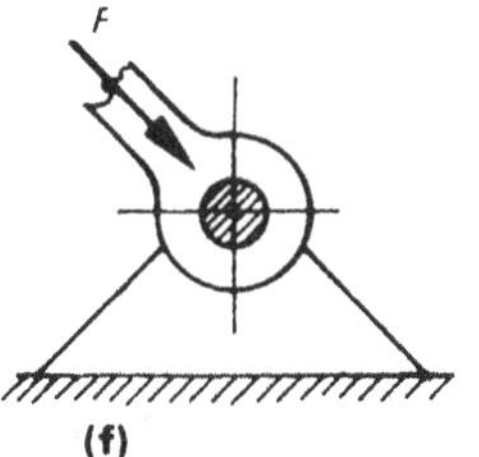

(*g*) *Idler pulley*

An idler pulley may usually be regarded as a frictionless bearing. Therefore the pulley merely changes the direction of the belt tension which is the same on both sides. The force transmitted to the bearing is the resultant of the belt tensions and must pass through the centre of the bearing at the mid-angle of the belt.

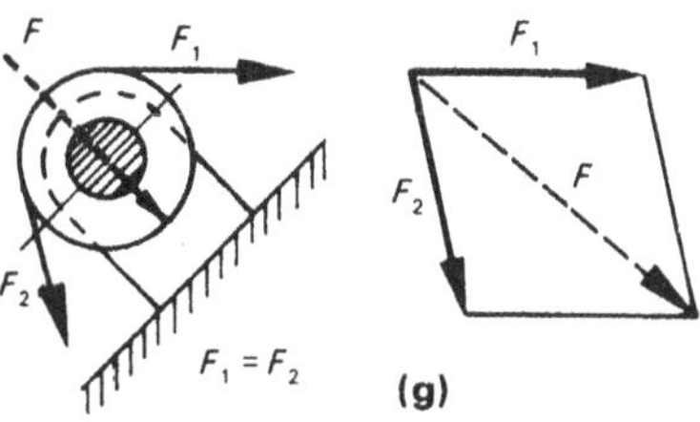

Fig. 2.18

Members

(*a*) *Flexible*

Rope, wire rope (cable), chain, vee belt, flat belt—flexible members may have a force only axially, i.e. in the direction of the member itself. Furthermore the force must be a tension force (pulling out from the support). Flexible members cannot transmit compression, shear force or moment.

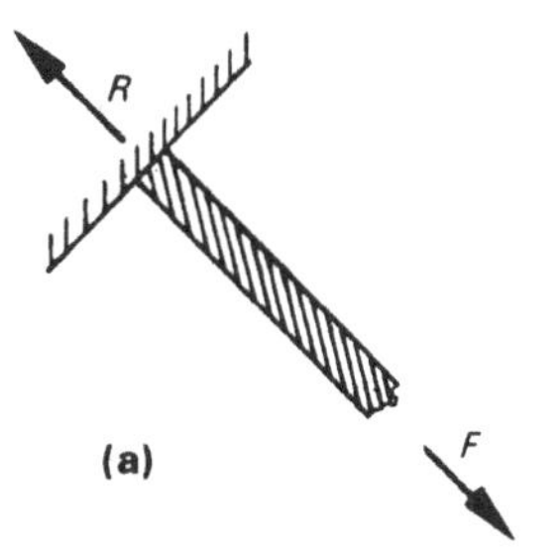

(*b*) *Rigid*

A rigid member may support tension or compression forces and also moment.

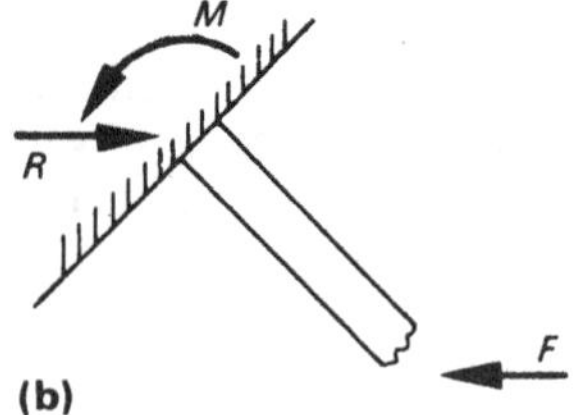

(*c*) *Pin jointed*

Of great importance in mechanics is the pin jointed rigid member. This may have tension or compression force and also moments. However two important conclusions follow:

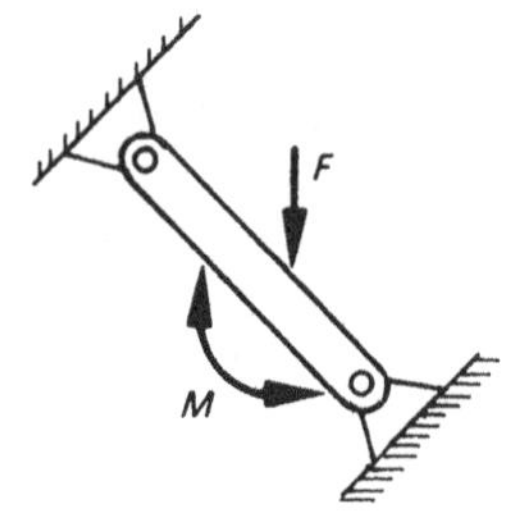

1. Force in any direction may be transmitted through the pin connection but no moment can be transmitted through the pin.

2. If forces are applied to the member only at the *pin joints* then the direction of the force in the member must be along the centre line joining the pins; that is either pure tension (tie) or compression (strut) and there will be no shear force or bending moment in the member.

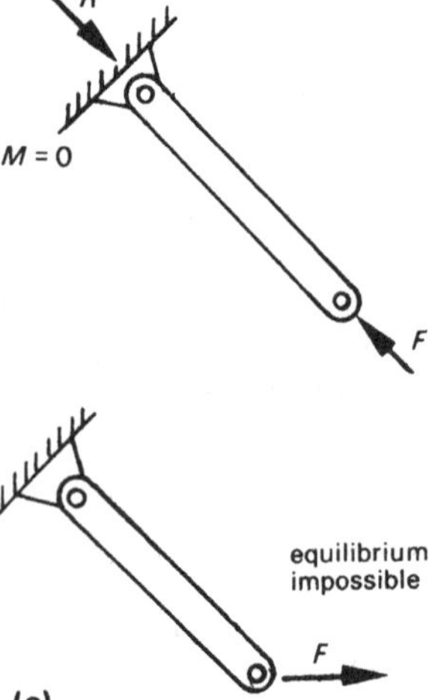

Fig. 2.19

2.7 Determination of unknown forces

In practice it is seldom necessary to prove that a force system is in equilibrium from knowledge of all the forces acting in the system. Usually the problem is to determine some unknown force or forces in a system which is known to be in equilibrium. This type of problem may be solved using the principles of force equilibrium (statics) as follows.

One unknown force

Where a single force is unknown it may be determined in both magnitude and direction.

Two unknown forces

Where two forces in the system are unknown in both magnitude and direction, a solution is impossible by the method of statics. However, solution is possible if some additional fact is known—usually the direction of one of the unknown forces. This is why the principle of concurrency of three non-parallel forces is so important since it enables unknown directions to be established.

In many cases the unknown forces are reactive forces, namely forces resulting from the application of other forces (the applied or active forces). That is the reactive forces arise in response to the applied forces—if the applied forces were removed the reactive forces would be zero.

Whenever the sense of an unknown force is not given nor obvious, but the line of action is known, the sense may be assumed. Then if a negative answer is obtained, the force must act in the opposite sense to that assumed.

Below are several examples showing how the principles of force equilibrium may be used to solve typical engineering problems involving unknown forces. In all cases the forces due to weight of members themselves are not included unless stated otherwise. Where the weight of the member is to be included, it is treated as a force acting vertically downward at the centre of mass of the member.

Example 2.6

A tug boat exerts a force of 25 kN on a towing cable which makes an angle of 20° to the forward direction of the ship being towed.

Determine the forward drag and side drag on the ship.

Solution

Since a cable is used to transmit the force, this force must be tensile and acting in the direction of the cable itself.

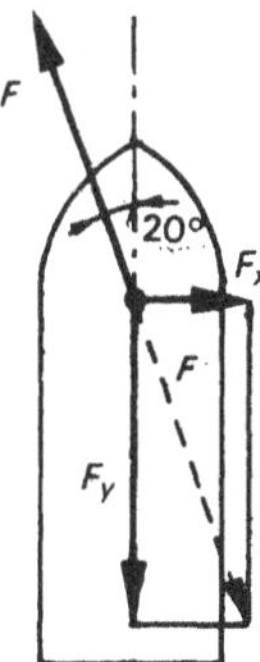

Fig. 2.20

Calling the forward drag F_y and the side drag F_x:

$F_y = F \cos 20°$ $\qquad$ $F_x = F \sin 20°$

$= 25 \cos 20°$ $\qquad$ $= 25 \sin 20°$

$= \mathbf{23.5\ kN}$ $\qquad$ $= \mathbf{8.55\ kN}$

Example 2.7

Determine the tension in the cable and the reactions at A and B for the beam of mass 20 kg/m shown in Figure 2.21. Use both analytical and graphical solutions.

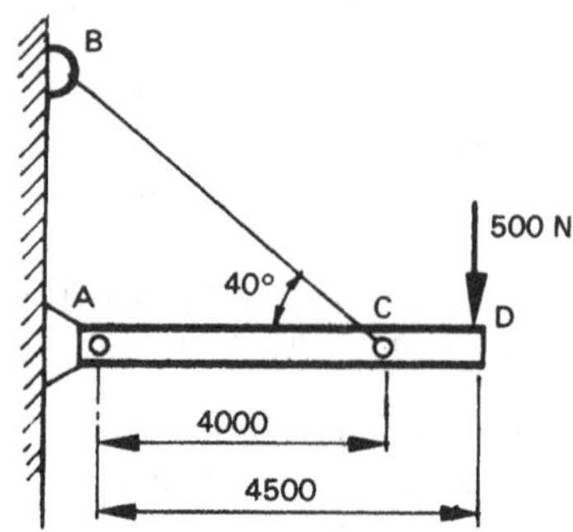

Fig. 2.21

Solution

The weight of the beam is

$$20 \times 4.5 \times 9.81 = 883 \text{ N}$$

and acts downward at the midpoint of the beam. Since BC is a cable, the force in the cable must be axial and therefore the reaction at B must also lie along the same line of action. However the direction of the reaction at A is unknown for, although the member AD is pin-jointed, the forces are applied at intermediate positions as well as at the pins.

Analytical solution

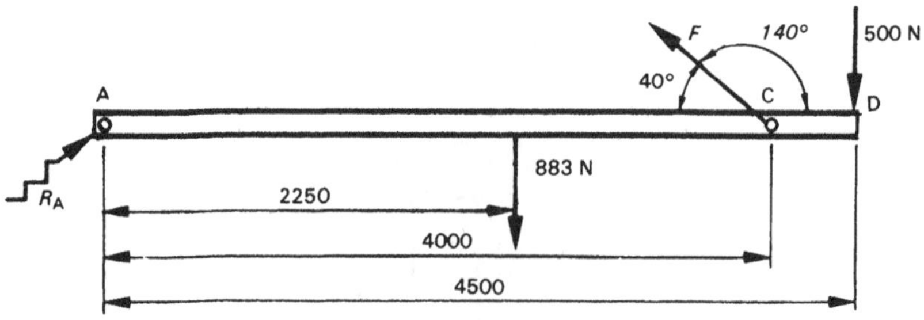

Fig. 2.22

Taking moments about A:

$$883 \times 2.25 + 500 \times 4.5 - F \sin 40° \times 4 = 0$$

$$\therefore F = \mathbf{1.65\ kN\ @\ 140°}\ (1648 \text{ N})$$

At joint B (see Fig. 2.23):

it is evident that $\boldsymbol{R_B} = \mathbf{1.65\ kN\ @\ 140°}$.

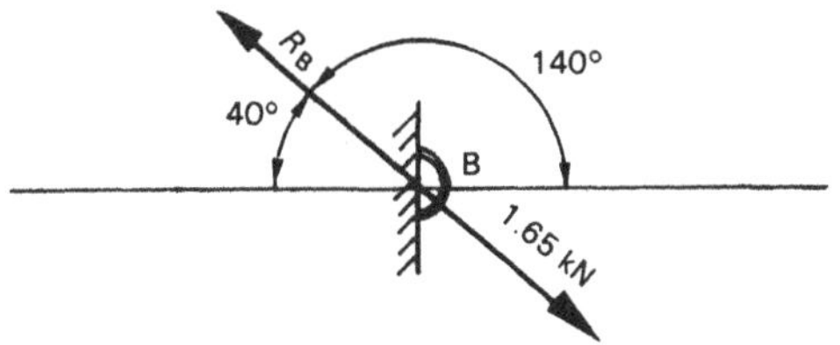

Fig. 2.23

In order to solve for the reaction at A, now tabulate the horizontal and vertical components of the known forces in the system shown in Figure 2.24.

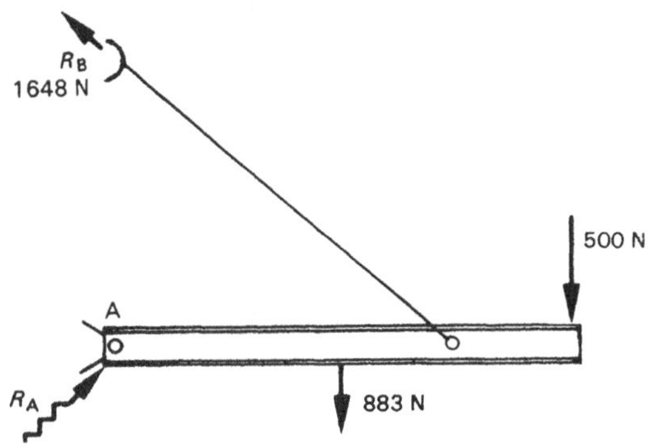

Fig. 2.24

Force (N)	*Vertical component (N)*		*Horizontal component (N)*	
500		−500		0
883		−883		0
R_B = 1648	1648 sin 140°	1059.3	1648 cos 140°	−1262.4
	Total	−323.7		−1262.4

Since $\Sigma F_V = 0$ and $\Sigma F_H = 0$:

$R_{AV} - 323.7 = 0$ $\quad$ $R_{AH} - 1262.4 = 0$

$\therefore R_{AV} = 323.7$ N $\quad$ $\therefore R_{AH} = 1262.4$ N

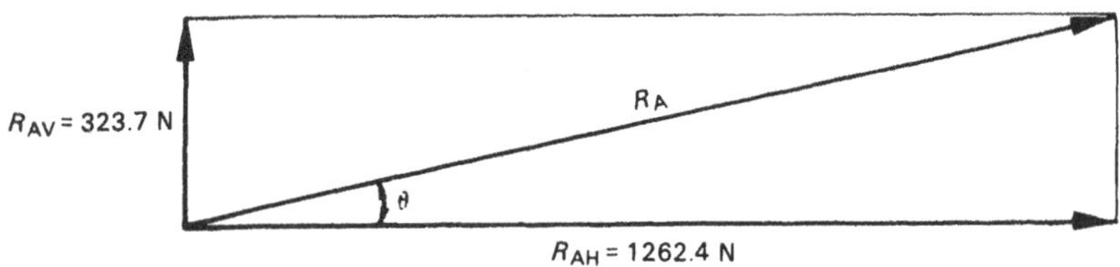

Fig. 2.25

$\therefore$ **R_A = 1.3 kN @ 14.4°.**

Graphical solution

One method of graphical solution is to use the three force principle, that is the system shown in Figure 2.24 will be a three force system if the 500 N and 883 N forces are combined. This may be done graphically or by simple calculation:

$$F = 1383 \text{ N}$$

and by taking moments about the 883 N force:

$$500 \times 2.25 = 1383 \times x$$

$$\therefore \quad x = 0.813$$

that is the resultant acts 3.063 m from A.

Now complete the graphical solution as shown in Figure 2.26 by determining the point of concurrency of the three forces and drawing the vector force diagram.

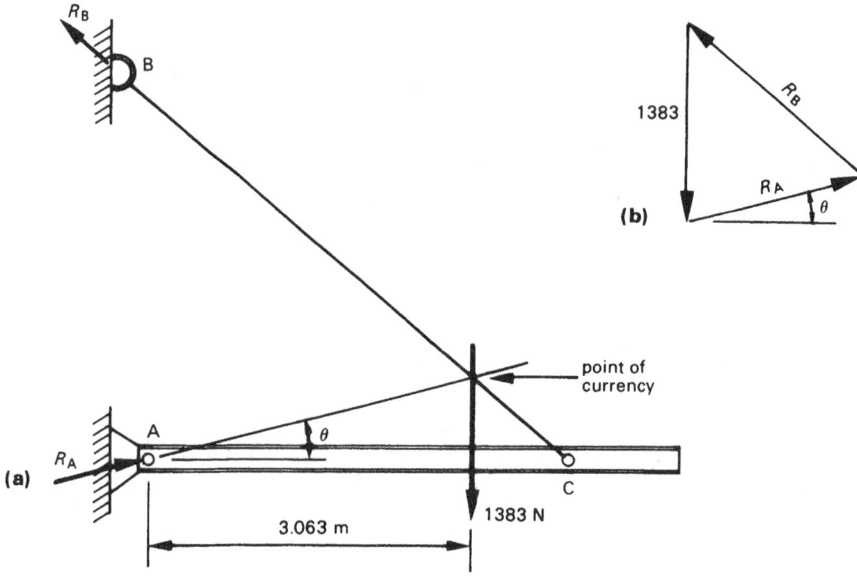

Fig. 2.26 *(a) Space diagram. (b) Vector force diagram*

By measurement: $\boldsymbol{R_A} = \mathbf{1.3\ kN}$

$\boldsymbol{R_B} = \mathbf{1.65\ kN}$

$\boldsymbol{\theta} = \mathbf{14.4°}$

Example 2.8

Determine the pin reactions (resultant forces) acting on pins A, B, C of the structure shown in Figure 2.27, using both analytical and graphical methods. Also draw a free body diagram for each member.

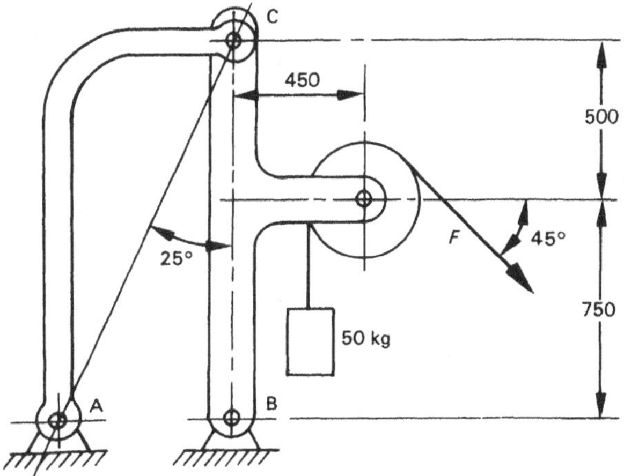

Fig. 2.27

Solution

The following observations may be made:

1. The force F must be equal to the weight force of the 50 kg load, namely 490.5 N. The resultant of these two forces passes through the centre of the pulley and the diameter of the pulley is immaterial.
2. Member AC is pinned at both ends and has forces transmitted to it only at the pin joints. Therefore the forces at pins A and C act along the line joining A and C.

Analytical solution

The distance between A and B is 1250 tan 25° = 582.9 mm.

Assuming positive directions for the components of the reaction at A and taking moments about B:

$$R_{AV} \times 0.5829 + 490.5 \times 0.45 + 490.5 \sin 45° \times 0.45 + 490.5 \cos 45° \times 0.75 = 0$$

$$\therefore R_{AV} = -1092.7 \text{ N}$$

Hence the vertical component of the reaction at A is downward as shown in Figure 2.28.

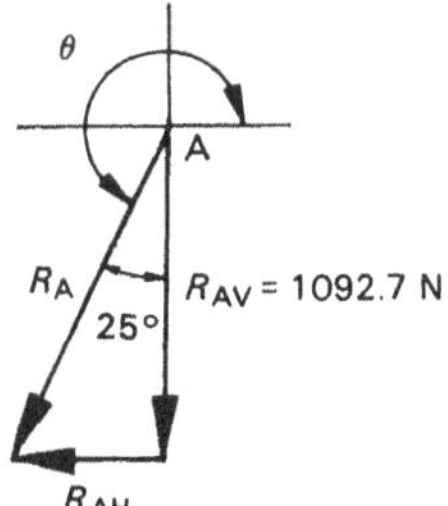

Fig. 2.28

By trigonometry:

$$R_{AH} = -509.5 \text{ N}$$

$$\mathbf{R_A = 1206 \text{ N @ } 245°}$$

Now obtain the reaction at B by taking horizontal and vertical components:

Force (N)	*Vertical component (N)*		*Horizontal component (N)*	
50 kg mass		−490.5		0
$F = 490.5$ N	−490.5 sin 45°	−346.8	490.5 cos 45°	346.8
$R_A = 1206$ N		−1092.7		−509.5
	Total	−1930		−162.7

Assuming a positive direction for the components at B:

$$R_{BV} - 1930 = 0 \qquad R_{BH} - 162.7 = 0$$

$$R_{BV} = 1930 \text{ N} \qquad R_{BH} = 162.7 \text{ N}$$

Hence the horizontal component of the reaction at B acts toward the right.

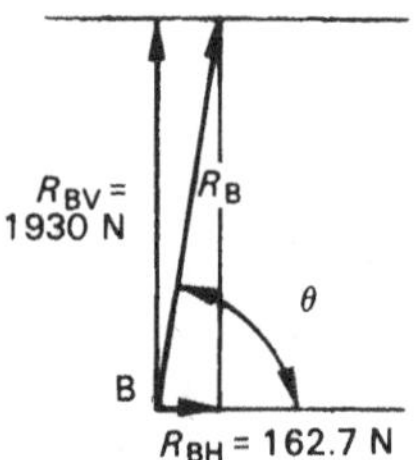

Fig. 2.29

By trigonometry, $\theta = 85.2°$ $\therefore$ $\boldsymbol{R_B}$ **= 1937 N @ 85.2°**

The reactions at pins A and B have now been determined. At pin C the reaction must be the same as at A only acting in the opposite sense, namely 1206 N @ 65°.

The free body diagrams of each member may be drawn as shown in Figure 2.30.

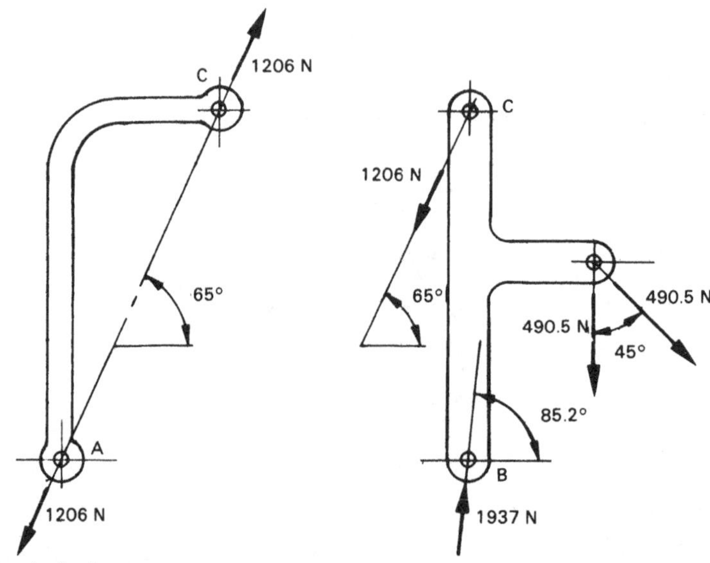

Fig. 2.30 *Free body diagrams*

Note that the direction of the force at pin C is opposite with respect to the other member.

Graphical solution

By combining the two forces at the pulley, the system may be reduced to a three force system and hence the principle of concurrency used to determine the direction of the reaction at B. Then a vector force diagram may be drawn to determine the magnitude of the reactions at A and B. This is shown in Figure 2.31.

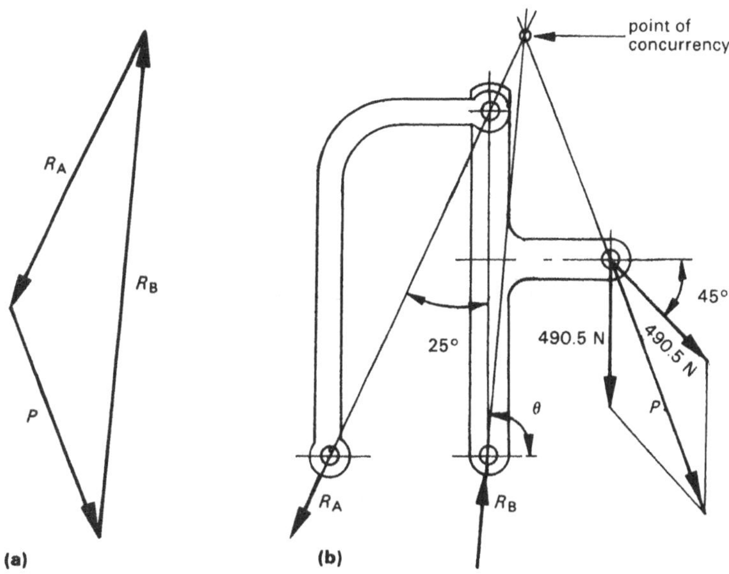

Fig. 2.31 *(a) Vector diagram. (b) Space diagram*

By scaling from the diagram, the values of R_A, R_B and θ are determined as before.

Problems

2.1 For the force system shown in Figure P2.1 determine the equilibrant force using an analytical method.

104 N @ 258.4°

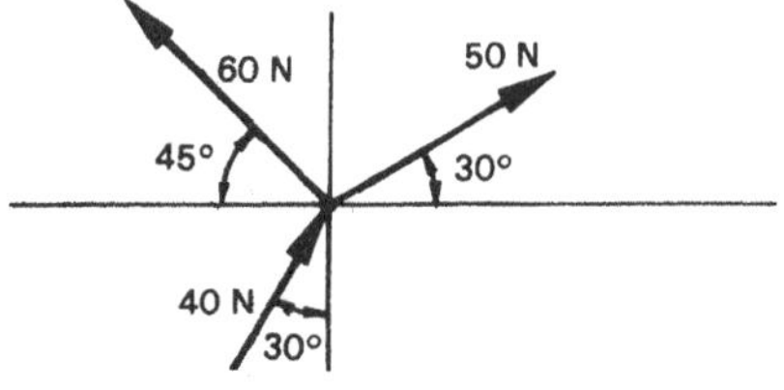

Fig. P2.1

2.2 Solve problem 2.1 graphically.

2.3 For the force system shown in Figure P2.3 determine, using an analytical method:

(a) equilibrant force

(b) location of the equilibrant force on the *x* axis

(c) equilibrant moment.

(a) 2.15 kN @ 15.7° (b) −3068 mm (c) 1.78 kNm

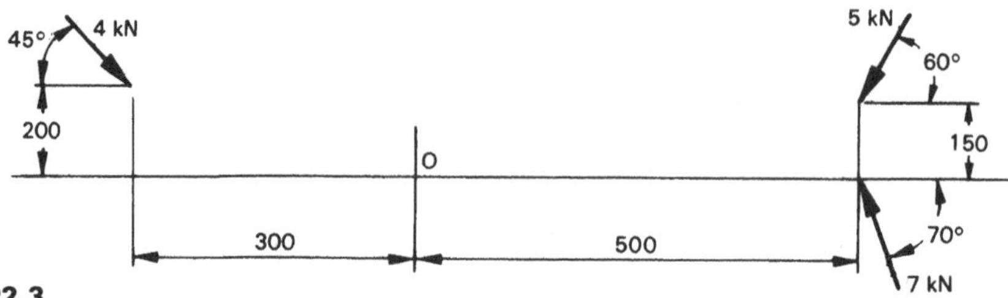

Fig. P2.3

2.4 Solve problem 2.3 (a) and (b) graphically.

2.5 In the yacht rigging shown in Figure P2.5, the mast pivots at the base and the backstay and forestay are tightened to a tension of 2.5 kN.

Determine the angle θ and the compressive force in the mast using an analytical method.

92.5°, 4.435 kN

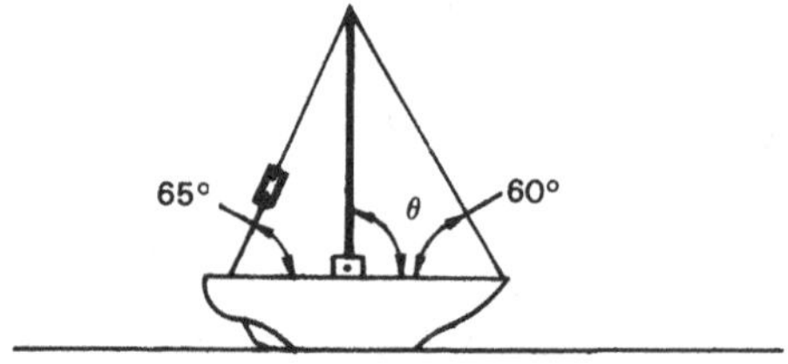

Fig. P2.5

2.6 Solve problem 2.5 graphically.

2.7 For the lever system shown in Figure P2.7, determine force *F* and the pin reaction at A using an analytical method.

357 N, 1.67 kN @ 245.8°

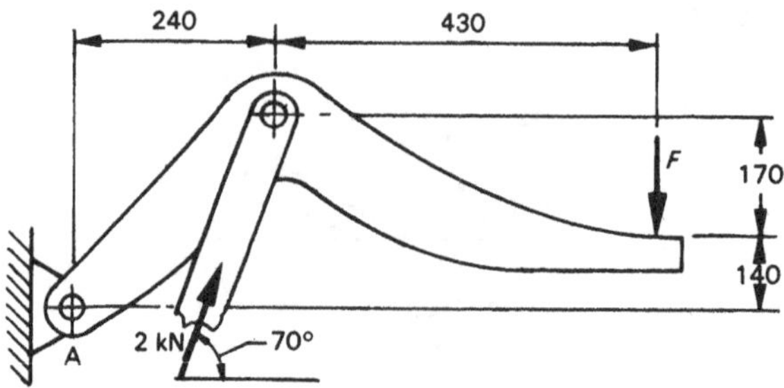

Fig. P2.7

2.8 Solve problem 2.7 graphically.

2.9 A glider mass 200 kg is towed by a cable which makes an angle of 25° to the horizontal. If the tension in the cable is 1.2 kN, determine the lift and drag forces on the glider.

2.47 kN, 1.09 kN

2.10 A glider mass 230 kg is towed by a cable which makes an angle of 55° with the horizontal. If the lift force is 20 per cent greater than the weight of the glider, determine the tension in the cable and the drag force on the glider.

551 N, 316 N

2.11 For the lever system shown in Figure P2.11, determine the force F and the bearing reaction at A using both analytical and graphical methods.

156 N, 222 N @ 138.7°

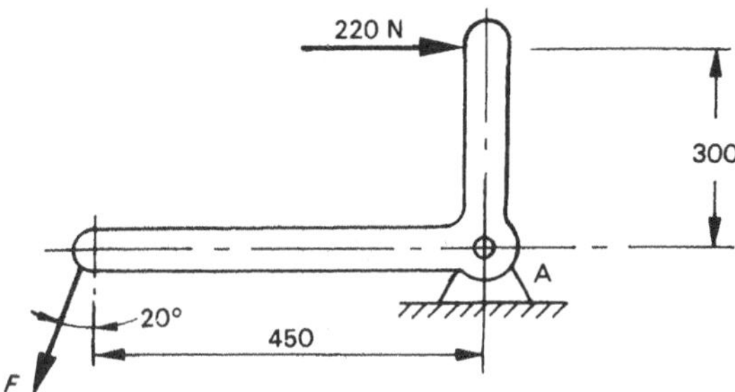

Fig. P2.11

2.12 The hold-down device for a boat is illustrated in Figure P2.12.

Determine the force in the top and bottom straps when the applied force F is 50 N.

193 N, 212 N

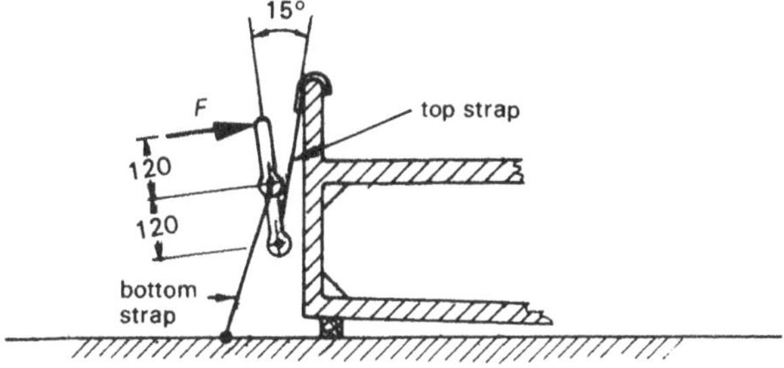

Fig. P2.12

2.13 A ladder mass 30 kg, length 6 m, rests against a smooth (frictionless) wall making an angle of 75° with the horizontal. Determine:

(a) wall reaction force

(b) ground reaction force

using both analytical and graphical methods.

(a) 39.4 N @ 180° (b) 297 N @ 82.4°

2.14 Solve problem 2.13 if a man mass 80 kg stands on a rung 1 m from the top of the ladder.

(a) 215 N @ 180° (b) 1100 N @ 78.75°

2.15 For the beam mass 30 kg/m shown in Figure P2.15, determine the tension in the cable and the reaction force at the hinge pin A using both analytical and graphical methods.

1.07 kN, 2.77 kN @ 68.7°

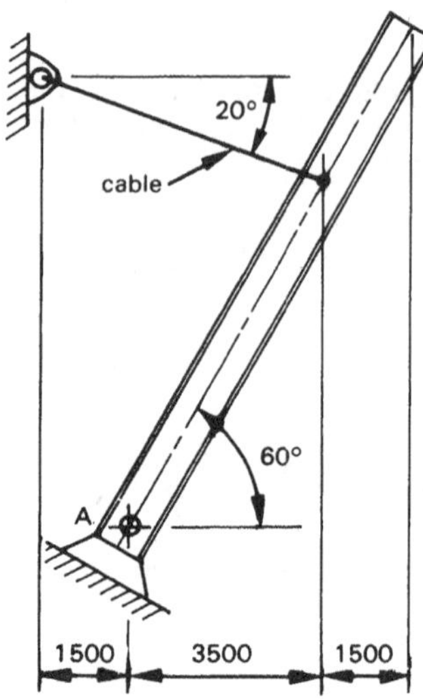

Fig. P2.15

2.16 Solve problem 2.15 if a force of 2 kN is applied downward perpendicular to the axis of the beam at the free end.

3.97 kN, 3.27 kN @ 52.3°

2.17 Two brackets as shown in Figure P2.17 are used to support equally a lifeboat of total mass 800 kg.

Assuming that all the vertical force is taken by the ground at A, determine the reaction force at A and B due to the weight of the lifeboat. Use both analytical and graphical methods.

5.28 kN @ 48°, 3.53 kN @ 180°

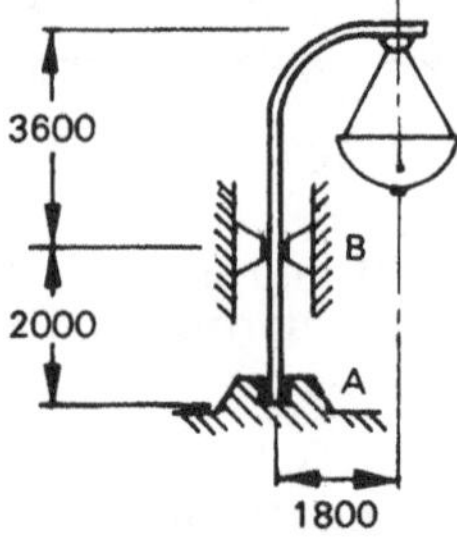

Fig. P2.17

2.18 A light pole has a mass of 120 kg and centre of mass located 250 mm to the right of the base. It supports a light fitting of mass 20 kg as shown in Figure P2.18.

Determine the reactive force and moment at the ground.

1.37 kN @ 90°, −451 Nm (anticlockwise)

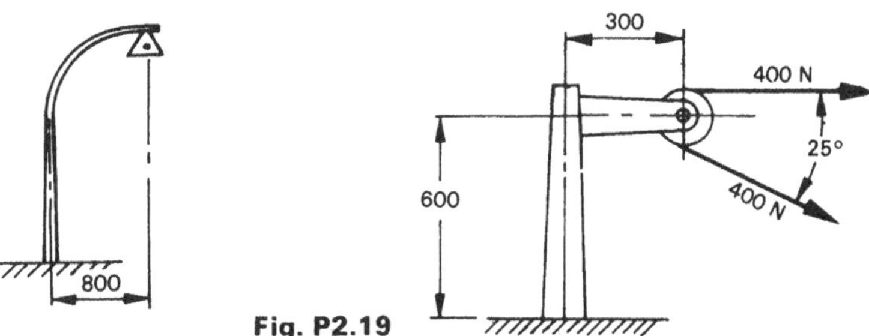

Fig. P2.18 **Fig. P2.19**

2.19 An idler pulley with belt tensions 400 N is supported by a bracket as shown in Figure P2.19.

Determine the reactive force and moment at the base of the bracket.

781 N @ 167.5°, −508 Nm

2.20 An idler pulley with belt tensions 500 N is supported by a bracket as shown in Figure P2.20.

Determine the force in the turnbuckle and the reaction at support pin A using both analytical and graphical methods.

1.08 kN, 1.15 kN @ 104.5°

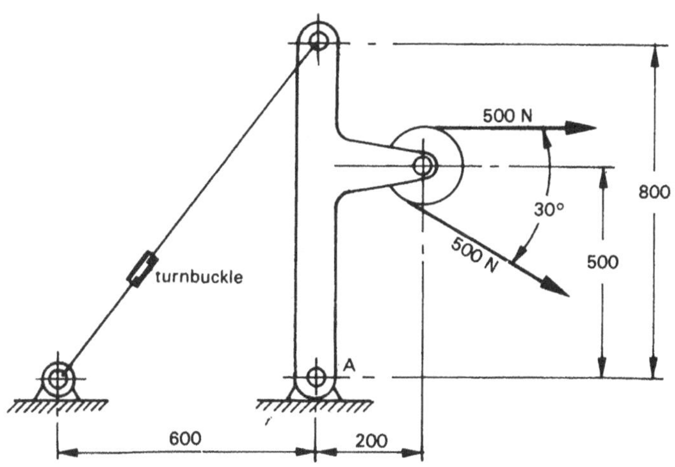

Fig. P2.20

2.21 The electric motor shown in Figure P2.21 has a mass of 18 kg.

Determine the belt tensions and the reaction force in pin A when the motor is at rest if the weight of the bracket itself and bearing friction are negligible.

353 N, 570 N @ 221.6°

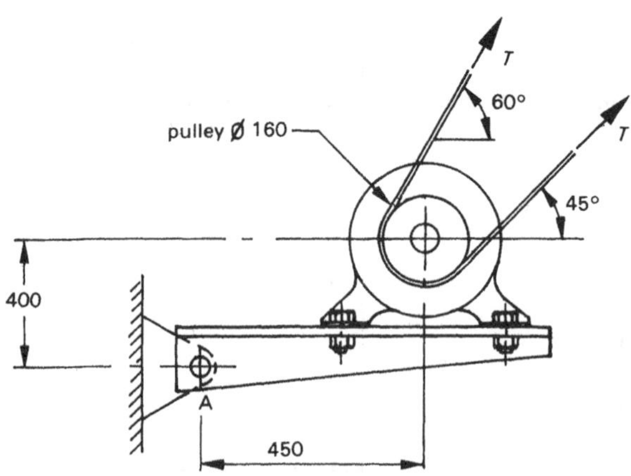

Fig. P2.21

2.22 The toggle clamping device for a machine in which the links have equal length is illustrated in Figure P2.22.

Determine the horizontal clamping force when the applied force P is 80 N if friction is negligible.

916 N

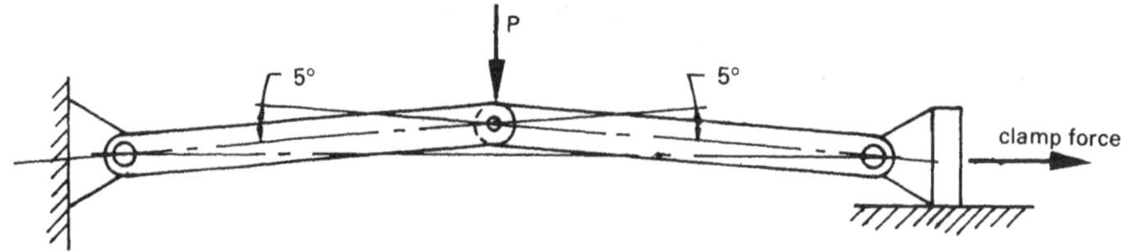

Fig. P2.22

2.23 Determine the reactive force in pins A and B for the hinged arch shown in Figure P2.23. Use both analytical and graphical methods.

1.39 kN @ 25.64°, 1.6 kN @ 141.3°

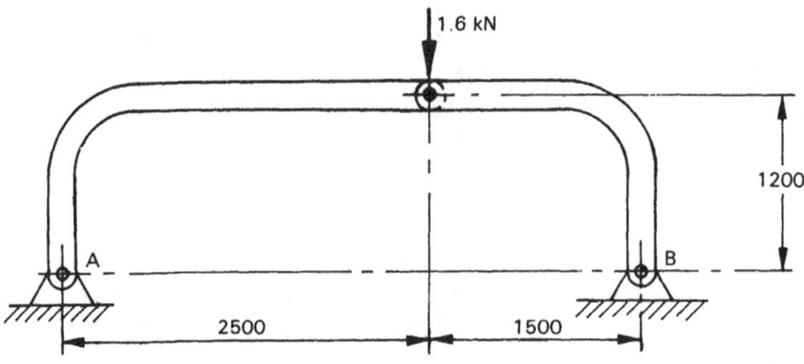

Fig. P2.23

2.24 Solve problem 2.23 if an additional load of 2.8 kN acts vertically downward 1.5 m to the right of A.

3.48 kN @ 42.5°, 3.28 kN @ 141.3°

2.25 At the position shown in Figure P2.25, the gas pressure acting on the top of the piston of the internal combustion engine of bore 100 mm and stroke 100 mm is 8 Mpa (gauge).

If friction and inertia forces are neglected determine:

(a) compressive force in the connecting rod

(b) side thrust between piston and bore

(c) turning moment (torque) on the crankshaft.

(a) 63.8 kN (b) 11.1 kN (c) 2.05 kNm

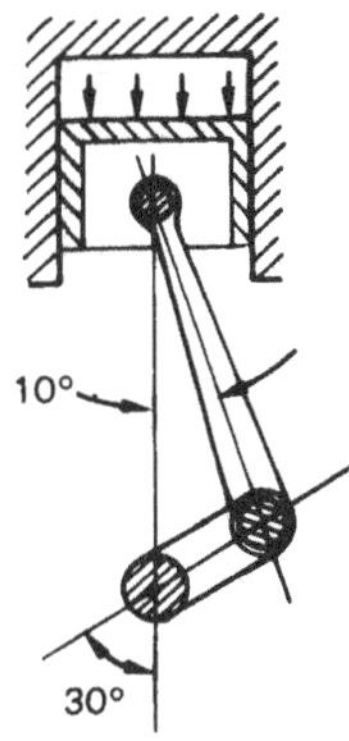

Fig. P2.25

2.26 For the yacht rigging shown in Figure P2.26, the mast pivots at the base and the backstay is tensioned with a turnbuckle.

If the tension in the backstay is 6 kN and the mast is perpendicular to the deck, determine:

(a) tension in the forestay

(b) pivot pin reaction.

Also draw a free body diagram showing all the forces acting on the mast.

(a) 8.67 kN (b) 13.4 kN @ 92.3°

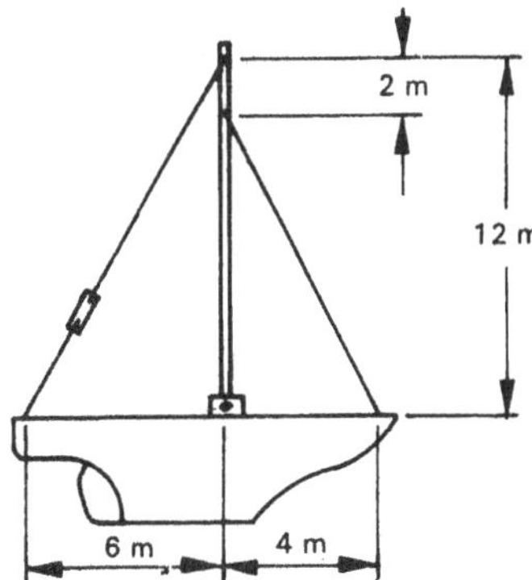

Fig. P2.26

2.27 For the pinned bracket shown in Figure P2.27, determine the reactions acting at:

(a) pin A

(b) pin B

(c) pin C.

Also draw a free body diagram of each member showing all the forces acting on it.

(a) 472 N @ 238° (b) 1226 N @ 78.2° (c) 472 N @ 58°

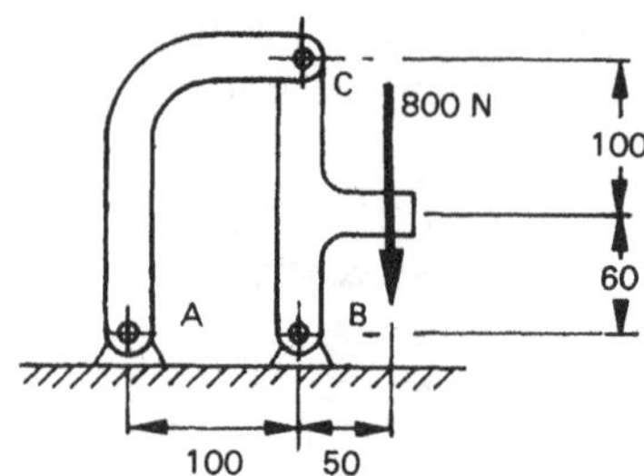

Fig. P2.27

3

Loads and support reactions

This chapter is concerned with loading situations often encountered in engineering applications involving beams, frames, cantilevers and other structures and the support reactions induced by the loads. The structure is considered as a complete unit, that is it is not necessary to take into account the internal forces or moments in the structure itself in order to determine the reactions. For example the structures A and B in Figure 3.1 have the same reactions.

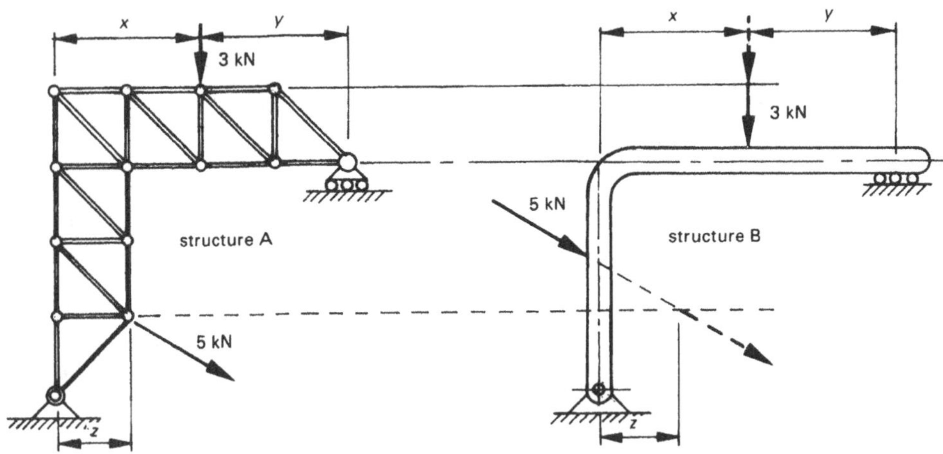

Fig. 3.1 *Support reactions are the same for both structures A and B which are therefore equivalent in this respect*

It will be noted that the principle of transmissibility has been used in order to obtain the equivalent structure, that is the point of application of a force along its line of action does not change the support reactions induced by that force. However, the internal forces induced in the structure itself may change if the point of application is changed, and therefore the point of application is important when considering the internal forces induced in the members and joints of a structure. For example it is clear from Figure 3.1 that the movement of the applied forces along their lines of action has no effect on the support reactions for structure A, but will affect the internal forces in the members and joints of this structure.

Determination of the reactions is usually the first step in a complete analysis of a loaded structure. It is a most important step because if an error is made, the subsequent calculations will be in error. There are no new principles involved and determination of the reactions is simply a particular application of the principles of force equilibrium in a system. To this extent this chapter could be considered merely as an extension of Chapter 2 and indeed many of the problems in Chapter 2 did involve the determination of reactive forces. However, because of the variety of loads and structures and because of the importance of correct determination of the reactions for subsequent calculations, a full chapter has been devoted to this topic.

Generally, the direction (sense) of a reactive force or moment will not be known. In calculations, it is good practice to assume positive directions: vertical reactive forces up, horizontal reactive forces acting to the right and reactive moments clockwise. Then if the calculation produces a negative result, the correct sense is opposite to the one assumed.

As a final note, remember that the structure is considered as being completely rigid—that is without deflection. It would be more accurate to say that the amount of deflection in rigid structures is usually so small that this deflection has negligible effect on the reactions.

3.1 Types of load

Load is simply another word for force. However just as "pipe" and "tube" are equivalent words, each has traditionally been used in different engineering applications. So the word "load" is used for the applied forces occurring in beams and structures and which these beams and structures have to resist.

Distinction may be made between "dead" loads and "live" loads. A **dead load** is one which does not vary appreciably throughout the life of the structure, such as the weight of the structure itself or fixed weights attached to the structure. A **live load** is one which may vary continuously throughout the life of the structure, such as wind load or the movement of vehicles on the structure.

Dead loads can usually be determined with accuracy whereas live loads are far more difficult to predict. For example, is it necessary to design a vehicular bridge for the eventuality of bumper-to-bumper fully laden cement-mixers at the time of a gale strength storm? This loading is possible but obviously highly unlikely and design for this eventuality would involve high unnecessary cost.

The determination of probable live loads in actual structures is beyond the scope of the subject of mechanics and in this book the magnitude of the load will always be given.

A **concentrated load** is a load which acts at a point such as the loads shown in Figure 3.1. In practice concentrated loads occur whenever the load is distributed over a small area compared to the size of the structure, for example loads applied through cables or similar members.

A **distributed load** is one which is spread over a considerable area compared to the size of the structure itself. Distributed loads may be uniformly distributed or non-uniformly distributed as shown in Figure 3.2.

For the purpose of determining the support reactions, a distributed load may be treated as a concentrated load of magnitude equal to the total load (area of distributed load) and acting through the centroid of the area of the distributed load. For a uniformly distributed load, this means the equivalent concentrated load acts at the centre of the distributed load and for the uniformly increasing distributed load it acts at the $\frac{2}{3}$, $\frac{1}{3}$ point.

Although distributed loads often act in a perpendicular direction to the axis of the member, this is not necessarily the case; for example, when the distributed load is due to the

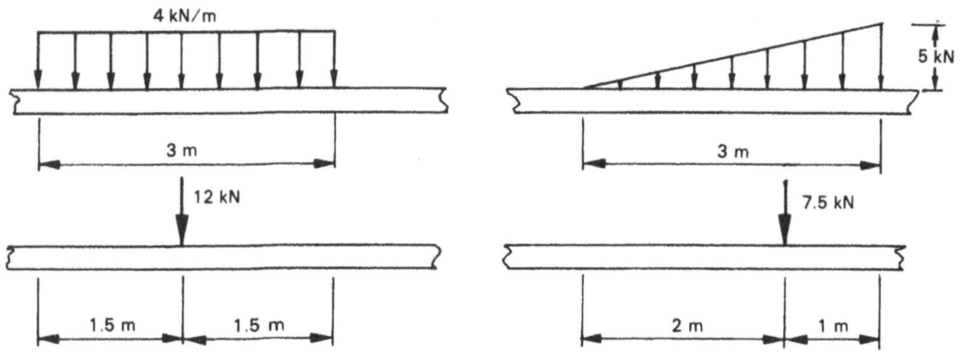

Fig. 3.2 *Uniformly and non-uniformly distributed loads and their equivalent concentrated loads*

weight of the member itself and the axis of the member is not horizontal. This is illustrated in Figure 3.3 where a beam of mass 204 kg/m is inclined at an angle to the horizontal.

The distributed load due to the weight of the beam is 9.81 × 204 N/m = 2 kN/m. The distributed and equivalent concentrated load for a 3 m length of this beam is shown in Figure 3.3.

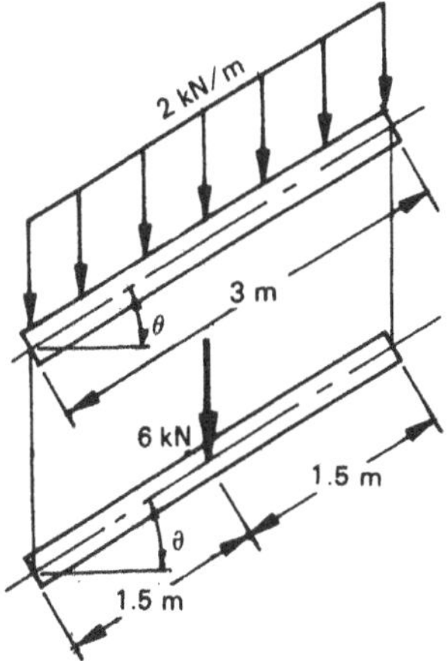

Fig. 3.3 *Uniformly distributed load and equivalent concentrated load for the weight of an inclined member*

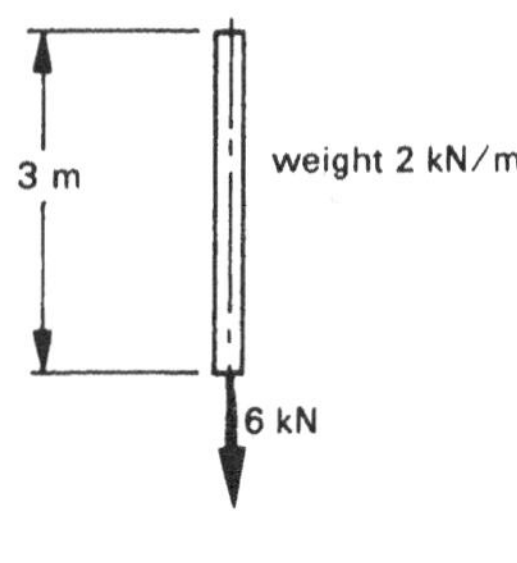

Fig. 3.4 *Vertical member—load due to weight of member*

Notes

1. The angle of the inclination does not affect the magnitude or direction of the load which is always vertical, and the equivalent concentrated load acts through the centrepoint of the beam.
2. When a member is vertical, the load due to the weight of the member itself is no longer distributed but acts as a concentrated load as shown in Figure 3.4.

3.2 Types of support

Any of the types of support described in section 2.6 may occur. These may be classified into three groups:

1. rigid or fixed
2. sliding or roller
3. bearing or pinned.

If a rigid support is used it is only necessary to support the structure at one position to ensure stability. With any other type of support used with a beam or frame, at least two support positions are required to ensure stability.

Depending upon the support system used the structure may be termed cantilever, simply supported or statically indeterminate.

Determination of the reactions in each case with both point and distributed loads will now be investigated in greater detail.

3.3 Cantilever beam reactions: Analytical solution

The cantilever is a beam which is rigidly supported at one end and free at the other end as illustrated in Figure 3.5.

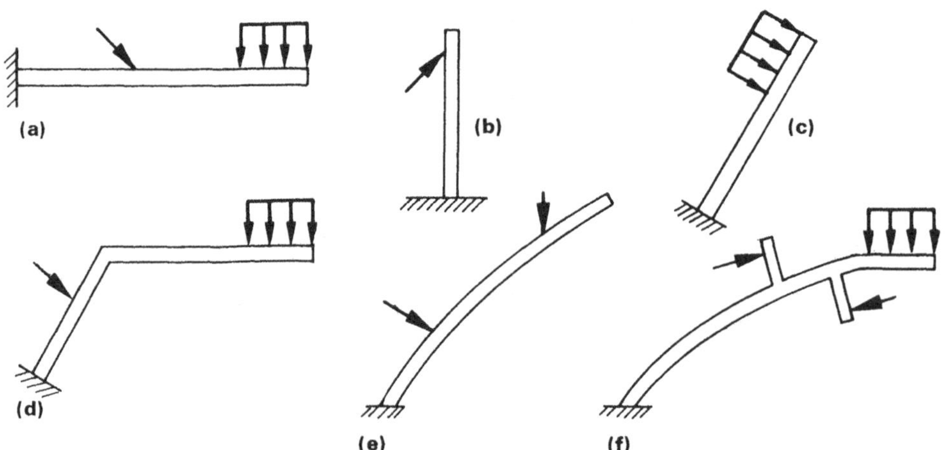

Fig. 3.5 *Cantilever beams*

The cantilever may be (a) horizontal, (b) vertical, (c) inclined, (d) cranked, (e) curved, (f) composite and loaded by any combination of concentrated or distributed loads (in any direction).

The reaction at the support may be a force (in any direction) and a moment. When there is a moment at the support, *the three force principle does not apply*. Hence graphical methods are not applicable and the reactive force and moment are determined using the equations of equilibrium for force systems:

$$\Sigma F_V = 0, \Sigma F_H = 0, \Sigma M = 0.$$

Example 3.1
Determine the reactive force and moment for the cantilever shown in Figure 3.6.

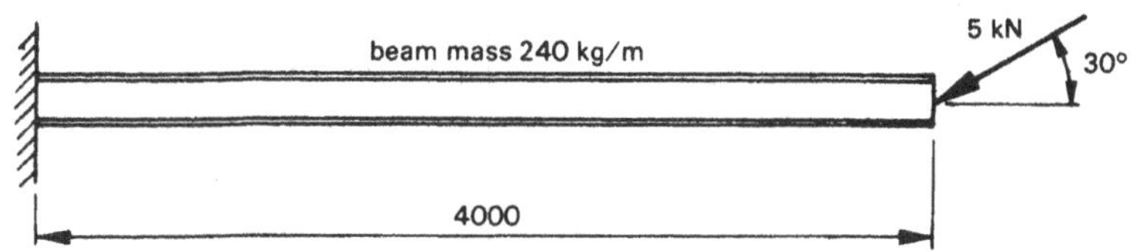

Fig. 3.6

Solution
The beam mass of 240 kg/m is equal to a distributed load of 240 × 9.81 = 2354 N/m which (for the purpose of determining the reactions) may be treated as an equivalent concentrated load of 9.42 kN at the centre of the span.

The 5 kN load may be broken into horizontal and vertical components of 4.33 kN and 2.5 kN.

The reactive force at the support is assumed to have positive components R_H and R_V and the reactive moment M is assumed positive (clockwise).

The actual loads, equivalent loads and free body diagrams for the beam are as shown in Figure 3.7.

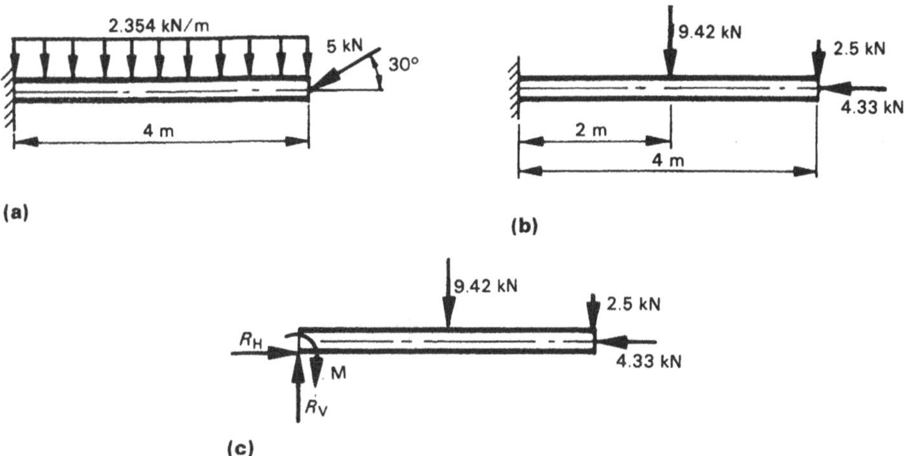

Fig. 3.7 *(a) Actual loads. (b) Equivalent loads. (c) Free body diagram*

Using the equations of equilibrium:

$\Sigma F_V = 0 \therefore R_V - 9.42 - 2.5 = 0, \qquad \therefore R_V = 11.92\text{ kN}$

$\Sigma F_H = 0 \therefore R_H - 4.33 = 0, \qquad \therefore R_H = 4.33\text{ kN}$

$\Sigma M = 0 \therefore M + 9.42 \times 2 + 2.5 \times 4 = 0, \therefore M = -28.8\text{ kNm}$

Note that the positive value of R_V and R_H indicates correct direction was assumed for them, whereas the negative value of M indicates that the reactive moment is anticlockwise and not clockwise as assumed.

From the components R_V and R_H the resultant reactive force R and angle made with the x axis may be determined by trigonometry. The result is as shown in Figure 3.8.

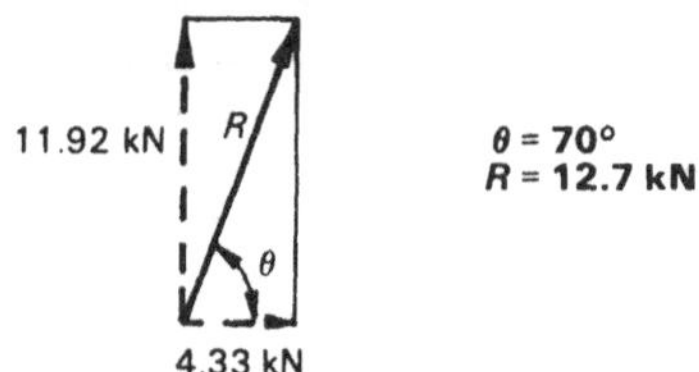

Fig. 3.8

The reactive force at the support is therefore **12.7 kN @ 70°** and the reactive moment is **−28.8 kNm** (anticlockwise).

Example 3.2

Determine the reactive force and moment for the cantilever beam shown in Figure 3.9. The beam itself has a mass of 120 kg/m.

Note: The pulley is shown diagrammatically but actually has negligible diameter.

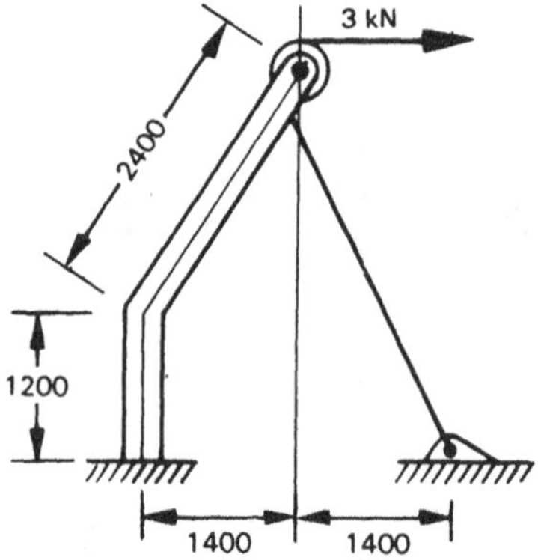

Fig. 3.9

Solution

The unknown angles and dimensions may be determined by trigonometry and the 3 kN tension in the cable resolved into horizontal and vertical components. Also the weight of each section of the beam may be calculated and converted into an equivalent concentrated load.

A free body diagram showing these forces may be drawn as shown in Figure 3.10. This diagram also shows the reactive force R broken into components R_V and R_H (assumed positive) and the reactive moment M (also assumed positive).

$\Sigma F_V = 0 \quad \therefore R_V - 1.413 - 2.825 - 2.741 = 0$

$\therefore R_V = 6.98 \text{ kN } (\uparrow)$

$\Sigma F_H = 0 \quad \therefore R_H + 1.22 + 3 = 0$

$\therefore R_H = -4.22 \text{ kN } (\leftarrow)$

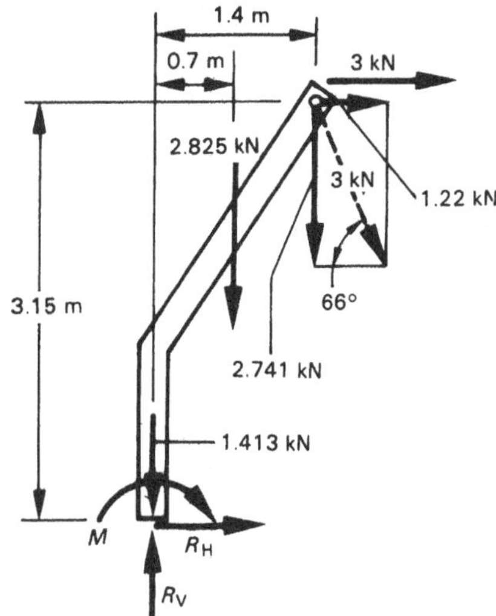

Fig. 3.10

Combining R_V and R_H:

$$\therefore \boldsymbol{R} = \mathbf{8.16\ kN\ @\ 121.2°}$$

$$\Sigma M = 0 \quad \therefore M + 2.825 \times 0.7 + 2.741 \times 1.4 + 4.22 \times 3.15 = 0$$

$$\therefore \boldsymbol{M} = \mathbf{-19.1\ kNm} \text{ (anticlockwise)}$$

3.4 Cantilever beam reactions: Computer solution

A computer program named CANTRE written in BASIC language for determining the reactive force and moment for a cantilever beam of any shape loaded with any number of concentrated loads in any direction or position on the beam is listed in Appendix 3.3.

This program is an extension of FORCE 2 and the forces acting on the beam are input in the same way, namely that each applied force is specified by magnitude, angle with the x axis, and x and y coordinates with reference to the support.

The program outputs are the horizontal and vertical components of the reactive force, the resultant and angle made by the resultant force with the x axis and the reactive moment.

Notes: The same notes are applicable as for FORCE 2 with these additional points:

1. If distributed loads act on the beam, the user must first convert these loads to equivalent concentrated loads.
2. The frame of reference is at the cantilever's support. Therefore if any forces act to the left of the support they will have a negative x coordinate.
3. If there is a moment load on the beam, this must be converted to an input force load by means of an equivalent couple.

The solution to example 3.2 is given below the program listing in Appendix 3.3.

The applied forces are:

F_1	−1.413, 90, 0, 0	F_3	3, 0, 1.4, 3.15
F_2	−2.825, 90, 0.7, 0	F_4	3, −66, 1.4, 3.15

Note that since F_1 and F_2 are vertical, their y coordinate is immaterial and can have any value including zero. Similarly since F_3 is horizontal, its x coordinate is immaterial.

3.5 Cantilever beam reactions: Graphical solution

Since there is a reactive moment as well as a reactive force, the cantilever beam does not lend itself to a graphical solution. Although the force polygon will close because the forces must balance, there is no convenient graphical method for determining the reactive moment. The other important graphical method, the concurrency method (three forces), does not apply because of the moment present.

3.6 Simply supported beams

The simply supported beam is a beam which is mounted on two supports, one of which provides constraint against movement in all directions except rotation (e.g. hinge) and the other which provides constraint only against lateral movement (e.g. roller). Types of simply supported beam are illustrated in Figure 3.11.

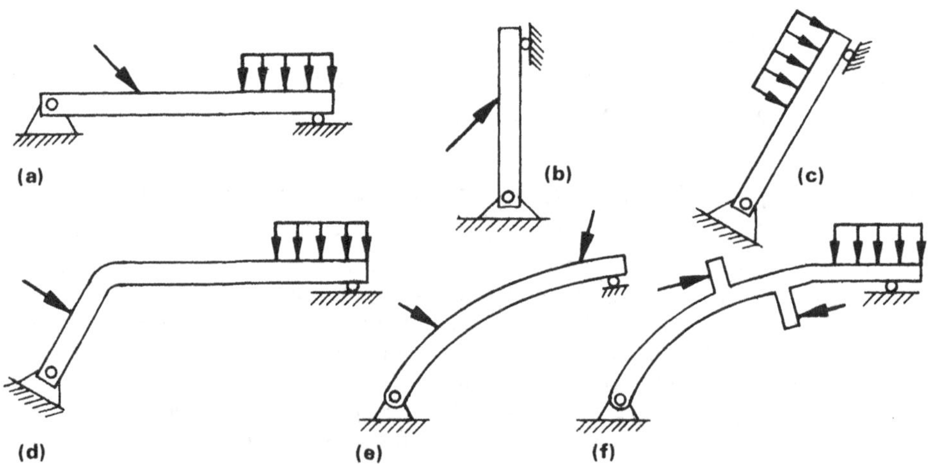

Fig. 3.11 *Simply supported beams*

The simply supported beam may be (a) horizontal, (b) vertical, (c) inclined, (d) cranked, (e) curved, (f) composite, and have the supports at the extreme ends or located inwards of either end (overhung). The load may be any combination of concentrated or distributed load in any direction within the span or outside the span (in the case of the overhung beam). However the load cannot be such as to require a downward reaction at the roller end since this is impossible (see Fig. 3.12).

By taking moments about the hinge, it is evident that the roller support would require a downward reaction since the clockwise moment is 24 kNm and the anticlockwise moment is 30 kNm; that is to say that in this case the beam would lift and rotate about the hinge.

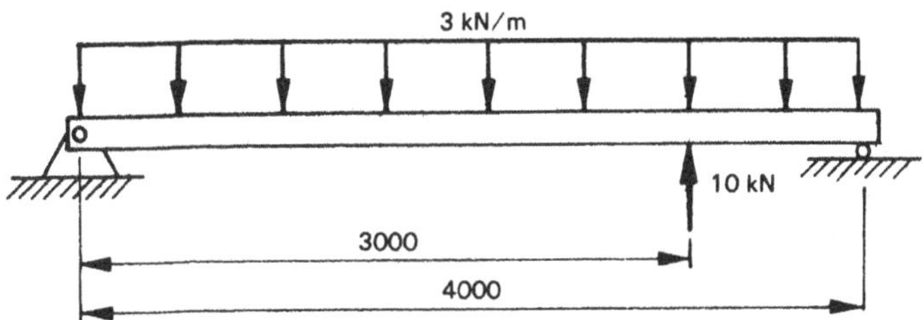

Fig. 3.12 *Impossible loading*

Variations

Although the hinge-roller combination is commonly encountered, other variations of support systems are possible. Some of these are illustrated in Figure 3.13. In (a) the roller has been replaced by a smooth surface, in (b) the roller has been replaced by a frictionless slot and in (c) the roller has been replaced by a smooth surface and the hinge by an indent in the surface.

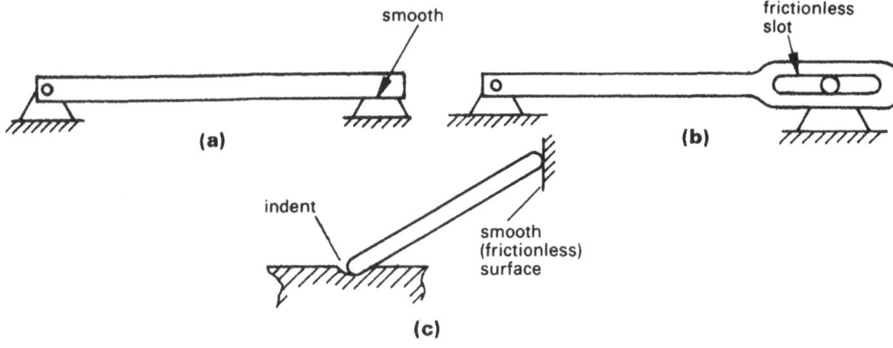

Fig. 3.13 *Variations in simply support*

Direction of the reactions

Neither support can withstand rotation so there is no moment reaction at either support but only a force reaction. In general, the direction of the reactive force is unknown at the hinge, but at the roller (or smooth surface) the line of action of the reactive force must be perpendicular to the surface as illustrated in Figure 3.14.

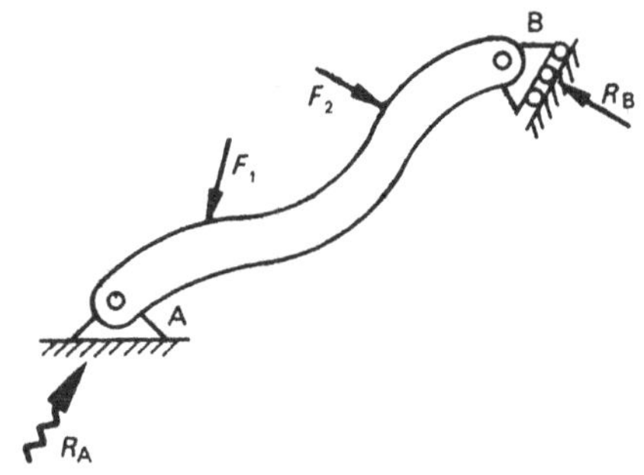

Fig. 3.14 *Direction of reactions*

The reaction direction is unknown at A, and this is indicated by the wavy line, whereas the reaction direction at B must be perpendicular to the surface and acting toward the roller.

3.7 Simply supported beam reactions: Analytical solution

The analytical solution is again an application of the equations for equilibrium of a force system:

$$\Sigma F_V = 0, \Sigma F_H = 0, \Sigma M = 0$$

Generally the most direct method is to take moments about the hinge and thus determine the reaction at the roller. Then the reaction at the hinge may be found by application of the equations:

$$\Sigma F_V = 0 \text{ and } \Sigma F_H = 0$$

This method is illustrated in example 3.3.

Example 3.3
Determine analytically the reactions for the beam illustrated in Figure 3.15.

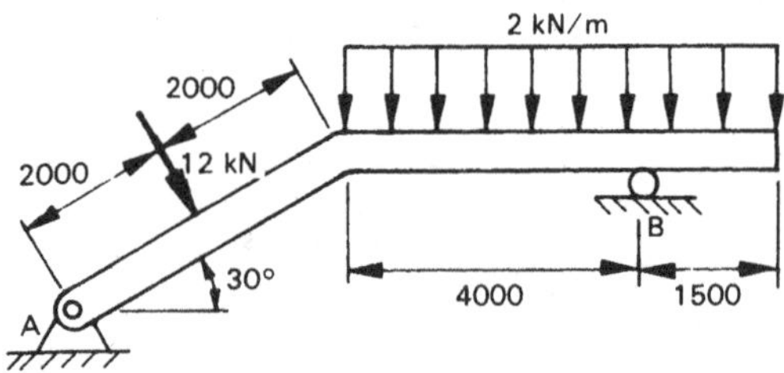

Fig. 3.15

Solution
A free body diagram of the beam is shown in Figure 3.16.

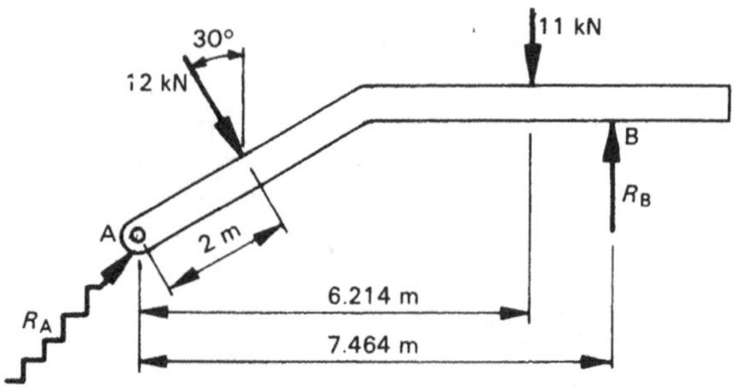

Fig. 3.16

Taking moments about A:

$$12 \times 2 + 11 \times 6.214 - R_B \times 7.464 = 0$$

$$\therefore R_B = \mathbf{12.37\ kN\ @\ 90^\circ}$$

$\Sigma F_V = 0$

$$\therefore -12 \cos 30^\circ - 11 + 12.37 + R_{AV} = 0$$

$$\therefore R_{AV} = 9.02 \text{ kN} \uparrow$$

$\Sigma F_H = 0$

$$\therefore 12 \sin 30^\circ + R_{AH} = 0$$

$$R_{AH} = -6 \text{ kN (the negative sign indicates} \leftarrow \text{direction)}$$

The reaction at A is as shown in Figure 3.17.

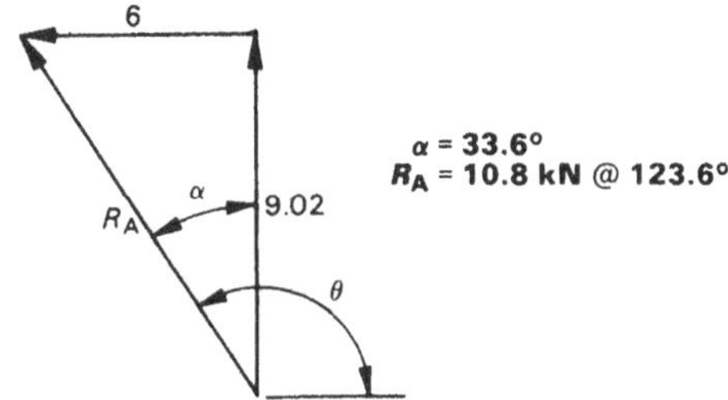

Fig. 3.17

3.8 Simply supported beam reactions: Computer solution

A computer program named REACT written in BASIC language for determining the reactions for a simply supported beam of any shape loaded with any number of concentrated loads in any direction or position on the beam is listed in Appendix 3.4.

This program is an extension of FORCE 2 and the forces acting on the beam are input in the same way: each applied force is specified by magnitude, angle with the x axis, and x and y coordinates with reference to the pinned joint named joint A.

The roller support is named B and the known direction of the reaction at this support, as well as the x and y coordinates of the support, are input. Thus the program inputs are the same as for FORCE 2 with these additional inputs applicable to the support reactions.

The program outputs are the same as for FORCE 2 in that the resultant force and moment of the applied forces is output. The additional outputs are for the vertical and horizontal components, resultant and angle of the reactions with the x axis.

Notes: The same notes are applicable as for FORCE 2 with these additional points:

1. If distributed loads act on the beam, the user must first convert these loads to equivalent concentrated loads.
2. The pin support must be named A and the roller support B. The frame of reference is located at joint A so that if joint A is to the right of joint B, then the x coordinate of joint B will be negative. Similarly if any forces act to the left of joint A they will have a negative x coordinate.
3. If there is an applied moment load, this must be converted to a force load by means of an equivalent couple.

The solution to example 3.3 is given below the program listing in Appendix 3.4.

The applied forces are:

F_1 −12, 120, 1.732, 1

F_2 −11, 90, 6.214, 2

The direction of the reaction at B in degrees is 90. The x and y coordinates of the reaction at B are 7.464, 2.

Note that since F_2 and R_B are vertical, their y coordinate is immaterial and could be input as zero or any other value without altering any of the computed output values.

3.9 Simply supported beam reactions: Graphical solution

There are several graphical methods for determination of simply supported beam reactions. One method is to combine all the applied forces acting on the beam into a single resultant. Since there are two support reactions, this means that the system has been reduced to a three force system and if the resultant applied force is not parallel to the direction of the roller reaction, the intersection of the lines of action of these forces is at the point of concurrency. Hence the direction of the pin reaction may be obtained and the force polygon (triangle) drawn. The use of this method to solve example 3.3 is shown in Figure 3.18.

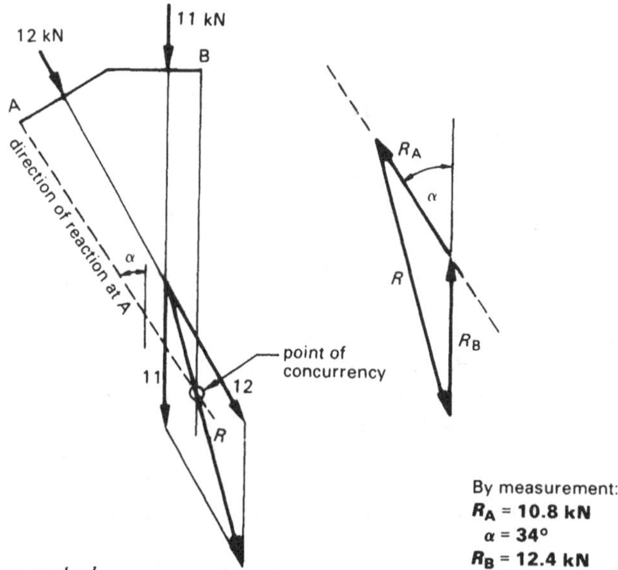

Fig. 3.18 *Three force method*

This method cannot be applied if the resultant of the applied forces is parallel to the direction of the roller reaction since the line of action of the pin reaction must also be parallel to this direction and no point of concurrency exists. In such cases, the string polygon method used previously to determine the resultant of parallel or nearly parallel forces may be modified to determine the reactions graphically. The use of the string polygon method to solve example 3.3 is shown in Figure 3.19.

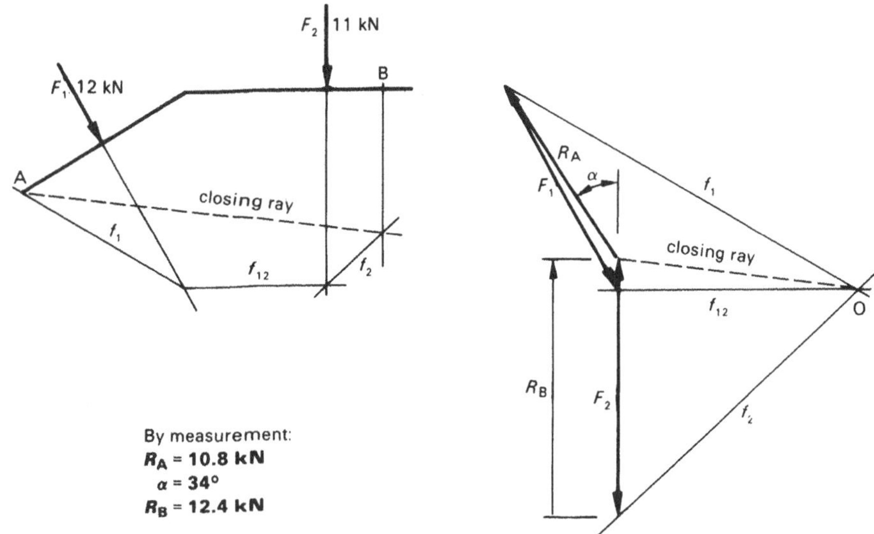

Fig. 3.19 *String polygon method for reactions*

Notes

1. The string polygon method can be applied to any system of forces and reactions but is most suitable where loads and reactions are parallel or nearly parallel.
2. Any convenient pole point O may be chosen; in this case component f_{12} to the pole point is horizontal but this is not necessary.
3. When drawing the components f_1, f_{12} and f_2 on the free body diagram, the starting point must be the pin reaction where the direction of the reaction is unknown.
4. The closing ray (shown dashed) joins the starting point (A) to where the final component (f_2) cuts the line of action of the known direction (R_B). The closing ray is then drawn in a parallel direction through the pole point on the string polygon to enable the polygon to be completed and the magnitude and direction of the reactions to be scaled off it.

3.10 Link or cable supported structures

In many cases the roller support may be replaced by a link or cable as shown in Figure 3.20. The variations shown in this figure are: (a) cable support, (b) link support, (c) frame with link support, (d) beam with curved link support.

The cable or link supported structure may be considered as a special case of a simply supported beam. Provided there is no applied load between the ends of the link or cable itself, then the line of action of the reaction at the link or cable must pass through the centreline drawn between the ends of the link or cable. That is, if the link or cable is straight, the reaction is axial to the link or cable. In Figure 3.20 in all cases there is no applied load between the end of the link or cable so that the line of action of R_B is known, whereas the line of action of R_A is unknown.

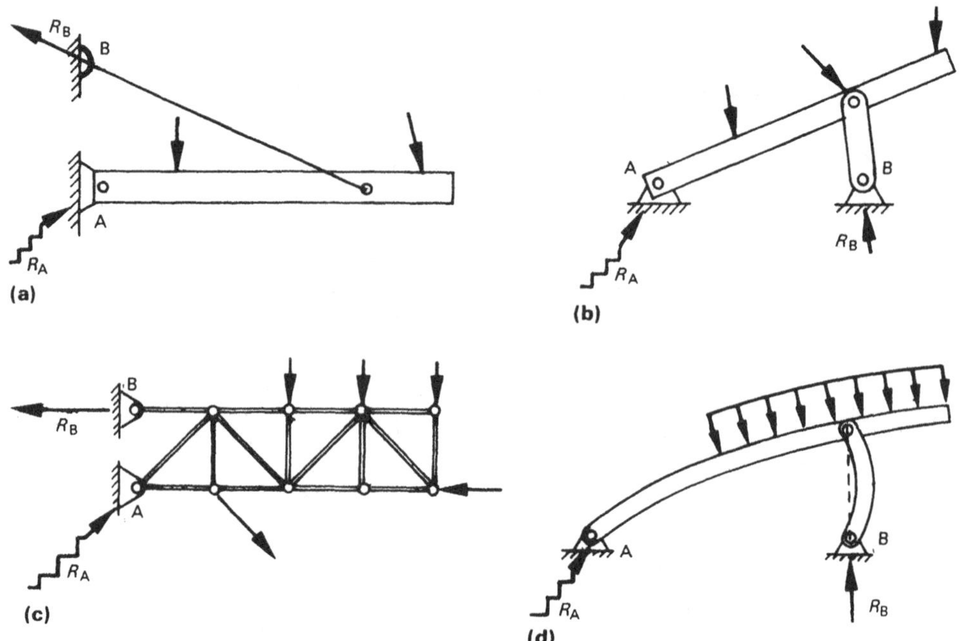

Fig. 3.20 *Cable and link supported structures*

Example 3.4

Determine the reactions at A and B for the frame illustrated in Figure 3.21.

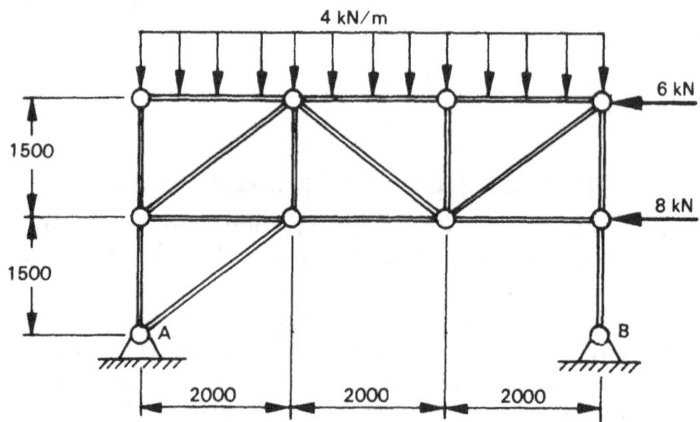

Fig. 3.21

Solution

The load of 4 kN/m may be replaced by a single, concentrated load of 24 kN at the midspan. The reaction at B must be vertical; the direction of the reaction at A is unknown.

Analytical method

Moments about A: $\Sigma M = 0$

$$\therefore 24 \times 3 - 6 \times 3 - 8 \times 1.5 - R_B \times 6 = 0$$

$$\therefore \quad \mathbf{R_B = 7\ kN} \uparrow$$

$$\Sigma F_V = 0$$

$$\therefore \quad R_{AV} - 24 + 7 = 0$$

$$\therefore \quad R_{AV} = 17\ \text{kN} \uparrow$$

$$\Sigma F_H = 0$$

$$\therefore \quad R_{AH} - 6 - 8 = 0$$

$$\therefore \quad R_{AH} = 14\ \text{kN} \rightarrow$$

$$\therefore \mathbf{R_A = 22\ kN\ @\ 50.5°}$$

Computer solution

Using computer program REACT the inputs are:

Number of applied forces = 3

Forces are: F_1 −24, 90, 3, 3
F_2 −6, 0, 6, 3
F_3 −8, 0, 6, 1.5

Reaction at B: R_B, 90, 6, 0

The output from the run of the program is as follows:

```
THE RESULTANT FORCE IS   27.78
THE ANGLE MADE BY THE RESULTANT IN DEGREES IS   239.74
THE SUM OF THE VERTICAL COMPONENTS IS   -24
THE SUM OF THE HORIZONTAL COMPONENTS IS   -14
THE RESULTANT MOMENT IS   42
THE RESULTANT FORCE CROSSES THE X AXIS AT   1.75
THE REACTION AT B IS AS FOLLOWS
VERTICAL COMPONENT   7   HORIZONTAL COMPONENT   0
RESULTANT   7   AT   90 DEGREES
THE REACTION AT A IS AS FOLLOWS
VERTICAL COMPONENT   17   HORIZONTAL COMPONENT   14
RESULTANT   22.02   AT   50.53 DEGREES
```

The computed values of the reactions at A and B therefore agree with those calculated analytically.

3.11 Statically indeterminate beams

Where a beam is supported in such a way that the reactive forces or moments cannot be determined by the methods of statics, the beam is said to be statically indeterminate. This is *not the case* with the double pinned beam illustrated in Figure 3.22.

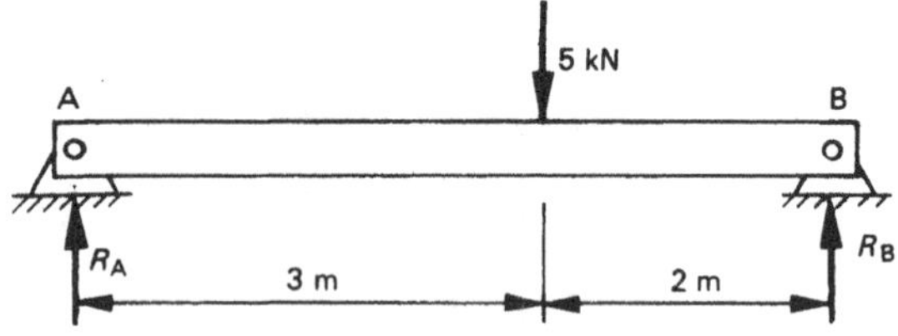

Fig. 3.22 *Double pinned beam— statically determinate*

It is possible to determine R_A and R_B by taking moments about either A or B. The result is:

$R_A = 2$ kN, $R_B = 3$ kN.

However, suppose the load acts at an angle as shown in Figure 3.23:

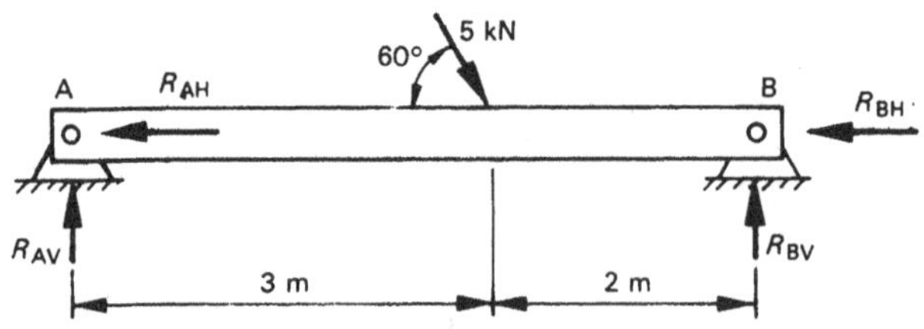

Fig. 3.23 *Double pinned beam— statically indeterminate*

By taking moments about A or B:

$R_{AV} = 1.73$ kN, $R_{BV} = 2.6$ kN.

However it is not possible to determine R_{AH} or R_{BH} by taking moments and there is only one equation with two unknowns:

$2.5 - R_{AH} - R_{BH} = 0$

There are no further equations or procedures of statics which will enable the horizontal components to be determined, hence this beam is statically indeterminate.

Other examples of statically indeterminate beams are illustrated in Figure 3.24.

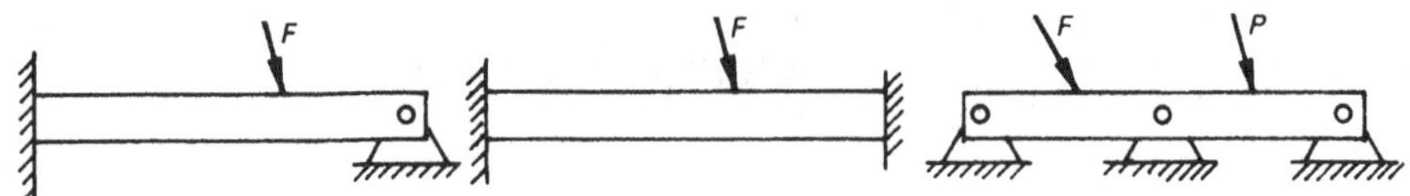

Fig. 3.24 *Statically indeterminate beams*

Statically indeterminate beam reactions may be determined by consideration of the deflection involved, as well as the principles of static equilibrium of forces and moments. Such cases are beyond the scope of this book and are an example of problems which cannot be solved by separate consideration of the principles of the mechanics of non-deformable bodies and those of deformable bodies.

Problems

Note: For these problems the weight of the beam itself does not have to be taken into account unless the beam mass is given.

3.1–3.8 Determine the reactive force and moment for the cantilever beams loaded as shown in Figures P3.1–P3.8.

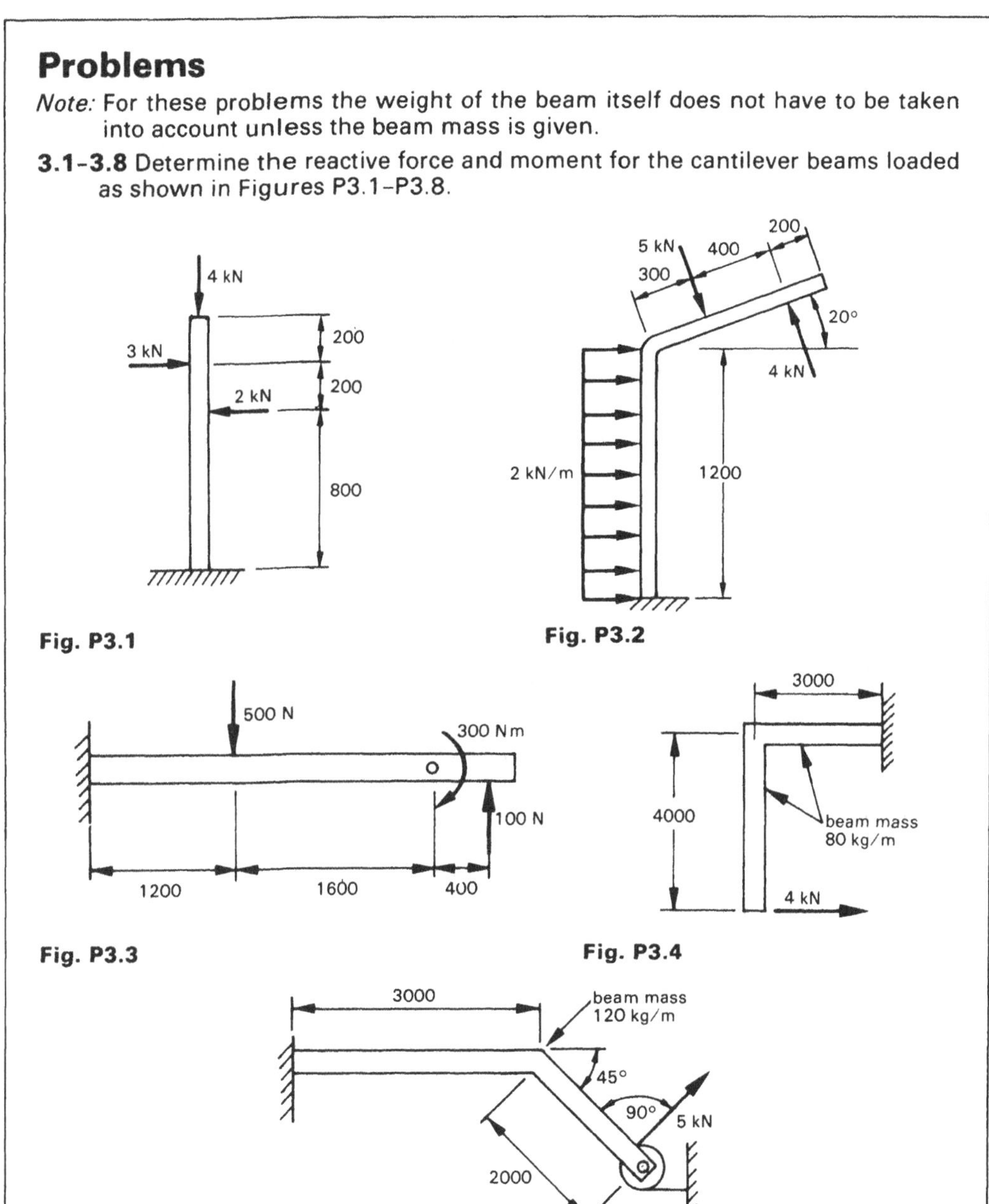

Fig. P3.1

Fig. P3.2

Fig. P3.3

Fig. P3.4

Fig. P3.5

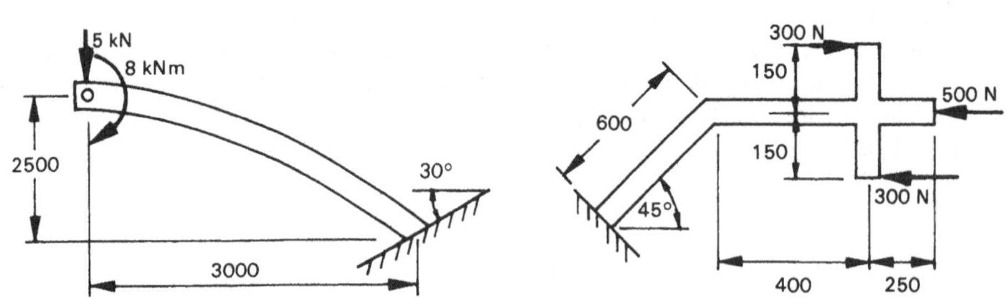

Fig. P3.6

Fig. P3.7

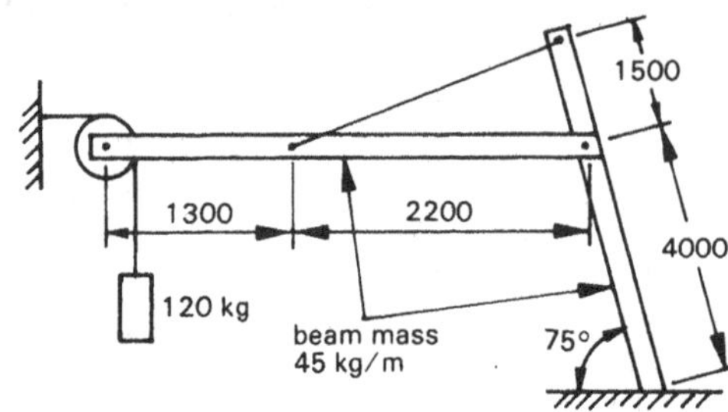

Fig. P3.8

Solutions

	Force	*Moment*
3.1	4.12 kN @ 104°	-1.4 kNm
3.2	2.9 kN @ 161°	-0.552 kNm
3.3	400 N @ 90°	-580 Nm
3.4	6.8 kN @ 126°	28.95 kNm
3.5	8.85 kN @ 164.6°	13.65 kNm
3.6	5 kN @ 90°	7 kNm
3.7	500 N @ 0°	122 Nm
3.8	5.28 kN @ 77.1°	15.9 kNm

3.9–3.26 Determine the reactions at A and B for the beams and frames loaded as shown in Figures P3.9–P3.26.

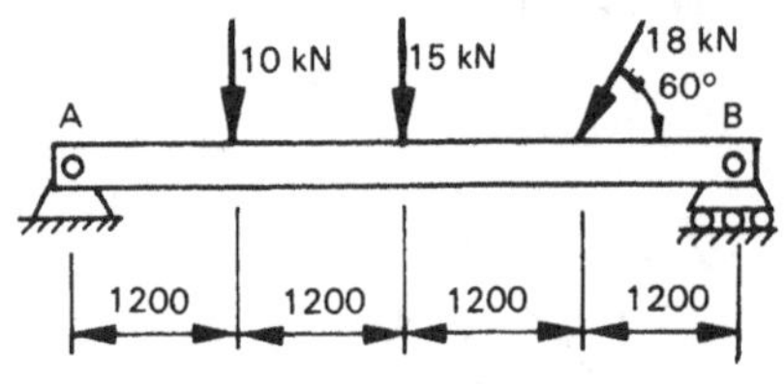

Fig. P3.9

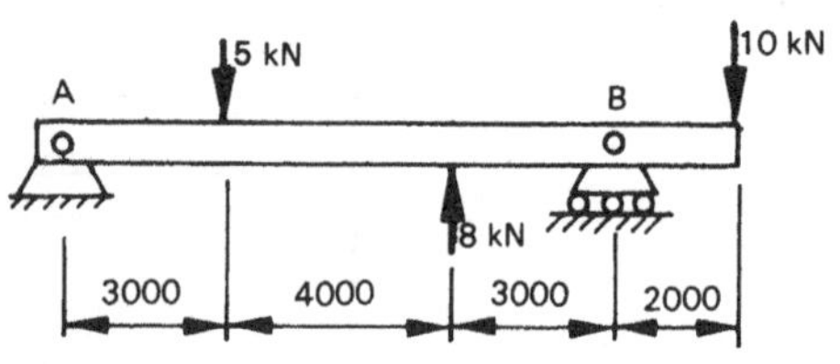

Fig. P3.10

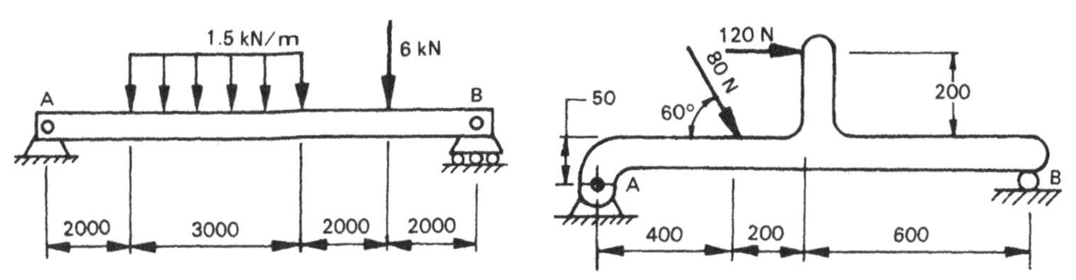

Fig. P3.11

Fig. P3.12

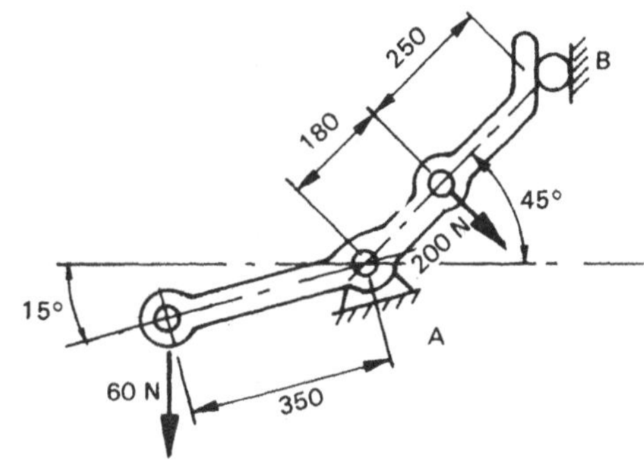

Fig. P3.13

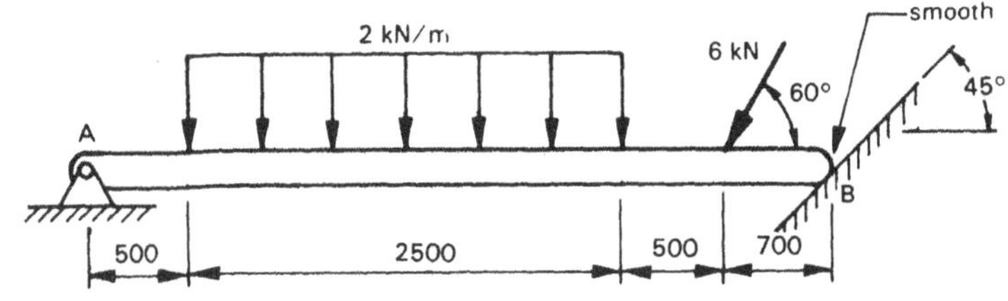

Fig. P3.14

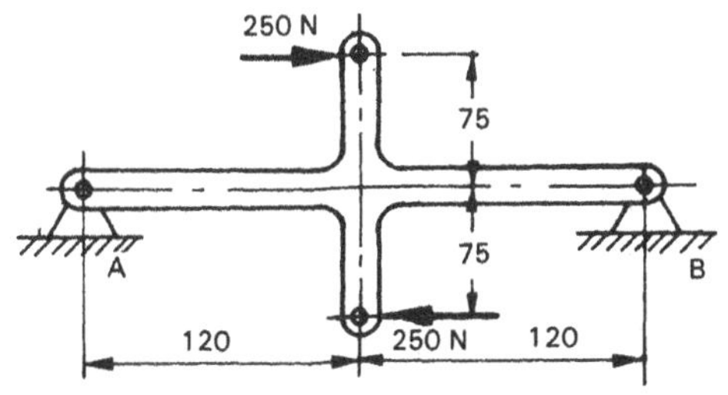

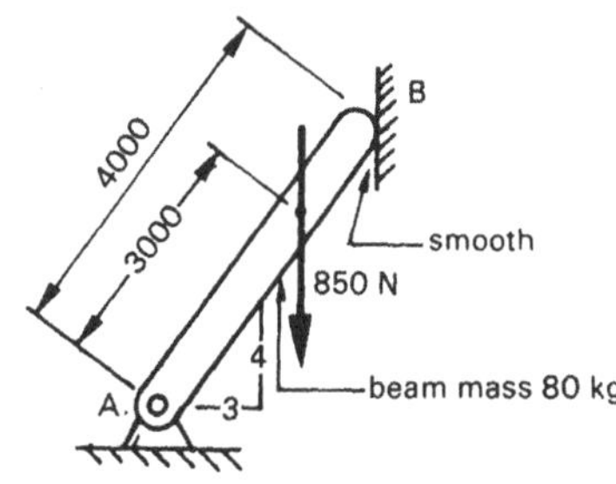

Fig. P3.15

Fig. P3.16

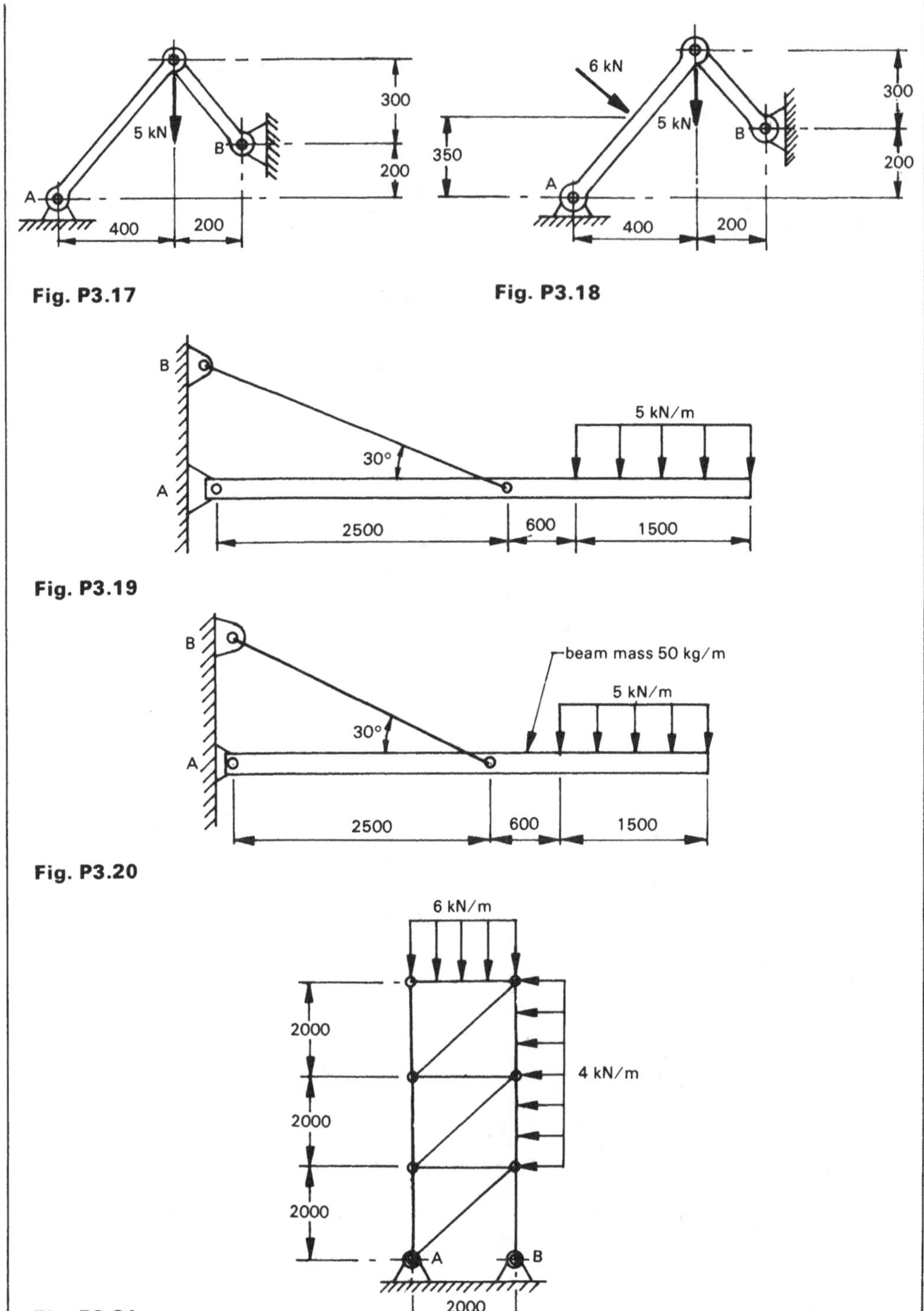

Fig. P3.17

Fig. P3.18

Fig. P3.19

Fig. P3.20

Fig. P3.21

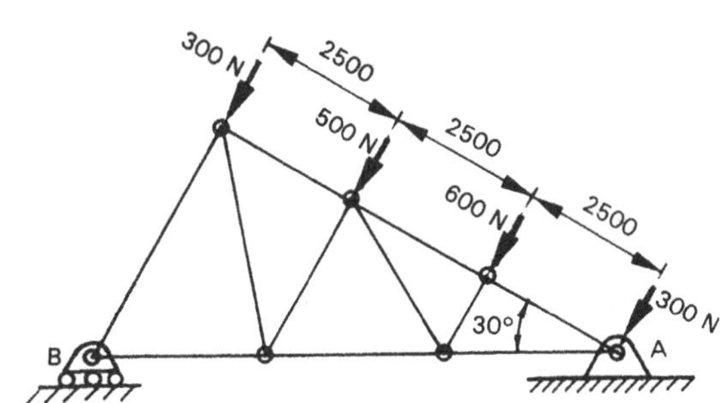

Fig. P3.22

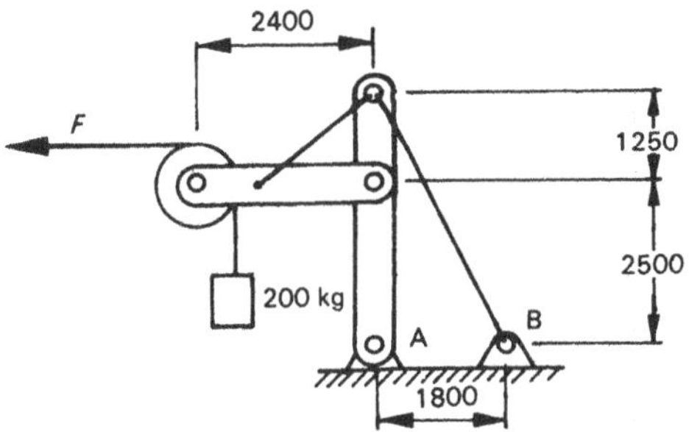

Fig. P3.23

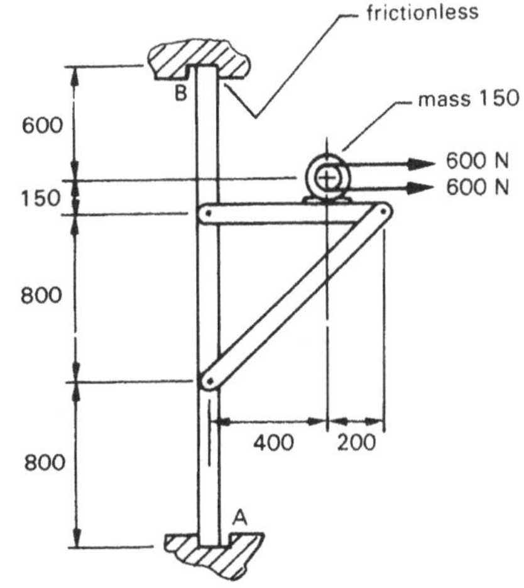

Fig. P3.24

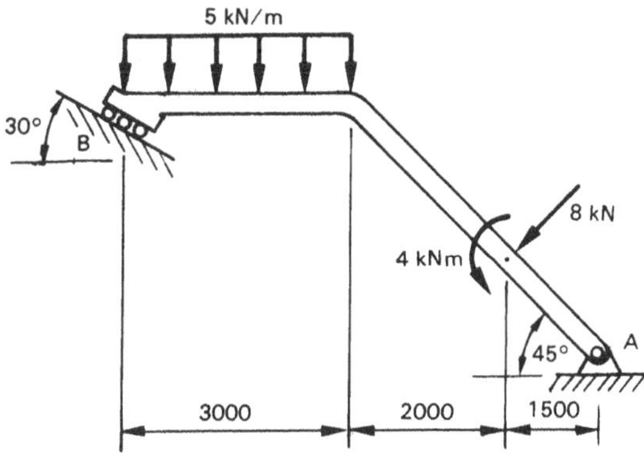

Fig. P3.25

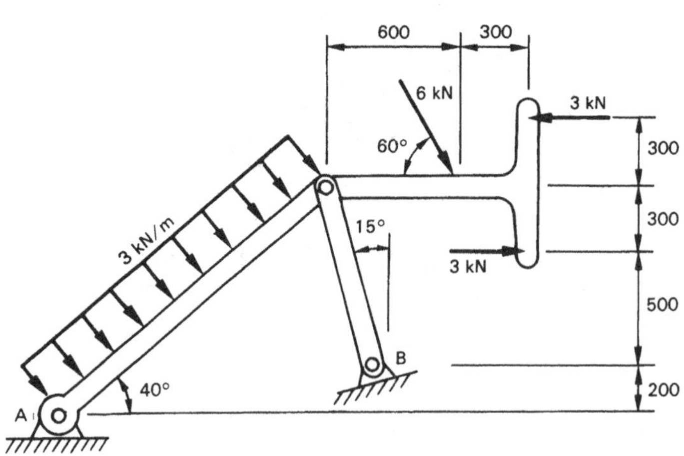

Fig. P3.26

Solutions

	Reaction at A	*Reaction at B*
3.9	20.9 kN @ 64.5°	21.7 kN @ 90°
3.10	0.9 kN @ 270°	7.9 kN @ 90°
3.11	4.08 kN @ 90°	6.42 kN @ 90°
3.12	161 N @ 173°	49.8 N @ 90°
3.13	221 N @ 114°	51.7 N @ 180°
3.14	10.15 kN @ 21.9°	9.07 kN @ 135°
3.15	156 N @ 270°	156 N @ 90°
3.16	1808 N @ 64.7°	772 N @ 180°
3.17	2.91 kN @ 51.3°	3.28 kN @ 124°
3.18	2.39 kN @ 100°	7.69 kN @ 124°
3.19	20.4 kN @ 349°	23.1 kN @ 150°
3.20	23.9 kN @ 351°	27.3 kN @ 150°
3.21	41.2 kN @ 67.2°	26 kN @ 270°
3.22	1134 kN @ 41.4°	722 N @ 90°
3.23	7.33 kN @ 94.7°	5.92 kN @ 296°
3.24	1.47 kN @ 92.2°	1.14 kN @ 180°
3.25	9.43 kN @ 95.2°	13 kN @ 60°
3.26	3.53 kN @ 195°	10 kN @ 105°

4

Forces in frame members

Until now a frame has been regarded as a unified structure, for example Chapter 3 dealt with methods for determining the reactions when a beam or frame was subject to various loading conditions by considering the frame to be a complete structural unit. It was not necessary to consider the internal forces or moments in the members of the frame in order to determine the reactions. However, when designing or analysing the design of a frame made up of several members, it is of course necessary to ascertain the individual forces and moments to which each member is subjected.

Various methods for determining the forces in frame members are presented in this chapter. Although several methods are outlined it is necessary to use only one method for a complete solution. However, an understanding of all methods is desirable since each has particular applications where it is best suited.

As is the case with most topics in statics, the determination of forces in frame members involves no new principles but rather the application of the principles already developed to solve this particular class of problems. Hence this chapter is really a combination and extension of Chapters 1, 2 and 3 into a particular area of application—the determination of forces in frame members.

4.1 General principles

Frame

A frame or truss is a rigid, fabricated structure which consists of members and joints. For example, consider the structures A and B shown in Figure 4.1 (previously considered as Fig. 3.1).

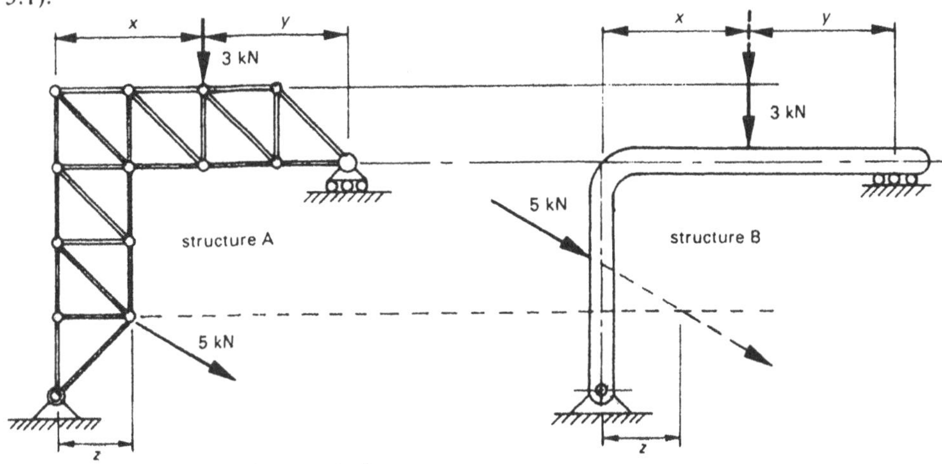

Fig. 4.1 *Distinction between a frame and a beam*

Structure A is clearly a frame, whereas structure B is clearly a continuous beam. Although both A and B are equivalent structures for determination of the reactions, they are not equivalent when it is necessary to determine the internal forces or moments present.

Pin-jointed frame

A pin-jointed frame is a frame or truss in which the joints are considered to be pinned, that is the joints themselves do not resist rotation (see sect. 2.6). This means that in order to be rigid the frame must consist of a series of triangles. For example consider the pin-jointed frames A and B shown in Figure 4.2.

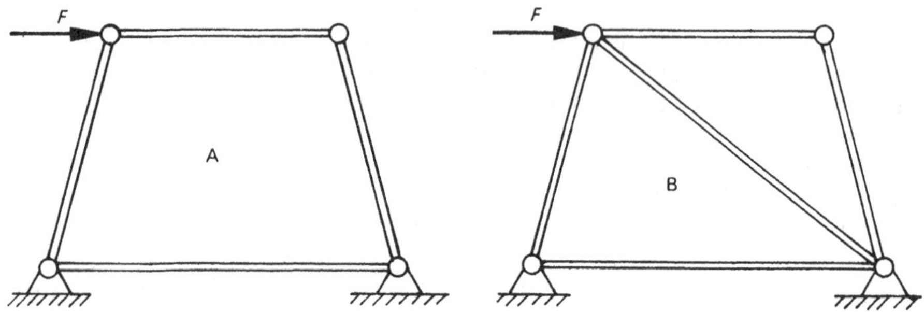

Fig. 4.2 *Distinction between non-rigid and rigid frames*

It is quite clear that frame A will collapse when force F is applied if the joints themselves do not offer any resistance to rotation. It is also clear that frame B is rigid even though the joints do not have any resistance to rotation. Therefore frame A cannot be considered as a pin-jointed frame, whereas frame B can be so considered. Frame A is in fact a four bar mechanism and used in various forms (inversions) to transmit motion.

In practice, bolted, welded and riveted joints are usually used in frame construction. Even though the joints may be strengthened by gussets and may in fact offer considerable resistance to rotation, the analysis of the frame as a pin-jointed structure is a widely adopted method and has been found to give acceptable results. In this chapter, joints will always be considered as pin connections.

Continuous member

A continuous member is longer than the length between the joints and hence, if a load is applied outside the joints, there may be a bending moment in the member at the joint. Consider Figure 4.3. It is evident that member AB must be continuous and will be subject to bending moment at joint C. In this chapter, all members will be assumed to be non-continuous, i.e. discrete.

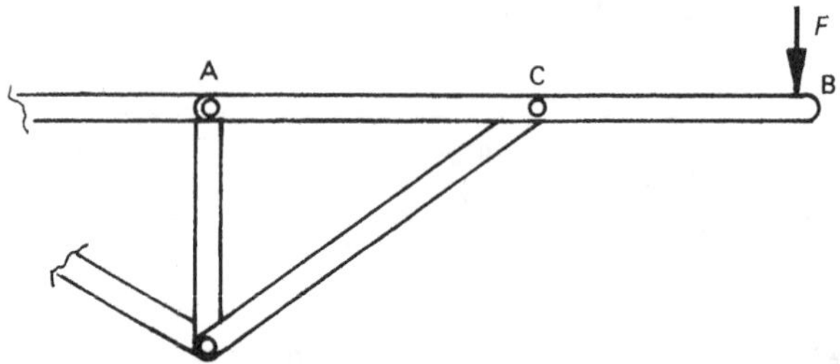

Fig. 4.3 *Member AB is continuous*

Principle of transmissibility

This principle states that a force may be moved along its line of action within a force system without changing the mechanics of that system. That is to say, changing the point of application of a force produces no change within the system.

This principle may be applied to frame analysis in order to determine the reactions because the force system involves only the external forces and reactions. The internal forces in the members are in balance and the frame may be considered as a rigid unit. However, the principle cannot be applied to the analysis of the forces in individual members of a frame if this involves transferring a force outside the system under consideration. For example, consider part of a frame shown in Figure 4.4 where force F has been moved from pin A to the centre of member AC.

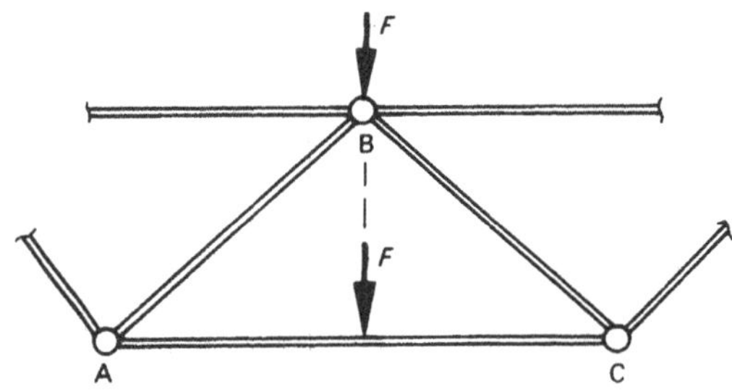

Fig. 4.4 *The principle of transmissibility may not be used in the analysis of the forces in members AB, BC or AC*

The force in the members AB, BC and AC will obviously be greatly affected by movement of force F, although the external members and the reactions are not affected by this movement. The reason for this is clear. When analysing the equilibrium of joint B in order to determine the force in members AB and BC, the force system is the force in each of the members pinned at B as well as force F. If force F is transferred as shown, it is no longer included in this force system so the equilibrium of this system is changed.

Determination of the reactions

Although it is not always necessary to determine the reactions before solving for the forces in the frame system, it is desirable that the reactions are determined first by considering the frame as a complete unit. Any of the methods outlined in Chapter 3 may be used but it is obviously best to use a method compatible with the one which will be used for subsequent determination of the forces.

Forces in members

In the analysis of frame structures, the following assumptions will be made:

1. All joints are pure pinned joints and have no rotational resistance.
2. All members are discrete (non-continuous).
3. All loads on the structure are point loads applied at the joints only.

Under these conditions the following conclusions follow:

1. There is no bending moment in any member.
2. There is no torsional load at any joint.
3. The force in each member acts axially, i.e. along the line joining the centres of the pinned joints at each end of the member.
4. Each member will carry only tension or compression force.

Tension and compression forces

The distinction between tension and compression force in a member may be indicated by sign (tension +, compression −) or by the use of arrowheads as shown in Figure 4.5.

Fig. 4.5 *Tension and compression force in a member*

Bow's notation

A convenient method of identifying the forces in structural members is by Bow's notation. In this notation the spaces between the forces and members are lettered using capital letters. The forces themselves are then referred to by lower case letters. The method is illustrated in Figure 4.6.

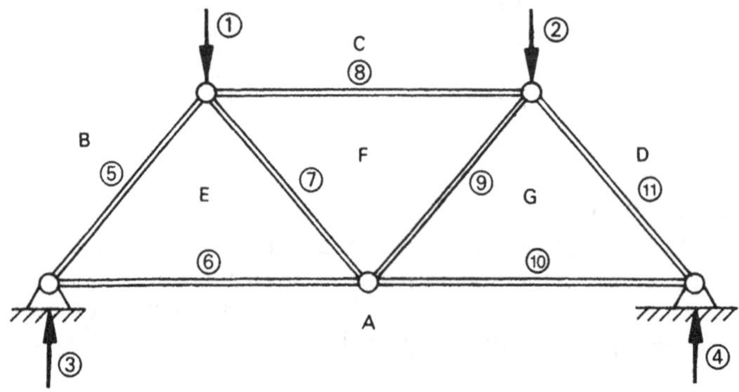

Fig. 4.6 *Bow's notation*

In this system there are eleven forces present, two applied forces, two reactions, and a force in each of the seven members which make up the frame. Numbering the forces 1 to 11 to illustrate Bow's notation:

Force number	*Bow's notation for the force*
1	*bc*
2	*cd*
3	*ab*
4	*da*
5	*be*
6	*ae*
7	*ef*
8	*cf*
9	*fg*
10	*ag*
11	*dg*

Note: The sequence of lettering the spaces is customarily down clockwise, first moving around the outside of the structure taking into account the applied forces and reactions, then inside the structure taking in the member forces.

4.2 Method of joints: Analytical solution

In the method of joints, each joint is isolated from the rest of the frame and treated as a separate force system. This is done by taking a *closed* boundary around the joint which must cut each member acting at the joint and each applied force acting at the joint. This system then becomes a concurrent force system in equilibrium under the action of the forces in it. There is no turning moment and the sum of the forces in any direction must be zero. Two different directions may be used thus giving two equations which may be solved for two unknown forces. If there are more than two unknown forces at a joint, this joint cannot be solved immediately and another joint must be selected where there are no more than two unknown forces. A suggested procedure is as follows:

1. Draw a neat diagram of the frame showing all the forces. This diagram does not necessarily have to be to scale.
2. Mark the diagram using Bow's notation and number the joints progressively. Show the reactions on the diagram by means of arrows.
3. Calculate the reactions and mark their magnitude and sense on the frame diagram.
4. Identify a joint with no more than two unknown forces. Draw a separate diagram of this force system. Write the equations for the equilibrium of the forces and solve for the unknown forces at this joint.
5. Mark the appropriate members of the frame diagram with the magnitude of the forces found in step 4. Show the sense of the forces by means of arrowheads at the joint, thus identifying whether the members are in tension or compression.
6. Repeat steps 4 and 5 until the entire frame has been solved. The order in which the joints are solved is immaterial provided that there are no more than two unknown forces at a joint.
7. Check for error by ensuring that the forces at the final joint balance.

Notes

1. Each member does not necessarily carry a force.

2. The direction of the force in a member will be opposite at the joint located at each end of the member. (Refer to Fig. 4.5.)
3. When the force in a member is unknown, assume tension. If the calculation yields a negative result, the member must be in compression.

Example 4.1

Determine the forces in all the members of the frame shown in Figure 4.7 using the method of joints (analytically).

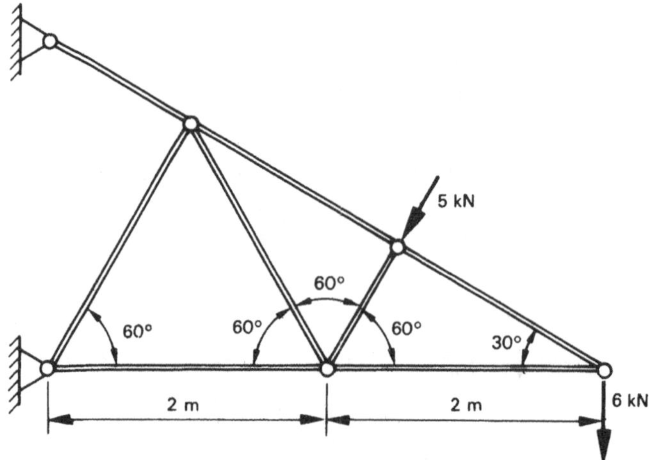

Fig. 4.7

Solution

Following the method step by step:

1. and 2. The frame diagram is redrawn, the reactions are shown, the joints are numbered and Bow's notation used (see Fig. 4.8). Note that the direction of the reaction R_2 may be determined by inspection but the direction of the reaction R_1 is unknown and is shown by a wavy arrow.

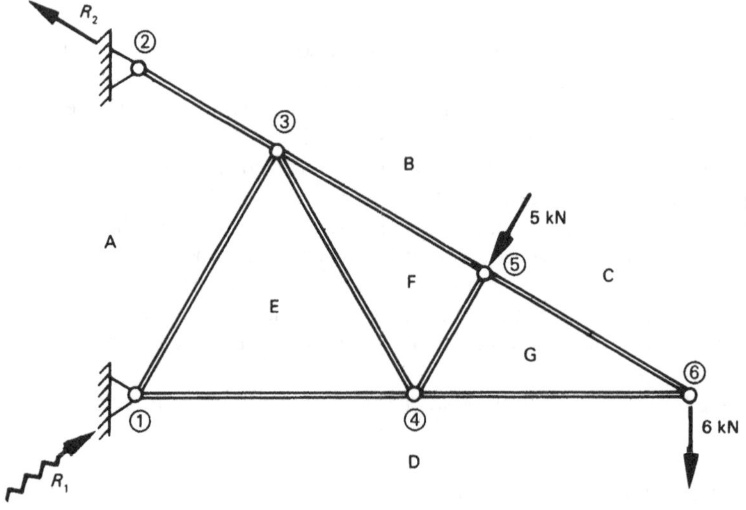

Fig. 4.8

3. In order to calculate the reactions it is first necessary to use trigonometry to determine some unknown dimensions. These are shown in Figure 4.9.

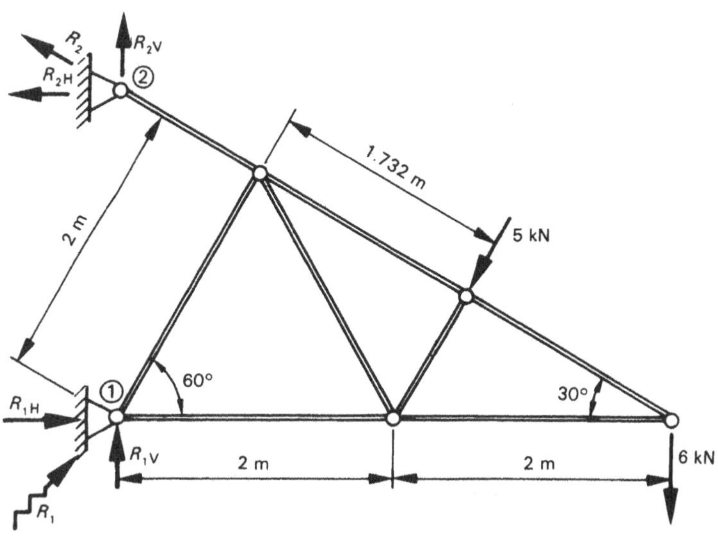

Fig. 4.9

Taking moments about ①:

$$-R_2 \times 2 + 5 \times 1.732 + 6 \times 4 = 0$$

$$\therefore \boldsymbol{R_2} = \textbf{16.33 kN},\ R_{2V} = 8.165 \text{ kN},\ R_{2H} = 14.14 \text{ kN}$$

$$\Sigma F_V = 0 \quad \therefore R_{1V} + R_{2V} - 5 \sin 60° - 6 = 0$$

$$\therefore R_{1V} = 2.17 \text{ kN}$$

$$\Sigma F_H = 0 \quad \therefore R_{1H} - R_{2H} - 5 \cos 60° = 0$$

$$\therefore R_{1H} = 16.64 \text{ kN}$$

$$\therefore \boldsymbol{R_1} = \textbf{16.78 kN @ 7.41°}$$

The reactions may now be shown on the frame diagram (see Fig. 4.11).

4. A joint with no more than two unknown forces may now be identified. Chosing joint ⑥ and assuming tension in the unknown members, the system is as shown in Figure 4.10.

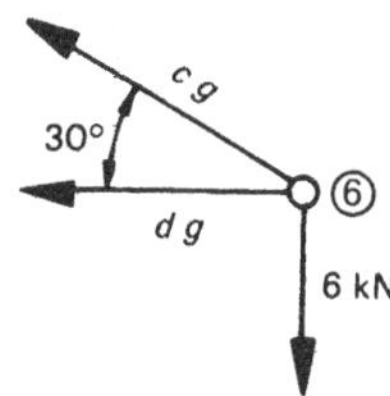

Fig. 4.10

$\Sigma F_V = 0 \quad \therefore cg \sin 30° - 6 = 0 \quad \therefore \mathbf{cg = 12\ kN}$

$\Sigma F_H = 0 \quad \therefore cg \cos 30° - dg = 0 \quad \therefore \mathbf{dg = -10.4\ kN}$

Since force dg is negative, it is evident that the wrong direction was chosen and member dg must be in compression.

5. The frame diagram may now be marked with the magnitude and direction (sense) of the forces found thus far. This is shown in Figure 4.11.

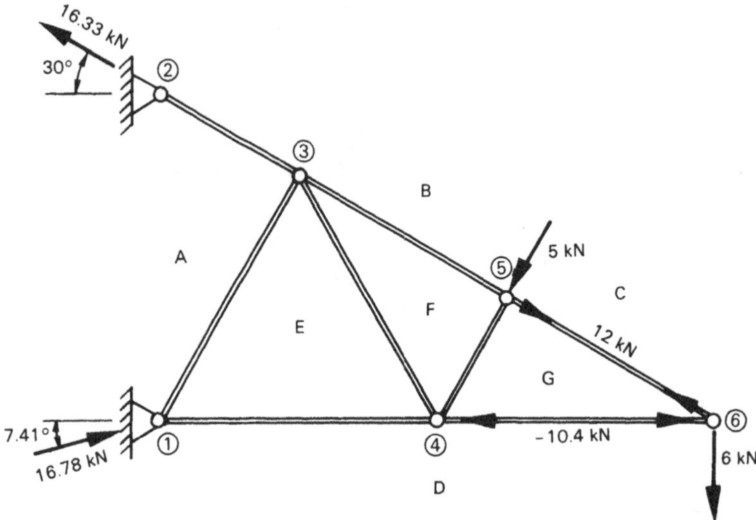

Fig. 4.11

6. Steps 4 and 5 may now be repeated at another joint where there are no more than two unknown forces. Such a joint is joint ⑤. The force system for this joint is shown in Figure 4.12.

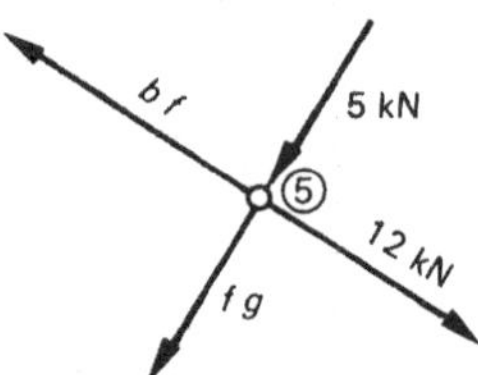

Fig. 4.12

Since the forces are mutually perpendicular:

$bf = 12$ kN (tension), $fg = -5$ kN (compression)

These values may now be marked on the frame diagram.
Steps 4 and 5 may now be repeated at joint ④ (Fig. 4.13).

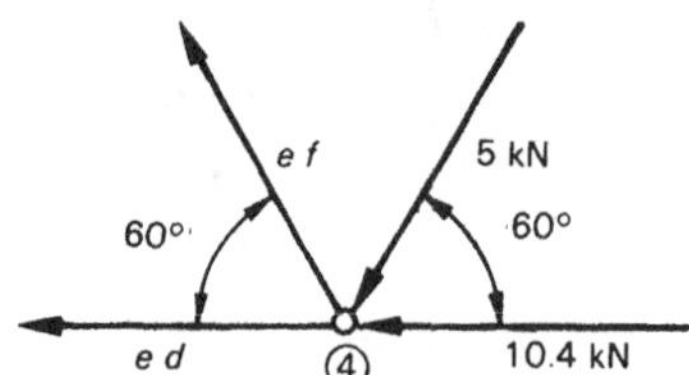

Fig. 4.13

$\Sigma F_V = 0 \quad \therefore -5 \sin 60° + ef \sin 60° = 0$

$\therefore$ ***ef* = 5 kN (tension)**

$\Sigma F_H = 0 \quad \therefore -10.4 - 5 \cos 60° - ef \cos 60° - ed = 0$

$\therefore$ ***ed* = −15.4 kN (compression)**

These values may be marked on the frame diagram.
Now moving to joint ③ (Fig. 4.14).

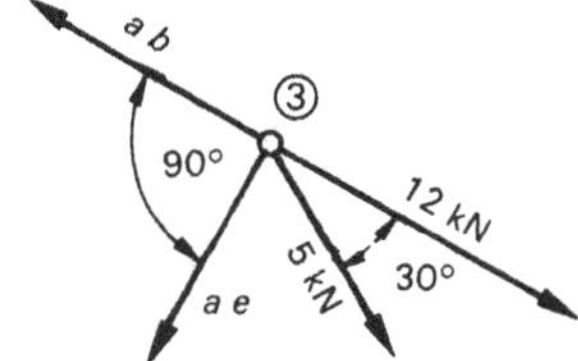

Fig. 4.14

It is evidently more convenient to rotate the frame of reference so that the x axis coincides with the inclined angle of the frame.

$\Sigma F_V = 0 \quad \therefore -5 \sin 30° - ae = 0$

$\therefore$ ***ae* = −2.5 kN (compression)**

$\Sigma F_H = 0 \quad \therefore 12 + 5 \cos 30° - ab = 0$

$\therefore$ ***ab* = 16.33 kN (tension)**

The magnitude of force *ab* is therefore correct since from joint ② it is evident that $ab = R_2 = 16.33$ kN

7. The entire frame has now been solved. However as a final check, consider joint ①. The force system is:

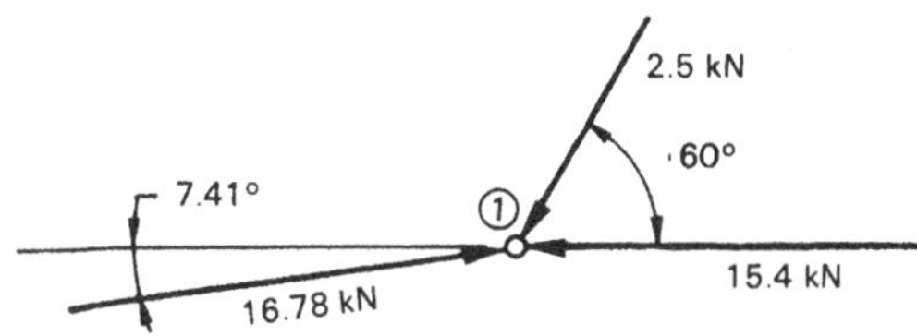

Fig. 4.15

$\Sigma F_H = 0 \quad \therefore -2.5 \sin 60° + 16.78 \sin 7.41° = 0$

which checks.

$\Sigma F_H = 0 \quad \therefore -15.4 - 2.5 \cos 60° + 16.78 \cos 7.41° = 0$

which checks.

The completed frame diagram showing all forces is as shown in Figure 4.16.

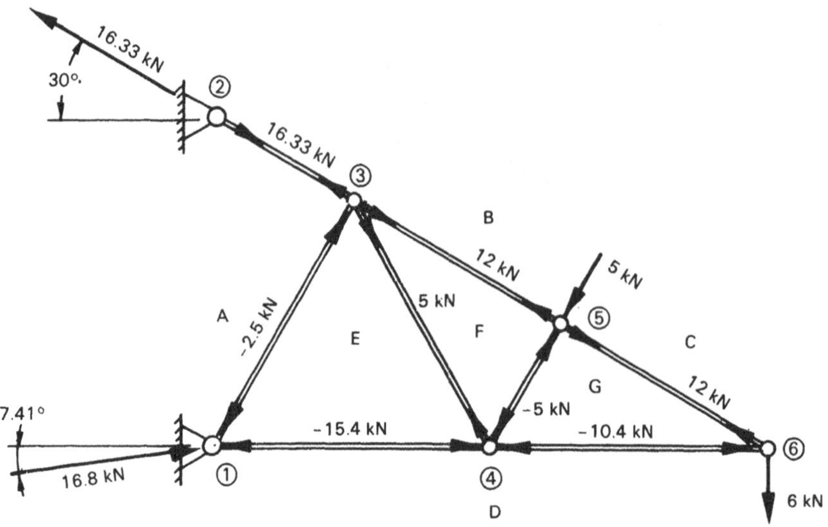

Fig. 4.16

4.3 Method of joints: Computer solution

A computer program for the method of joints named FORCE 3 written in BASIC language is listed in Appendix 3.5. This program may be used to determine the magnitude and sign (tension or compression) of two unknown forces in a concurrent force system, given the magnitude and direction of the other forces in the system. It is particularly suitable for the method of joints but is a general program which may be used for concurrent force systems other than those occurring in frames.

The program inputs are:

1. the number of known forces at the joint
2. the magnitude of each known force
3. the angle made by the known force with the x axis (in degrees anticlockwise)
4. the direction of the two unknown forces (angles in degrees anticlockwise with the x axis).

2 and 3 must be input for each known force in turn.

The program outputs are:

the magnitude and sign of the two unknown forces at each of the two directions specified.

Notes

1. The convention for specifying a known force and its direction as illustrated in Figure 1.5 and used in previous computer programs must be used.
2. There may be any number of known forces at the joint but only two unknown forces.
3. Joints may be rotated clockwise or anticlockwise for convenience in specification of angles.
4. Consistent units for the magnitude of the forces (N or kN) must be used throughout.
5. Output forces and angles are rounded off to two decimal places.

6. Correct sign (+ tension, − compression) will be obtained automatically from the computed output.
7. The program may also be used to resolve any known force into two components in any two directions. The easiest way of doing this is to reverse the direction of the known force; the components will then be in the correct sense.
8. A complete computer solution to a frame or truss may be obtained by:
   (a) first applying computer program REACT to determine the reactions
   (b) then using the computer program FORCE 3 moving around the frame progressively from joint to joint as in the analytical solution.
9. The programs REACT and FORCE 3 could in fact be combined into one program but have been kept separate for simplicity and in order to maintain their general utility.
10. A frame diagram should be drawn and progressively marked with the value of the forces obtained as each joint is solved and each pair of unknown forces is determined. It is a good idea to use arrowheads to denote the sense of the forces as well as the sign convention (+ tension, − compression).

The solution to example 4.1 using both programs (REACT and FORCE 3) is also given in Appendix 3.5 below the listing of FORCE 3 and should be self-explanatory.

However the following points may assist the user:

1. When using REACT, the known direction is 150° (R_2) and the coordinates of joint ② (using joint ① as reference) are 0, 2.3094.
2. When using FORCE 3, joints ⑤ and ③ have been rotated for convenience. This is not essential of course.
3. The entire frame is solved without using joint ①. However this joint has been used as a check for error by making force *ad* the known force (R_1) and forces *ae* and *de* the unknown forces and then checking their value with that obtained previously.

4.4 Method of joints: Graphical solution

Each joint in the frame may be isolated from the remainder of the frame and treated as a system of concurrent forces in equilibrium. A polygon of forces drawn for each joint force system must therefore close. The polygon may be drawn provided that there are no more than two unknown force magnitudes at the joint. Once the polygon is drawn, the unknown magnitudes may be determined by measurement and the sense of the unknown forces determined by inspection.

Notes

1. Each member does not necessarily carry a force.
2. The direction of the force in a member is opposite at the joint at each end of the member.
3. It is not necessary to assume tension or compression for the unknown forces. The correct sense will automatically follow from the head-tail sequence which must be observed when drawing the polygon.

Suggested procedure

1. Draw the frame to suitable scale around the top section of a reasonably sized sheet of drawing paper. If a drawing machine is used, plain paper may be used, otherwise graph paper is better. Show all applied forces acting on the frame.

2. Mark the frame diagram using Bow's notation and number the joints progressively. Identify the reactions.
3. Use a suitable method to determine the reactions. A graphical method will usually be most applicable but in simple cases it may be possible to determine the reactions more conveniently by inspection or short calculation. Typical graphical methods are the concurrency method (three force system) or the string polygon. Note that the concurrency method may not be used if the resultant of the applied forces is parallel to the known direction of one of the reactions. When the reactions have been determined, show them on the frame diagram.
4. Identify a joint with no more than two unknown forces. Using a suitable scale, draw the polygon of forces for the joint under consideration. Measure the magnitude of the unknown forces and determine their sense (tension or compression) by inspection of the force polygon.
5. Mark the appropriate members of the frame diagram with the magnitude of the forces found in step 4, and use arrowheads to show whether the members are in tension or compression.
6. Repeat steps 4 and 5 until the entire frame has been solved. The order in which the joints are solved is not of any consequence provided it is done so that no more than two unknown forces exist at a joint at the time that the polygon is drawn for that joint.
7. Check for error by ensuring that the forces at the final joint constitute a vector polygon which closes.

Example 4.2

Solve example 4.1 using the graphical method of joints. The frame diagram is reproduced here for convenience as Figure 4.17.

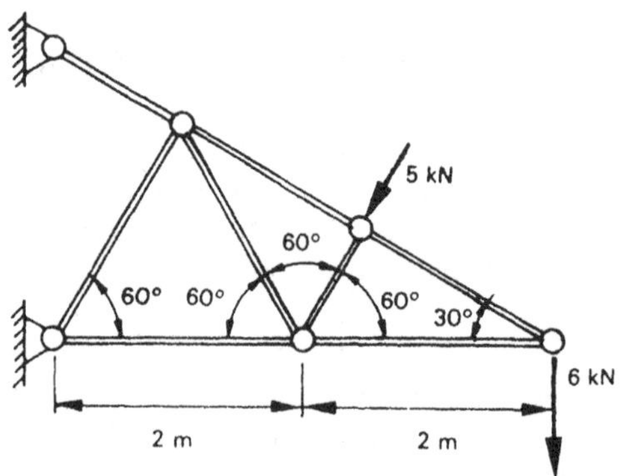

Fig. 4.17

Solution

The entire solution (using the three force method for the reactions) is shown in Figure 4.18. Note that each unknown point on the force polygon has been shown by means of double arrowheads along the known directions of the force to obtain a point of intersection. For example at joint ⑥ the unknown point is point *g*, at joint ⑤ it is point *f*, and so on.

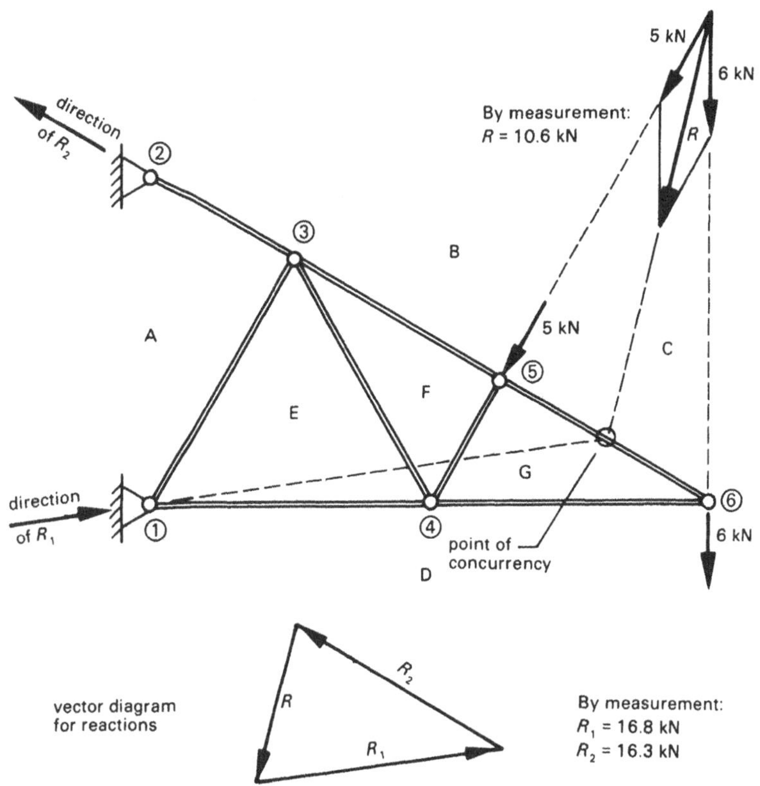

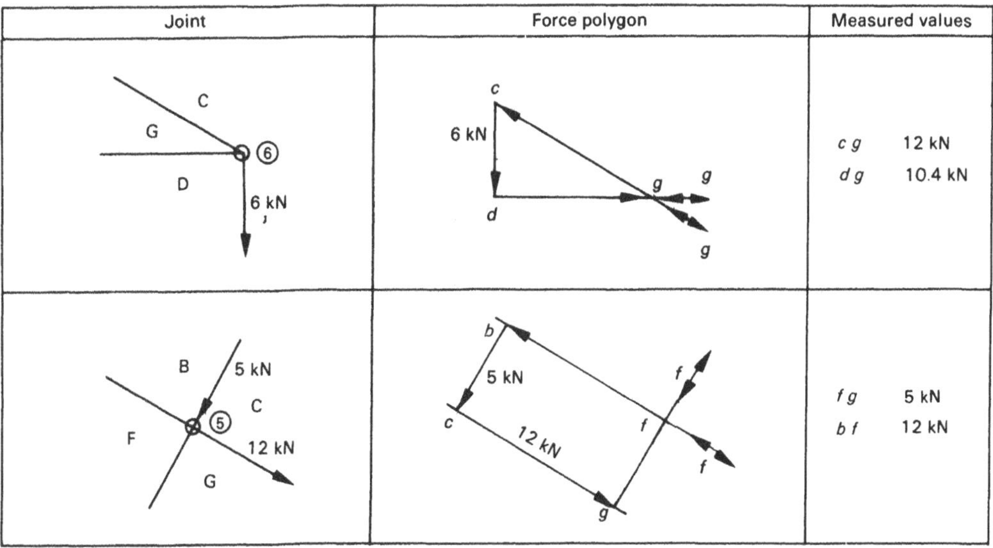

Joint	Force polygon	Measured values
		cg 12 kN dg 10.4 kN
		fg 5 kN bf 12 kN

Fig. 4.18 *(continued on next page)*

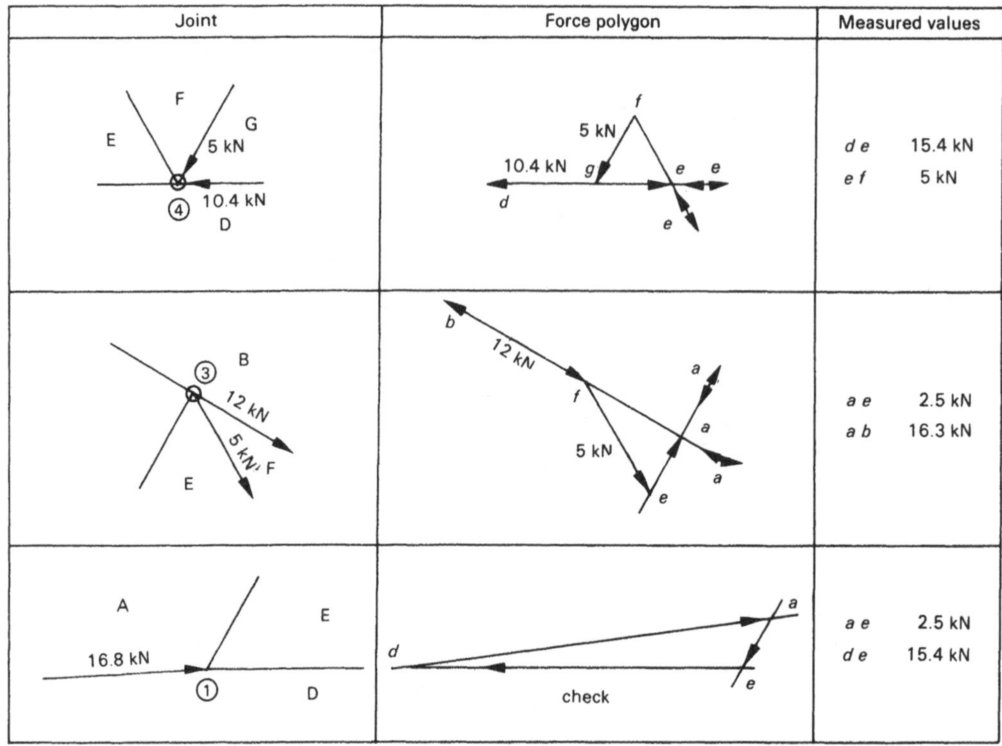

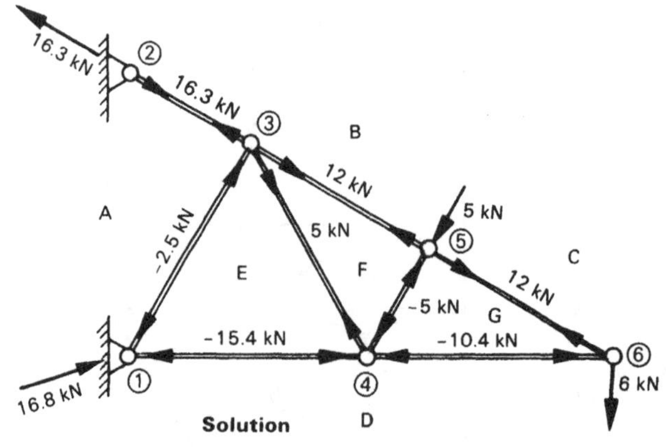

Fig. 4.18 *(contd)*

4.5 Combined force polygon

It will be evident that, when using the method of joints graphically, the vector for the force in each frame member must be drawn twice—when considering the joint at each end of the member. This unnecessary duplication may be avoided by drawing only one force polygon for the entire frame. This combined force polygon is also known as a Maxwell diagram.

The Maxwell diagram avoids duplication and is more compact but has one disadvantage: arrowheads showing the sense of the force (tension or compression) cannot be used because there is a reversal of direction at each end of the member. The sense of the forces must be determined by examination of each joint individually and by knowing the direction of at least one of the forces at the joint. Then since heads must follow tails in the vector polygon, the direction of all other forces at the joint may be ascertained.

The procedure for solving a frame using the combined force polygon is essentially the same as that used previously with the method of joints, except that only one force polygon is drawn and this polygon does not use arrowheads to indicate the direction of the forces.

Example 4.3

Determine the forces in the symmetrical frame shown in Figure 4.19 using the combined force polygon method (Maxwell diagram).

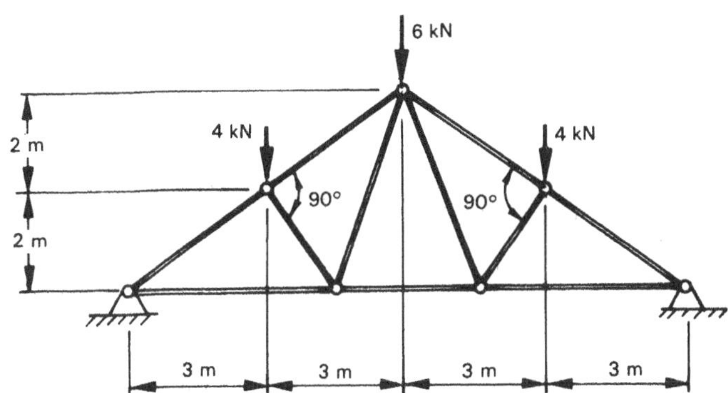

Fig. 4.19

Solution

The reactions may be determined by inspection since only symmetrical vertical forces act on the frame. Each reaction must be vertical and of magnitude 7 kN.

The development of the Maxwell diagram is shown in Figure 4.20 below the scale drawing of the frame in which the reactions have been shown, Bow's notation used and the joints numbered. So that the procedure can more readily be followed, the development of the diagram at each joint has been shown, but of course in practice only one diagram is drawn.

6 kN
C
④
D
4 kN
4 kN
B
②
G
⑥
I
E
H
①
F
J
③
⑤
⑦
7 kN
A
7 kN
known direction
Joint ①
Joint ②
Joint ③

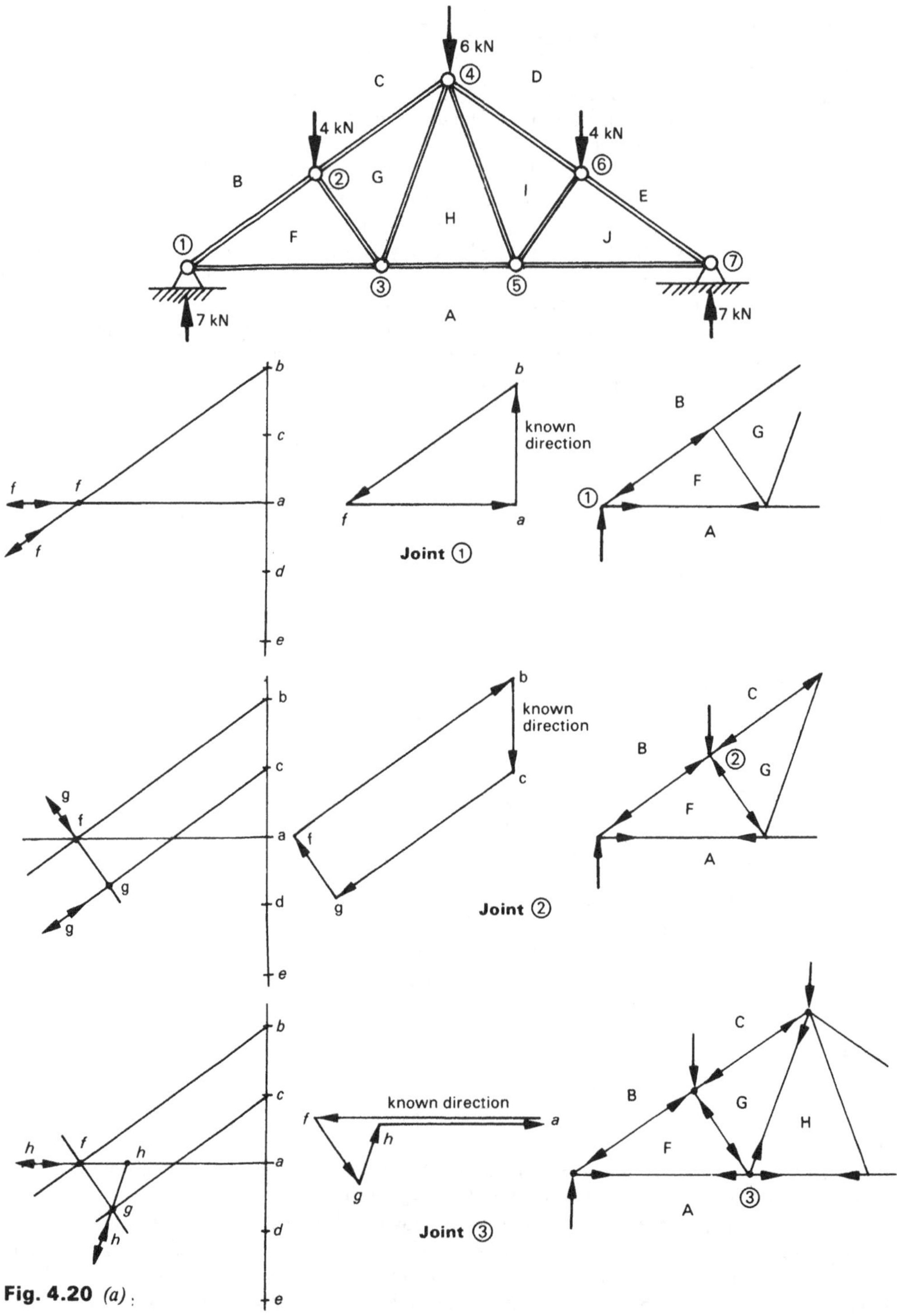

Fig. 4.20 *(a)*

b
c
a
d
e
f
g
h
i
j
known direction
Joint ④
Joint ⑤
Joint ⑥
A
B
C
D
E
F
G
H
I
J
6
4
4
7
7
-10.4
-10.4
-12.6
-12.6
-3.3
-3.3
3.0
3.0
10.5
7.5
10.5
Solution complete
(measured value of forces shown in kN)

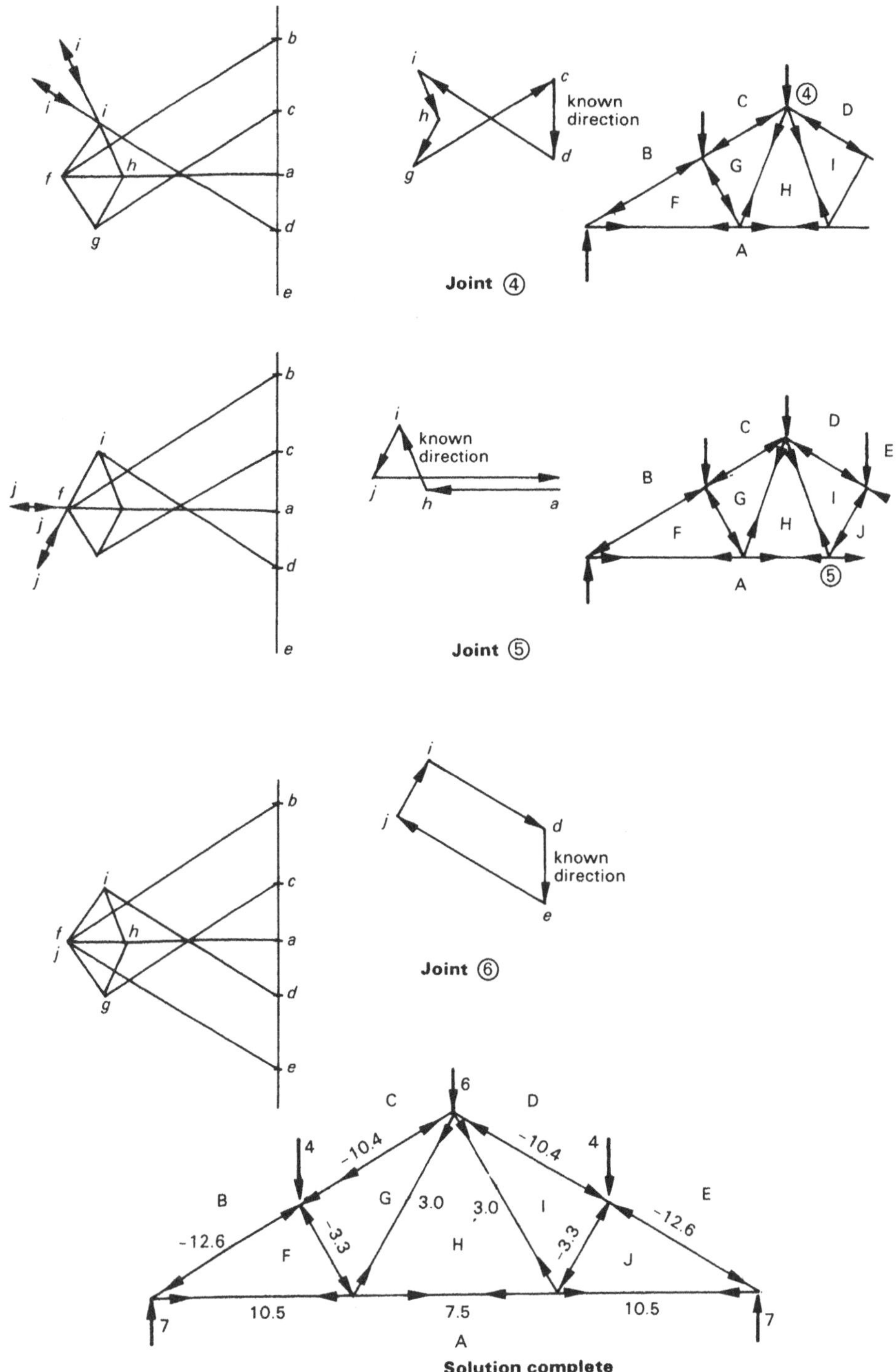

Fig. 4.20 *(b)*

Example 4.4

Determine the forces in the frame shown in Figure 4.21 by drawing a Maxwell diagram.

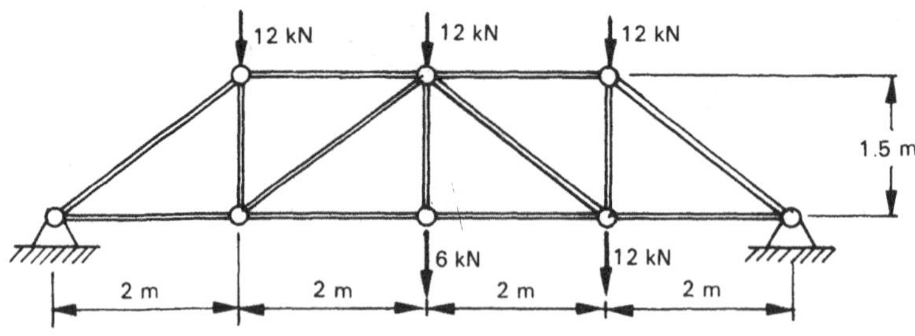

Fig. 4.21

Solution

The reactions are vertical and may be obtained by simple calculation (moments about either end and $\Sigma F_V = 0$). The diagram with Bow's notations is shown in Figure 4.22(a), the Maxwell diagram in Figure 4.22(b) and the frame diagram marked with measured values and directions in Figure 4.22(c).

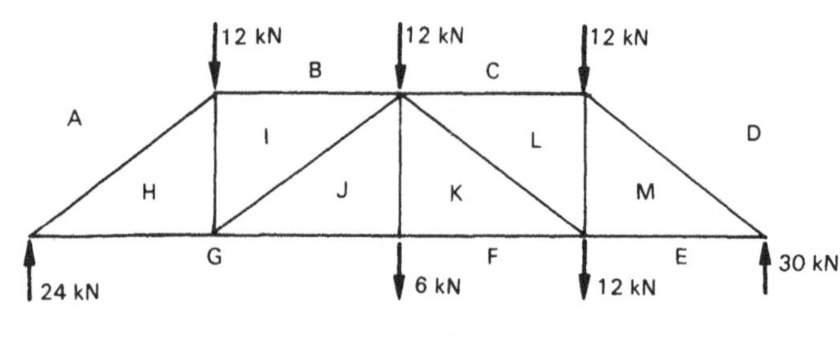

(a)

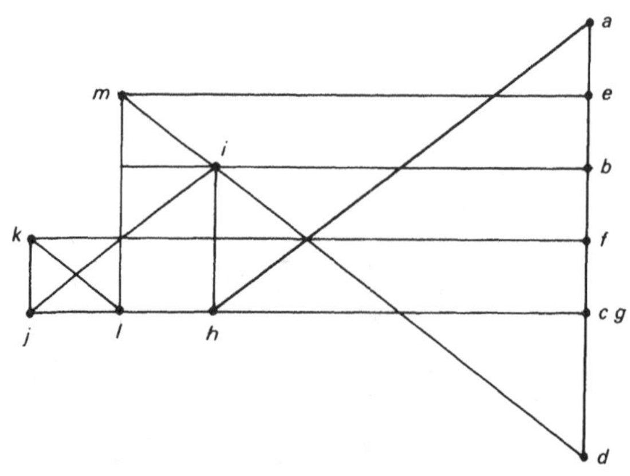

(b)

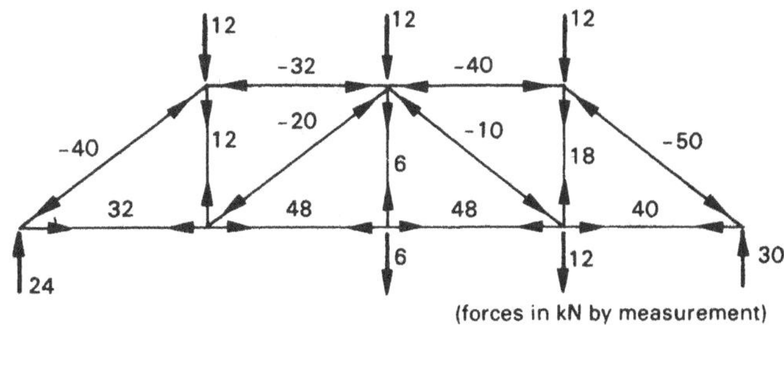

Fig. 4.22 *(a) Frame diagram. (b) Maxwell diagram. (c) Completed solution*

4.6 Method of sections

The method of sections is based upon the fact that, for a beam or frame in equilibrium, members may be cut and the forces in the cut members together with the external forces acting on either side of the cut section, form two new force systems, each of which must be in equilibrium.

Consider the simple structure shown in Figure 4.23. It is evident that the reactions at A and B must be 14.14 kN acting in the direction of each member. This is also the force (compression) in each member. Since the structure is pin-jointed with forces acting only at the pins, there is no bending moment in either of the members.

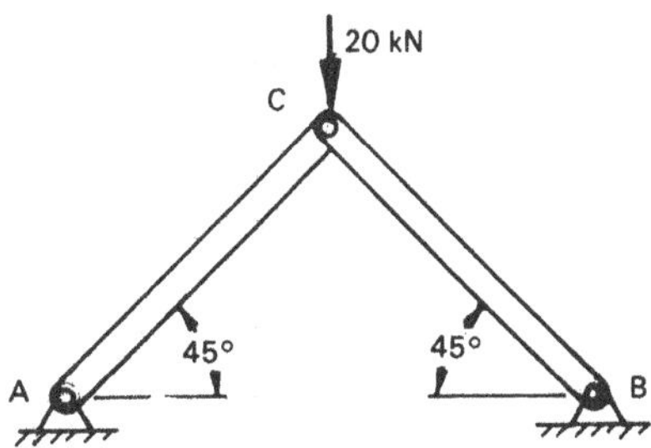

Fig. 4.23 *Simple frame structure*

Imagine member CB to be cut thus forming two force systems as shown in Figure 4.24.

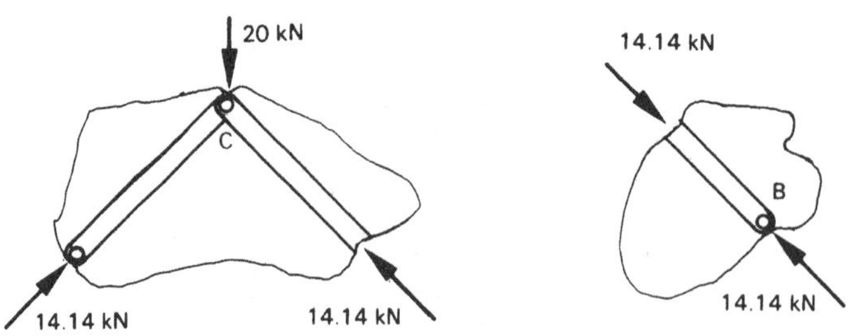

Fig. 4.24 *Force systems for a cut frame structure*

Each of the force systems shown must also be in equilibrium.

The method of sections enables unknown forces in frame members to be determined without an analysis of the entire structure by suitable choice of cutting plane. However the entire frame may also be solved by use of sufficient cutting planes to take in each unknown force, but this is not the usual application of the method.

Notes

1. The method may only be used to determine unknown forces in frame members provided that no bending moment or shear force acts in the members. That is, the applied forces (or reactions) act only at the pin joints.
2. The cutting plane must form a complete force system around the cut section with no uncut members still supporting the structure. All applied forces (and reactions) acting in the force system must be shown but only the internal forces in cut members are considered.
3. It is only necessary to analyse one of the force systems formed by the cutting plane. Usually it is more convenient to consider the one which has the least number of forces acting.
4. Unknown forces may be determined by taking moments about a suitable point which may be located either inside or outside the force system. If three forces are unknown, direct solution to one of the forces may be obtained by taking moments about a point where two unknown forces intersect, since these forces will have no moment about that point.
5. Unknown forces may also be determined by use of the equations $\Sigma F_V = 0$, $\Sigma F_H = 0$, provided that there are no more than two unknown forces in the force system.
6. The direction (sense) of the forces in cut members is generally unknown and must be assumed. As a rule is it convenient to assume tension (+), i.e. the force is pulling away from the joint being considered.

Example 4.5

Use the method of sections to determine the forces in the members marked ①, ②, ③, ④ and ⑤ of the frame shown in Figure 4.25.

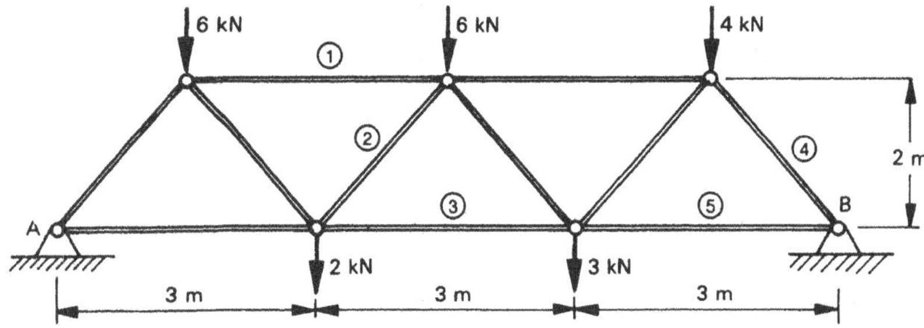

Fig. 4.25

Solution

1. Determine the reactions.

   Moments about A:

   $$6 \times 1.5 + 2 \times 3 + 6 \times 4.5 + 3 \times 6 + 4 \times 7.5 - R_B \times 9 = 0$$

   $$\therefore R_B = 10 \text{ kN}$$

   $$\Sigma F_V = 0, \therefore R_A = 11 \text{ kN}$$

2. Choose a suitable cutting plane and isolate the force system. In order to determine F_1, F_2 and F_3, the most suitable cutting plane and the simplest force system is the left-hand section as shown in Figure 4.26.

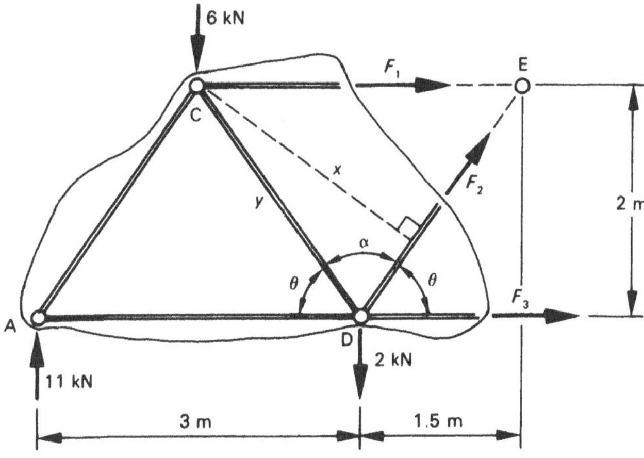

Fig. 4.26

3. Force F_3 may now be determined directly by taking moments about point E, namely:

   $$11 \times 4.5 - 6 \times 3 - 2 \times 1.5 - F_3 \times 2 = 0$$

   $\therefore F_3 =$ **14.25 kN** (+ hence assumed tension is correct)

   Force F_1 may be determined directly by taking moments about point D, namely:

   $$11 \times 3 - 6 \times 1.5 + F_1 \times 2 = 0$$

   $\therefore F_1 =$ **−12 kN** (− hence compression)

Force F_2 may be determined by taking moments about point C, after having determined the unknown distance x.
By trigonometry:

$$\tan\theta = \frac{2}{1.5} \quad \therefore \theta = 53.13°, \alpha = 73.74°$$

The length of the sloping side y is given by:

$$\sin\theta = \frac{2}{y} \quad \therefore y = 2.5 \text{ m}$$

And finally:

$$\frac{x}{y} = \sin\alpha \quad \therefore x = 2.4 \text{ m}$$

Moments about C:

$$11 \times 1.5 + 2 \times 1.5 - 14.25 \times 2 - F_2 \times 2.4 = 0$$

$$\therefore F_2 = \mathbf{-3.75\ kN} \text{ (– hence compression)}$$

It is easy to check F_2 by $\Sigma F_V = 0$.

$$\therefore 11 - 6 - 2 + F_2 \sin 53.13° = 0$$

$$\therefore F_2 = \mathbf{-3.75\ kN} \text{ (which checks)}$$

4. To determine forces F_4 and F_5, choose the cutting plane and isolate the force system as shown in Figure 4.27.

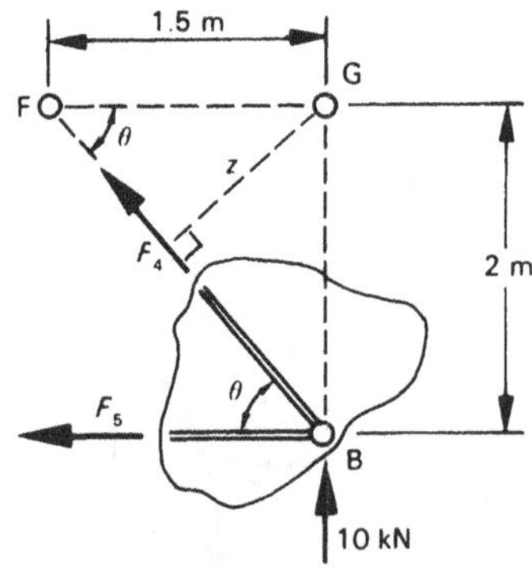

Fig. 4.27

Force F_5 may be determined by taking moments about point F.

$$-10 \times 1.5 + F_5 \times 2 = 0$$

$$\therefore F_5 = \mathbf{7.5\ kN} \text{ (tension)}$$

Similarly F_4 may be determined by taking moments about point G, after first determining distance z.

$$\sin\theta = \frac{z}{1.5} \quad \therefore z = 1.5 \sin 53.13°$$

$$= 1.2 \text{ m}$$

Moments about G:

$7.5 \times 2 + F_4 \times 1.2 = 0$

$\therefore F_4 = \mathbf{-12.5\ kN}$ (compression)

Alternatively F_4 and F_5 could have been determined by the equations for force equilibrium at joint B, namely:

$\Sigma F_V = 0 \quad \therefore F_4 \sin\theta + 10 = 0$

$$\therefore F_4 = \frac{-10}{\sin 53.13°}$$

$= \mathbf{-12.5\ kN}$ (as before)

$\Sigma F_H = 0 \quad \therefore -F_5 - F_4 \cos\theta = 0$

$\therefore F_5 = 12.5 \cos 53.13°$

$= \mathbf{7.5\ kN}$ (as before)

Problems

Note: In the solution to these problems, the sense of the force in each member is indicated by + tension or - compression. All forces are in kN.

4.1 For the frame shown in Figure P4.1, determine the reactions. Then determine the force in each member using:

(a) method of joints analytically

(b) method of joints graphically

(c) combined force polygon (Maxwell diagram).

Reactions

a b	*c d*
5.12 @ 90°	2.88 @ 90°

Forces

a e	*b e*	*c f*	*d f*	*e f*
3.84	-6.4	-4.8	3.84	3.0

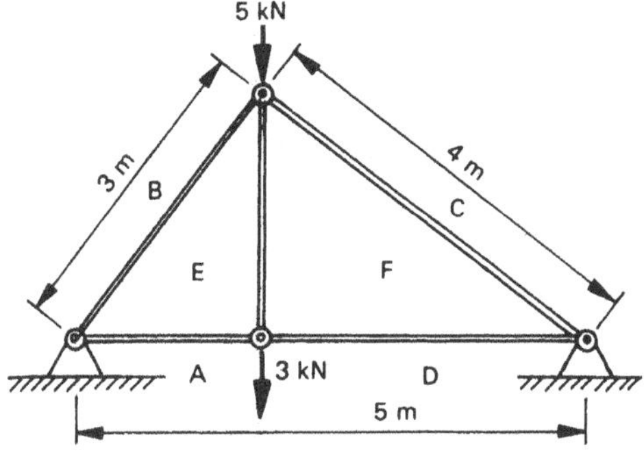

Fig. P4.1

4.2 Repeat problem 4.1 for the frame shown in Figure P4.2.
Reactions

a b	*a c*
5.55 @ 205.6°	2.4 @ 90°

Forces

a d	*a e*	*b d*	*c e*	*d e*
3.2	3.2	3.0	-4.0	0

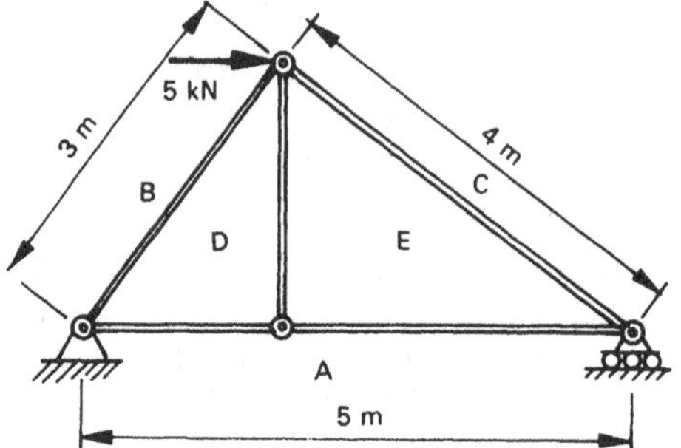

Fig. P4.2

4.3 Repeat problem 4.1 for the frame shown in Figure P4.3.
Reactions

a b	*a e*
10.4 @ 180°	16.6 @ 51.4°

Forces

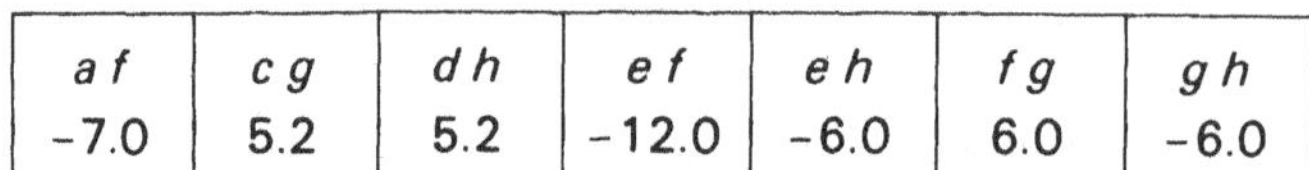

a f	*c g*	*d h*	*e f*	*e h*	*f g*	*g h*
-7.0	5.2	5.2	-12.0	-6.0	6.0	-6.0

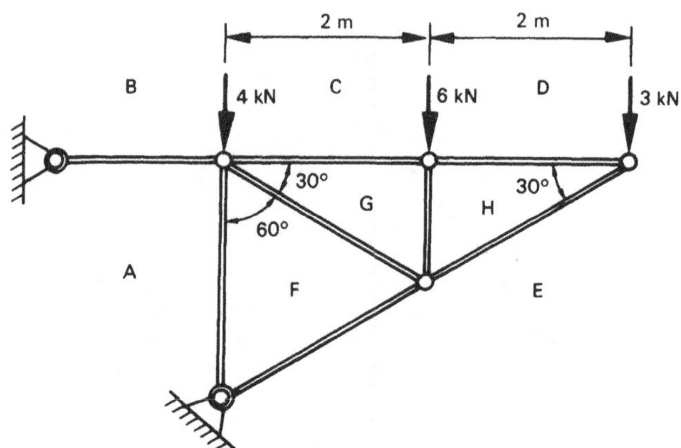

Fig. P4.3

4.4 Repeat problem 4.1 for the frame shown in Figure P4.4.

Reactions

a b	*a e*
16.8 @ 150°	9.65 @ 351.7°

Forces

a f	*b g*	*b h*	*d h*	*e f*	*f g*	*g h*
6.9	11.9	11.9	-8.0	-11.0	3.0	0

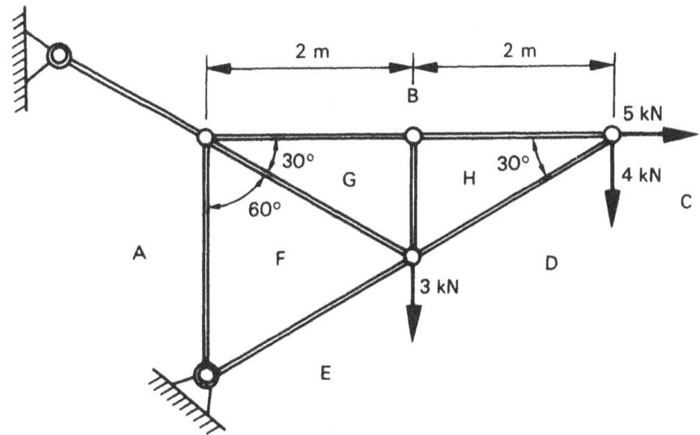

Fig. P4.4

4.5 Repeat problem 4.1 for the frame shown in Figure P4.5.

Reactions

a b	*a c*
10 @ 180°	11.2 @ 26.6°

Forces

a b	*a d*	*b e*	*c d*	*c e*	*d e*
10	-7.07	5.0	-5.0	-7.07	5.0

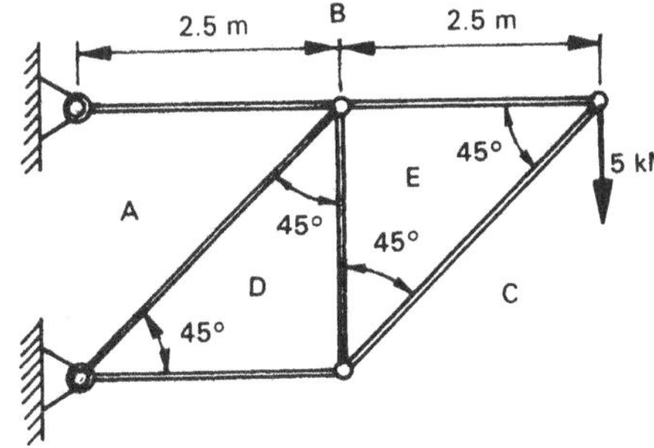

Fig. P4.5

4.6 The frame shown in Figure P4.6 consists of three equilateral triangles. Determine the reactions and the force in each member. Check the force in members BG and FG by the method of sections.

Reactions

a e	*c d*
20.5 @ 90°	19.5 @ 90°

Forces

a f	*b g*	*c h*	*d h*	*e f*	*f g*	*g h*
-23.7	-15.0	-22.5	11.3	11.8	6.35	7.51

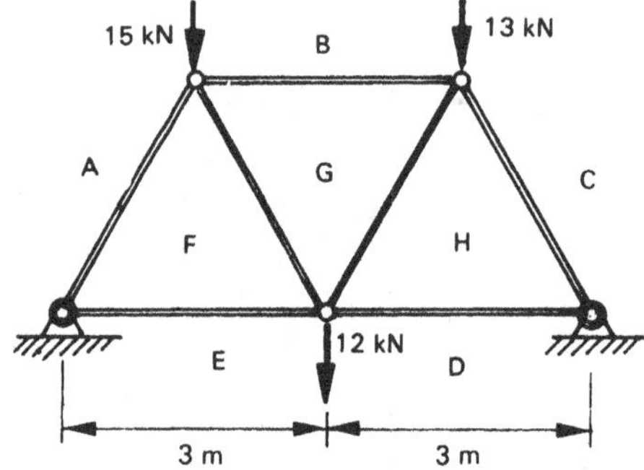

Fig. P4.6

4.7 Repeat problem 4.6 for the frame of the same design shown in Figure P4.7.

Reactions

a e	*c d*
17.4 @ 115.5°	22.2 @ 90°

Forces

a f	*b g*	*c h*	*d h*	*e f*	*f g*	*g h*
-18.2	-18.2	-25.7	12.8	16.6	3.18	10.7

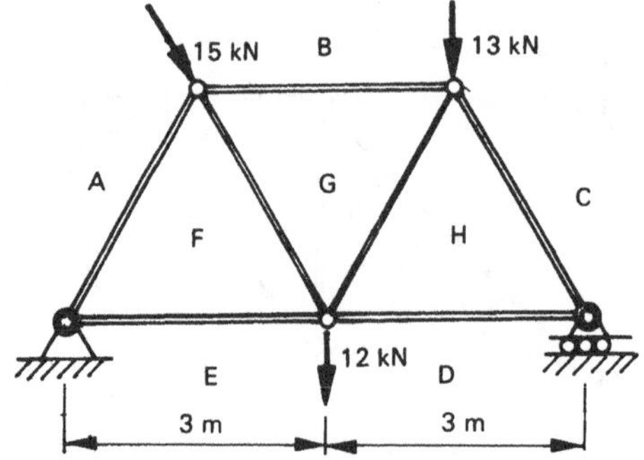

Fig. P4.7

4.8 For the frame shown in Figure P4.8, determine:

(a) reactions using a graphical method

(b) reactions using an analytical method

(c) force in members marked C, D, E using the method of sections.

Reactions

B	A
35 @ 143.1°	36.3 @ 24.5°

Forces

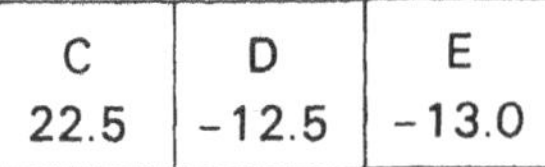

C	D	E
22.5	-12.5	-13.0

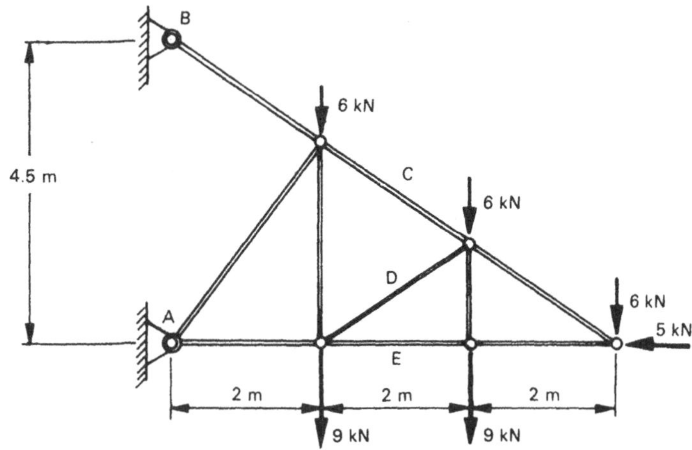

Fig. P4.8

4.9 For the frame shown in Figure P4.9, determine the force in members marked A, B, C.

Forces

A	B	C
110	76.4	-164

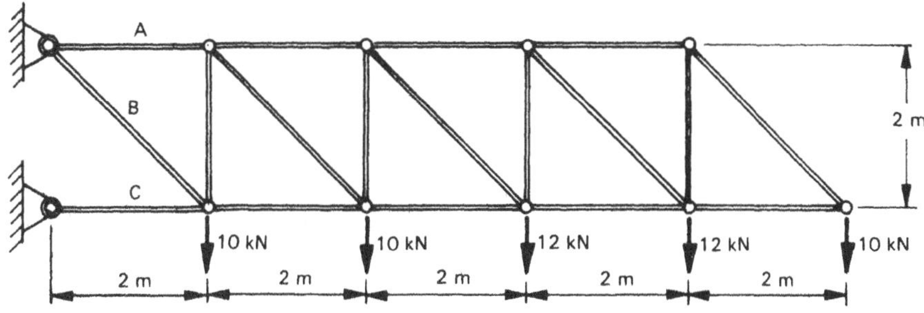

Fig. P4.9

4.10 Determine the force in all the members of the frame shown in Figure P4.10.

Forces

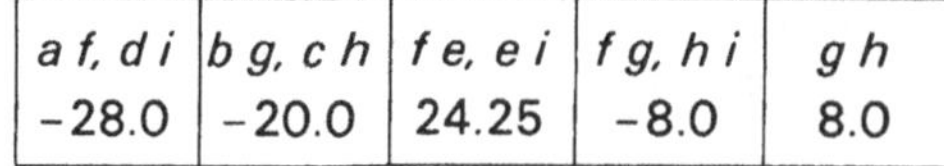

a f, d i	*b g, c h*	*f e, e i*	*f g, h i*	*g h*
-28.0	-20.0	24.25	-8.0	8.0

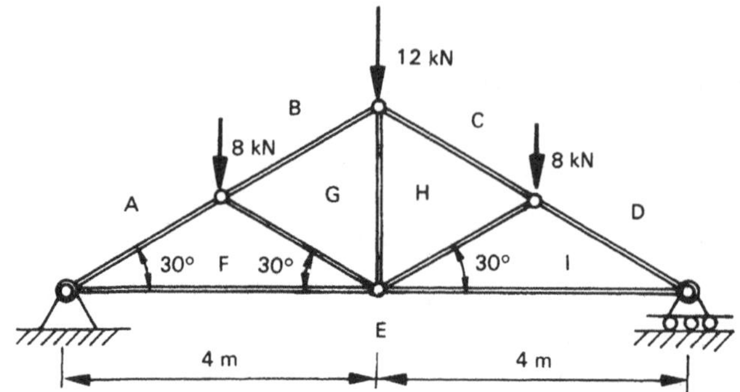

Fig. P4.10

4.11 Determine the force acting in each member of the symmetrical frame shown in Figure P4.11.

Forces

a f, d j	*b g, c i*	*e f, e j*	*e h*	*f g, i j*	*g h, h i*
-40.0	-30.0	34.6	23.1	-10.0	5.77

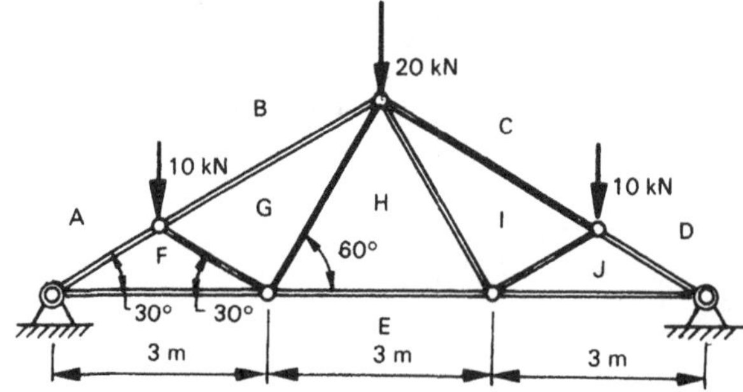

Fig. P4.11

4.12 For the frame shown in Figure P4.12, determine the reactions and the force in each member.

Reactions

a g	*d e*
26.6 @ 55.7°	18.0 @ 90°

Forces

a h	*b j*	*b k*	*d m*	*e m*	*f i*	*f l*	*g h*
-46.75	-25.5	-25.5	-38.25	33.75	26.25	33.75	26.25
h i	*i j*	*j k*	*k l*	*l m*			
12.0	-21.3	24.0	-29.75	12.0			

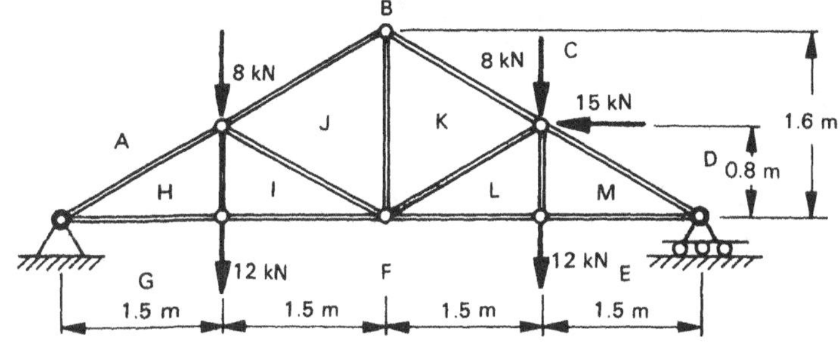

Fig. P4.12

5

Shear force, bending moment and torque distribution diagrams

When a shaft, beam or other structural member is loaded, it is essential to ensure that the stresses and strains caused by the shear force, bending moment and torque do not exceed allowable values otherwise failure could occur. A necessary first step is to determine their values at any section in the member which is the subject of this chapter.

The most convenient way of showing the variation in shear force, bending moment and torque in a member is by means of a diagram. These diagrams are usually drawn to scale and from the diagram the required values at any location may be read.

Of course, once the principles for drawing the diagrams have been established, the shear force, bending moment or torque at any section may be determined without drawing the full diagram and this is more convenient when values are required at only one section. However, the usual problem in mechanical design is to determine the plane of highest stress and the location of this plane in the member may be difficult to determine without a diagram.

5.1 Shear force at any section

The shear force at any section of a structural member is the sum of the components of the forces acting in a perpendicular direction to the axis of the member at *either* side of the section. The forces may be applied or reactive forces, so that in order to calculate the shear force at any section it is first necessary to determine the reactive forces at the supports of the member. The shear force at any section may then be determined from the total of the perpendicular components of the forces at either side of the section being considered.

Sign convention for shear force

The usual sign convention for shear force is that upward force is positive when considering the left-hand side of the section. Hence downward force will be positive when considering the right-hand side of the section. This is illustrated in Figure 5.1.

This sign convention may also be remembered by imagining the member to be cut and a small roller placed at the section. A clockwise rotation of the roller would indicate positive shear, whereas anticlockwise rotation would indicate negative shear. This is also shown in Figure 5.1.

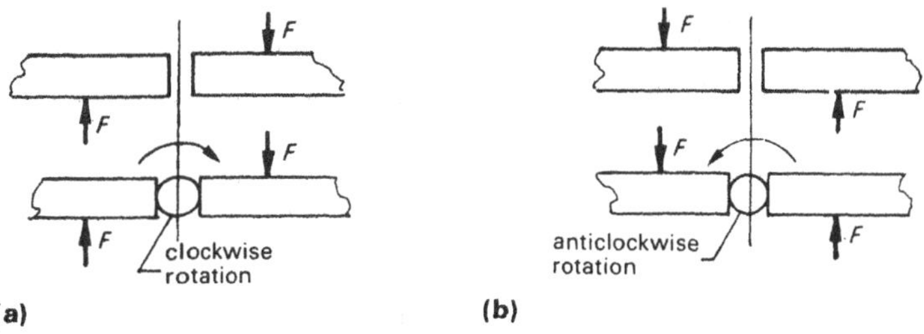

Fig. 5.1 *Sign convention for shear force: (a) positive shear force, (b) negative shear force*

5.2 Bending moment at any section

The bending moment at any section of a structural member is the sum of the moments acting on *either* side of the section. The total moment may be due to applied or reactive moment as well as moment due to applied or reactive forces. In order to calculate the bending moment at any section, it is first necessary to determine reactive forces and moments at the supports of the member. The moment at any section may then be determined from the total moment on either side of the section being considered.

Sign convention for bending moment

The usual sign convention for bending moment is that clockwise moment is positive when considering the left-hand side of the section. Hence anticlockwise moment will be positive when considering the right-hand side of the section. This is illustrated in Figure 5.2.

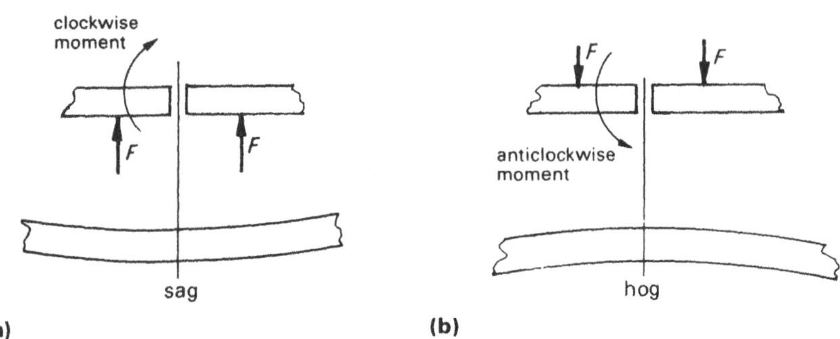

Fig. 5.2 *Sign convention for bending moment: (a) positive bending moment, (b) negative bending moment*

This sign convention may also be remembered by imagining the beam to be flexible. A positive moment is then a sag shape (downward) whereas a negative moment is a hog shape (upward). This is also shown in Figure 5.2.

5.3 Shear force and bending moment: Concentrated loads

Consider the left-hand portion of a member with concentrated loads as shown in Figure 5.3. In (a), section A is located a distance x from F_1 and is to the left of F_2.

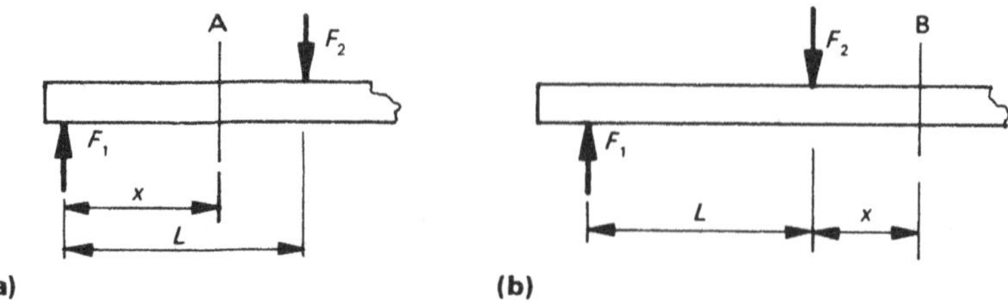

Fig. 5.3 *Concentrated loads on part of a member*

The shear force at section A is $F = +F_1$

The bending moment at section A is $M = +F_1x$

It is evident that the bending moment due to a concentrated load varies directly (linearly) with the distance x whereas the shear force is independent of x.

In Figure 5.3(b), section B is located a distance x from F_2 and is to the right of F_2.

The shear force at section B is $F = F_1 - F_2$

The bending moment at section B is $M = F_1(L + x) - F_2x$

$$= F_1L + (F_1 - F_2)x$$

Again it is seen that the bending moment varies directly with x whereas the shear force is independent of x.

Hence it is clear that at the point of application of a concentrated load the shear force undergoes a sudden change in value equal to the magnitude of the load. Therefore it is impossible to state the value of shear force in the plane of the load, rather it is necessary to state the value a little to the left or a little to the right of the plane. Thus at the plane or the load F_2, the shear force is F_1 a little to the left and $F_1 - F_2$ a little to the right.

The bending moment does not change suddenly in the plane of the load and its value may be stated at this plane. At the plane of the load F_2, the bending moment is F_1L.

5.4 Shear force and bending moment diagrams: Concentrated loads on a simply supported beam

When a beam is simply supported, there is no bending moment at the supports unless the beam is overhung. For an overhung beam, there will be no bending moment at the extreme ends but there may be a moment at the overhung support. The shear force and bending moment diagrams in each case for a beam with a single load is shown in Figure 5.4.

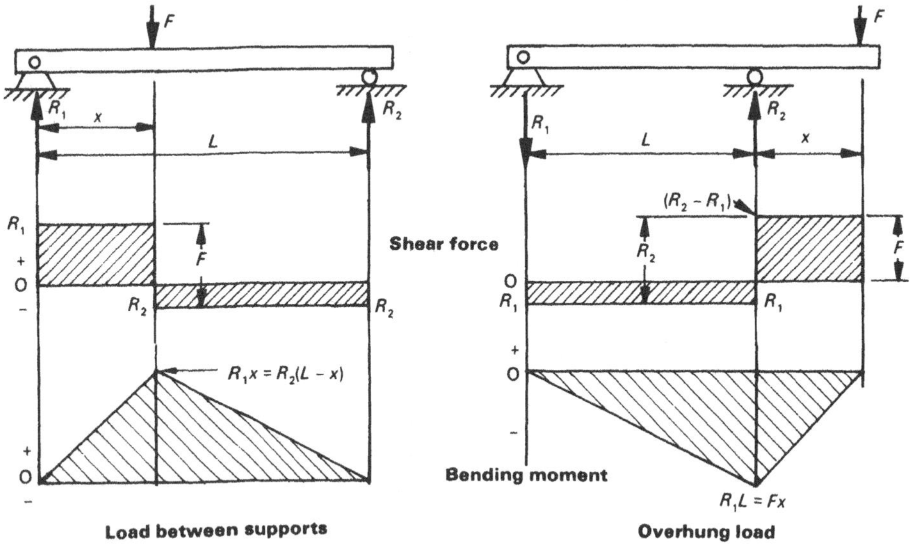

Fig. 5.4 *Shear force and bending moment diagrams for a simply supported beam with a single concentrated load*

Example 5.1

Draw shear force and bending moment diagrams for the simply supported beam shown in Figure 5.5.

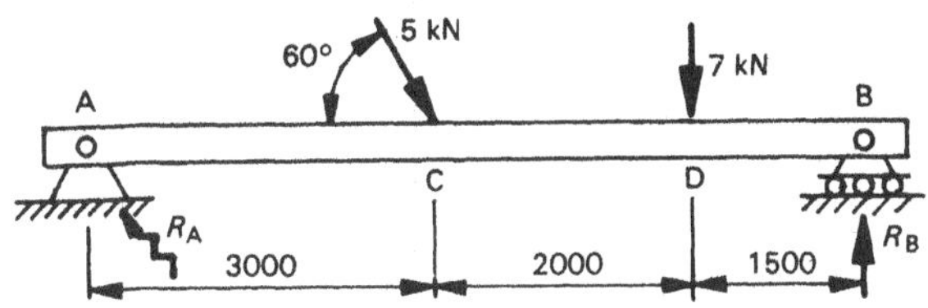

Fig. 5.5

Solution

First it is necessary to calculate the reactions.

Taking moments about A:

$$\Sigma M = 0 \quad \therefore 5 \sin 60° \times 3 + 7 \times 5 - R_B \times 6.5 = 0$$

$$\therefore R_B = 7.38 \text{ kN}$$

$$\Sigma F_V = 0 \quad \therefore R_{AV} + 7.38 - 5 \sin 60° - 7 = 0$$

$$\therefore R_{AV} = 3.95 \text{ kN}$$

$\Sigma F_H = 0 \quad \therefore R_{AH} + 5 \cos 60^\circ = 0$

$\therefore R_{AH} = -2.5 \text{ kN}$

$\therefore R_A = 4.67 \text{ kN @ } 122.35^\circ$

A free body diagram may now be drawn showing all the forces acting on the beam (Fig. 5.6, all forces in kN).

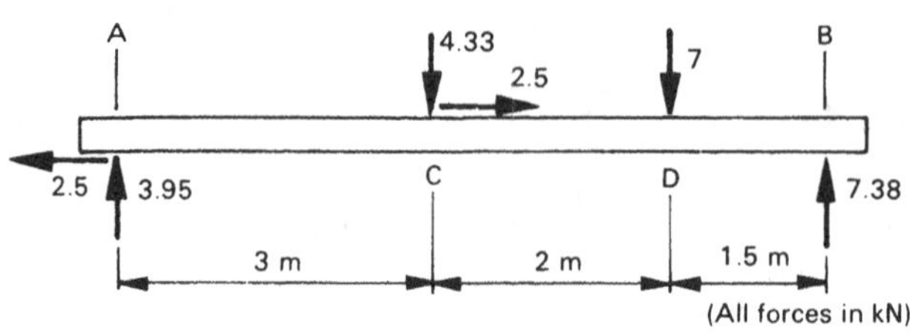

Fig. 5.6

Working from the left-hand side of the beam (discarding the right-hand side of each section), the shear force and bending moment may now be determined at each section. Since shear force cannot be specified directly in the plane of the section, subscripts L and R will be used to denote points just to the left and just to the right of the section.

Section		*Shear force (kN)*	*Bending moment (kNm)*
A	A_L	$F = 0$	$M = 0$ (pin joint)
	A_R	$F = +3.95$ **Fig. 5.7**	
C	C_L	Same as A_R $\quad F = +3.95$	$M = 3.95 \times 3 - 4.33 \times 0$ $= +11.8$
	C_R	$F = 3.95 - 4.33$ $= -0.38$ **Fig. 5.8**	

Section		*Shear force (kN)*	*Bending moment (kNm)*
D	D_L	Same as C_R $F = -0.38$	$M = 3.95 \times 5 - 4.33 \times 2 - 7 \times 0$
	D_R	$F = -0.38 - 7$ $= -7.38$ Fig. 5.9	$= +11.1$
B	B_L	Same as D_R $F = -7.38$	$M = 0$ (roller support)
	B_R	$F = 0$	

It is a good idea to check for error by determining the shear force and bending moment at section D by considering the right-hand side of the section. This has been done below and the same answer is obtained, thus indicating a correct result.

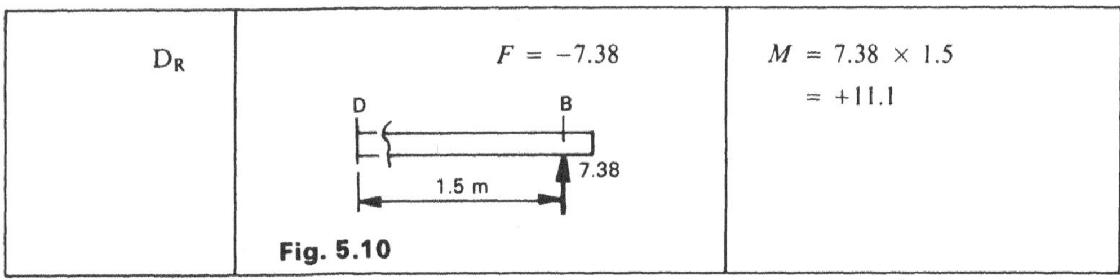

D_R	$F = -7.38$ Fig. 5.10	$M = 7.38 \times 1.5$ $= +11.1$

The shear force and bending moment diagrams may now be drawn to scale below the beam diagram as shown in Figure 5.11.

Notes

1. When checking values by working from the right-hand side of the section rather than the left, remember the opposite sign applies, that is upward forces will be negative and anticlockwise moments positive.
2. The shear force may be developed progressively working along the beam by adding or subtracting each new vertical force component as it occurs. It is generally better to develop the bending moment diagram independently at each section rather than progressively.
3. Components of forces parallel to the section (horizontal components in this case) have no effect on the shear force or bending moment provided the depth of the beam is small in relationship to its length (as is usually the case.)
4. When drawing the shear force and bending moment diagrams, any convenient scale may be used. The scale on the y axis should be clearly shown or else key values marked (as has been done in Fig. 5.11).
5. Whereas the shear force diagram can have discontinuities (sudden changes in value), this

cannot occur on the bending moment diagram. There can be a change in slope at the position of a concentrated force but not a sudden change in value.

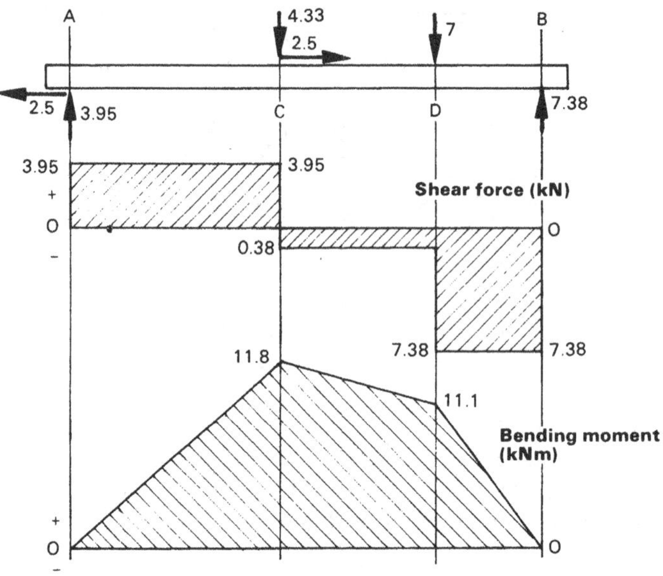

Fig. 5.11

5.5 Relationship between shear force and bending moment

It is possible to develop the shear force and bending moment diagrams independently of one another—as was done in example 5.1. However there is a definite relationship between the two diagrams which is useful when drawing and checking for error.

Consider a single concentrated force F acting at distance x from a section as shown in Figure 5.12.

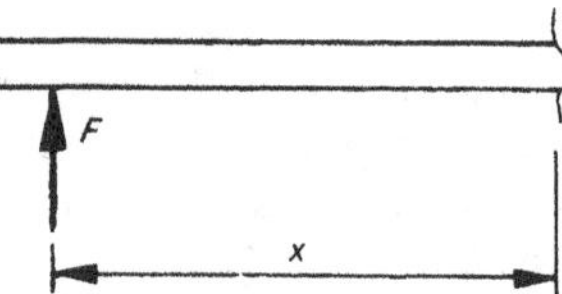

Fig. 5.12 *Concentrated load*

The shear force at the section is $F = F$.

The bending moment at the section is $M = Fx$.

If the bending moment is differentiated with respect to x, the result is:

$$\frac{dM}{dx} = F \quad \text{or} \quad M = \int F dx$$

$\frac{dM}{dx}$ is the slope of the bending moment diagram.

$\int F dx$ is the area of the shear force diagram.

Hence it may be stated that:

1. the slope of the bending moment diagram at any section is equal to the shear force at that section;
2. the area under the shear force diagram between any two sections is equal to the change in bending moment between the two sections.

Several important principles follow:

1. If the shear force is positive, the bending moment diagram has a positive slope, whereas if the shear force is negative, the bending moment diagram has a negative slope.
2. The greater the value of the shear force, the greater the slope of the bending moment diagram.
3. If the shear force is constant between any two sections, the bending moment diagram is an inclined straight line between these sections.
4. If the shear force is zero, the bending moment diagram is changing slope on either side of the section (being momentarily zero slope at the section). This will often represent the point of maximum or minimum bending moment.

These principles are clearly seen with reference to Figures 5.4 and 5.11. For example, consider section A to C in Figure 5.11. For this section of the beam, the shear force is positive and the bending moment diagram has a positive slope of $\frac{11.8}{3} = 3.93$, which is the value of the shear force in this part of the beam (slight difference due to rounding off). Similarly, the area of the shear force diagram

$$= 3.95 \times 3 = 11.85,$$

which is the change in bending moment between the two sections.

Similarly for section D to B, the shear force is negative, the bending moment diagram has a negative slope of $\frac{11.1}{1.5} = 7.4$, which is the value of the shear force in this part of the beam. Also the area of the shear force diagram

$$= 7.38 \times 1.5 = 11.1,$$

which is the change in bending moment between the two sections.

5.6 Shear force and bending moment: Uniformly distributed loads

Consider a uniformly distributed load F N/m acting on a horizontal section of beam as shown in Figure 5.13. In (a), the section A is located inside the span of the distributed load, a distance x m from the left-hand end of the beam. At this section the equivalent concentrated load is Fx acting at distance $\frac{x}{2}$ m from A.

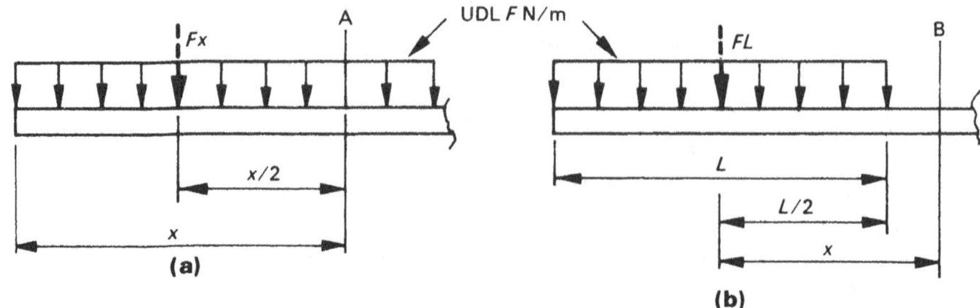

Fig. 5.13 *Uniformly distributed load*

The shear force at section A is $F = -Fx$

The bending moment at section A is $M = -Fx \times \frac{x}{2} = \frac{-Fx^2}{2}$

Hence it will be seen that the bending moment varies parabolically with x and the shear force varies linearly with x. This is also confirmed by differentiating M with respect to x:

$$\frac{dM}{dx} = -Fx$$

which is the same shear force value as that obtained above.

For Figure 5.13(b), the section B is located outside the span of the distributed load, a distance x m from the centre of the distributed load.

The shear force at section B is $F = -FL$

The bending moment at section B is $M = -FL \times x$

Since L is constant and does not change as x changes, the shear force is constant and the bending moment varies directly with x. That is, for any section outside the span of the distributed load, the uniformly distributed load may be regarded as an equivalent concentrated load of magnitude FL acting at the centre span of the distributed load.

5.7 Shear force and bending moment diagrams: Uniformly distributed load on a simply supported beam

If a uniformly distributed load (UDL) acts over the entire span of a simply supported beam, the shear force and bending moment diagrams will be as shown in Figure 5.14.

Note that at the centre span of the beam, the shear force is zero and the bending moment is:

$$M = R_A \times \frac{L}{2} - \frac{FL}{2} \times \frac{L}{4} = \frac{FL}{2} \times \frac{L}{2} - \frac{FL}{2} \times \frac{L}{4} = \frac{FL^2}{8}$$

The method of developing the shear force and bending moment diagrams for a combination of concentrated and distributed loads on a simply supported beam is illustrated in example 5.2. This beam has the additional complication of being overhung.

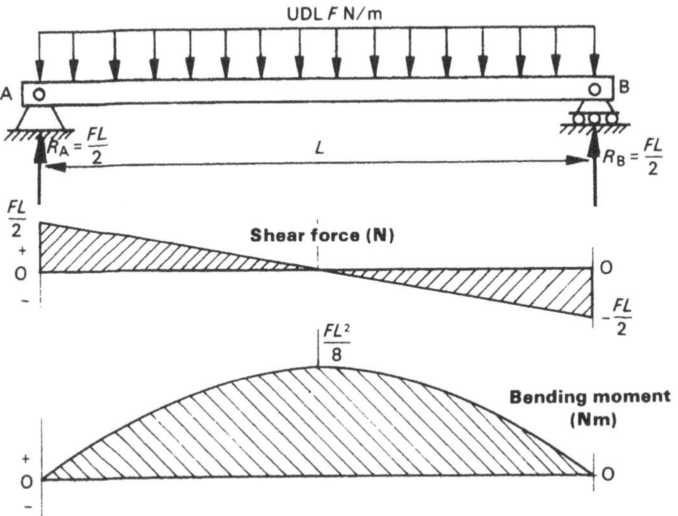

Fig. 5.14 *Shear force and bending moment diagram for UDL on a simply supported beam*

Example 5.2

Draw the shear force and bending moment diagrams for the beam shown in Figure 5.15.

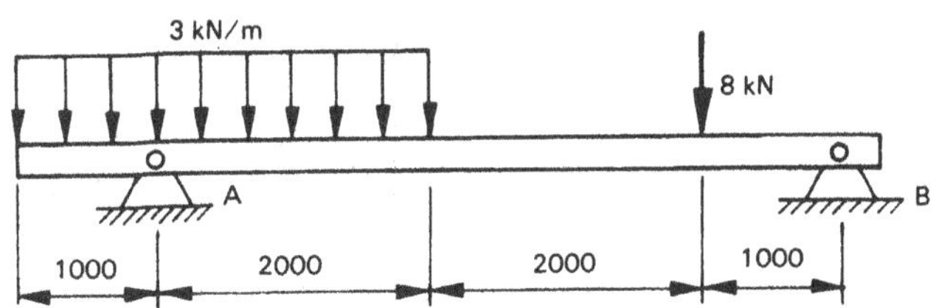

Fig. 5.15

Solution

Since only vertical loads act on the horizontal beam, the reactions at A and B must be vertical.

Taking moments about A:

$$3 \times 3 \times 0.5 + 8 \times 4 - R_B \times 5 = 0$$

$$\therefore R_B = 7.3 \text{ kN}$$

$\Sigma F_V = 0 \quad \therefore 7.3 - 8 - 9 + R_A = 0$

$\therefore R_A = 9.7 \text{ kN}$

Draw a free body diagram and choose several key sections (Fig. 5.16).

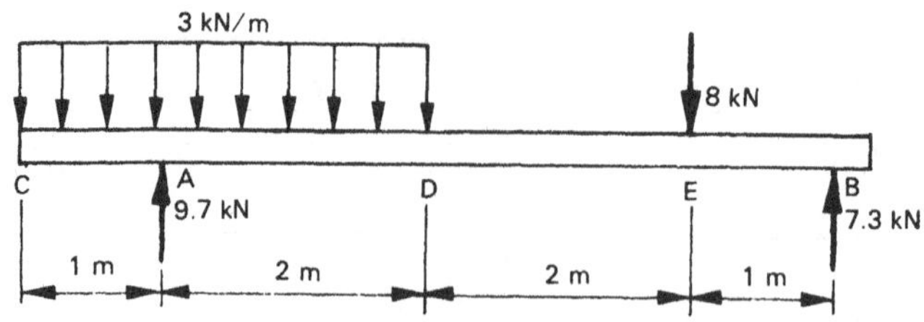

Fig. 5.16

Working from the left-hand end of the beam:

Section		*Shear force (kN)*	*Bending moment (kNm)*
C		O	O
A	A_L	$F = -3$	$M = -3 \times \frac{1^2}{2}$ $= -1.5$
	A_R	$F = -3 + 9.7$ $= 6.7$ **Fig. 5.17**	
D		$F = -3 \times 3 + 9.7$ $= 0.7$ **Fig. 5.18**	$M = -3 \times \frac{3^2}{2} + 9.7 \times 2$ $= 5.9$

Section		*Shear force (kN)*	*Bending moment (kNm)*
E	E_L	Same as D $F = 0.7$	
	E_R	$F = 0.7 - 8$ $= -7.3$	$M = 9.7 \times 4 - 9 \times 3.5 - 8 \times 0$ $= 7.3$
		Fig. 5.19	
B	B_L	Same as E_R $F = -7.3$	$M = 0$ (pin support)
	B_R	$F = 0$ check $F = -7.3 + 7.3$ $= 0$	

The shear force and bending moment diagrams may now be drawn to scale below the beam diagram as shown in Figure 5.20.

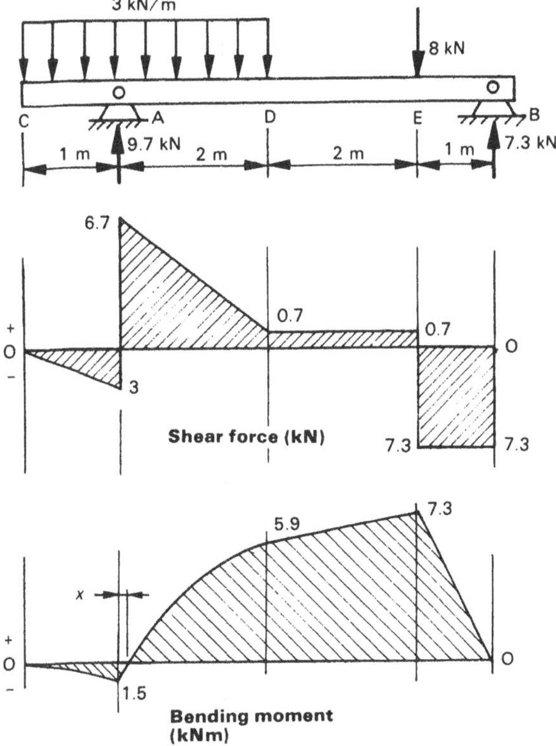

Fig. 5.20

Note: It is interesting to calculate the distance x to the right of A within the distributed load where the bending moment is zero. Writing the bending moment in terms of x and setting equal to zero:

$$\frac{-3(x+1)^2}{2} + 9.7x = 0 \quad \text{or} \quad 1.5x^2 - 6.7x + 1.5 = 0$$

$$\therefore x = \frac{6.7 \pm \sqrt{-6.7^2 - 4 \times 1.5 \times 1.5}}{2 \times 1.5} = 4.23 \text{ or } 0.236$$

The result of 4.23 m is invalid as it lies outside the distributed load. Hence the result of 0.236 m to the right of A is the position of zero bending moment.

5.8 Shear force and bending moment diagrams: Cantilever beam

When developing the shear force and bending moment diagrams for a cantilever beam, exactly the same principles apply as those for a simply supported beam. The fact that there is a reactive moment at the support as well as a reactive force has no effect on the shear force at any section of the beam. However, the reactive moment must be taken into account when drawing the bending moment diagram.

Consider the simple cantilever shown in Figure 5.21. In (a), the reactive force at the support is F and the reactive moment at the support is $-Fx$. The shear force has a constant value, $+F$, over the span x, whereas the bending moment varies linearly from $-Fx$ at the support to 0 at distance x from the support.

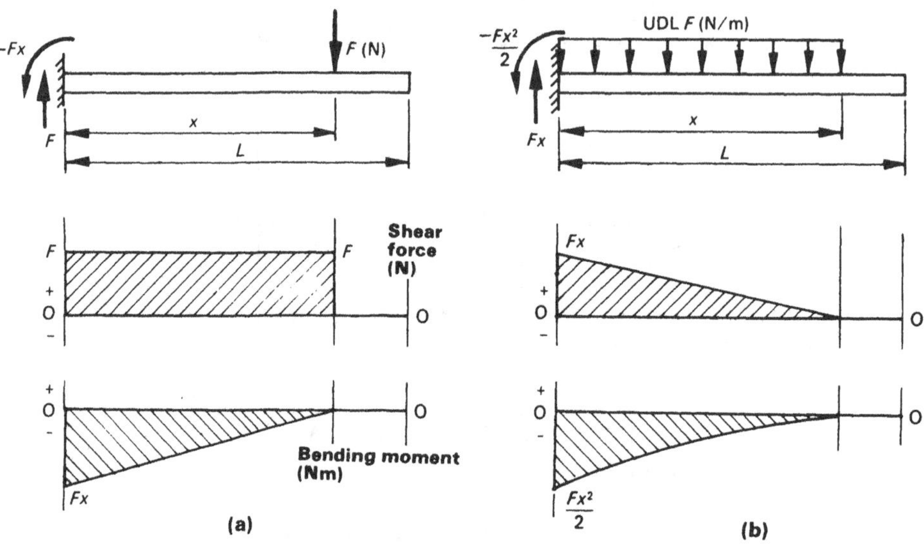

Fig. 5.21 *Shear force and bending moment diagrams for a cantilever beam: (a) concentrated load, (b) uniformly distributed load*

In Figure 5.21(b), the reactive force at the support is Fx. The shear force varies linearly from Fx at the support to 0 at distance x from the support, whereas the bending moment varies parabolically from $-\frac{Fx^2}{2}$ at the support to 0 at distance x from the support.

Note that in both cases there is no shear force or bending moment at positions to the right of the beam beyond x.

Example 5.3

Draw the shear force and bending moment diagrams for the cantilever beam illustrated in Figure 5.22.

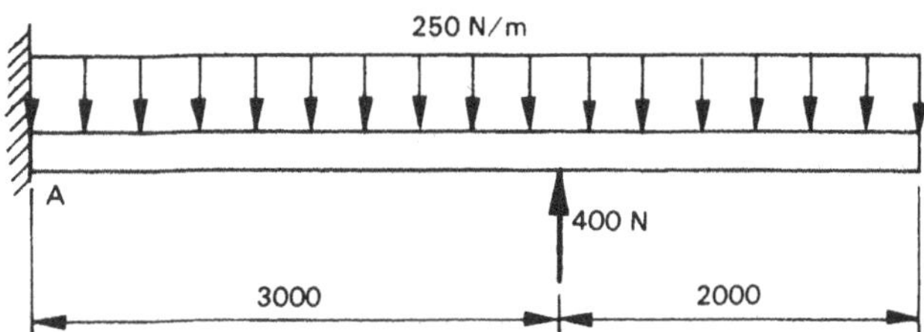

Fig. 5.22

Solution

Taking moments about A:

$$M_A + \frac{250 \times 5^2}{2} - 400 \times 3 = 0$$

$$\therefore M_A = -1925 \text{ Nm (anticlockwise)}$$

$$\Sigma F_V = 0$$

$$\therefore R_A + 400 - 250 \times 5 = 0$$

$$\therefore R_A = 850 \text{ N}$$

Now a free body diagram is drawn and the key points labelled as shown in Figure 5.23:

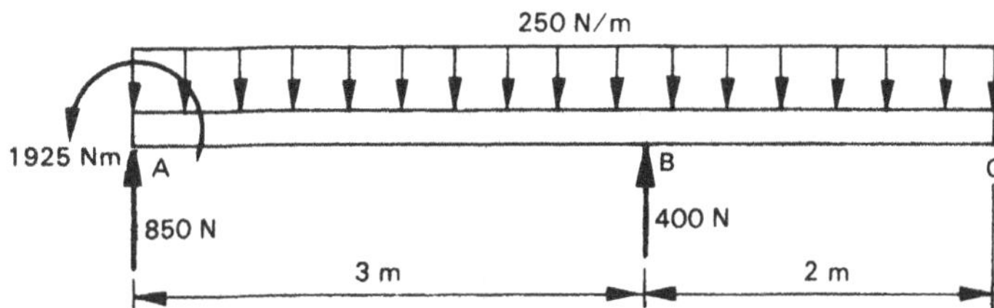

Fig. 5.23

Working from the left-hand end of the beam:

Section		*Shear force (N)*	*Bending moment (Nm)*
A	A_L	$F = 0$	$M = -1925$
	A_R	$F = 850$	
B	B_L	$F = 850 - 250 \times 3$ $= 100$ 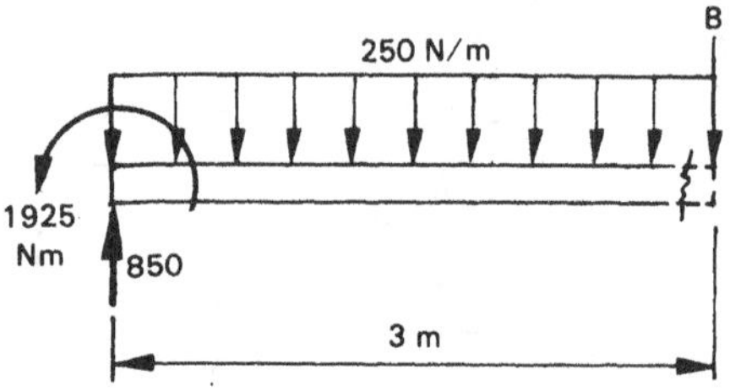	$M = -1925 - \dfrac{250 \times 3^2}{2}$ $+ 850 \times 3$ $= -500$ Check from RHS: $M = -\dfrac{250 \times 2^2}{2}$ $= -500$
	B_R	$F = 100 + 400$ $= 500$ 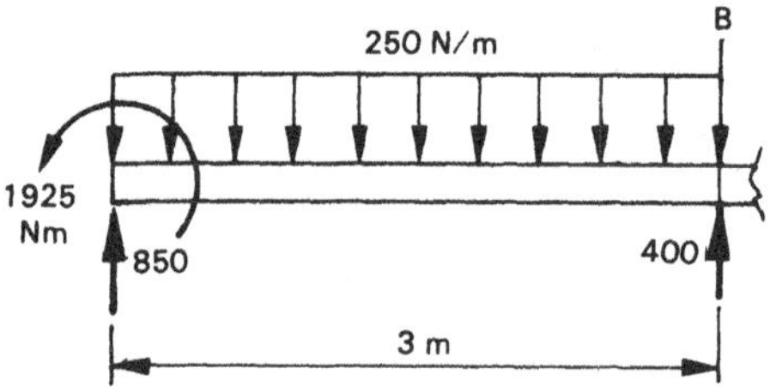 Fig. 5.24	
C		$F = 0$ Check: $F = 500 - 250 \times 2$ $= 0$	$M = 0$ Check: $M = -1925 - \dfrac{250 \times 5^2}{2}$ $+ 850 \times 5 + 400 \times 2$ $= 0$

The shear force and bending moment diagrams may now be drawn to scale as shown in Figure 5.25.

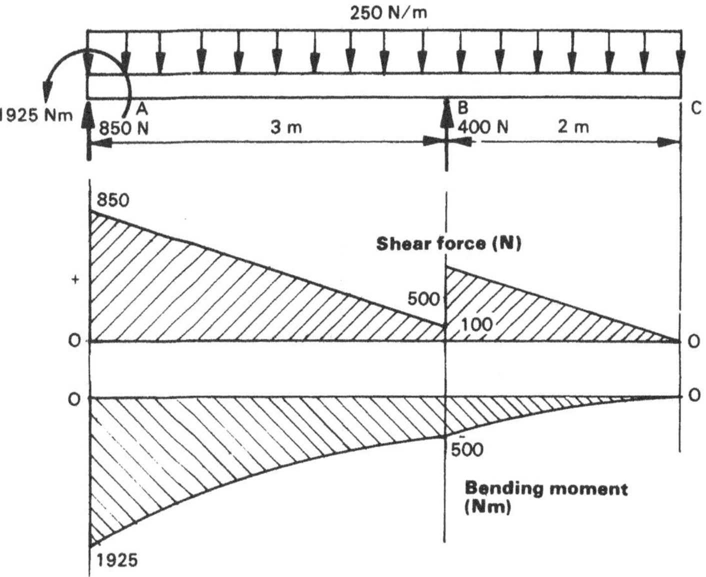

Fig. 5.25

5.9 Turning moment or torque (τ)

Turning moment or torque is defined as the product of a force and the perpendicular distance of that force from the point about which the moment is taken. Hence by definition, torque or turning moment is exactly the same as bending moment. The feature which distinguishes torque from bending moment is that torque is always specified with relationship to an axis of rotation and causes the member to twist about its central axis, whereas bending moment causes the member to bend about its longitudinal axis; that is, the reference planes are 90° apart as shown in Figure 5.26.

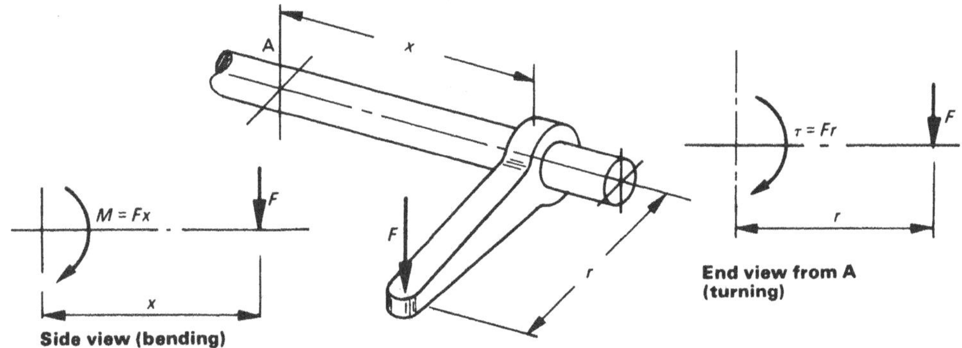

Fig. 5.26 *Distinction between torque and bending moment caused by a force*

It will be seen that a single force may cause simultaneous torque and bending moment as in this case about point A:

Bending moment $M = Fx$ ↻
Torque (turning moment) $\tau = Fr$ ↻

Typical engineering examples where torque is of importance are: nuts and bolts, shafts, flywheels, flanges, pulleys, couplings and similar devices.

Notes

1. There is also a shear force at A of magnitude F.
2. Torque has units Nm.
3. The sign convention used here is that clockwise torque is positive. For example, when viewed from A, force F tends to cause clockwise rotation of the shaft and is therefore positive.
4. If a couple occurs about the centre of rotation, the result is pure torque and there is no resultant force or bending moment about the longitudinal axis (see Fig. 5.27). Examples of this in practice are electric motors, turbines and rotating flywheels.

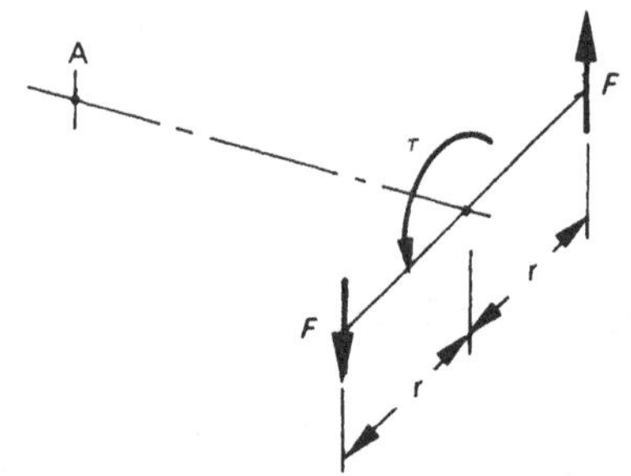

Fig. 5.27 *Pure torque caused by a couple*

About A: torque $\tau = 2Fr$
bending moment $M = 0$
shear force $F = 0$

5. The relationship between torque and power is:

$$P = \tau\omega \quad \text{or} \quad P = \tau \times \frac{2\pi N}{60} = \frac{\tau \times \pi N}{30}$$

where P = power (W)
τ = torque (Nm)
ω = rotational speed (rad/s)
N = rotational speed (rpm)

5.10 Torque distribution diagrams

When a member in static equilibrium (at rest or rotating with constant velocity) is subject to one or more torques applied along the length of the member, the same conditions of static equilibrium apply to torque as also apply to force or bending moment, namely that the sum of the torques must be zero. That is:

$$\Sigma \tau = 0$$

When there is only a single input or output torque, the distribution of torque along the member is usually obvious, but where there are several input or output torques the distribution of torque may not be so obvious and in such cases it is advisable to draw a torque distribution diagram. Such a diagram is particularly useful when read in conjunction with a shear force and bending moment diagram for the member.

The same procedure will be followed when drawing torque distribution diagrams, namely the diagrams will be developed from left to right, and the sign convention will be as previously stated, namely that torque will be positive when clockwise (viewed from the left-hand end of the shaft).

Example 5.4

Draw a torque distribution diagram for the shaft shown in Figure 5.28. The shaft is fixed to a rigid member at A and is supported by a bearing of negligible friction at B.

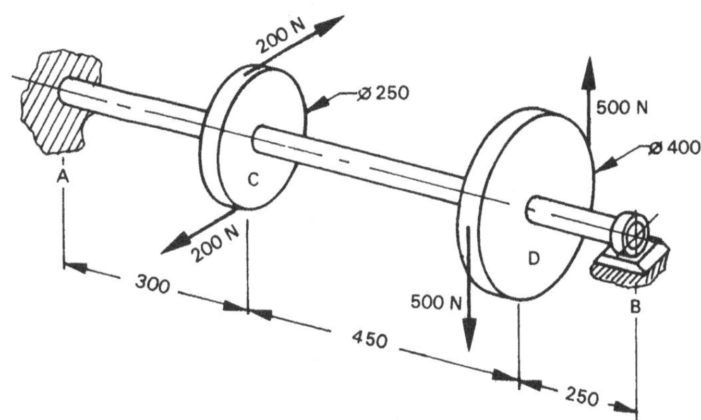

Fig. 5.28

Solution

Working from the left-hand end of the shaft:

Applied torque at C: $\tau_C = -200 \times 0.25 = -50$ Nm (anticlockwise)

Applied torque at D: $\tau_D = 500 \times 0.4 = 200$ Nm (clockwise)

$\Sigma \tau = 0 \quad \therefore \tau_A + \tau_B + \tau_C + \tau_D = 0$

τ_A = reactive torque at A, $\tau_B = 0$ (bearing friction negligible)

$\therefore \tau_A + 0 - 50 + 200 = 0$

$\therefore \tau_A = -150$ Nm

The torque distribution diagram may now be drawn:

It will be seen that the torque distribution diagram is similar in appearance to a shear force diagram for concentrated loads.

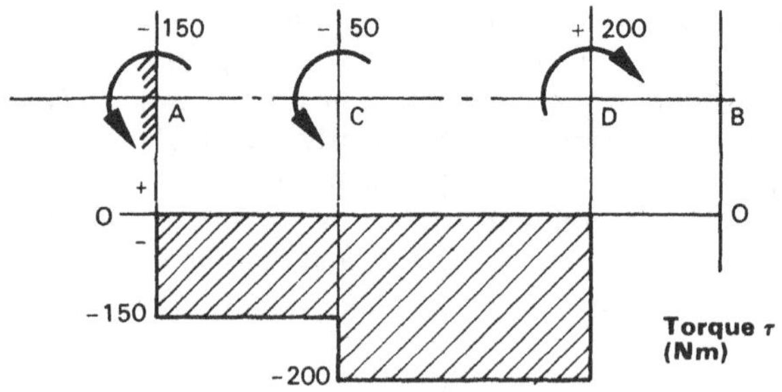

Fig. 5.29

Example 5.5

Input power of 2 kW is supplied to pulley D in the shafting system shown in Figure 5.30, power being taken off at pulleys C and E. The shaft rotates at 720 rpm and the frictional power absorbed in each of the bearings A and B is 60 W.

Draw the torque distribution diagram and determine the maximum torque and its position.

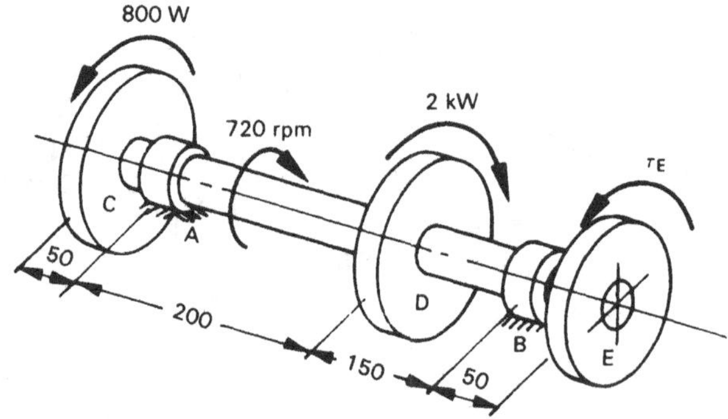

Fig. 5.30

Solution

The input power is 2 kW at D. The torque at D is negative (because anticlockwise viewed from the left-hand end).

$$\omega = \frac{2\pi N}{60} = \frac{2\pi \times 720}{60} = 75.4 \text{ rad/s}$$

$$P = \tau\omega \quad \therefore \tau = \frac{P}{\omega}$$

$$\tau_D = \frac{2000}{75.4} = -26.5 \text{ Nm}$$

$$\tau_C = \frac{800}{75.4} = 10.6 \text{ Nm (power take off at C = 800 W)}$$

$$\tau_A = \tau_B = \frac{60}{75.4} = 0.8 \text{ Nm}$$

$$\Sigma \tau = 0$$

$$\therefore \tau_A + \tau_B + \tau_C + \tau_D + \tau_E = 0$$

$$\therefore 0.8 + 0.8 + 10.6 - 26.5 + \tau_E = 0$$

$$\therefore \tau_E = 14.3 \text{ Nm}$$

Alternatively, the power take off at pulley E is:

$$P_E = 2 - 0.06 - 0.06 - 0.8 = 1.08 \text{ kW}$$

$$\tau_E = \frac{P_E}{\omega} = \frac{1080}{75.4} = 14.3 \text{ Nm (which checks)}$$

Working from left to right the torque in each section of the shaft may now be tabulated and the torque distribution diagram drawn (see Fig. 5.31).

Position	*Torque (Nm)*
C_L	0
$C_R - A_L$	10.6
$A_R - D_L$	10.6 + 0.8 = 11.4
$D_R - B_L$	11.4 − 26.5 = −15.1
$B_R - E_L$	−15.1 + 0.8 = −14.3
E_R	−14.3 + 14.3 = 0 (checks)

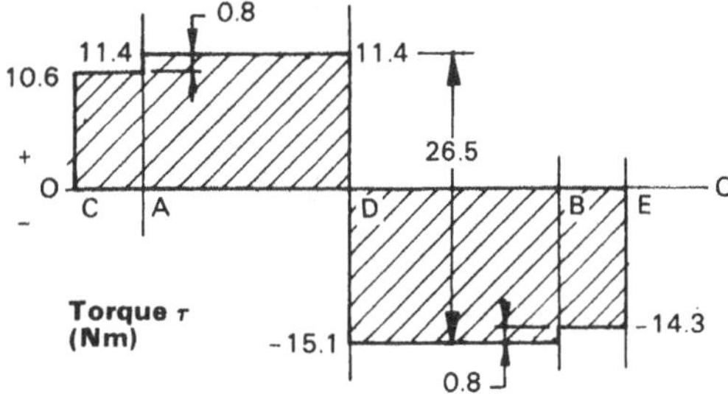

Fig. 5.31

It will be seen that the maximum torque in the shaft is −15.1 Nm and occurs between D_R and B_L.

5.11 Combined shear force, bending moment and torque distribution diagrams

In the design of rotating shafts, it is often the case that shear force, bending moment and torque occur simultaneously. For example, even if there is only a single force applied between the bearings, if this force is offset from the centre of rotation it will give rise to shear force, bending moment and torque. A combined diagram showing all three effects is the best way of giving the engineer a complete picture of the factors which cause stress in the shaft.

Example 5.6

The power supply and take off to the shaft in example 5.5 is by means of belt drives. The combined belt tensions are as follows:

Pulley C: 150 N vertically down

Pulley D: 160 N vertically up

Pulley E: 140 N vertically down

The bearings A and B are self-aligning and do not support bending moment. The weight of the pulleys and shaft is negligible.

Draw shear force, bending moment and torque distribution diagrams.

Solution

The side view of the shaft with forces acting is shown in Figure 5.32.

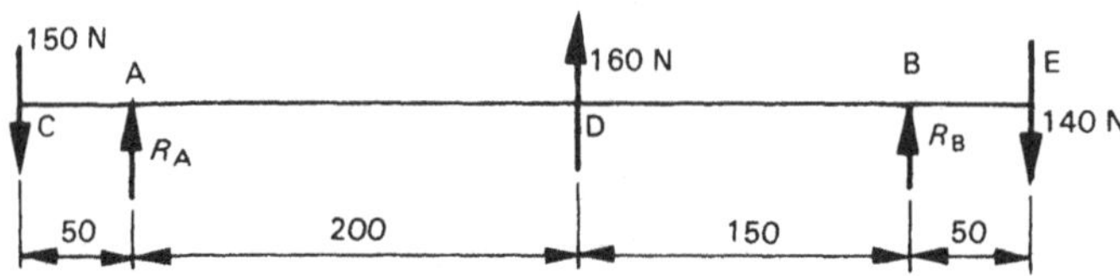

Fig. 5.32

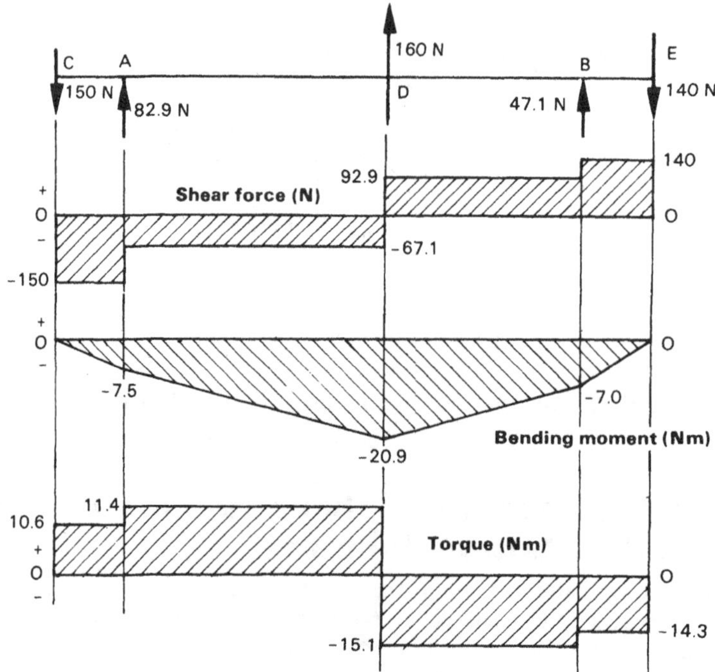

Fig. 5.33

Taking moments about B:

$$\Sigma M_B = 0$$

$$\therefore -150 \times 0.4 + R_A \times 0.35 + 160 \times 0.15 + 140 \times 0.05 = 0$$

$$\therefore R_A = 82.9 \text{ N}$$

$$\Sigma F_V = 0$$

$$\therefore -150 + 82.9 + 160 + R_B - 140 = 0$$

$$\therefore R_B = 47.1 \text{ N}$$

The calculation of shear force and bending moment at each key position is straightforward and is not reproduced here. The torque distribution diagram has already been drawn (Fig. 5.31) and is shown below the shear force and bending moment diagrams in Figure 5.33.

Problems

Notes

1. All key points are denoted by capital letters.
2. All diagrams should be drawn to scale which should be clearly marked.
3. All calculations should be shown.
4. Unless otherwise stated, the answers given for shear force are in kN, bending moment in kNm and torque in Nm.
5. Answers given for shear force and torque are values to the left of the key point. Values to the right of the key point are indicated by use of the subscript R.
6. Not all values are given in the answers.

5.1-5.12 Draw shear force and bending moment diagrams for each of the beams shown in Figures P5.1–P5.12. For one section of the beam, check the shear force from the slope of the bending moment diagram and the change in bending moment from the area under the shear force diagram.

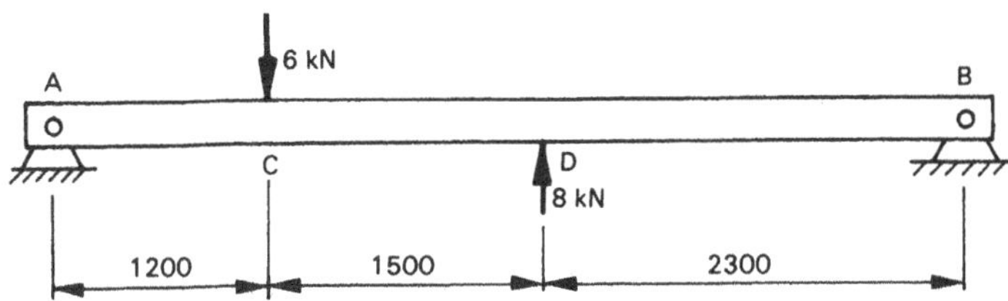

Fig. P5.1

	C	D	B
F	0.88	-5.12	2.88
M	1.06	-6.62	0

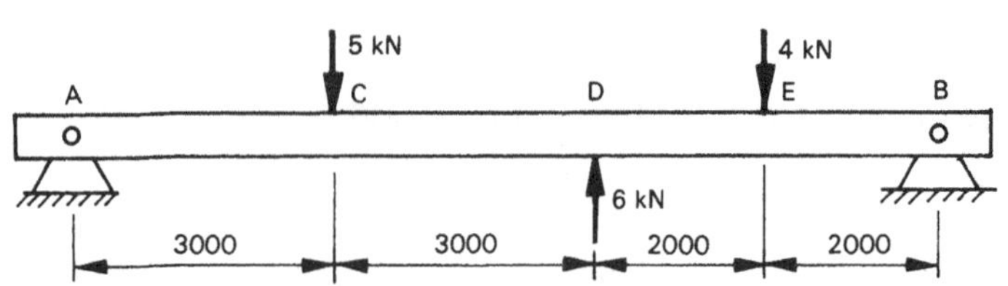

Fig. P5.2

	C	D	E	B
F	1.9	-3.1	2.9	-1.1
M	5.7	-3.6	2.2	0

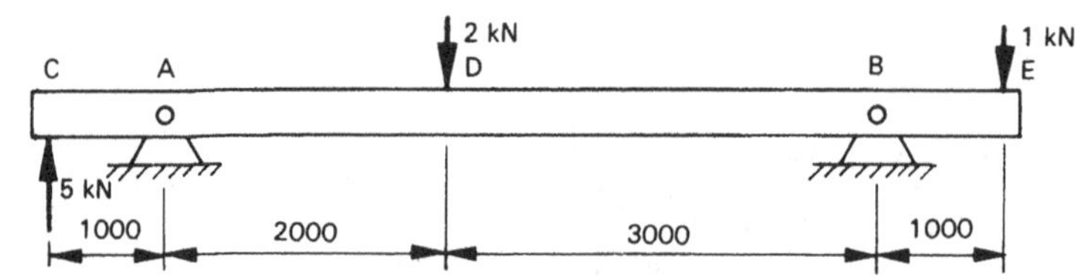

Fig. P5.3

	A	D	B	E
F	5	0	-2	1
M	5	5	-1	0

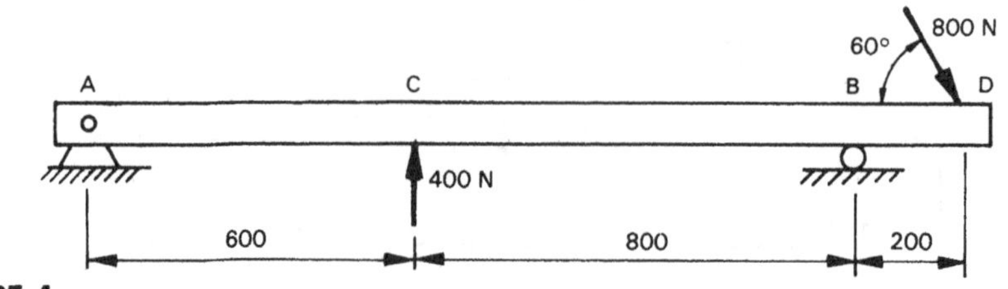

Fig. P5.4

	C	B	D
F	-0.327	0.072	0.693
M	-0.196	-0.138	0

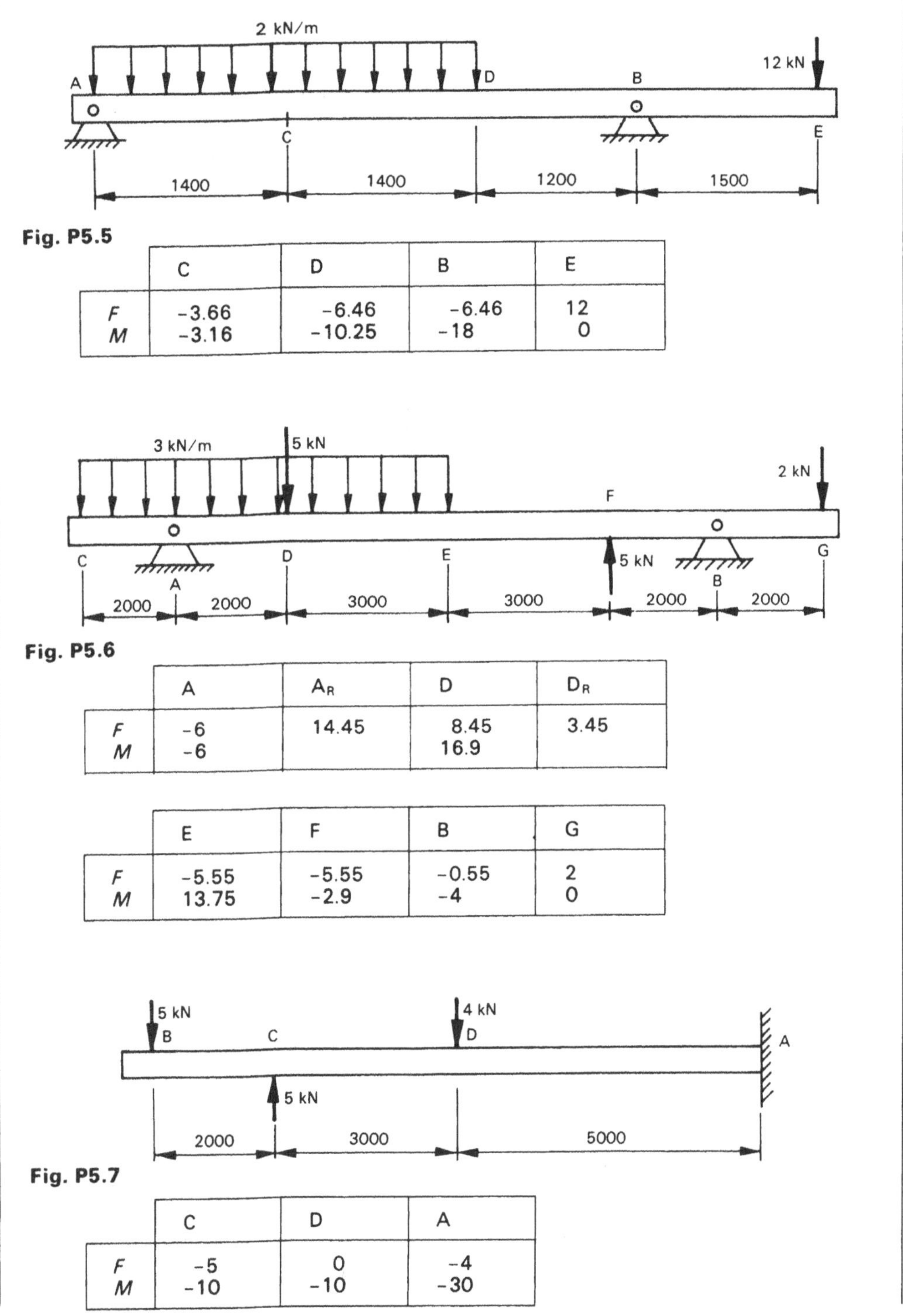

Fig. P5.5

	C	D	B	E
F	-3.66	-6.46	-6.46	12
M	-3.16	-10.25	-18	0

Fig. P5.6

	A	A_R	D	D_R
F	-6	14.45	8.45	3.45
M	-6		16.9	

	E	F	B	G
F	-5.55	-5.55	-0.55	2
M	13.75	-2.9	-4	0

Fig. P5.7

	C	D	A
F	-5	0	-4
M	-10	-10	-30

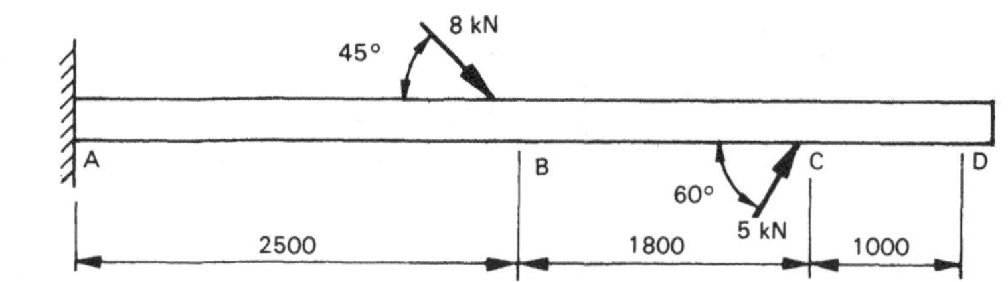

Fig. P5.8

	A	B	C	D
F		1.33	-4.33	0
M	4.48	7.8	0	0

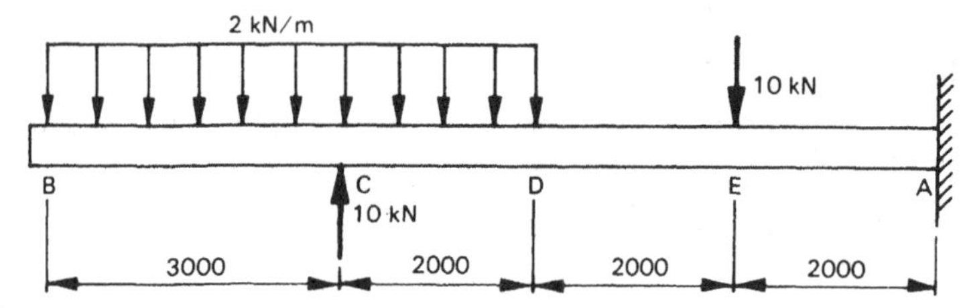

Fig. P5.9

	C	C_R	D	E	A
F	-6	4	0	0	-10
M	-9		-5	-5	-25

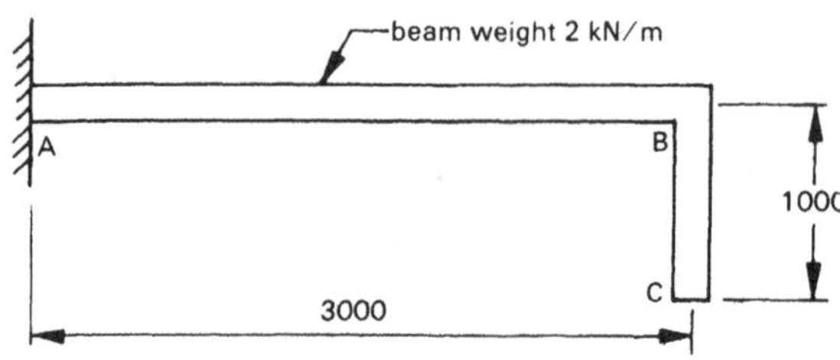

Fig. P5.10

	A	B	C
F	8	2	0
M	-15	0	0

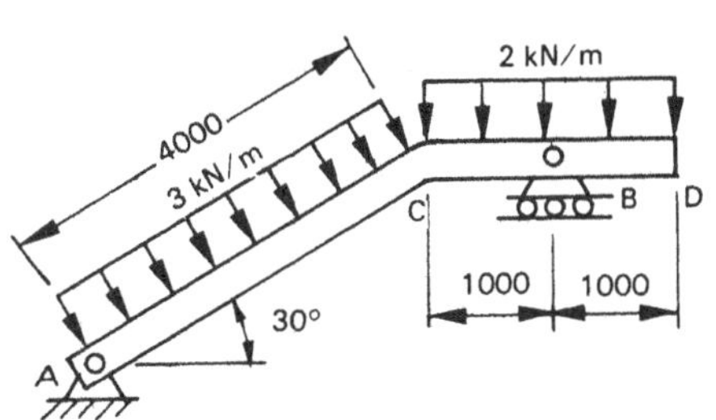

Fig. P5.11

	AR	C	C_R	B	B_R	D
F	7.34	-4.66	-5.38	-7.38	2	0
M	0	5.38		-1		0

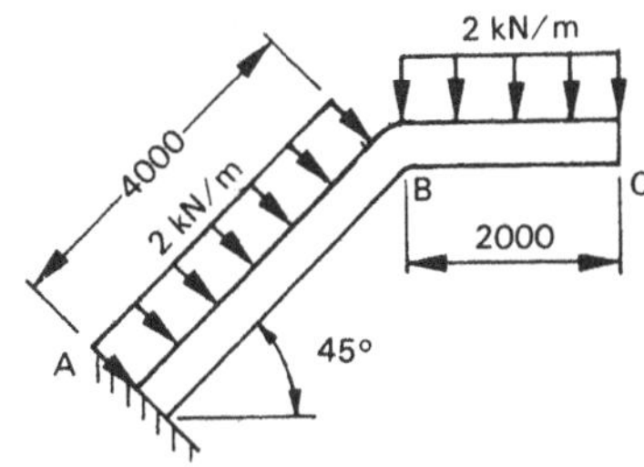

Fig. P5.12

	AR	B	B_R	C
F	10.8	2.83	4	0
M	-31.3	-4		0

5.13 Draw the torque distribution diagram for the shaft shown in Figure P5.13. The shaft is rigidly fixed at A and is supported by a bearing of negligible friction at B.

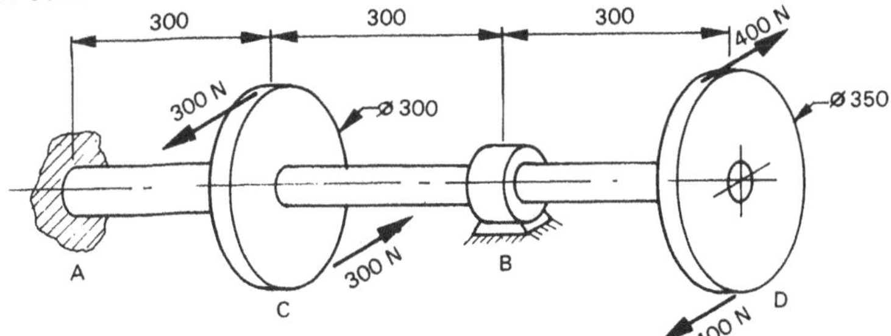

Fig. P5.13

	C	B	D
T	-50	-140	-140

5.14 Input power of 5 kW is supplied at pulley C to the shaft rotating at 1440 rpm shown in Figure P5.14. The power take off at pulley D is 2 kW and at pulley E 2.8 kW. Assuming that each of the bearings A and B absorb equal power, draw the torque distribution diagram and determine the maximum torque and where it occurs.

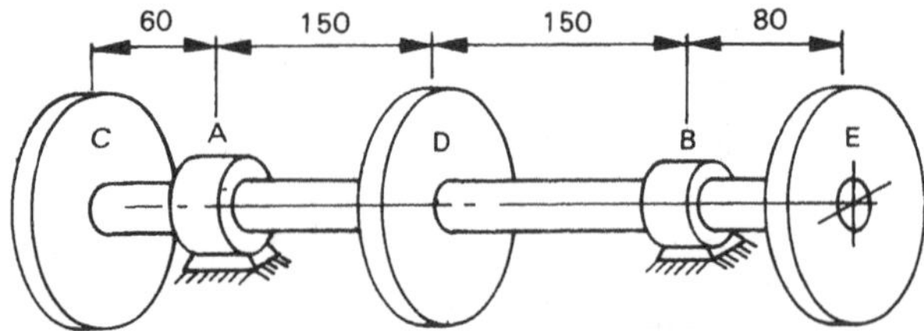

Fig. P5.14

A	D	B	E
33.2	32.5	19.2	18.6

5.15 The power to the shafting system given in problem 5.14 is by means of vee belts at the pulleys. The sum of the belt tensions are: C 280 N vertically up, D 150 N vertically down, E 180 N vertically down.

Draw complete shear force, bending moment and torque distribution diagrams and determine maximum values of each and where they occur in the shaft.

The weight of the pulleys and the shaft are negligible.

	A	D	B	E
F (N) M (Nm)	280 16.8	-29 12.45	-179 -14.4	180 0
T	(see problem 5.14)			

5.16 Draw shear force, bending moment and torque distribution diagrams for the shaft shown in Figure P5.16 which is rigidly supported at A. Determine maximum values of each and where they occur.

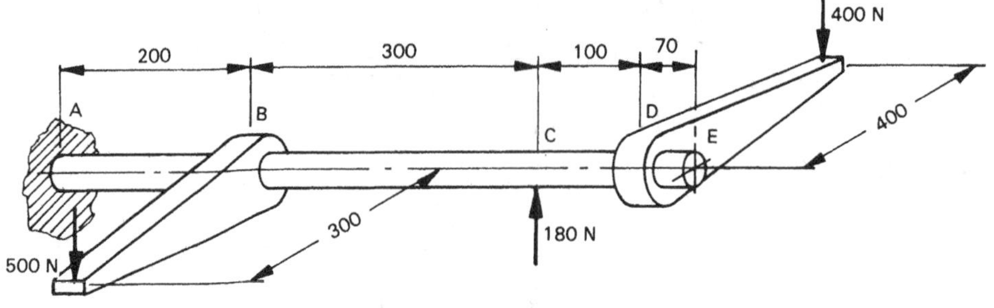

Fig. P5.16

	AR	B	C	D
F (N)	720	720	220	400
M (Nm)	-250	-106	40	0
τ (Nm)	-10	-10	-160	-160

5.17 For the shafting system shown in Figure P5.17, A and B are self-aligning bearings with negligible friction. The diameter and tensions for pulleys C, D and E are given in the table below.

Draw shear force, bending moment and torque distribution diagrams and determine the maximum values of each and where they occur.

The weight of the pulleys and shaft are negligible.

Pulley	*Pulley dia.* (mm)	T_1 (N)	T_2 (N)
C	250	50	200
D	200	40	160
E	225	40	100

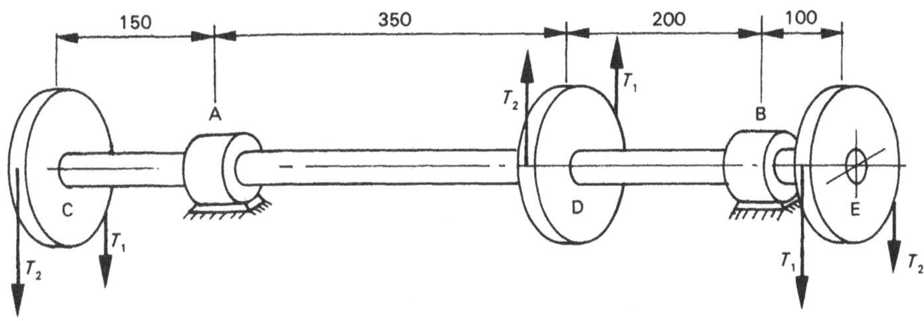

Fig. P5.17

	AR	D	B	E
F (N)	-250	-30	170	140
M (Nm)	-37.5	-48	-14	0
τ (Nm)	-18.75	-18.75	-6.75	-6.75

6 Friction

Friction occurs when two bodies or materials are in contact and one body moves or attempts to move relative to the other. Mechanics is generally concerned only with dry friction, that is the friction resulting from two solids in contact. When one material is a fluid, analysis of frictional effects is complex and falls into the realm of fluid mechanics.

Friction results in a force which resists the motion or impending motion and the direction of the frictional force is always opposite to the direction of the motion or impending motion, that is friction attempts to oppose the motion. Without friction, life as we know it would be impossible: simple tasks like walking could not be done without spiked shoes, and all objects would have to be fastened together to stop movement between them. On the other hand, in many cases life would be much easier without friction, and in most machines and mechanisms friction is undesirable as it causes a loss in efficiency as well as inducing side-effects such as heating; this results in a far greater energy input than would be necessary if there were no friction.

In some cases, the effects of friction may be ignored without serious loss of accuracy, for example in the analysis of pin-jointed frames or when a shaft rotates in a rolling element bearing. Often in theoretical analysis, friction is neglected in the first instance in order to simplify the theory and the effect of friction is considered later on. In most engineering applications, friction should be considered, otherwise appreciable difference can occur between theoretical and actual values.

The precise analysis of friction is very complex, particularly if lubricant or fluid is present between the surfaces. This chapter considers only the simplified analysis which is, however, sufficiently accurate for most problems in mechanics. In the simplified analysis, the coefficient of friction is taken to have a constant value which does not vary with other factors such as the relative velocity or temperature of the bodies.

6.1 Review of friction

Below is a summary of the key points necessary in the analysis of mechanical friction.

1. Friction occurs when the surfaces of two bodies capable of relative sliding motion are in contact and there is a motion or impending motion between the surfaces. The distinction between bodies capable of relative sliding motion and those which are not is clarified in Figure 6.1.

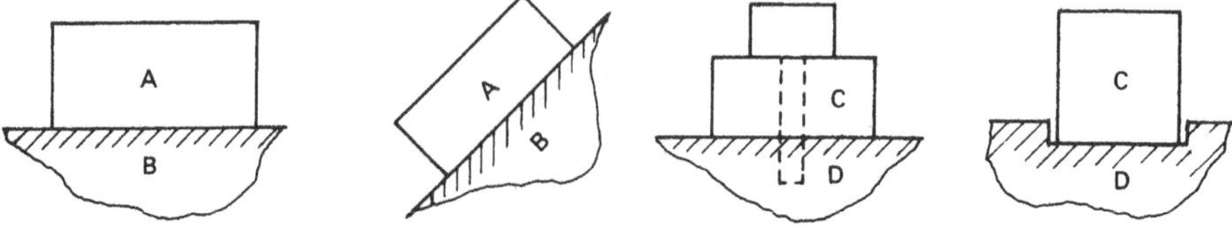

Fig. 6.1 *Distinction between bodies capable of relative sliding motion and those which are not*

It is clear that bodies A and B are capable of relative sliding motion, whereas C and D are not and may be considered to be interlocked. The distinction is not always so clear cut, particularly when there is appreciable deformation between the surfaces, for example when a person walks on soft ground with grooved shoes or when soft rubber tyres are in contact with a rough road surface. In such cases, the coefficient of friction can be greater than 1, but in this book it will be assumed that if the bodies are not interlocked they are capable of relative sliding motion so that the coefficient of friction is always less than 1.

2. Friction results in a force which acts parallel to the surfaces at the point of contact and in the opposite direction to the motion or impending motion, that is opposite to the direction of the velocity or impending velocity. This means that the friction force is in the opposite direction for each body considered as a free body since the direction of the relative motion is opposite for each body. This is clarified in Figure 6.2.

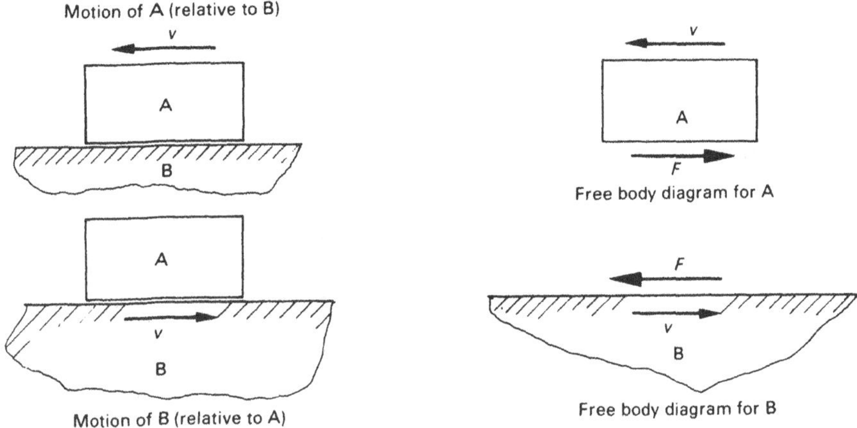

Fig. 6.2 *Free body diagrams for bodies in contact*

It is clear that the direction of motion is opposite when considering the motion of A relative to B rather than B relative to A. Hence the direction of the friction force is opposite for each body considered as a free body.

3. The maximum value of the friction force occurs when motion impends (that is the bodies are just on the point of sliding relative to one another) and is assumed constant thereafter. This is clarified in Figure 6.3 which shows the friction force plotted against the applied force for dry surfaces in contact.

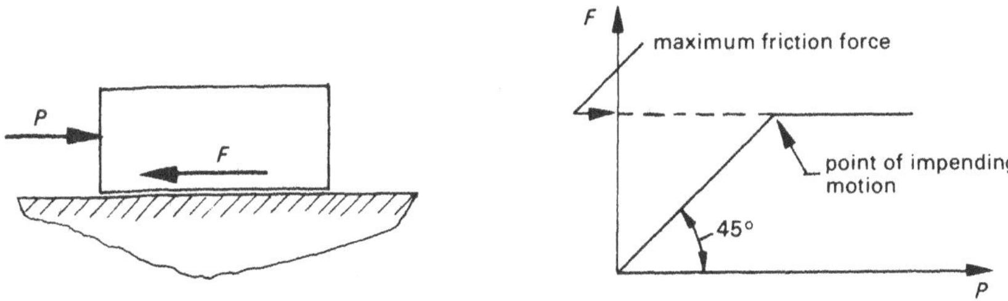

Fig. 6.3 *Friction force and applied force for bodies in contact*

4. The maximum value of the friction force is given by:

$$F = \mu N \qquad \textbf{(6.1) friction force}$$

where F = maximum friction force (N)
N = normal force (reaction) between the two surfaces (N)
μ = coefficient of friction (dimensionless)

The reason why the normal reaction is used in this formula rather than the weight of the body is because the friction force is proportional to the normal force between the bodies which is not necessarily equal to the weight of the body. For example, when surfaces are inclined, the normal reaction is not the same as the weight and varies with the angle of inclination. On the other hand, the weight remains the same at any angle of inclination. This is clarified in Figure 6.4.

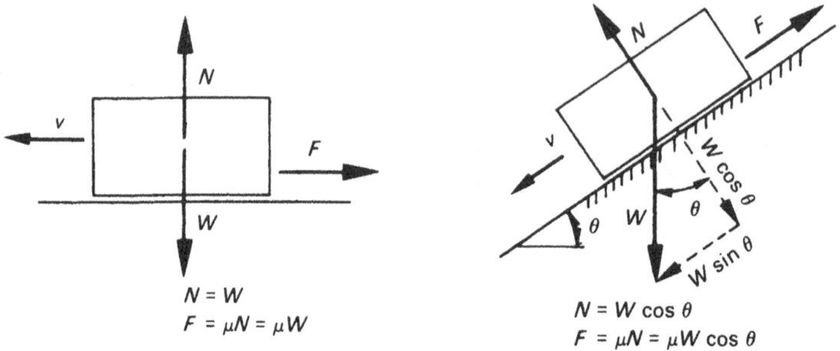

Fig. 6.4 *Normal reaction and friction force for blocks on horizontal and inclined surfaces*

5. The coefficient of friction depends primarily upon the nature of the surfaces in contact, that is the material and the surface finish, and may be considered independent of the velocity or the surface area. This is a good approximation for dry surfaces but does not hold for lubricated surfaces. The coefficient of friction must lie in the range 0 to 1.
6. The value of the coefficient of friction may be determined experimentally by inclining the contact surface gradually as shown in Figure 6.5.

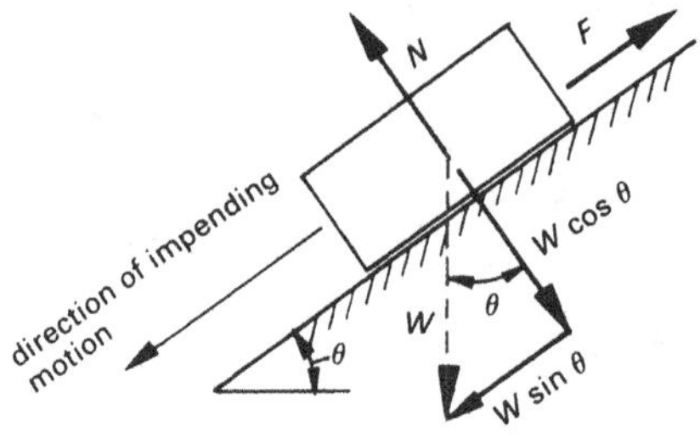

Fig. 6.5 *Forces on a block on an inclined surface*

The normal reaction $N = W\cos\theta$.
Sliding impends when $F = W\sin\theta$.
$\therefore \mu N = \mu W\cos\theta = W\sin\theta$

$$\therefore \quad \mu = \frac{\sin\theta}{\cos\theta} = \tan\theta$$

Hence the coefficient of friction may be determined by measuring the angle at which sliding impends. This angle is also known as the **angle of friction ϕ**.

That is

$$\boxed{\phi = \tan^{-1}(\mu)} \qquad \textbf{(6.2) angle of friction}$$

6.2 Friction on the inclined plane

Motion up the plane

Consider a block weight W being moved at constant velocity up an inclined plane inclined at angle θ to the horizontal by an applied force P acting at angle α to the block as shown in Figure 6.6.

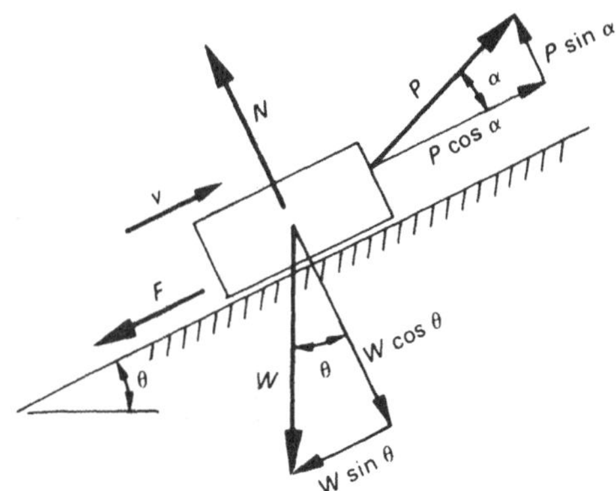

Fig. 6.6 *General case of constant velocity motion up an inclined plane*

Resolving perpendicular to the plane:

$$N - W\cos\theta + P\sin\alpha = 0$$

$$\therefore \quad N = W\cos\theta - P\sin\alpha$$

Resolving parallel to the plane:

$$P\cos\alpha - F - W\sin\theta = 0$$

But $F = \mu N$:

$$\therefore P\cos\alpha - \mu(W\cos\theta - P\sin\alpha) - W\sin\theta = 0$$

$$\therefore \quad P(\cos\alpha + \mu\sin\alpha) = W(\sin\theta + \mu\cos\theta)$$

$$\therefore \quad \boxed{P = \frac{W(\sin\theta + \mu\cos\theta)}{\cos\alpha + \mu\sin\alpha}} \qquad \textbf{(6.3) motion up plane}$$

Notes

1. Equation 6.3 applies to constant velocity motion up the plane or to a static block with motion impending up the plane.
2. In the above analysis, angle α is measured anticlockwise above the plane angle. If force P acts below the plane angle so that α is measured clockwise below the plane angle, α will be negative and equation 6.3 is still valid.
3. Angle α may be zero in which case $\sin \alpha$ will be zero and $\cos \alpha$ will be 1. Equation 6.3 is still valid but reduces to:

$$P = W(\sin \theta + \mu \cos \theta).$$

Motion down the plane

The same principles apply with two variations:

1. The direction of the friction force is up the plane.
2. Force P may act up or down the plane according to the relationship between θ and ϕ (angle of friction).
   (a) If $\theta < \phi$, the block will not slide down the plane of its own accord and hence P must act down the plane.
   (b) If $\theta > \phi$, the block will slide down the plane of its own accord and hence P must act up the plane (otherwise the block will accelerate down the plane).

Considering case (a) the forces are as shown in Figure 6.7.

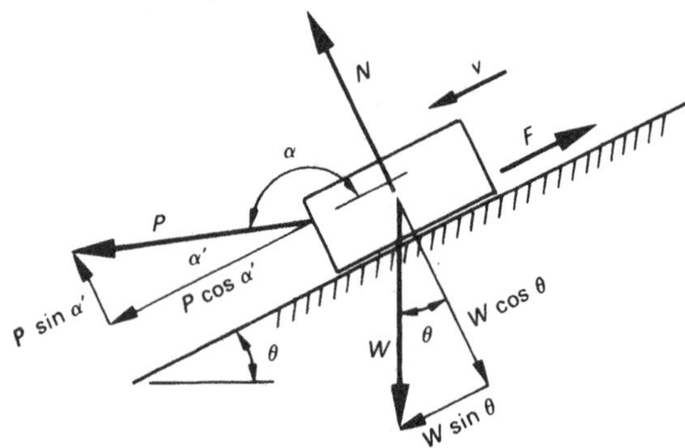

Fig. 6.7 *General case of constant velocity motion down an inclined plane*

Resolving perpendicular to the plane:

$$N + P\sin \alpha' - W\cos \theta = 0$$

$$\therefore N = W\cos \theta - P\sin \alpha'$$

Resolving parallel to the plane:

$$F - P\cos \alpha' - W\sin \theta = 0$$

$$\therefore \mu(W\cos \theta - P\sin \alpha') - P\cos \alpha' - W\sin \theta = 0$$

$$\therefore \quad \boxed{P = \frac{W(\mu \cos \theta - \sin \theta)}{\cos \alpha' + \mu \sin \alpha'}} \qquad \textbf{(6.4(a) motion down plane } (\theta < \phi))$$

Considering case (b), the forces are the same as shown in Figure 6.6 except that, since motion is down the plane, F acts up the plane. The equation thus becomes:

$$P = \frac{W(\sin\theta - \mu\cos\theta)}{\cos\alpha - \mu\sin\alpha}$$

(6.4(b) motion down plane ($\theta > \phi$))

Notes

1. In case (a) angle α' is measured with P pointing outward (away from the block) *clockwise* above the plane. If α' acts below the plane, equation 6.4(a) is still valid with α' negative. In case (b), angle α is measured *anticlockwise* above the plane. If α acts below the plane, equation 6.4(b) is still valid with α negative.
2. Angle α or α' may be zero, in which case equation 6.4(a) reduces to:

   $$P = W(\mu\cos\theta - \sin\theta)$$

   and equation 6.4(b) reduces to:

   $$P = W(\sin\theta - \mu\cos\theta)$$

3. It is best that the equations are not committed to memory but derived as required.
4. An alternative approach is to use equation 6.3 to cover motion down the plane as well as up the plane. This may be done if the following convention is used.

   Angle α is measured with P pointing outward (away from the block) anticlockwise from the plane angle in all cases.

   If motion is down the plane instead of up the plane, the coefficient of friction is negative.

   These principles are clarified in example 6.1.

Example 6.1

Determine force P necessary to cause the 5 kg block shown in Figure 6.8 to move with constant velocity (a) up the plane, (b) down the plane. The coefficient of friction is 0.4.

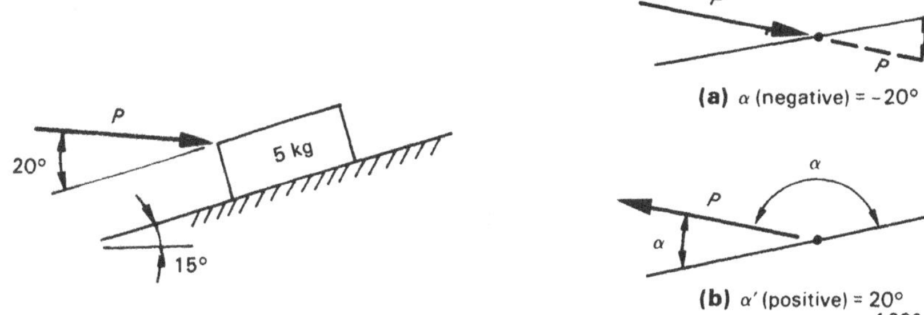

Fig. 6.8

Solution

(a) Motion up the plane.

Here $\alpha = -20°$, $\theta = 15°$, $W = 5 \times 9.81$ N, $\mu = 0.4$.

Applying equation 6.3:

$$P = \frac{W(\sin\theta + \mu\cos\theta)}{\cos\alpha + \mu\sin\alpha}$$

$$= \frac{5 \times 9.81(\sin 15° + 0.4 \cos 15°)}{\cos(-20°) + 0.4 \sin(-20°)}$$

$$= \mathbf{39.4\ N}$$

(b) Motion down the plane.

First it is necessary to check if the block will move down the plane of its own accord.

$$\phi = \tan^{-1}(\mu) = \tan^{-1}(0.4) = 21.8°.$$

Since $\theta < \phi$, the block will not slide down the plane of its own accord, therefore equation 6.4(a) should be applied.

$$P = \frac{W(\mu \cos \theta - \sin \theta)}{\cos \alpha' + \mu \sin \alpha'}$$

where force P acts down the plane and α' is 20°. Substituting:

$$P = \frac{5 \times 9.81(0.4 \cos 15° - \sin 15°)}{\cos 20° + 0.4 \sin 20°}$$

$$= \mathbf{5.81\ N}$$

Note that the same result is achieved by substituting in equation 6.3 with:

$$\alpha = 160°;\ \mu = -0.4$$

Graphical solution

An alternative solution to inclined plane problems is by use of a graphical method. The method involves replacing the friction force and the normal reaction by a single force (R), using the angle of friction to determine the direction of R. The block on the plane may then be considered as a free body acted upon by three forces: R, W and P. Generally R and P are unknown in magnitude but known in direction, whereas W is known both in magnitude and direction. Hence a vector force diagram (triangle of forces) may be drawn and the magnitude of the unknown forces determined.

The solution to example 6.1(a) and (b) using this method is shown in Figure 6.9. In this case:

angle of friction $\phi = \tan^{-1}(0.4) = 21.8°$

weight $W = 5 \times 9.81 = 49.05$ N

Note that in Figure 6.9 the friction force has been drawn above the block for clarity. Of course, since there are three forces acting on the block they must be concurrent, but their actual location on the block is not of importance in solving the problem.

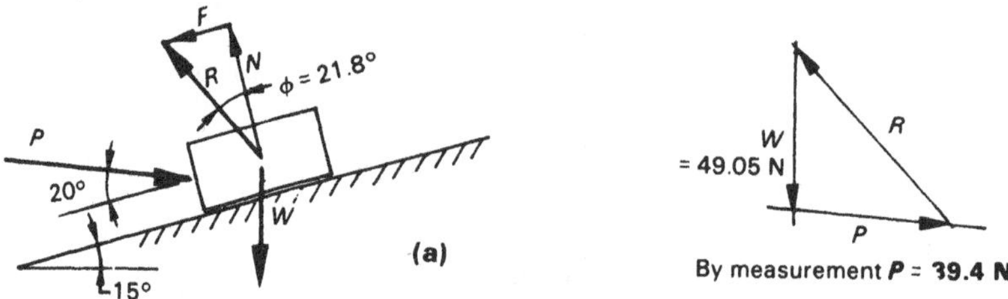

Fig. 6.9 *(a) Motion up the plane*

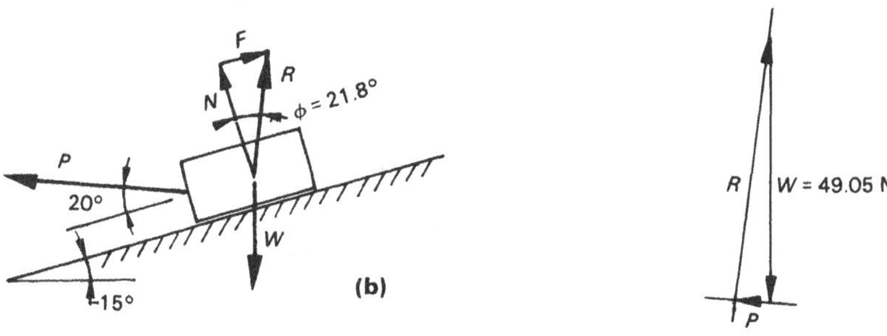

Fig. 6.9 *(b) Motion down the plane*

Computer solution

The computer program INPLANE 1 was written to solve problems involving blocks on the inclined plane, with motion up or down the plane. The program inputs are:

1. Whether motion is up or down the plane.
2. The angle of inclination of the plane to the horizontal.
3. The angle of inclination of the force with the plane. If the force acts parallel to the plane, this is zero. The angle is measured anticlockwise with force P pointing outward away from the block.
4. The coefficient of friction. It is not necessary to use a negative coefficient of friction for motion down the plane as the program automatically takes this into account.
5. The mass of the block.

The program then outputs the magnitude of the force.

Note: If the force is known and the mass of the block is unknown, the program may be run with a nominal block mass (say, 1 kg) and the force determined. The unknown mass for the given force may then be obtained by ratio.

The program is listed in Appendix 3.6. The solution to example 6.1 is given below the program listing and should be self-explanatory.

6.3 Wedges

A wedge is a tapered block in which an axial force is used to balance a side thrust force as shown in Figure 6.10.

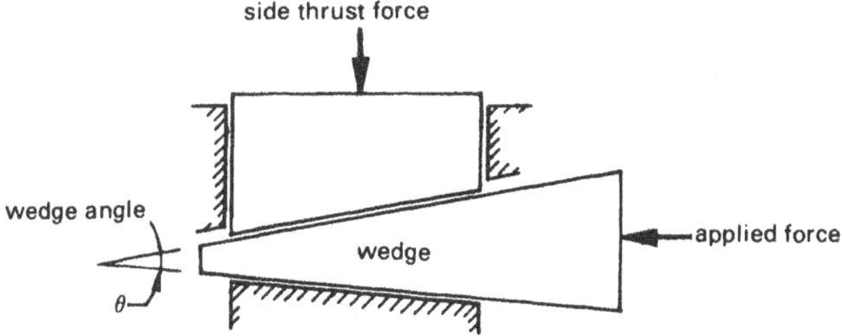

Fig. 6.10 *Wedge configuration*

The magnitude of the side thrust force for a given applied force depends upon the wedge angle and the coefficient of friction. The wedge is often self-locking, depending upon the same two factors. A self-locking wedge is one which will not slip back when the applied force is removed and, indeed, an applied force acting in the opposite direction to that shown in Figure 6.10 will be necessary to remove the wedge. If the coefficient of friction is the same on both the top and bottom wedge faces, self-locking occurs when $\theta < 2\phi$, where ϕ is the angle of friction. This may be proved by examining the forces acting on a self-locking wedge (considered as a free body) when the wedge is removed.

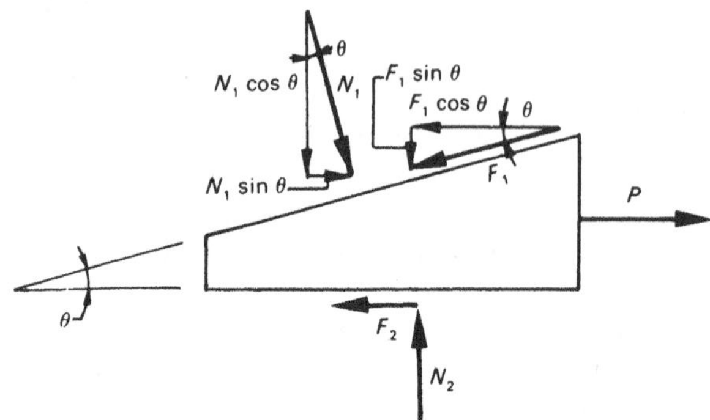

Fig. 6.11 *Wedge as a free body (removing wedge)*

The limiting angle for self-locking occurs when these forces are in equilibrium with force $P = 0$. Under these conditions:

$\Sigma F_H = 0 \quad \therefore N_1 \sin\theta = F_1 \cos\theta + F_2$

but $\quad F_1 = \mu N_1,\ F_2 = \mu N_2$

$\therefore N_1 \sin\theta = \mu N_1 \cos\theta + \mu N_2$ **(1)**

$\Sigma F_V = 0 \quad \therefore N_1 \cos\theta + F_1 \sin\theta = N_2$ **(2)**

Substituting in (1),

$N_1 \sin\theta = \mu N_1 \cos\theta + \mu N_1 \cos\theta + \mu F_1 \sin\theta$

$\therefore N_1 \sin\theta = 2\mu N_1 \cos\theta + \mu^2 N_1 \sin\theta$

Dividing by $N_1 \sin\theta$,

$$1 = \frac{2\mu}{\tan\theta} + \mu^2$$

$$\therefore \tan\theta = \frac{2\mu}{1 - \mu^2} \qquad \textbf{(3)}$$

$$\text{Now } \tan(a + b) = \frac{\tan a + \tan b}{1 - \tan a \tan b}$$

$$\text{or} \quad \tan(\phi + \phi) = \frac{\tan\phi + \tan\phi}{1 - \tan\phi \tan\phi}$$

but $\quad \mu = \tan\phi \quad (\phi = \text{angle of friction})$

$$\therefore \quad \tan(2\phi) = \frac{\mu + \mu}{1 - \mu \times \mu} = \frac{2\mu}{1 - \mu^2} \qquad \textbf{(4)}$$

Comparing equations (3) and (4), it will be seen that:

$$\tan\theta = \tan(2\phi)$$

$$\text{or} \quad \theta = 2\phi$$

which is the maximum value of the wedge angle θ in order that the wedge be self-locking.

Example 6.2

A dry metal wedge is used to lock a component on a machine tool. If the coefficient of friction is 0.2, determine the maximum wedge angle in order that the wedge be self-locking.

If the wedge surface were greasy, so that the coefficient of friction was 0.05, what maximum wedge angle would now be necessary?

Solution

If $\mu = 0.2$, $\phi = \tan^{-1}(0.2) = 11.3°$

Since $\theta = 2\phi$ (maximum for self-locking)

$\boldsymbol{\theta = 22.6°}$

If $\mu = 0.05$, $\phi = 2.86°$

$\boldsymbol{\theta = 5.72°}$

Analysis of wedge problems

Wedge problems are readily solved by graphical methods, although of course analytical or computer solutions are also possible. An outline of the graphical procedure is as follows:

1. Draw a configuration diagram. The dimensions of the wedge do not have to be known provided the angles are given.
2. Draw each wedge or block to scale as a free body and show the forces acting on it. There are generally forces acting on three faces of each block or wedge. The exception is the self-locking wedge in which there may be forces on two faces only.
3. On each face, combine the friction force and the normal reaction into a single force whose direction relative to the face is determined by the angle of friction only. This is shown in Figure 6.12. Note that when two faces are in contact, the friction force is opposite on each face and the combined force (resultant) is equal and opposite on each block. Also note that the direction of the friction force depends upon the direction of motion or impending motion of each block and may be determined by inspection. The normal reaction force in each block acts perpendicular to the surface and in toward the block.

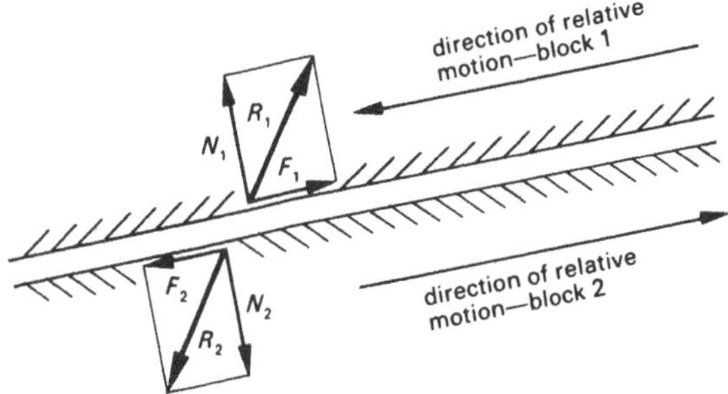

Fig. 6.12 *When two faces are in contact, the forces at each face act in opposite directions*

4. Start from a block or wedge in which the force on one face is known in both magnitude and direction. The forces on the other faces will be known in direction but not in magnitude. Draw a vector force diagram. Since the forces are in equilibrium, the vector diagram must close and the magnitude of the two unknown forces may be found.
5. Transfer the common known force (found in step 4) to the next block (where the force will be opposite in direction) and draw the vector diagram for this block as in step 4. In this way, moving from block to block, the entire system may be solved.

Example 6.3

An applied force $P = 1$ kN is applied to wedge A shown in Figure 6.13.

Determine the side thrust force Q acting on block B if:

(a) the vertical side face on block B may be considered frictionless and the coefficient of friction on the other faces in contact is 0.25
(b) all faces in contact have a coefficient of friction 0.25.

The weight of the blocks themselves may be ignored.

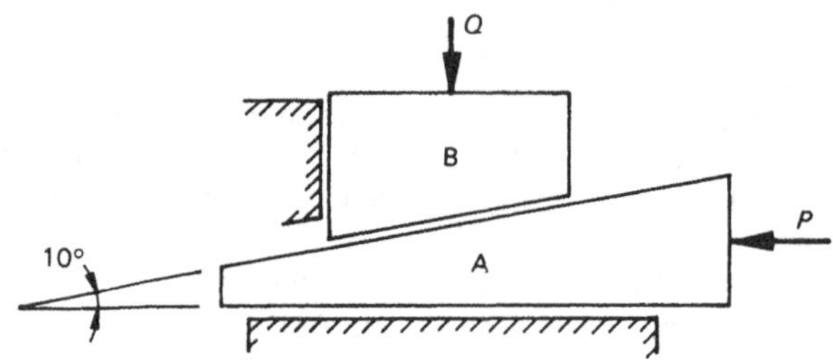

Fig. 6.13

Solution

(a) No friction on the vertical side face

The free body diagrams and vector force diagrams for each block are shown in Figure 6.14.

Notes

1. The vector diagram is started from wedge A which is solved before block B.
2. Force R_3 is equal and opposite to R_2.
3. Angle ϕ is $\tan^{-1}(0.25) = 14°$.
4. From the vector diagram for block B, Q may be scaled off and its magnitude measured and found to be **1.44 kN**.

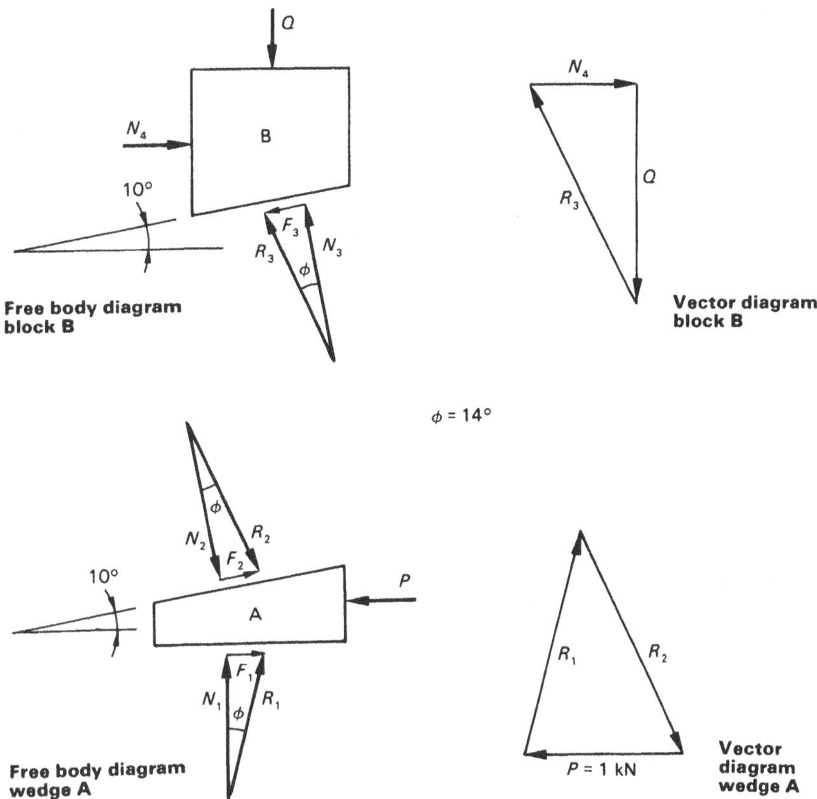

Fig. 6.14

(b) Friction on all faces in contact

The vector diagram for wedge A does not change. However, there will now be a friction force on the side face of block B which changes the direction of the resultant force on this face as shown in Figure 6.15.

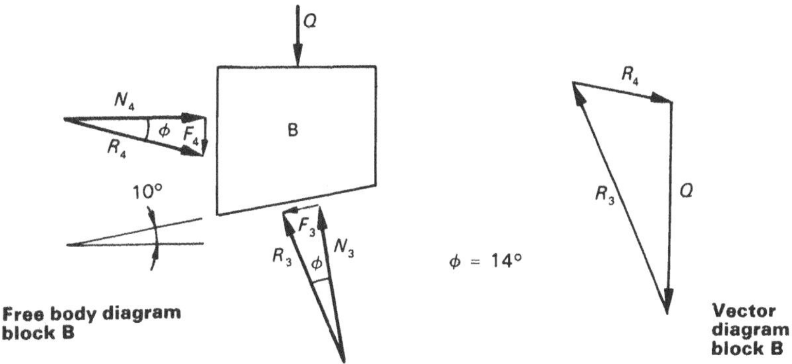

Fig. 6.15

The magnitude of Q may now be scaled off the vector force diagram and is now found to be **1.28 kN** (which is less than the value found in (a) as should be the case).

Example 6.4

For the wedge system shown in Figure 6.16, determine the mass M which must be placed on the wedge A of mass 2 kg in order to cause block B of mass 10 kg just to slide to the right. The coefficient of friction on all faces in contact is 0.3.

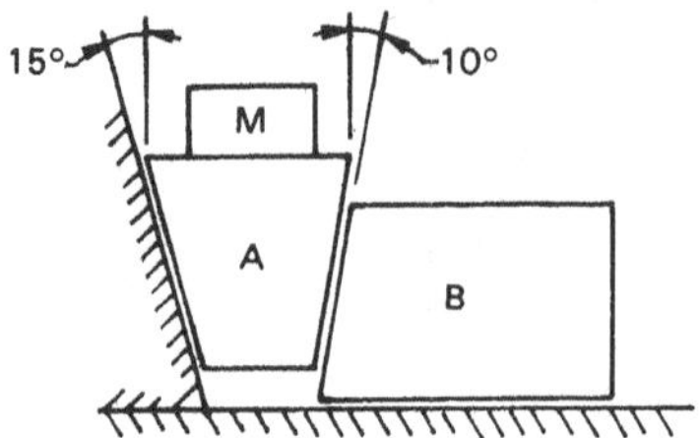

Fig. 6.16

Solution

$\phi = \tan^{-1}(0.3) = 16.7°$

$W_B = 10 \times 9.81 = 98.1$ N, $W_A = 2 \times 9.81 = 19.6$ N

The free body diagrams and vector diagrams are shown in Figure 6.17. The vector diagrams are drawn starting with block B. Note that R_2 and R_3 are equal and opposite.

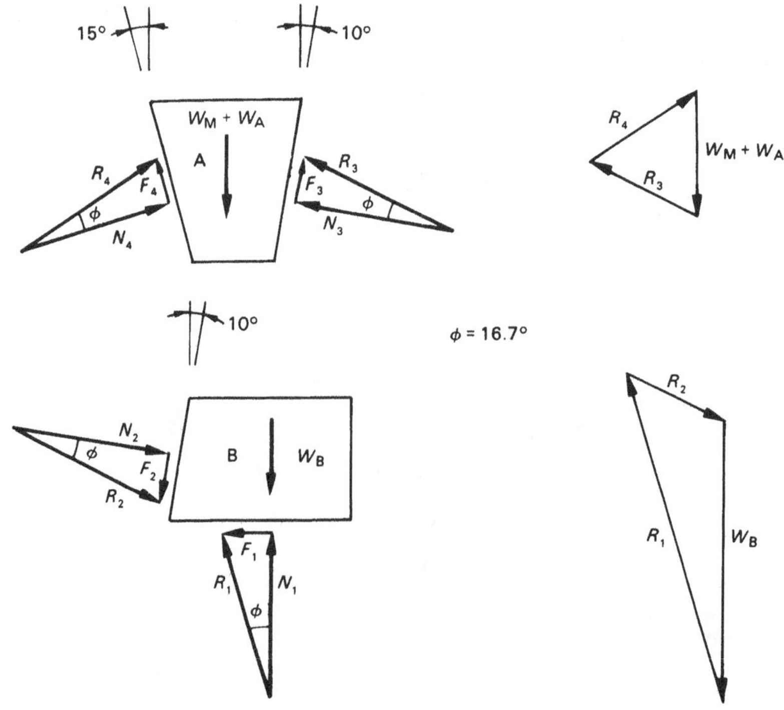

Fig. 6.17

By scaling off the vector diagram for block A, it is found that $W_M + W_A = 39$ N. Since $W_A = 19.6$ N, therefore $W_M = 19.4$ N.

Hence M = **2 kg.**

Computer solution

The computer program WEDGE was written to solve standard problems involving a block and wedge where the coefficient of friction is the same on all faces and the face angles are given. The applied force P is then input and the program calculates and outputs the corresponding side thrust force Q. The face angles are defined as shown in Figure 6.18.

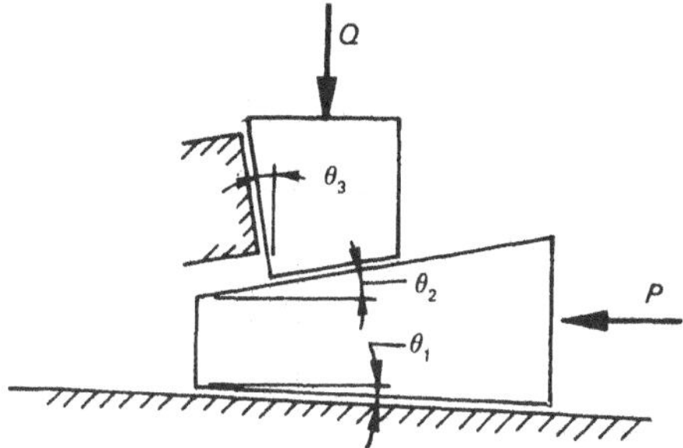

Fig. 6.18 *Wedge configuration for computer solution*

The program is listed in Appendix 3.7 and the solution to example 6.3(b) is given below the program listing and should be self-explanatory.

6.4 Screw threads

The screw thread is an application of the inclined plane in which the plane is wrapped around a cylinder. The three-dimensional form traced out by the plane is known as a **helix** and the plane angle of inclination is the helix angle.

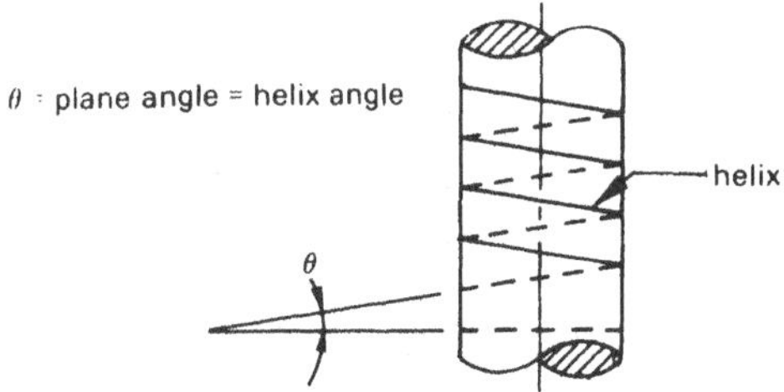

Fig. 6.19 *Obtaining a helix from an inclined plane*

A screw thread is obtained when a helix is cut into a solid rod. If a mating thread is then cut into a hollow block, a nut is formed. If the screw is held and the nut rotated, the nut will move axially along the screw. Similarly, if the nut is held and the screw rotated, the screw will move axially along the nut.

The helix angle is usually small (less than 5°), so that the nut is self-locking and axial force will not cause it to slip back along the thread after it is tightened. However, if torsional forces or vibration can occur, a locking device should be used. With small helix angles, there is a large velocity ratio and hence a large mechanical advantage. This means that large axial forces may be obtained and is the main reason why threaded fasteners and power screws find such wide application in mechanical engineering.

When the helix angle is large, the thread is not used for fastening but rather as a mechanism for conversion of rotary motion into axial motion or vice versa. In such cases, the thread is often made multi-start, i.e. two threads are cut out of phase 180° to one another or three threads out of phase by 120°. Multi-start threads are relatively uncommon, and in this book threads will always be single-start.

Thread form

The vast majority of screws used are for the purpose of fastening. In engineering, threads are also used for power transmission, for example in lathe saddle screws, screw jacks and worm-steering systems. Power screws usually are a square thread form or slightly angled thread form (ACME) as this gives lower frictional losses than the vee form. Threaded fasteners require friction for self-locking and therefore usually are a vee pattern (with about 60° included angle) which also gives greater strength to the thread.

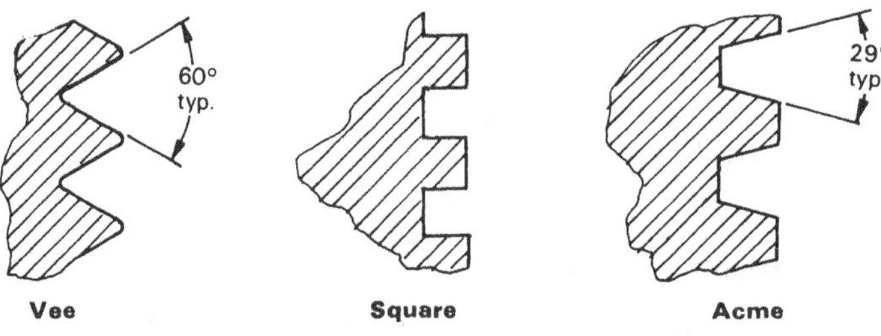

Fig. 6.20 *Common thread forms*

Thread pitch (*p*)

The axial distance between corresponding points on adjacent threads on one side of a screw or nut is known as the thread pitch.

Lead

The lead is the axial distance advanced for one revolution of the thread or nut. For a single-start thread the lead is the same as the pitch, but for a multi-start thread the lead is the pitch multiplied by the number of starts.

Pitch diameter (*d*)

If the thread depth were made increasingly smaller, so that the root diameter and outside diameter approached one another and eventually were equal, then the pitch diameter would be the diameter of the resulting cylinder which would have the same pitch as the original thread. The pitch diameter is approximately equal to the mean diameter between outside and root diameter but not exactly equal to it because most actual thread forms are truncated unequally.

It is standard practice to define the helix angle based upon the pitch diameter. In an actual thread, the helix angle varies slightly from the inside (root) to the outside.

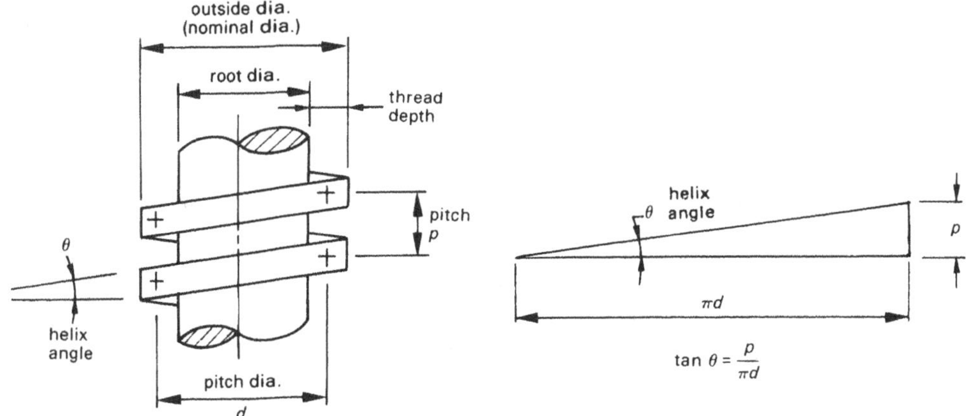

Fig. 6.21 *Thread geometry*

It will be seen that the relationship between the helix angle, the pitch diameter and the pitch is:

$$\tan\theta = \frac{p}{\pi d}$$ **(6.5) helix angle**

Example 6.5

From the table for the SI metric thread (standard or coarse pitch), the following data are obtained:

thread diameter (outside)	36 mm
pitch	4 mm
pitch diameter	33.4 mm

Determine the helix angle for this thread and the number of turns necessary to advance the nut 20 mm along the screw.

Solution

From equation 6.5:

$$\tan\theta = \frac{p}{\pi d} = \frac{4}{\pi \times 33.4} = 0.0381$$

$$\therefore \ \theta = 2.18°$$

To advance 20 mm, the number of turns required is:

$$N = \frac{20}{p} = \frac{20}{4} = 5$$

6.5 Forces on a square screw thread

When a torque is applied to a screw thread or nut, a force is transmitted along the axis of the thread. The forces present on a square thread may be analysed by imagining the helix to be unwrapped around the pitch diameter. The inclined plane which results from this may be analysed using the methods previously used for forces on an inclined plane. This is shown in Figure 6.22.

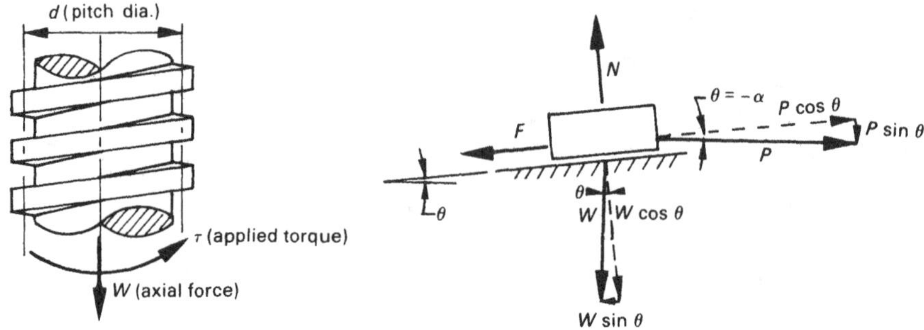

Fig. 6.22 *Forces acting on an equivalent inclined plane (thread tightened)*

If the thread is tightened, this is equivalent to motion of the block up the plane so that the friction force (F) acts down the plane.

The applied torque (τ) results in an equivalent applied force (P) on the inclined plane and the relationship between them is:

$$\tau = P \times \frac{d}{2}$$

Now applying equation 6.3:

$$P = \frac{W(\sin\theta + \mu\cos\theta)}{\cos\alpha + \mu\sin\alpha}$$

In this case $\alpha = -\theta$, so that $\cos\alpha = \cos\theta$, but $\sin\alpha = -\sin\theta$.

$$\therefore P = \frac{W(\sin\theta + \mu\cos\theta)}{\cos\theta - \mu\sin\theta}$$

Dividing top and bottom of the right-hand side of the fraction by $\cos\theta$, the result is:

$$P = \frac{W(\tan\theta + \mu)}{1 - \mu\tan\theta}$$

or

$$\boxed{\tau = W\frac{d}{2}\left[\frac{\tan\theta + \mu}{1 - \mu\tan\theta}\right]}$$ **(6.6) tightening a screw thread**

This equation gives the relationship between the axial load W and the applied torque τ for a screw thread which is being tightened.

If the thread is loosened instead, the direction of the applied force and the friction force is opposite and is equivalent to motion down the inclined plane as shown in Figure 6.23.

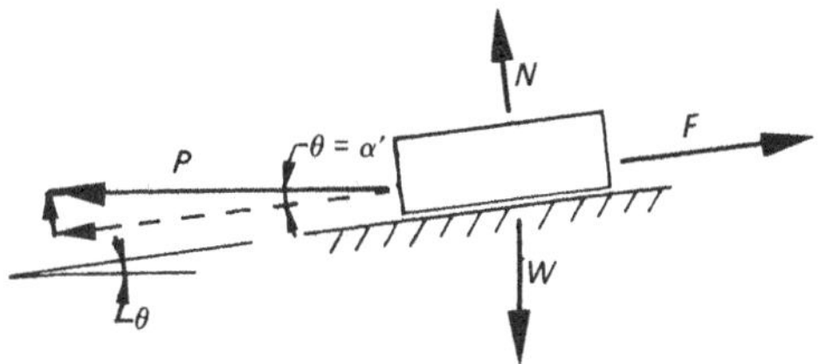

Fig. 6.23 *Forces acting on an equivalent inclined plane (thread loosened)*

For this case, equation 6.4(a) is applicable since $\theta < \phi$. The helix angle $\theta = \alpha'$ and therefore:

$$P = \frac{W(\mu \cos\theta - \sin\theta)}{\cos\theta + \mu \sin\theta}$$

$$\therefore P = \frac{W(\mu - \tan\theta)}{1 + \mu \tan\theta}$$

or

$$\tau = W\frac{d}{2}\left[\frac{\mu - \tan\theta}{1 + \mu\tan\theta}\right]$$

(6.7) loosening a screw thread

Self-locking

The limiting angle for self-locking occurs when no force is required to loosen the thread.

i.e. $P = 0 \quad \therefore 0 = W(\mu - \tan\theta)$

$$\text{or } \mu = \tan\theta$$

$$\therefore \theta = \tan^{-1}\mu = \phi$$

That is, if the helix angle is less than the angle of friction, the screw is self-locking. Note that this is one-half the value of θ for a self-locking wedge. This is because the wedge binds on two surfaces (upper and lower) whereas a screw mates only on one surface of the thread, there being clearance or backlash on the other side.

Alternative analysis of forces on a square screw thread

A simpler analysis is obtained by use of the angle of friction to combine the friction force and the normal reaction into a single force (as was done in the graphical wedge analysis). The equivalent free body diagrams and vector diagrams are shown in Figure 6.24.

There are now three forces acting on the block, namely R, P and W.

It is clear from the vector diagrams that:

Tightening

$$\frac{P}{W} = \tan(\theta + \phi)$$

$$\text{but } \tau = P \times \frac{d}{2}$$

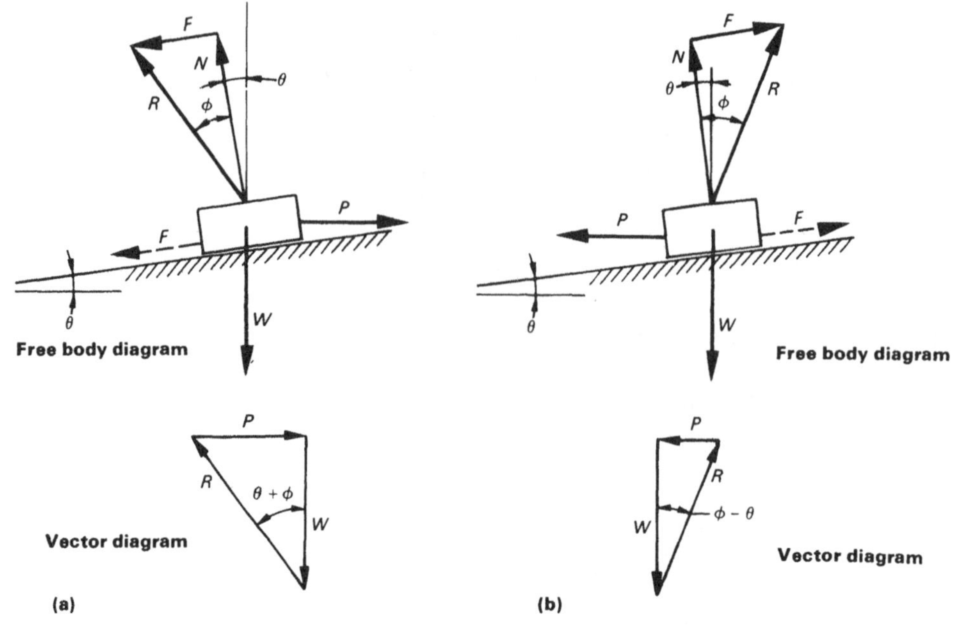

Fig. 6.24 *Free body diagrams and vector force diagrams for the equivalent inclined plane: (a) motion up the plane (tightening), (b) motion down the plane (loosening)*

$$\therefore \quad \boxed{T = W\frac{d}{2}\tan(\theta + \phi)} \qquad \textbf{(6.8) tightening a screw thread}$$

Loosening

$$\frac{P}{W} = \tan(\phi - \theta)$$

$$\therefore \quad \boxed{T = W\frac{d}{2}\tan(\phi - \theta)} \qquad \textbf{(6.9) loosening a screw thread}$$

Equations 6.8 and 6.9 give the same answer as equations 6.6 and 6.7 but are simpler.

Example 6.6

A vertical square screw thread, pitch diameter 100 mm and pitch 25 mm, is used with a load of 1 t. The coefficient of friction is 0.2.

Determine:

(a) screw torque required to raise the load
(b) screw torque required to lower the load
(c) maximum possible pitch in order that the thread is still self-locking.

Solution

From equation 6.5,

$$\tan\theta = \frac{p}{\pi d} = \frac{25}{\pi \times 100} \quad \therefore \theta = 4.55°$$

$$\phi = \tan^{-1}\mu = \tan^{-1}(0.2) = 11.31°$$

(a) Using equation 6.8,

$$\tau = W\frac{d}{2}\tan(\theta + \phi)$$

$$= 1000 \times 9.81 \times \frac{0.1}{2}\tan(4.55° + 11.31°)$$

$$= \mathbf{139\ Nm}$$

Alternatively, using equation 6.6,

$$\tau = W\frac{d}{2}\left[\frac{\tan\theta + \mu}{1 - \mu\tan\theta}\right]$$

$$= 1000 \times 9.81 \times \frac{0.1}{2}\left[\frac{\tan 4.55° + 0.2}{1 - 0.2\tan 4.55°}\right]$$

$$= \mathbf{139\ Nm} \qquad \text{(which is the same result)}$$

(b) From equation 6.9,

$$\tau = W\frac{d}{2}\tan(\phi - \theta)$$

$$= 1000 \times 9.81 \times \frac{0.1}{2}\tan(11.31° - 4.55°)$$

$$= \mathbf{58.1\ Nm}$$

Note: Equation 6.7 will give the same result.

(c) $\tan\theta = \dfrac{p}{\pi d}$ or $p = \pi d\tan\theta$

For self-locking, $\theta = \phi = 11.31°$

$$\therefore p = \pi \times 100 \times \tan 11.31° = \pi \times 100 \times 0.2$$

$$= \mathbf{62.8\ mm}$$

6.6 Efficiency of a screw thread

For a screw being tightened it is possible to obtain a simple expression of the efficiency.

Now efficiency $\eta = \dfrac{\text{MA}}{\text{VR}}$

$$\text{MA} = \frac{\text{load}}{\text{effort}} = \frac{W}{P} = \frac{1}{\tan(\theta + \phi)}$$

$$\text{VR} = \frac{\pi d}{p} \qquad \text{(single start thread)}$$

$$\therefore \quad \eta = \frac{p}{\pi d\tan(\theta + \phi)}$$

But $p = \pi d \tan\theta$

$$\therefore \quad \boxed{\eta = \frac{\tan\theta}{\tan(\theta + \phi)}} \qquad \text{(6.10) efficiency of a screw thread}$$

Example 6.7

Determine the efficiency of the screw thread given in example 6.6 when the load is being raised.

Solution

From equation 6.10,

$$\eta = \frac{\tan\theta}{\tan(\theta + \phi)}$$

$$= \frac{\tan 4.55°}{\tan 15.86°}$$

$$= \mathbf{0.28 \text{ or } 28\%}$$

It is evident, therefore, that a single start screw thread is not a very efficient way of converting rotary to linear motion because of the high frictional loss. For greater efficiency the frictional loss must be reduced (for example, with recirculating balls, or by making the thread multi-start so that the helix angle is large).

6.7 Forces on a vee screw thread

Consider a vee screw thread as shown in Figure 6.25 in which the included angle of the thread is α.

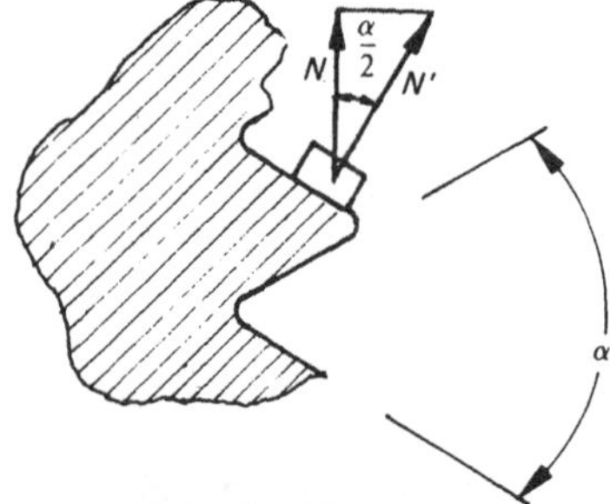

Fig. 6.25 *Equivalent normal force on a vee screw thread*

Let N = normal reaction for a square thread
N' = normal reaction of vee thread
F' = friction force on the vee thread

Then it is clear that:

$$N' = \frac{N}{\cos(\alpha/2)}$$

$$F' = \mu N' = \mu\frac{N}{\cos(\alpha/2)}$$

It is more convenient to rewrite this formula:

$$F' = \mu' N \quad \text{where} \quad \boxed{\begin{aligned} \mu' &= \frac{\mu}{\cos(\alpha/2)} \\ \phi' &= \tan^{-1}\mu' \end{aligned}}$$

(6.11) vee thread

Using equation 6.11 any of the formulas valid for a square screw thread may be used for a vee thread by substituting μ' for μ and ϕ' for ϕ.

Example 6.8

A horizontal power screw driving a lathe tailstock of mass 50 kg is ACME thread, included angle 29°, pitch diameter 31 mm, pitch 6.4 mm.

Determine the torque required to move the tailstock at constant velocity and the efficiency of the screw if the coefficient of friction is 0.1 on all sliding surfaces and bearing friction is negligible.

Solution

From equation 6.11 with $\alpha = 29°$, $\mu = 0.1$.

$$\phi' = \tan^{-1}\left(\frac{0.1}{\cos(29°/2)}\right) = 5.9°$$

$$\tan\theta = \frac{p}{\pi d} = \frac{6.4}{\pi \times 31}$$

$$\therefore \quad \theta = 3.76°$$

Axial load $W = 0.1 \times 50 \times 9.81 = 49.05$ N

From equation 6.8,

$$\tau = W\frac{d}{2}\tan(\theta + \phi')$$

$$= 49.05 \times \frac{0.31}{2} \times \tan(3.76° + 5.9°)$$

$$= \mathbf{0.129\ Nm}$$

From equation 6.10,

$$\eta = \frac{\tan\theta}{\tan(\theta + \phi')} = \frac{\tan 3.76°}{\tan(3.76° + 5.9°)} = \mathbf{38.6\%}$$

6.8 Band friction

It is a common observation that when several turns of a rope are taken around a circular post fixed to the ground, a considerable load on one end of the rope may be balanced by a relatively small pull on the other end. This is an example of the mechanism of band friction and occurs whenever a flexible member is partially or fully wrapped around a circular member. Band friction is used to apply braking torque in engineering applications such as band brakes, or to transfer torque in belt drives.

Relationship between the band tensions

Consider the flexible member (band) passed around a fixed circular member (drum) as shown in Figure 6.26.

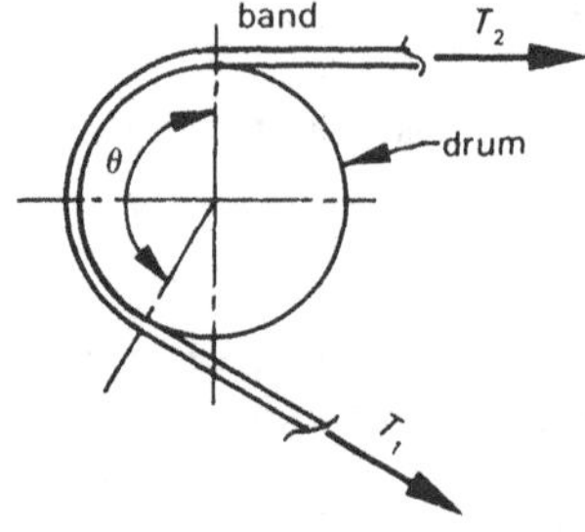

Fig. 6.26 *Band tensions*

Using the symbols T_2 for the larger tension and T_1 for the smaller tension, it can be shown (by application of calculus) that on the point of slipping, i.e. when T_1 is just sufficient to balance T_2, that T_2 and T_1 are related by the formula:

$$\boxed{\frac{T_2}{T_1} = e^{\mu\theta}}$$ **(6.12) band friction**

where T_2 = larger band tension (N)
T_1 = smaller band tension (N)
μ = coefficient of friction between band and drum
θ = angle of contact or angle of wrap (radians)
e = constant = 2.7183 and $\ln(e) = 1$ (natural logarithm)

Note: θ (radians) $= \theta° \times \frac{\pi}{180}$ or $360°$ (1 turn) $= 2\pi$ rads.

Example 6.9

A rope is wrapped around a bollard and used to hold a ship fast. The rope is capable of supporting a tension of 50 kN. Determine how many turns must be taken around the bollard so that a man can hold the ship when the rope is at maximum tension by exerting a pull of 200 N on the rope. Assume the coefficient of friction is 0.2.

Solution

Here $T_2 = 50 \times 10^3$ N

$T_1 = 200$ N

$\mu = 0.2$

From equation 6.12,

$$\frac{T_2}{T_1} = e^{\mu\theta}$$

Denoting $x = \mu\theta$,

$$\frac{T_2}{T_1} = e^x$$

$$\therefore \quad e^x = \frac{50 \times 10^3}{200} = 250$$

Taking natural logarithms of both sides:

$$x \ln(e) = \ln(250)$$

$$\therefore \quad x \times 1 = 5.5215 \quad \therefore x = 5.5215$$

$$\therefore \quad \mu\theta = 5.5215 \quad \theta = \frac{5.5215}{0.2} = 27.61 \text{ rads}$$

But 2π rads $= 360°$ (1 turn)

$$\therefore N \text{ (number of turns)} = \frac{27.61}{2\pi}$$

$$= \mathbf{4.4\ turns}$$

It is seen therefore that a force reduction factor of 250 is obtained with about four and a half turns around the bollard.

6.9 Band brakes

A band brake is a particular application where band friction is used to apply braking torque to a rotating drum. The band is normally manufactured from high tensile steel with friction material facing.

Some band brake configurations are shown in Figure 6.27. Note that in configuration (b), the larger band tension is used to assist braking by anchoring the band on the other side of the pivot point.

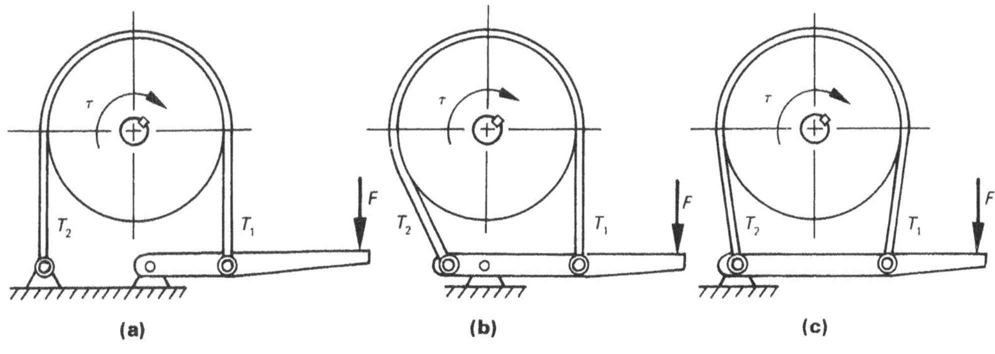

Fig. 6.27 *Some band brake configurations*

Braking torque and power

Consider the section through a band brake shown in Figure 6.28.

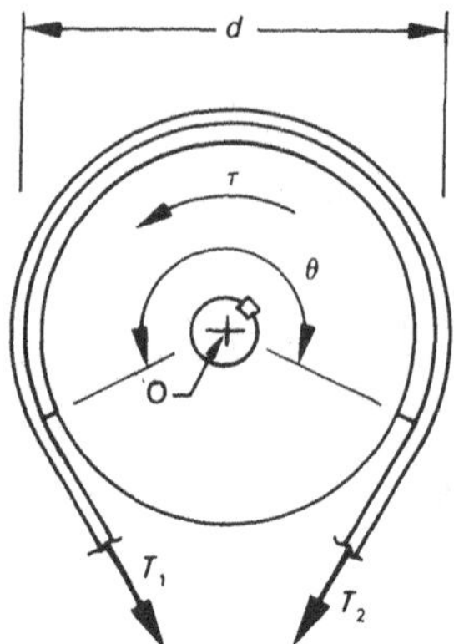

Fig. 6.28 *Section through a band brake*

Taking moments about the centre O at the point of slipping:

$$T_2 \times \frac{d}{2} - T_1 \times \frac{d}{2} - \tau = 0$$

$$\therefore \quad \boxed{\tau = (T_2 - T_1)\frac{d}{2}} \qquad \textbf{(6.13) braking torque}$$

If the brake is rotating at constant angular velocity, the power dissipated by the brake may be determined by the standard formula relating torque and power,

$$\boxed{P = \tau\omega} \qquad \textbf{(6.14) braking power}$$

In these formulas:

τ = braking torque (Nm)
P = power dissipated by the brake (W)
d = mean diameter over the centres of the band (m)
T_2 = larger belt tension (N)
T_1 = smaller belt tension (N)
ω = angular velocity of the drum (rad/s)
$= \frac{\pi N}{30}$ (where N is the angular velocity of the drum in rpm)

Method of solution of band brake problems

Band brake problems cannot generally be solved directly by application of equation 6.13 because there are two unknowns (T_2 and T_1). Another equation is required relating T_2 and T_1 and this may be obtained by application of equation 6.12. This gives two equations in two unknowns which may be solved simultaneously. The method is illustrated in example 6.10.

Example 6.10

For the band brake shown in Figure 6.29 determine:

(a) braking force F in order that the 60 kg load can be lowered at constant velocity 1.5 m/s

(b) power dissipated during braking.

Assume the coefficient of friction is 0.3.

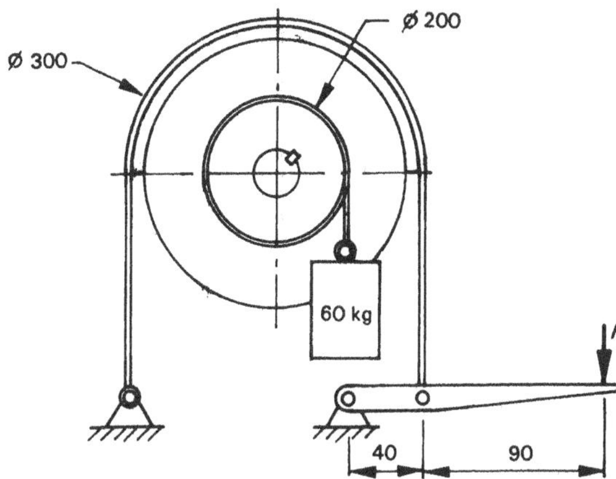

Fig. 6.29

Solution

Braking torque $\tau = 60 \times 9.81 \times \frac{0.2}{2} = 58.86$ Nm

(a) From equation 6.13,

$$\tau = (T_2 - T_1)\frac{d}{2}$$

$$\therefore\ 58.86 = (T_2 - T_1)\frac{0.3}{2}$$

$$T_2 - T_1 = 392.4 \qquad (1)$$

From equation 6.12,

$$\frac{T_2}{T_1} = e^{\mu\theta}$$

But $\theta = 180° = \pi$ rads

$$\therefore \frac{T_2}{T_1} = e^{0.3\pi}$$

$$\therefore T_2 = T_1 \times 2.566 \qquad (2)$$

Substituting in (1):

$$2.566T_1 - T_1 = 392.4$$

$$\therefore \quad T_1 = 250.5 \text{ N}$$

$$\therefore T_2 - 250.5 = 392.4$$

$$\therefore \quad T_2 = 643 \text{ N}$$

Consider the forces acting on the brake lever (Fig. 6.30).

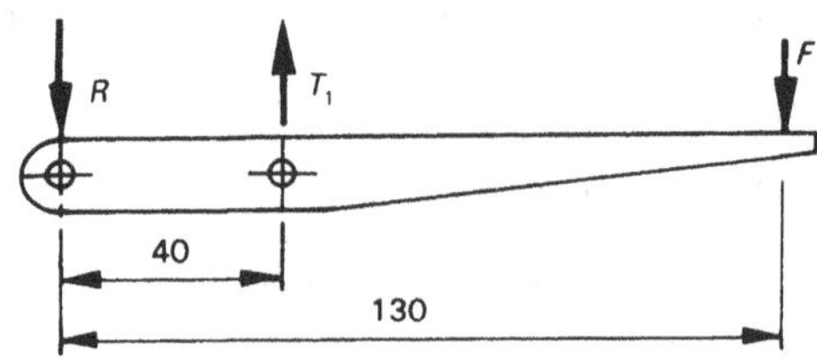

Fig. 6.30

Taking moments about the pivot point,

$$F \times 0.13 - T_1 \times 0.04 = 0$$

$$\therefore F = \frac{0.04 \times 250.5}{0.13} = \mathbf{77.1\ N}$$

(b) The angular velocity of the drum may be determined from the linear velocity by using the relationship:

$$v = r\omega$$

$$\therefore \omega = \frac{v}{r} = \frac{1.5}{0.1} = 15 \text{ rad/s}$$

Using equation 6.14,

$$P = \tau\omega$$
$$= 58.86 \times 15$$
$$= \mathbf{883\ W}$$

Alternatively:

$$P = Fv$$
$$= 60 \times 9.81 \times 1.5$$
$$= \mathbf{883\ W} \quad \text{(which checks)}$$

Example 6.11

Determine the braking effort, F, required to brake the 750 N load shown in Figure 6.31 if the coefficient of friction is 0.35.

Solution

$$\text{Braking torque } \tau = 750 \times \frac{0.15}{2} = 56.25 \text{ Nm}$$

$$\therefore \quad 56.25 = (T_2 - T_1)\frac{0.2}{2}$$

$$\therefore T_2 - T_1 = 562.5 \qquad \textbf{(1)}$$

$$\theta = 180 + 45 = 225° = 3.927 \text{ rads}$$

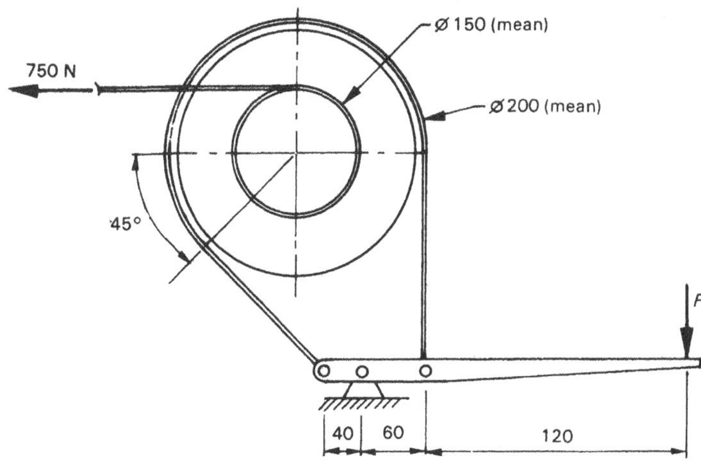

Fig. 6.31

$$\frac{T_2}{T_1} = e^{\mu\theta} = e^{0.35 \times 3.927}$$

$$\therefore \quad T_2 = 3.953\ T_1 \qquad \textbf{(2)}$$

Substituting in (1),

$$2.953\ T_1 = 562.5$$

$$\therefore \quad T_1 = 190.5$$

$$\therefore \quad T_2 = 753\ \text{N}$$

Taking moments about the pivot point:

$$F \times 0.18 + T_1 \cos 45^\circ \times 0.04 - T_2 \times 0.06 = 0$$

$$\therefore F = \frac{753 \times 0.06 - 190.5 \cos 45^\circ \times 0.04}{0.18}$$

$$= \mathbf{221\ N}$$

6.10 Flat belt drives

A flat belt drive is used to transfer power from one rotating shaft to another, the centrelines of the two shafts being parallel but displaced from one another. The mechanics of the flat belt drive at the point of slipping is the same as that of the band brake except that, because the belt itself is rotating, centrifugal force tends to lift the belt off the pulley and reduce the torque and power which can be transmitted. This effect is of importance in high speed belt drives but is neglected here.

Example 6.12

Determine the maximum torque and power which can be transmitted to the driven shaft of the flat belt drive shown in Figure 6.32. The coefficient of friction between belt and pulley is 0.3 and the slack side tension is 100 N. Neglect centrifugal tension.

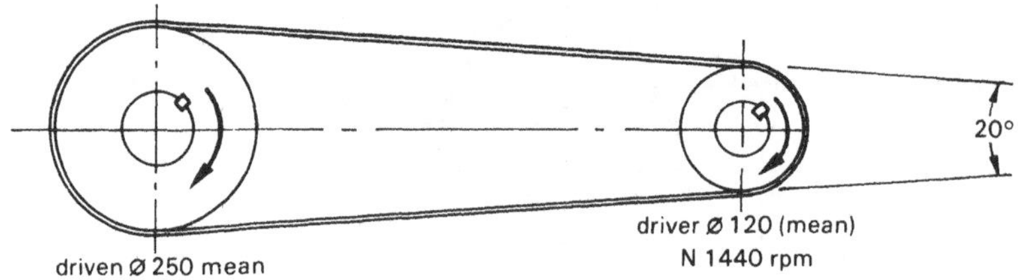

Fig. 6.32

Solution

The belt tensions are the same at both pulleys. However, because the angle of contact is less for the driver, slipping will occur first at the driver pulley and the point of slipping of the driver pulley determines the maximum belt tensions and power which the drive can transmit.

For the driver pulley:

$$\theta = 160° = 2.7925 \text{ rads}$$

$$T_1 = 100 \text{ N}$$

From equation 6.12,

$$\frac{T_2}{T_1} = e^{\mu\theta}$$

$$\therefore T_2 = 100 \times e^{0.3 \times 2.7925}$$

$$T_2 = 231 \text{ N}$$

At the driven pulley, $T_2 = 231$ N, $T_1 = 100$ N (same as for the driver)

$$\tau = (T_2 - T_1) \times \frac{d}{2} = (231 - 100) \times \frac{0.25}{2} = \mathbf{16.4\ Nm}$$

The rotational speed of the driven pulley is:

$$1440 \times \frac{120}{250} = 691 \text{ rpm}$$

From equation 6.14, the power is:

$$P = \tau\omega$$

$$= 16.4 \times \frac{\pi}{30} \times 691 \text{ (W)}$$

$$= \mathbf{1.19\ kW}$$

Note that the torque at the driver pulley is:

$$\tau = (231 - 100) \times \frac{0.12}{2} = 7.86 \text{ Nm}$$

$$P = 7.86 \times \pi \times \frac{1440}{30} = 1.19 \text{ kW}$$

which is the same since it has been assumed that there is no power loss between driver and driven pulleys.

6.11 Vee belt drives

In a vee belt drive, a vee belt rather than a flat belt is used. The vee belt usually has an included angle between 30° and 40° and there is a wedging action of the belt in the pulley groove which increases the friction and allows a greater difference between belt tensions before slipping occurs. A vee belt drive will therefore transmit considerably more torque and power than a flat belt drive of the same geometry.

Equations 6.13 and 6.14 are valid for a vee belt drive as well as a flat belt drive. However equation 6.12 must be modified to include the effect of the wedging action and becomes:

$$\frac{T_2}{T_1} = e^{\frac{\mu\theta}{\sin(\beta/2)}}$$ **(6.15) vee belt drive**

where β = included angle of the belt, and other terms are the same as in equation 6.12.

Example 6.13

Recalculate the maximum torque and power which can be transmitted by the belt drive given in example 6.12 if a vee belt (included angle 40°) were used rather than a flat belt, all other data being the same.

Solution

$$\frac{T_2}{T_1} = e^{\frac{\mu\theta}{\sin(\beta/2)}}$$

$$\therefore T_2 = 100 \times e^{\frac{0.3 \times 2.7925}{\sin(40°/2)}}$$

$$= 1158 \text{ N}$$

At the driven pulley:

$$\tau = (T_2 - T_1) \times \frac{d}{2} = (1158 - 100) \times \frac{0.25}{2} = \mathbf{132\ Nm}$$

$$P = \tau\omega = 132 \times \frac{\pi}{30} \times 691 = \mathbf{9.57\ kW}$$

Hence it is evident that the torque and power have been increased by about 800 per cent by using a vee belt rather than a flat belt.

6.12 Disc or collar friction

When an axial force such as that exerted by a screw acts on a rotating member, means must be provided for transmitting the thrust while the rotation occurs. If a rolling element thrust bearing is provided, the frictional torque is usually negligible. However, if the thrust is transmitted through a disc or collar as shown in Figure 6.33, the frictional torque may be significant.

Determination of the frictional torque depends upon the nature of the pressure distribution between the faces in contact. When the surfaces are newly machined there is likely to be even contact over the entire contact area so that the distribution of pressure is uniform. However, after a period the outer surfaces will wear more than the inner surfaces because of the higher velocity and distance moved. When this occurs, the pressure at the inner surfaces will be

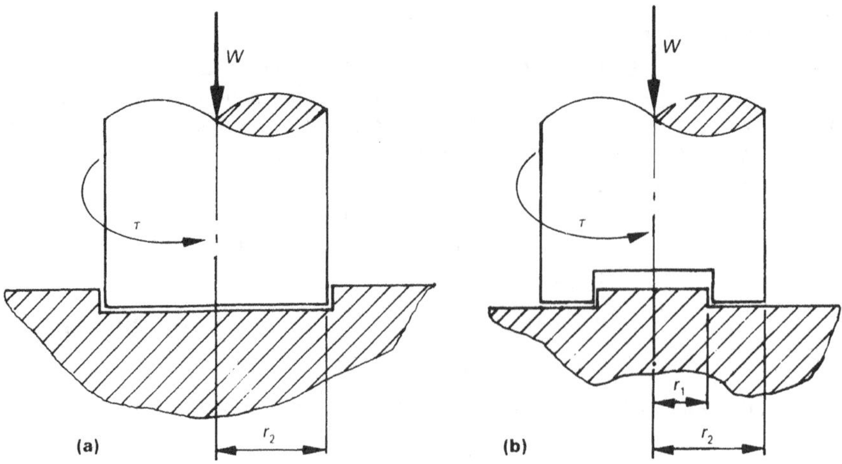

Fig. 6.33 *(a) Disc and (b) collar friction*

higher than that at the outer surfaces and it is usually assumed that the pressure varies with the radius such that:

$$pr = c \quad \text{(a constant)}$$

Consider a small annular ring of width dr located at radius r. The load on this ring is:

$$dW = p \times 2\pi\, r\, dr$$
$$= 2\pi\, c\, dr$$

The total load W is:

$$W = \int_{r_1}^{r_2} 2\pi\, c\, dr$$
$$= 2\pi\, c(r_2 - r_1)$$
$$\therefore c = \frac{W}{2\pi(r_2 - r_1)}$$

The torque on the small annular ring is:

$$d\tau = \mu\, r\, dW$$
$$= \mu\, r\, 2\pi\, c\, dr$$

The total torque is:

$$\tau = \mu\, 2\pi\, c \int_{r_1}^{r_2} r\, dr$$
$$= \mu\, 2\pi\, c \frac{(r_2^2 - r_1^2)}{2}$$
$$= \mu\, \pi\, c(r_2 - r_1)(r_2 + r_1)$$
$$= \mu\, \pi \frac{W(r_2 - r_1)(r_2 + r_1)}{2\pi(r_2 - r_1)} \qquad = \mu\, W \frac{(r_2 + r_1)}{2}$$

But $\frac{r_2 + r_1}{2} = r_m$ (mean radius),

$$\therefore \quad \boxed{\tau = \mu W r_m} \qquad \textbf{(6.16) disc or collar friction (uniform wear)}$$

Notes

1. In the case of a disc, $r_1 = 0$ and $r_m = r_2/2$.
2. In this book, uniform wear will be always assumed because in most engineering applications this is the most likely situation.

Example 6.14

For the screw thread in example 6.6, determine the total torque required to raise and lower the load if the load is supported on a thrust collar as shown in Figure 6.34. The coefficient of friction on the collar is the same as in the thread, 0.2.

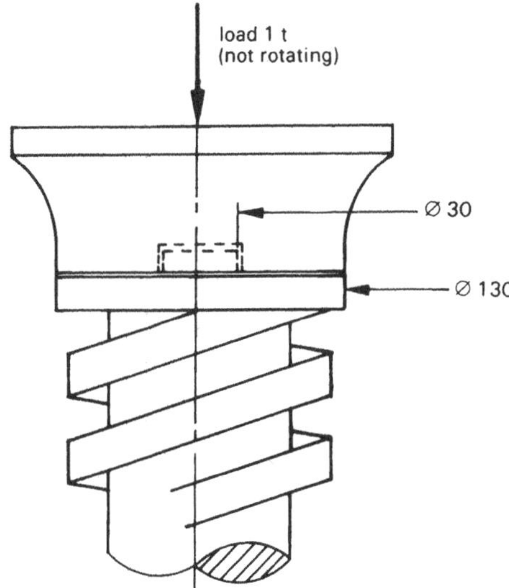

Fig. 6.34

Solution

In this case $r_m = \frac{15 + 65}{2} = 40 \text{ mm} = 0.04 \text{ m}$

$$W = 1000 \times 9.81 = 9810 \text{ N}$$

From equation 6.16, the frictional torque on the collar is:

$$\begin{aligned} \tau &= \mu W r_m \\ &= 0.2 \times 9810 \times 0.04 \\ &= 78.5 \text{ Nm} \end{aligned}$$

The solution to example 6.6 was:

screw torque to raise load = 139 Nm
screw torque to lower load = 58.1 Nm

Therefore the total torque including collar friction is:

raising load: 139 + 78.5 = **218 Nm**
lowering load: 58.1 + 78.5 = **137 Nm**

Problems

6.1 A block of mass 50 kg rests on an inclined plane as shown in Figure P6.1. The coefficient of friction is 0.4.
Determine force *P* in order to move the block up the plane with constant velocity.
451 N

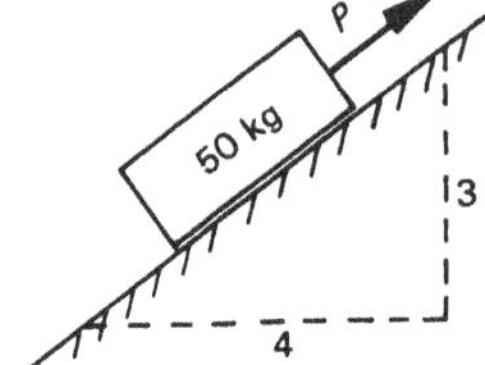

Fig. P6.1

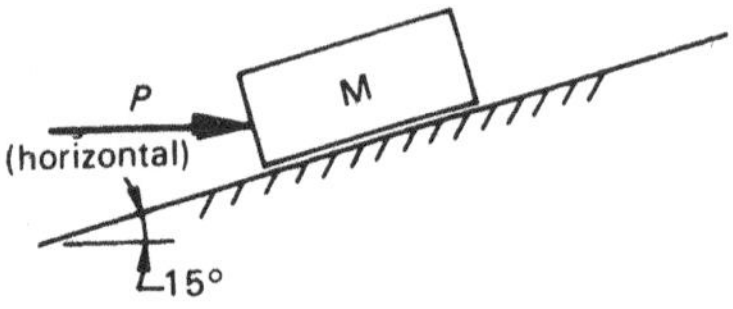

Fig. P6.3

6.2 Repeat problem 6.1 if the block is to move down the plane with constant velocity.
137 N

6.3 A block rests on an inclined plane as shown in Figure P6.3.
Determine the force *P* necessary to cause motion up the plane if *M* is 20 kg and the coefficient of friction is 0.35.
134 N

6.4 Repeat problem 6.3 if motion is to impend down the plane instead of up the plane.
14.7 N (down plane)

6.5 A block rests on an inclined plane as shown in Figure P6.5.
Determine force *P* necessary to cause motion up the plane if the coefficient of friction is 0.25 and mass *M* is 40 kg.
273 N

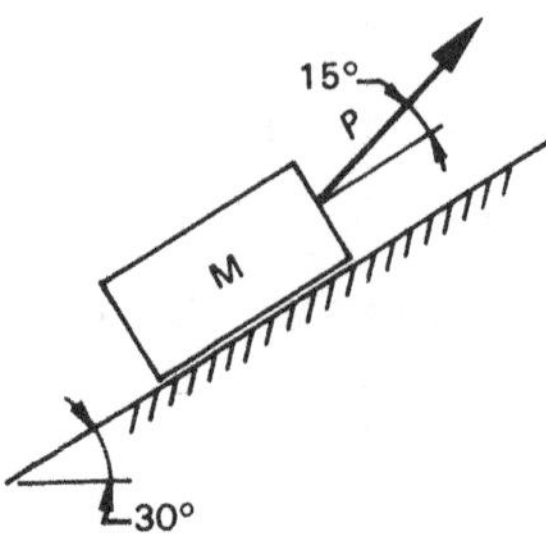

Fig. P6.5

6.6 Repeat problem 6.5 if motion is to impend down the plane instead of up the plane.

123 N

6.7 A block mass 80 kg rests on an inclined plane as shown in Figure P6.7.

Determine the force *P* necessary to move the block down the plane with constant velocity if the coefficient of friction is 0.35.

82 N

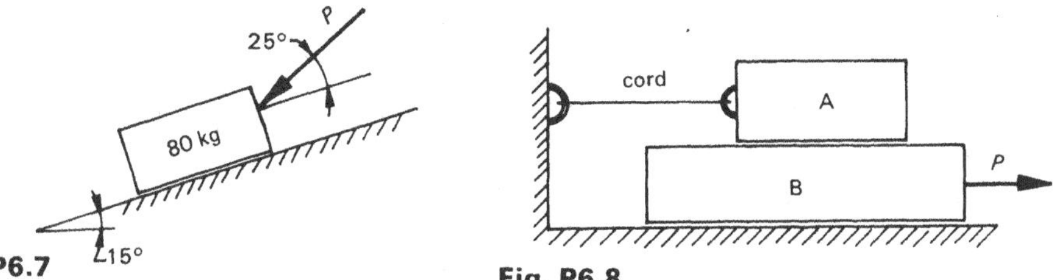

Fig. P6.7

Fig. P6.8

6.8 In the system of blocks shown in Figure P6.8, block A is 20 kg and block B 30 kg.

Determine the force *P* necessary to move block B if the coefficient of friction is 0.3 on all surfaces in contact.

206 N

6.9 In the system of blocks shown in Figure P6.9, block A is 5 kg, block B is 8 kg and the coefficient of friction 0.35 on all surfaces in contact.

Determine the tension in the cord and the force *P* necessary to move block B.

3.68 N, 48.3 N

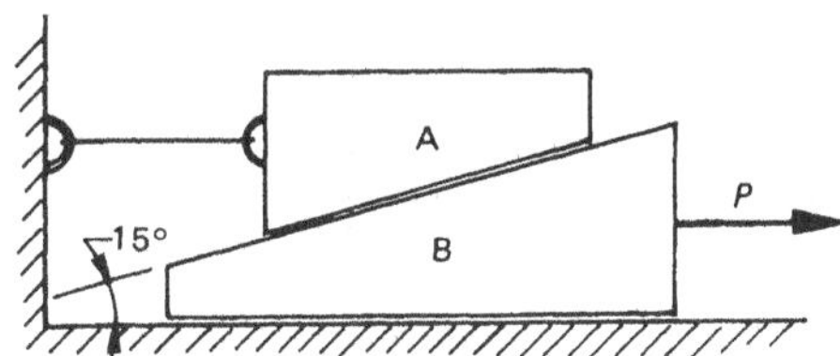

Fig. P6.9

6.10 In the system of blocks shown in Figure P6.10, block A is 2 kg, block B is 10 kg and the coefficient of friction is 0.3 on all surfaces in contact.

Determine mass *M* necessary to cause block B just to move to the right.

1.02 kg

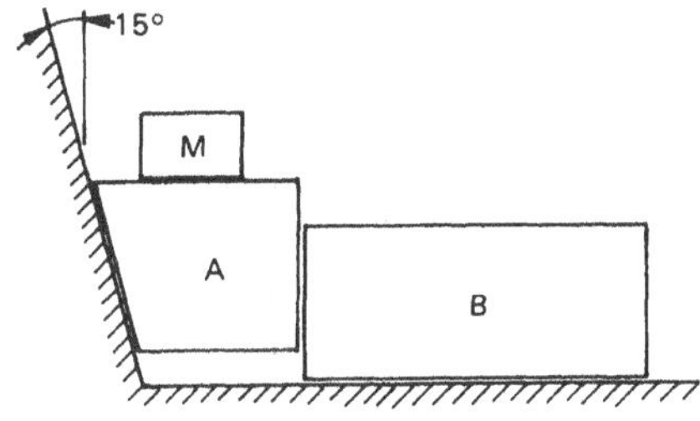

Fig. P6.10

6.11 In the system shown in Figure P6.11, the block is 10 kg, the wedge is 12 kg and the coefficient of friction is 0.3 on all surfaces in contact.

Determine force P necessary to cause the block just to lift.

306 N

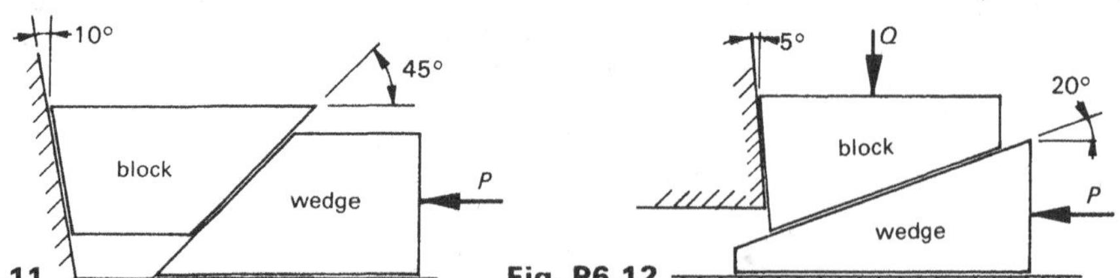

Fig. P6.11 **Fig. P6.12**

6.12 For the system shown in Figure P6.12, the block has a mass of 10 kg and the mass of the wedge is negligible. The coefficient of friction on all surfaces is 0.3.

Determine the side thrust force Q for an applied force P of 400 N.

225 N

6.13 Repeat problem 6.12 if the wedge itself has a mass of 8 kg, all other data being the same.

206 N

6.14 For the system shown in Figure P6.14, the block has a mass of 4 kg and the mass of the wedge is negligible. The coefficient of friction is 0.25 on all surfaces in contact.

Determine:

(a) force P necessary just to raise the block ($Q = 0$)

(b) side thrust force Q if P is 200 N

(c) side thrust force Q if P is 200 N and the mass of the wedge is 4 kg.

(a) 30.7 N (b) 216 N (c) 199 N

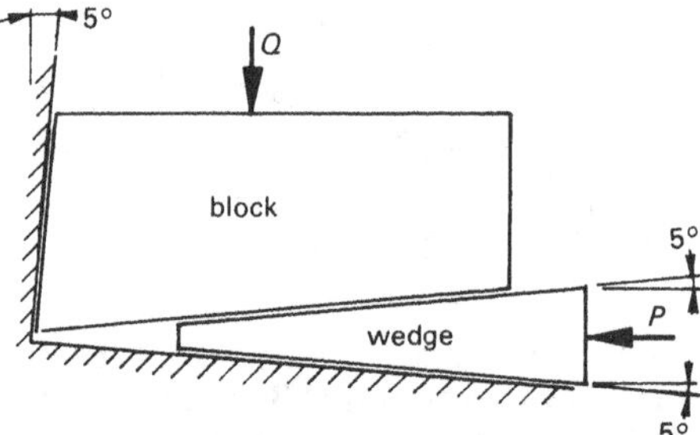

Fig. P6.14

6.15 Determine the maximum pitch for a 20 mm pitch diameter square screw thread to be self-locking if the coefficient of friction is 0.18.

11.3 mm

6.16 A screw jack with square thread, pitch 12 mm and pitch diameter 50 mm is used with a load of 700 kg. The coefficient of friction is 0.15.

Determine:

(a) torque necessary to raise the load

(b) efficiency

(c) torque necessary to lower the load.

(a) 39.3 Nm (b) 33.3% (c) 12.5 Nm

6.17 A vertical square screw thread, pitch diameter 75 mm and pitch 15 mm is used with a load of 500 kg. The coefficient of friction is 0.15.

Determine:

(a) helix angle
(b) torque required to raise the load
(c) torque required to lower the load
(d) maximum pitch in order that the thread is self-locking
(e) efficiency when raising the load.

(a) 3.64° (b) 39.7 Nm (c) 15.7 Nm (d) 35.3 mm (e) 29.5%

6.18 Repeat problem 6.17 if the thread were a vee thread rather than a square thread with included angle 60°.

(a) 3.64° (b) 44 Nm (c) 19.9 Nm (d) 40.8 mm (e) 26.6%

6.19 The horizontal table of a machine has a mass of 300 kg and rests on slides. It is driven at 0.15 m/s by a square screw of pitch diameter 40 mm and pitch 12 mm. The coefficient of friction between the table and slides is 0.15 and between the table nut and the screw is 0.22.

Determine:

(a) rotational speed, torque and power of the screw
(b) power of the table and hence the efficiency of the power drive system.

(a) 750 rpm, 2.84 Nm, 223 W (b) 66.2 W, 29.7%

6.20 A turnbuckle with left- and right-hand threads is used to tighten the rigging of a yacht. The screw has a vee form with 60° included angle, pitch diameter 9 mm and pitch 1.5 mm.

Determine the tightening force which can be exerted by the turnbuckle if a force of 50 N is applied to a lever 200 mm from the centre of the turnbuckle. Take the coefficient of friction to be 0.25.

3.2 kN

6.21 A power screw is used to move a mass of 120 kg at constant velocity of 0.2 m/s up an incline of 20° to the horizontal as shown in Figure P6.21. The screw has a single start square thread, pitch diameter 50 mm, pitch 10 mm and is supported by bearings with negligible friction.

If the coefficient of friction on all contact surfaces is 0.2, determine the motor torque and power.

4.17 Nm, 524 W

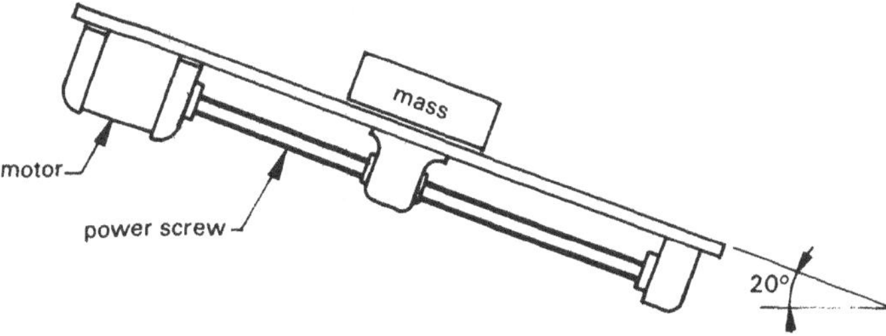

Fig. P6.21

6.22 A boat of mass 100 kg rests on support rails and is held by a cable attached to a winch drum as shown in Figure P6.22.

If the coefficient of friction between the boat and the rails is 0.3, and friction in the winch bearings is negligible, determine:

(a) force necessary at the winch handle to prevent the boat slipping back

(b) force necessary at the winch handle to raise the boat with constant velocity
(c) power expended in raising the boat if the winch drum is rotated at a constant speed of 40 rpm.

(a) 7.37 N (b) 76.5 N (c) 76.9 W

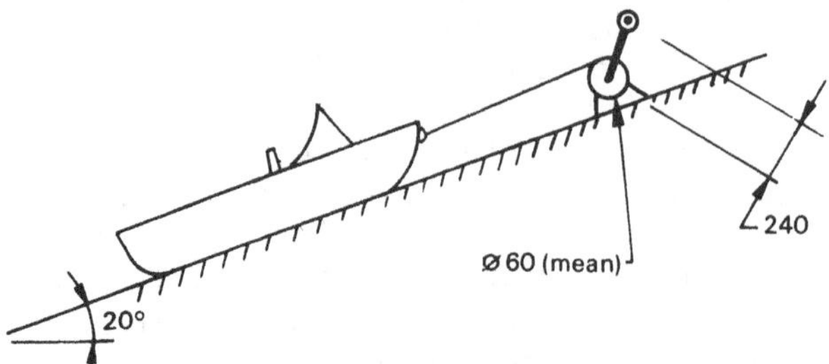

Fig. P6.22

6.23 A mass M kg is raised by means of a rotating screw and wedge arrangement as illustrated in Figure P6.23. The screw is single-start square thread, mean diameter 40 mm, pitch 6 mm. The coefficient of friction at the threads is 0.15 and the coefficient of friction on all other sliding faces is 0.3.

Determine:
(a) maximum mass M which may be raised by an applied force P of 500 N
(b) maximum torque T required to rotate the screw.
Friction in the thrust bearing and the weight of the wedge blocks themselves is negligible.

(a) 25.5 kg (b) 1.99 Nm

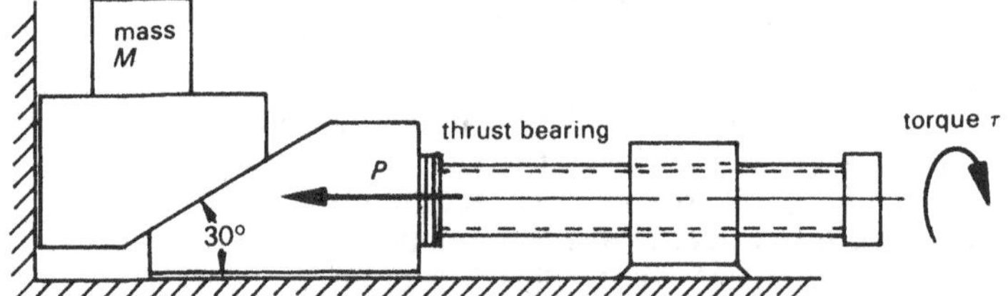

Fig. P6.23

6.24 For the band brake shown in Figure P6.24, the applied force P is 200 N and the coefficient of friction is 0.3.

Determine the braking torque when the drum rotation is:
(a) clockwise
(b) anticlockwise.

(a) 94 Nm (b) 36.6 Nm

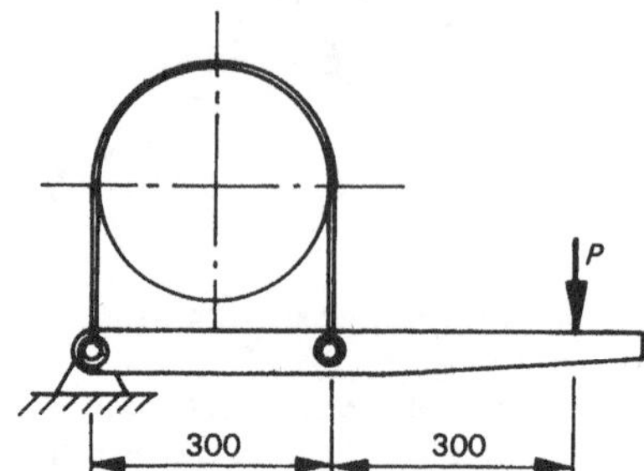

Fig. P6.24

6.25 The band brake shown in Figure P6.25 is required to lower the 200 kg mass with uniform velocity.

If the coefficient of friction is 0.5, determine the force P.

85.8 N

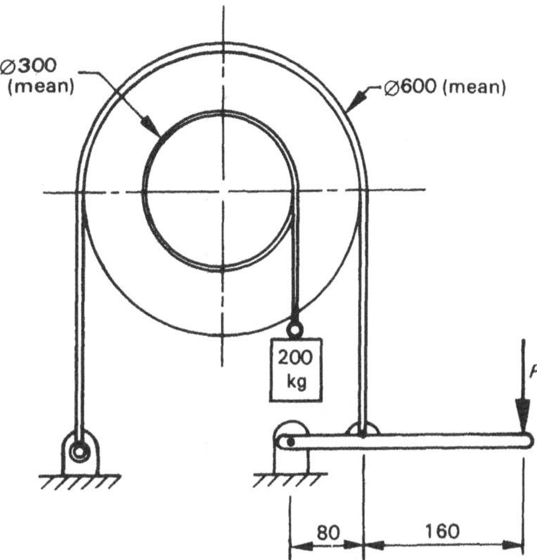

Fig. P6.25

6.26 For the band brake illustrated in Figure P6.26 determine:

(a) maximum load mass so that a braking force P of 250 N will just hold the load

(b) power dissipated by the brake if the load is lowered at a constant velocity of 1.8 m/s.

The coefficient of friction is 0.3.

(a) 160 kg (b) 2.83 kW

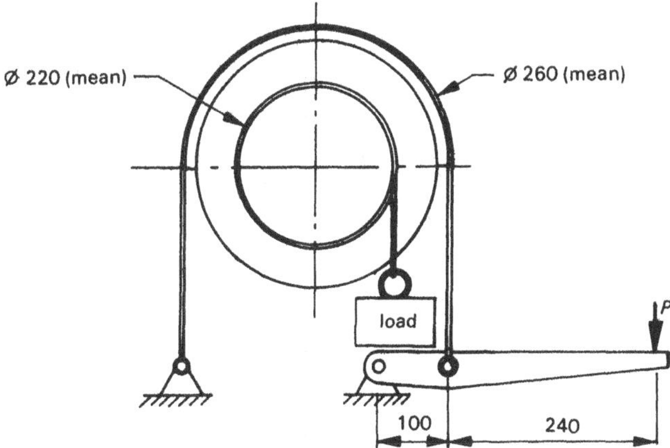

Fig. P6.26

6.27 Determine the braking effort P required to brake the load of 1.2 kN for the band brake shown in Figure P6.27. The coefficient of friction is 0.33.

371 N

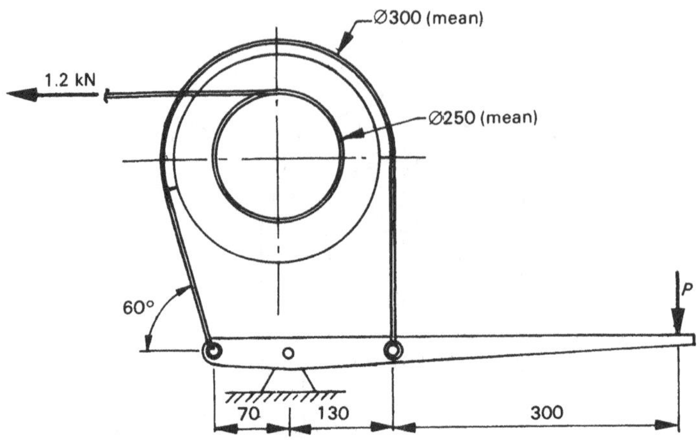

Fig. P6.27

6.28 A load of 500 kg is being lowered down an incline of 30° with velocity 1.2 m/s by a band brake as illustrated in Figure P6.28. Determine:

(a) braking torque
(b) braking power
(c) required value of the braking effort P.

The coefficient of friction between the load and the slide is 0.2 and between the brake and the drum is 0.35.

(a) 144 Nm (b) 1.92 kW (c) 124 N

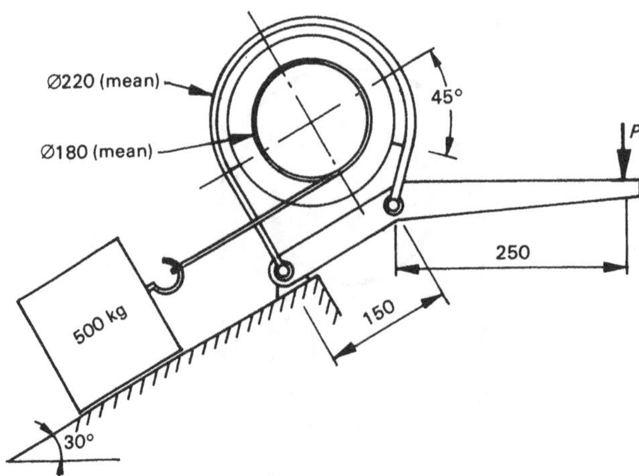

Fig. P6.28

6.29 A belt drive pulley has a mean diameter of 300 mm, an angle of contact of 180° and rotates at 920 rpm.

Determine the maximum torque and power which may be transmitted by the belt if it is (a) flat and (b) vee belt with included angle 40°. The smaller belt tension is 50 N and the coefficient of friction is 0.3. Neglect centrifugal tension.

(a) 11.75 Nm, 1.13 kW (b) 110 Nm, 10.6 kW

6.30 For the flat belt drive illustrated in Figure P6.30, determine the maximum torque and power which may be transmitted at 800 rpm if the coefficient of friction is 0.35 and the slack side tension is 20 N. Neglect centrifugal tension.

2.25 Nm, 188 W

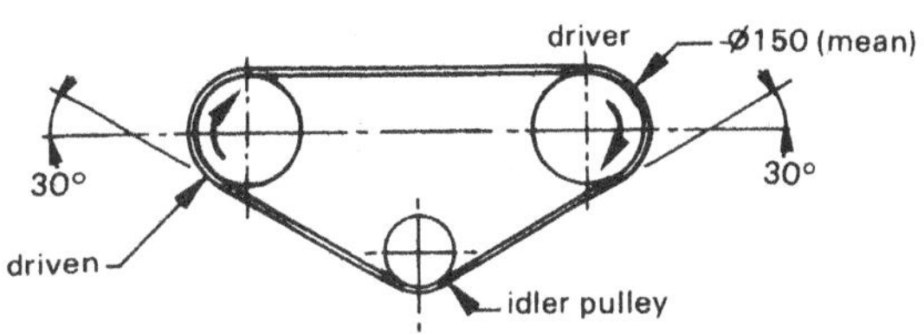

Fig. P6.30

6.31 Repeat the calculation for problem 6.30 if a vee belt, included angle 40°, is used rather than a flat belt, all other data being the same.

20.4 Nm, 1.7 kW

6.32 A vee belt pulley is designed for a T_2/T_1 ratio = 10. The pulley has a mean diameter of 200 mm, included angle 30° and rotates at 960 rpm. The coefficient of friction is 0.4.

If the smaller belt tension is 40 N, and centrifugal tension is neglected, determine:

(a) minimum angle of wrap to avoid slipping
(b) maximum power that may be transmitted.

(a) 85.4° (b) 3.62 kW

6.33 A thrust of 2 kN occurs axially in a shaft rotating at 120 rpm.

Determine the frictional torque and power if the coefficient of friction is 0.15 and the thrust is taken by:

(a) a flat face, diameter 50 mm
(b) a relieved face, outside diameter 50 mm, relief diameter 20 mm.

(a) 3.75 Nm, 47.1 W (b) 5.25 Nm, 66 W

6.34 Repeat problem 6.21 if the thrust is taken by a plain thrust bearing consisting of a collar, outside diameter 60 mm and inside diameter 40 mm, all other data being the same.

7.29 Nm, 916 W.

PART II

DYNAMICS

7

Kinematics of motion

Kinematics is the analysis of the motion of bodies in which the mass of the body and the forces acting on it are not considered. Kinematics is similar to navigation which is concerned with the position of bodies on or above the earth's surface, and the principles of navigation apply equally well to a small yacht or a large ship, the flight of a bird or an aircraft. It is evident, therefore, that the analysis of the motion or position of bodies may be a separate study to the analysis of the forces producing that motion.

Kinematics is concerned chiefly with three important variables: displacement, velocity and acceleration. The relationship between these variables will be considered separately for linear motion (translation) or curved motion (rotation). In fact, curved motion may follow any trajectory but the most important case is the case of rotation because, at any instant, any trajectory can be considered to be pure rotation about the instantaneous centre of curvature. Indeed, even linear motion can be considered as a special case of rotation where the centre of curvature is located at infinity.

7.1 Displacement (s)

When a body changes position relative to an initial position it has undergone a displacement. The shortest line drawn between the two positions will be a straight line, the length of which gives the magnitude of the linear displacement. The inclination of this line relative to a datum line or frame of reference indicates the direction of the linear displacement. For example, consider the motion of the body shown in Figure 7.1(a) which moves from point ① to point ② along the dashed line.

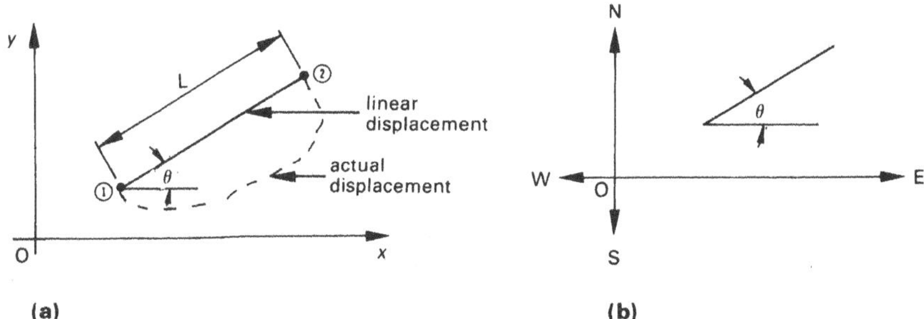

Fig. 7.1 *Actual and linear displacement: (a) x–y reference frame, (b) compass directions reference frame*

It is clear that the resultant linear displacement has magnitude L (m) and direction θ relative to the x–y frame of reference. In some instances, compass directions are used for the frame of reference as shown in Figure 7.1(b); in this case the direction of north is the $+y$ direction and the direction of east is the $+x$ direction. The angle θ is specified as $\theta°$ N of E; if $\theta = 45°$, this is known as NE.

7.2 Velocity (v)

Velocity is defined as the rate of change of displacement with respect to time. If a body moves in a straight line such that the displacement is constant for each increment of time, the body will have constant (or uniform) velocity v (m/s) given by:

$$v = \frac{s}{t}$$ **(7.1) constant (uniform) velocity**

where v = velocity (m/s)
s = displacement (m)
t = time interval over which the displacement occurs (s)

For example, if a body is moving in a straight line such that in every minute it is displaced a further 120 m, then the velocity will be $\frac{120}{60} = 2$ m/s.

Although the SI units of velocity are m/s, in practice the velocity is often given in km/h. To convert from km/h to m/s, multiply by 1000 and divide by 3600 (number of seconds in an hour), i.e. divide by 3.6. For example, a velocity of 72 km/h is 20 m/s (72/3.6).

If the body is not moving so that the displacement is constant with each increment of time, equation 7.1 gives only the *average* or *mean* velocity over the period of time. The actual velocity at any given instant is given by:

$$v = \frac{ds}{dt}$$ **(7.2) actual (instantaneous) velocity**

Therefore if a graph of displacement against time is drawn, the slope of the graph at any point represents the velocity at that point.

Equation 7.2 may also be written:

$$s = \int_{t_1}^{t_2} v\, dt$$ **(7.3) displacement**

Therefore if a graph of velocity against time is drawn, the area under the graph between any two time intervals is the displacement between those time intervals.

7.3 Acceleration (a)

Acceleration is defined as the rate of change of velocity with respect to time. If a body moves in a straight line such that the change in velocity is constant for each time increment, the body will have constant or uniform acceleration given by:

$$a = \frac{v_2 - v_1}{t}$$ **(7.4) constant (uniform) acceleration**

where a = acceleration (m/s^2)
v_2 = final velocity (m/s)
v_1 = initial velocity (m/s)
t = time interval over which the change in velocity occurs (s)

For example, if a body is moving in a straight line with constant acceleration, and changes velocity from 5 m/s to 8 m/s in 6 s, then the acceleration is:

$$\frac{8 - 5}{6} = 0.5 \text{ m/s}^2$$

In equation 7.4, if v_2 is greater than v_1 the acceleration is positive. If, however, v_2 is less than v_1, the acceleration is negative. This is also known as deceleration or retardation.

If the body is not moving so that the change in velocity is the same with each increment of time, the acceleration is not constant and equation 7.4 gives only the average or mean acceleration over the period of time. The actual or instantaneous acceleration at any point in time is given by:

$$a = \frac{dv}{dt}$$ **(7.5) actual (instantaneous) acceleration**

Therefore if a graph of velocity against time is drawn, the slope of the graph at any point represents the acceleration at that point.

Equation 7.5 may also be written:

$$v_2 - v_1 = \int_{t_1}^{t_2} a\,dt$$ **(7.6) change in velocity**

Therefore if a graph of acceleration against time is drawn, the velocity change between any two time intervals, t_1 and t_2, is the area under the graph between the two time intervals.

In practice, very large accelerations are possible in mechanisms; for example, the piston of a motor vehicle may experience maximum accelerations of the order of 10 000 m/s^2. However, the human body can only withstand much smaller accelerations—of the order of about 10 g or about 100 m/s^2 before there is unconsciousness or permanent physical damage. It is interesting to note that vehicles which rely upon friction for traction (as the conventional motor car) cannot accelerate forward at more than g (9.81 m/s^2) or decelerate (brake) at greater than this value and any attempt to do so only causes skidding. This limitation does not apply to jet or rocket-powered vehicles which may accelerate at a faster rate. Higher magnitude deceleration may be obtained by the use of special devices such as reverse thrust or parachutes.

Notes on displacement, velocity and acceleration

1. Displacement, velocity and acceleration are all vector quantities since the direction as well as the magnitude must be known in order that they be fully specified. The scalar version of displacement is called **distance** or **length** and is equal to the magnitude of the displacement. The scalar version of velocity is called **speed** and is equal to the magnitude of the velocity. The scalar version of acceleration is not given a special name.
2. In most instances, both in practice and in problems in mechanics, displacement, velocity and acceleration are taken to be linear unless otherwise specified. Hence "velocity" unqualified means linear velocity, and so on.
3. The word "uniform" is often used instead of "constant" when describing velocity or acceleration. Hence "uniform acceleration" means constant acceleration and "uniform velocity" means constant velocity. The term "uniformly increasing" is often used to describe a linear increase in velocity or acceleration with time.
4. When either the velocity or displacement follow a parabolic law (second order polynomial), the area under the curve may be determined simply from the fact that the

area enclosed by a parabola is 2/3 or 1/3 of the surrounding rectangle. This is demonstrated in example 7.3.

5. When either the velocity or acceleration is not constant, the average or mean value may be obtained by division of the total change by the total time. For example, a change in velocity from 10 to 20 m/s in 5 s, gives an average or mean acceleration of $(20 - 10)/5 = 2\ \text{m/s}^2$. However, it is incorrect to average different instantaneous values of the velocity or acceleration to obtain an average or mean value because the time base is different. This is made clear in example 7.1.

Example 7.1

A vehicle travels 120 km from A to B at 60 km/h and makes the return journey from B to A at 90 km/h.

Determine the average speed.

Solution

$$\text{Outward journey } t = \frac{120}{60} = 2\text{ h}$$

$$\text{Return journey } t = \frac{120}{90} = 1.\dot{3}\text{ h}$$

$$\text{Total time } t = 3.\dot{3}\text{ h}$$

$$\text{Total distance } s = 120 + 120 = 240\text{ km}$$

$$\text{Therefore average speed } \bar{v} = \frac{s}{t} = \frac{240}{3.\dot{3}} = \mathbf{72\ km/h}$$

Notice that by averaging an incorrect result is obtained, i.e.

$$\bar{v} = \frac{60 + 90}{2} = 75\text{ km/h}$$

7.4 Sign convention for displacement, velocity and acceleration

Since displacement must be specified relative to a frame of reference, displacement may be positive or negative according to the change in position. For example, consider the oscillating spring shown in Figure 7.2.

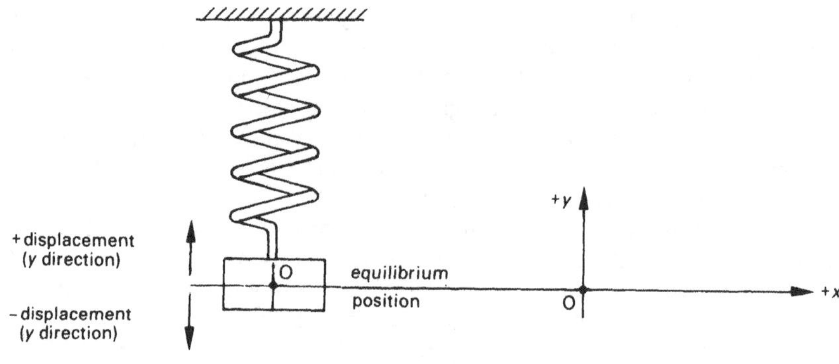

Fig. 7.2 *Displacement may be positive or negative*

In the usual convention, the upward vertical displacement from the equilibrium position is positive and downward vertical displacement below this position is negative. It is important to note that the **direction of movement** does not influence the sign of the displacement, and although the mass may be moving down it will still have a positive displacement if above the equilibrium position.

The direction of movement does influence the sign of the velocity. If the mass is moving in an upward direction, the velocity will be positive regardless of the position of the mass relative to the equilibrium position. Similarly, if the mass is moving downward, the velocity is negative.

The sign of the acceleration depends upon the direction of the change in velocity and *does not depend* upon the position of the mass or the direction of the velocity (direction of movement). When the velocity is decreasing, the acceleration is negative (deceleration).

A summary of the sign convention for displacement, velocity and acceleration is shown in Figure 7.3.

A final point: It is not necessary to maintain the same frame of reference throughout provided that consistency is maintained in specifying signs relative to the frame of reference and the frame of reference is clearly indicated.

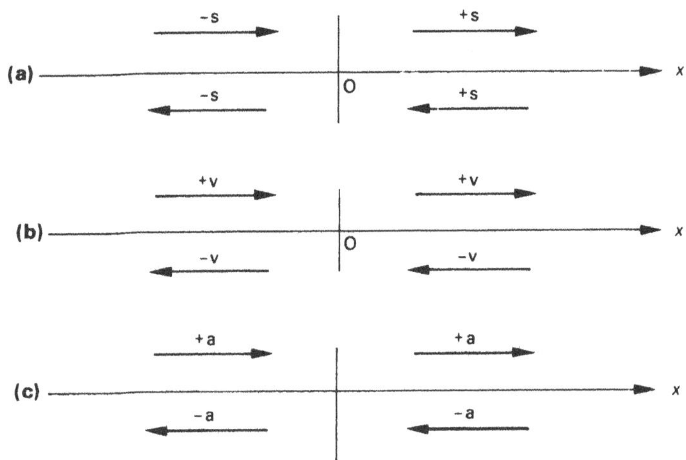

Fig. 7.3 *Signs of (a) displacement, (b) velocity and (c) acceleration*

7.5 Changes in velocity and acceleration

Velocity and acceleration may have negative or zero values but they cannot be infinite. The upper limit of velocity in the universe is that of a beam of light, about 300 000 km/s, but for motion with projectiles, rockets and jet aircraft, the maximum that has been achieved in the earth's atmosphere is about 10 000 km/h. There is no theoretical upper limit to acceleration but, because of Newton's law $F = ma$, the larger the acceleration, the larger the force, and the magnitude of the force which can be generated per unit mass places the restriction on the magnitude of the acceleration which can be achieved in practice. It is, of course, impossible for a mass to change velocity instantaneously since this would require an infinite acceleration as is illustrated in Figure 7.4.

Somewhat paradoxically it will be noted from Figure 7.4 that it is possible to have an instantaneous change in acceleration. A practical illustration of this is when a body is held in the earth's gravitational field and released. At the moment of release, the velocity is zero but the acceleration changes instantaneously from zero to 9.81 m/s^2. Furthermore, there is no reason why a mass cannot have velocity when the displacement is zero. It all depends upon the location of the frame of reference and occurs frequently with vibratory motion where the maximum velocity usually occurs at the point of zero displacement and maximum acceleration at the point of zero velocity.

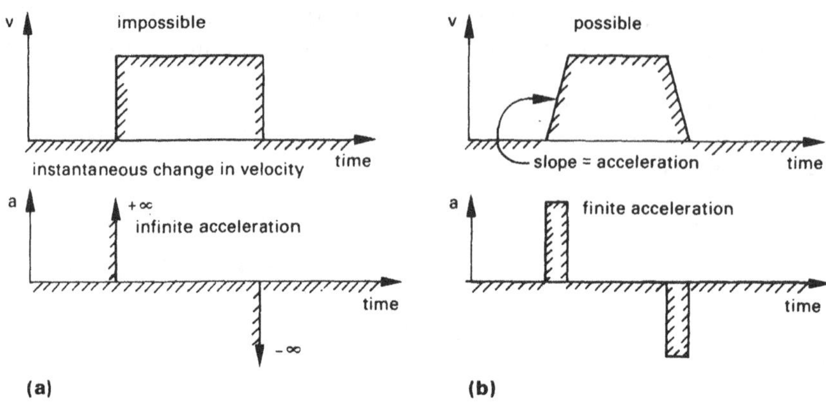

Fig. 7.4 *Instantaneous change in velocity: (a) is impossible*

Example 7.2

The displacement of a vehicle travelling in a straight line is given by:

$$s = 2t^2 + 5t + 2$$

where s is in m and t in s.

Draw motion diagrams between time $t = 0$ and $t = 5$ s and hence determine:

(a) velocity at time $t = 3$ s

(b) displacement between times $t = 1$ s and $t = 4$ s

(c) acceleration at time $t = 4$ s

(d) change in velocity between time $t = 1$ s and $t = 4$ s.

Solution

Since $v = \frac{ds}{dt} = \frac{d}{dt}(2t^2 + 5t + 2)$

$\therefore v = 4t + 5$ (m/s)

Since $a = \frac{dv}{dt}$, $a = 4$ m/s^2 (for any t)

Tabulating:

t (s)	0	1	2	3	4	5
s (m)	2	9	20	35	54	77
v (m/s)	5	9	13	17	21	25

The motion diagrams may now be drawn (Fig. 7.5).

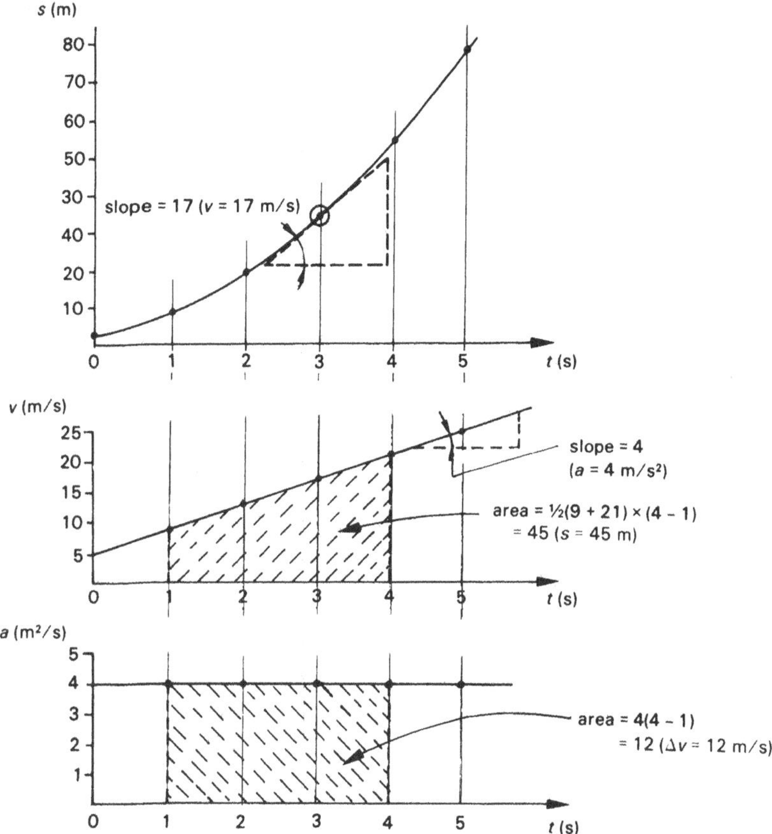

Fig. 7.5

Hence it may be seen:

(a) At time $t = 3$ s, $v =$ **17 m/s** (which is also the slope of the displacement-time diagram at this point.

(b) At time $t = 1$ s, $s = 9$ m and at time $t = 4$ s, $s = 54$ m. Therefore displacement between times $t = 1$ s and $t = 4$ s, = **45 m** (which is also the area under the velocity-time diagram between times $t = 1$ s and $t = 4$ s).

(c) The acceleration at time $t = 4$ s is the same as at any other time since a = constant and is equal to **4 m/s²**, (which is also the slope of the velocity-time diagram).

(d) At time $t = 1$ s, $v = 9$ m/s and at time $t = 4$ s, $v = 21$ m/s. Therefore the change in velocity between times $t = 1$ s and $t = 4$ s, = **12 m/s** (which is also equal to the area under the acceleration-time diagram between times $t = 1$ s and $t = 4$ s). Since the acceleration is constant, $a = \dfrac{v_2 - v_1}{t} = \dfrac{12}{3} = 4\ \text{m/s}^2$ which checks.

Example 7.3

A vehicle starts from rest at A and accelerates at a uniformly increasing rate from zero at A to 3.6 m/s^2 at B. The elapsed time is 4 s. The vehicle then moves at constant velocity for 36 m to C where the brakes are applied and it decelerates uniformly at 2.4 m/s^2 until it stops at D.

Draw motion diagrams and determine the total elapsed time and displacement.

Solution

The key concepts are "uniformly increasing" acceleration which means that the acceleration is increasing at a linear rate and "uniform" deceleration which means the deceleration is constant. The motion diagrams are shown in Figure 7.6 with the calculations shown to the right of the figure.

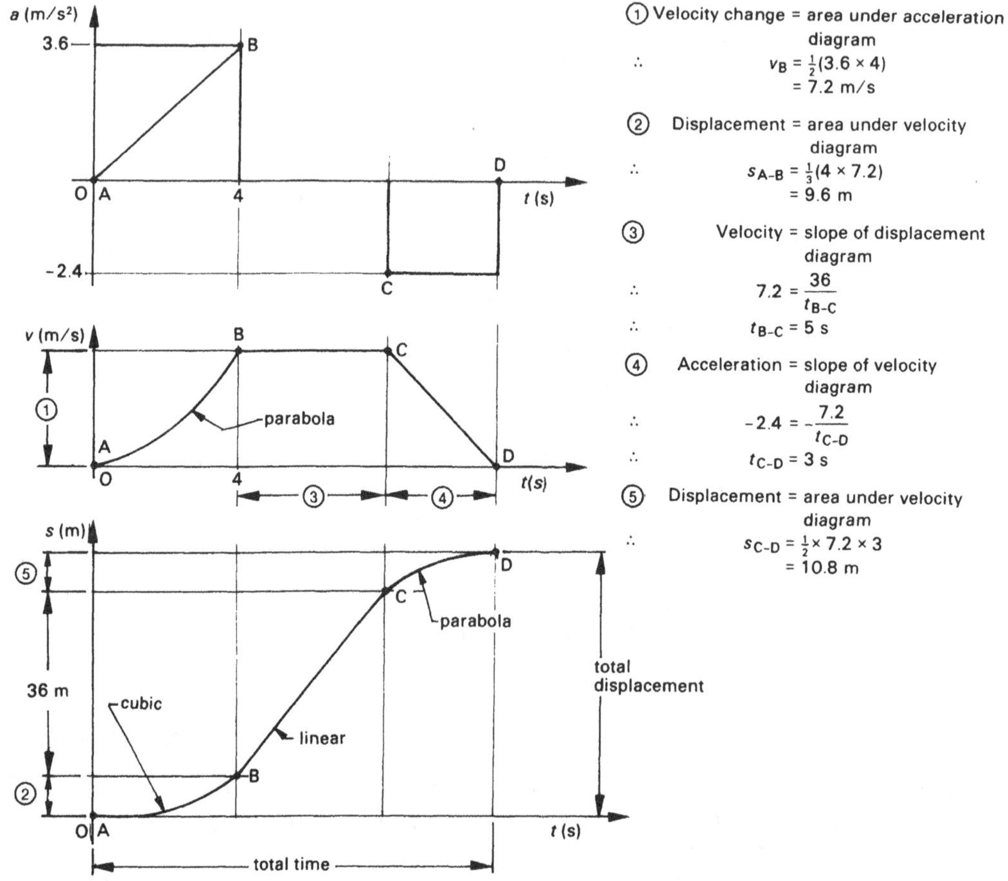

Fig. 7.6

Therefore it is seen that:

total elapsed time is 4 + 5 + 3 = **12 s**

total displacement is 9.6 + 36 + 10.8 = **56.4 m.**

7.6 Equations for common modes of motion

The motion diagram method is a general method and may be used for any problems involving the motion of bodies. However, several modes of motion are very common and are often more conveniently dealt with by the use of equations. These motion modes are:

1. constant velocity
2. constant acceleration
3. simple harmonic motion.

Equations for constant velocity and constant acceleration will now be developed. The case of simple harmonic motion is most applicable to the mechanics of vibrating bodies and is dealt with separately in Chapter 10.

The motion diagrams for constant velocity and constant acceleration motion are shown in Figure 7.7.

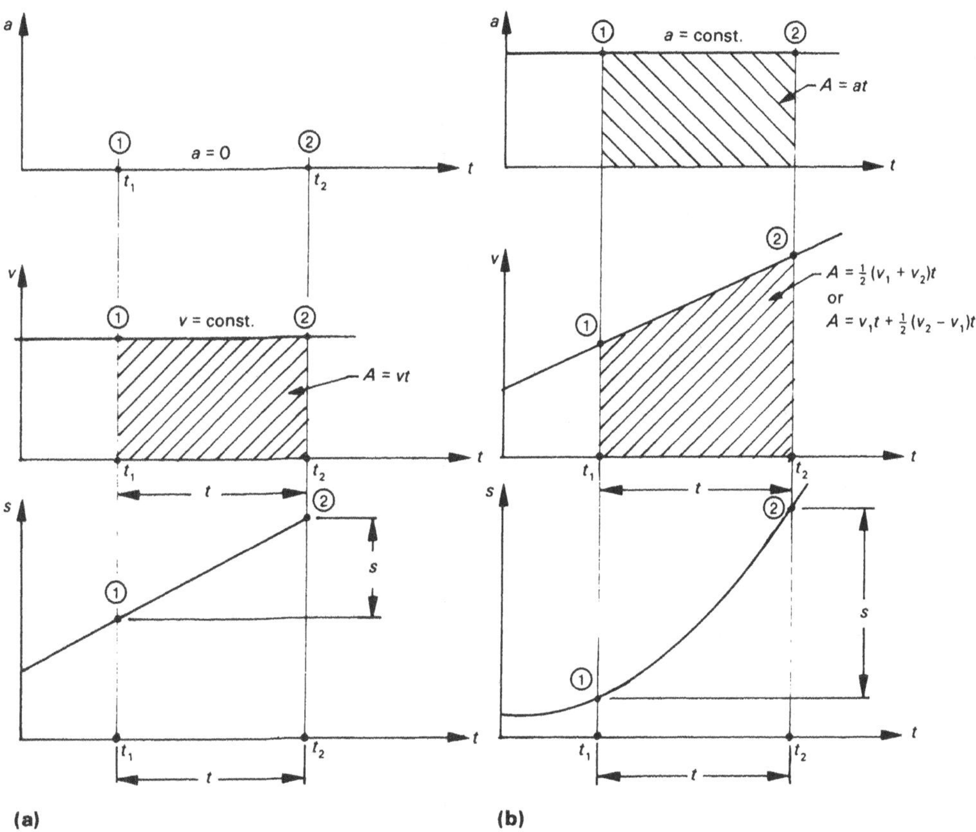

Fig. 7.7 *Motion diagrams for motion with (a) constant velocity and (b) constant acceleration*

Constant velocity motion (Fig. 7.7(a))

Velocity:

$$v_1 = v_2 = v \text{ (constant)}$$

Acceleration:

$a = 0$

Displacement:

Equation 7.1 may be transposed:

$$\boxed{s = vt}$$ **(7.1) displacement**

Constant acceleration motion (Fig. 7.7(b))

Acceleration:

$a_1 = a_2 = a$ (constant)

Velocity:

Equation 7.4 may be transposed:

$$\boxed{v_2 = v_1 + at}$$ **(7.4) velocity**

Displacement:

Displacement = area under velocity diagram (between t_1 and t_2)

$$A = \tfrac{1}{2}(v_1 + v_2)\,t$$

$\therefore$

$$\boxed{s = \tfrac{1}{2}(v_1 + v_2)\,t}$$ **(7.7) displacement**

An alternative equation may be derived for displacement since the area under the velocity diagram between t_1 and t_2 may also be written:

$$A = v_1 t + \tfrac{1}{2}(v_2 - v_1)\,t$$

But from equation 7.4,

$$v_2 - v_1 = a\,t$$

$\therefore$

$$\boxed{s = v_1 t + \tfrac{1}{2}at^2}$$ **(7.8) displacement**

A further equation may be derived by squaring equation 7.4:

$$v_2^2 = v_1^2 + 2\,v_1 at + a^2 t^2$$
$$= v_1^2 + 2\,a\,(v_1 t + \tfrac{1}{2}at^2)$$

$$\boxed{v_2^2 = v_1^2 + 2\,as}$$ **(7.9) velocity**

Notes

1. It will be seen that there is only one equation for motion with constant velocity and four equations for motion with constant acceleration. These four equations allow solution to all problems.
2. There are five variables involved, three of which must be known in order to solve for the other two which are unknown. The variables are: v_1, v_2, a, t, s.
3. Constant velocity motion may be treated as a special case of constant acceleration motion with $a = 0$. The equations for constant acceleration motion are still valid.

Example 7.4

A vehicle accelerates from rest with a constant acceleration of 2 m/s^2 for 5 s after which it travels with constant velocity for 10 s.

Determine the final velocity and total displacement.

Solution

The motion consists of two parts: (a) a period of constant acceleration and (b) a period of constant velocity.

(a) Constant acceleration

$$v_2 = v_1 + at = 0 + 2 \times 5 = 10 \text{ m/s}$$
$$s = v_1 t + \tfrac{1}{2}at^2 = 0 + \tfrac{1}{2} \times 2 \times 5^2 = 25 \text{ m}$$

(b) Constant velocity

$$s = vt = 10 \times 10 = 100 \text{ m.}$$

Therefore final velocity = **10 m/s**

total displacement = 100 + 25 = **125 m.**

Computer solution

The computer program MOTION 1 was written to solve problems involving motion with either constant acceleration or constant velocity (the latter being a special case with acceleration zero and final velocity = initial velocity).

The program operates with the five variables: initial velocity, final velocity, acceleration, elapsed time and displacement.

The program is listed in Appendix 3.8 as well as the solution to example 7.4, parts (a) and (b).

Notes

1. The two unknown variables are input as X.
2. Variables may have zero value.
3. The program computes the two unknown variables for every possible combination of variables (actually there are 10) and outputs all variable values.
4. Velocities are in m/s and must be converted (if necessary) from km/h by dividing by 3.6. Displacements are in metres and time in seconds.
5. Output values are rounded off to two decimal places.

7.7 Vector addition of motion

So far the motion of bodies has been considered only where there has been a single component of each motion variable. However, in many practical situations several components may act together. For example, a projectile launched at an angle to the horizontal will have a component of motion in the horizontal direction as well as a component of motion in the vertical direction.

Similarly, a boat moving across a river in which the tide is flowing also has two components of motion relative to the ground. In these cases two principles are evident:

1. The component of motion in any direction is not affected by the component of motion in a perpendicular direction. This follows because a vector has no component in a direction perpendicular to itself.

2. The resultant motion at any point is the vector sum of the motion components acting at that point. That is, displacement, velocity and acceleration may be treated the same way as forces have been treated in order to obtain the resultant.

Example 7.5

A boat moves in a NE direction at 15 km/h relative to a body of water flowing due west at 5 km/h.

Determine the velocity of the boat relative to the ground.

Solution

This problem may be solved by graphical vector addition of the two velocities as shown in Figure 7.8.

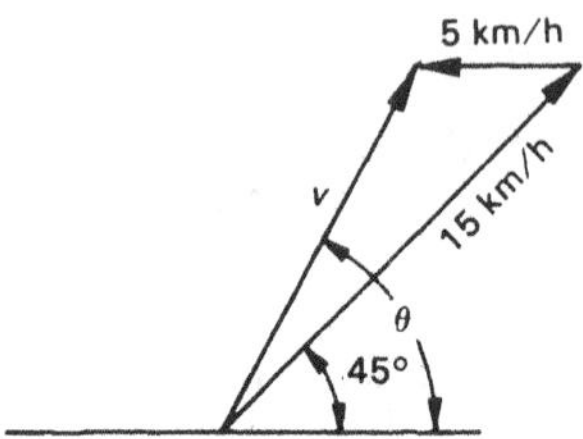

Fig. 7.8

By scaling from the diagram it is found that:

resultant velocity v = **12 km/h @ 62°.**

Alternatively an analytical method may be used:

Velocity	*Horizontal component*	*Vertical component*
15 km/h 5 km/h	15 cos 45° = 10.61 −5.0	15 sin 45° = 10.61 0
Resultant	5.61	10.61

Hence $\tan\theta = \dfrac{10.61}{5.61}$ $\therefore\ \theta =$ **62.1°**

and $\dfrac{10.61}{v} = \sin\theta$ $\therefore\ v =$ **12 km/h.**

Example 7.6

A projectile is launched horizontally and moves with constant horizontal velocity 120 m/s.

Determine the magnitude and direction of the resultant velocity of the projectile after a time lapse of 5 s if air resistance is neglected.

Solution

In the vertical direction after 5 s:

$$v_2 = v_1 + a\,t$$

$$\therefore\ v_2 = 0 + 9.81 \times 5$$

$$= 49 \text{ m/s}$$

The components of the velocity are shown in Figure 7.9.

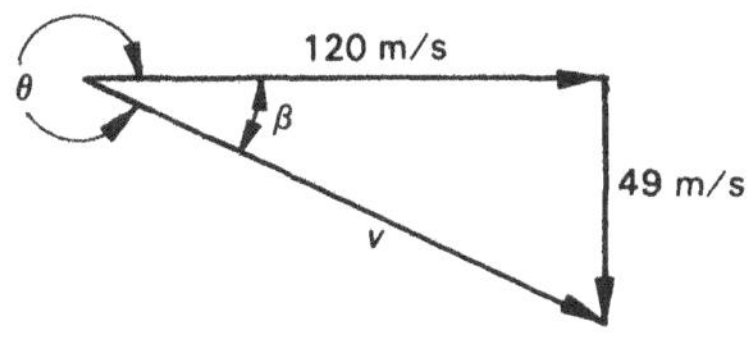

Fig. 7.9

Therefore v = **130 m/s @ 337.8°** (β = 22.2°)

7.8 Relative motion

The case of vector addition of motion just considered arises when the resultant motion of a body is *dependent* upon several motion components which act simultaneously upon it. That is, the frame of reference is *fixed* (usually the ground).

Call this type 1.

Now consider the case of motion where one body moves *independently* of another body which is also moving and it is desired to obtain the motion of one of the bodies relative to the other. In this case the frame of reference is *moving* and is the body to which the motion of the other body will be relative.

Call this type 2.

The difference between the two cases is illustrated in the following examples:

1. A man walks due west at 10 km/h in a train which is moving due east at 40 km/h The velocity of the man relative to the ground (fixed frame of reference) is obtained by vector addition and is:

   $v = +40 - 10 = +30$ km/h (+ due E)

2. The man is walking on the ground instead of in the train. The velocity of the train relative to the man (moving frame of reference is):

   $v = +40 + 10 = +50$ km/h (+ due E)

The two cases are illustrated in Figure 7.10(a) and (b).

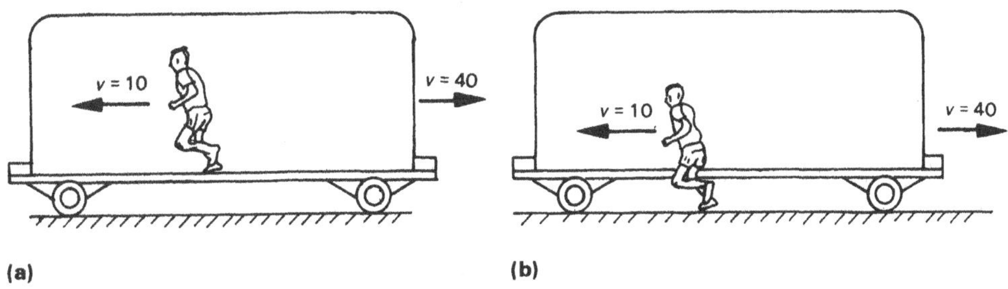

Fig. 7.10 *Relative motion*

A general rule may be stated for determining relative motion of one body to another (case 2):

In order to determine the motion of body B relative to body A, reverse the motion of A and then perform a vector addition.

If this rule is applied to case 2, the velocity of the man is reversed and thus becomes + 10 km/h which when added to the velocity of the train (+40 km/h) gives the correct result of +50 km/h (due E).

It is also clear that another general principle of relative motion is:

The motion of body A relative to body B is equal and opposite to the motion of body B relative to body A.

In case 2, the velocity of the man relative to the train is −50 km/h (that is due W). This result may also be obtained from the first principle of relative velocity, by reversing the velocity of the train (−40km/h) and adding to the velocity of the man (−10 km/h) gives −50 km/h (due W).

Example 7.7

Rain falls vertically at 50 km/h.

Determine the velocity of the rain relative to a car when the car is moving horizontally at (a) 40 km/h, (b) 80 km/h.

Solution

Because the velocity of the car is positive it must be moving in the x direction. The two situations are illustrated in Figure 7.11!

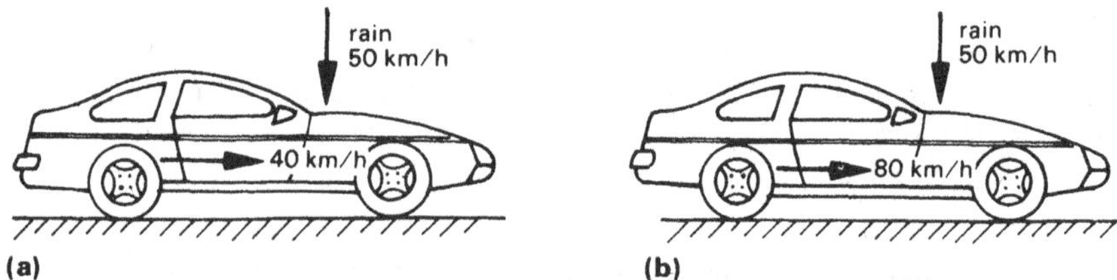

Fig. 7.11

Since it is desired to obtain the velocity of the rain relative to the car which is also moving. this is clearly an example of relative motion type 2.

Applying the principle of relative motion, reverse the velocity of the car and then add the velocities vectorially. This has been done in Figure 7.12.

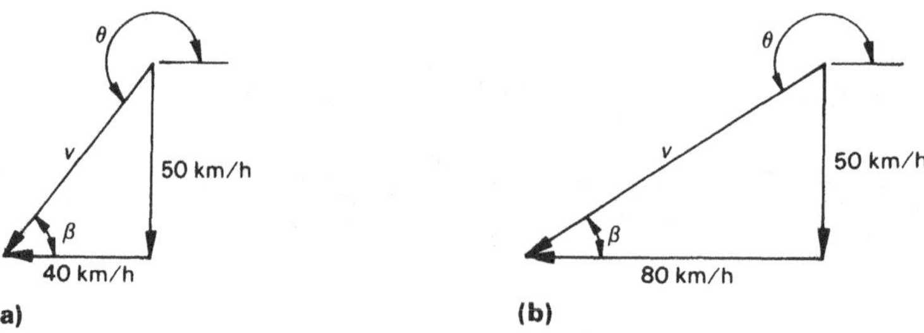

Fig. 7.12

Hence the velocity of the rain relative to the car is:

(a) v = **64 km/h @ 231.3°** (β = 51.3°)

(b) v = **94.3 km/h @ 212°** (β = 32°)

Example 7.8

A yacht moves NW at 10 km/h. The wind is blowing from the east at 25 km/h.

Determine the apparent wind (that is the velocity of the wind relative to the yacht).

Solution

The actual velocities are shown in Figure 7.13. These are the velocities relative to a fixed frame of reference (the earth).

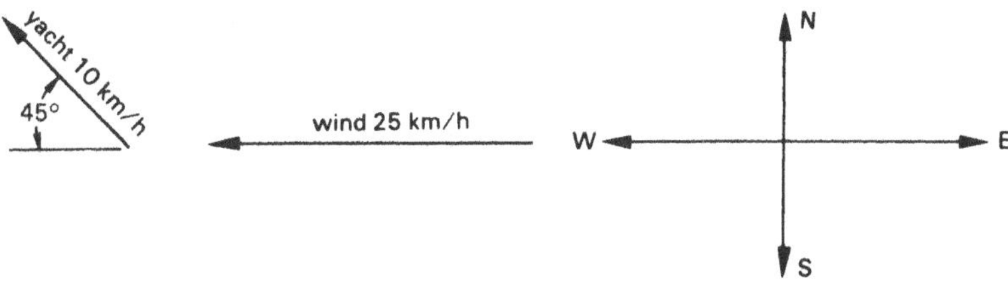

Fig. 7.13

Since the required frame of reference is the yacht which is moving, this is another example of relative motion with a moving frame of reference (type 2).

Reversing the velocity of the yacht and performing a vector addition, the result is shown in Figure 7.14.

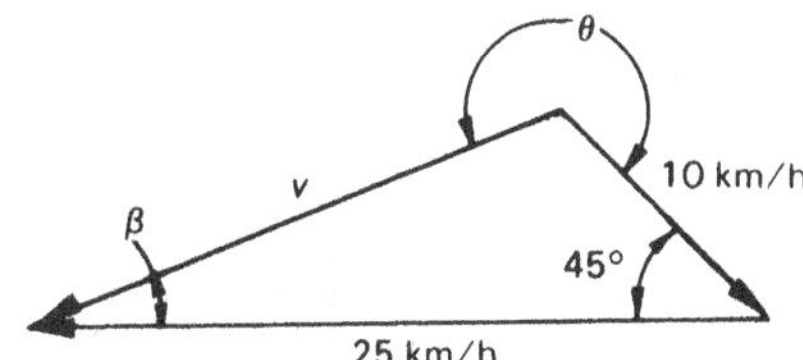

Fig. 7.14

Hence the relative velocity of the wind to the yacht (apparent wind) is:

v = **19.3 km/h @ 201.5°** (β = 21.5°)

Computer solution

Displacement, velocity and acceleration are all vector quantities in the same way that force is a vector quantity. Hence the computer program previously developed for vector addition of forces (FORCE 1) may be readily modified to solve problems involving vector addition of motion components or relative motion of one body with another. This program was named MOTION 2 and is listed in Appendix 3.9. The solution to examples 7.5, 7.6, 7.7 and 7.8 are given below the program listing and should be self-explanatory.

Notes

1. The same sign convention is used for motion components as was used for forces, i.e.
   positive magnitude: vector points outward from the origin
   positive direction: angle is measured anticlockwise from the x axis.

2. The program may be used with any of the three motion components, velocity, acceleration or displacement.
3. Any units may be used for any of the motion components provided that consistency is maintained.

7.9 Rotational motion

In mechanics, rotational motion (or angular motion) occurs very frequently. The rotational equivalents of displacement, velocity and acceleration are given below.

Angular displacement (θ)

The angular displacement may be measured in degrees, radians or revolutions. In mechanics formulas, radians are used. The conversion is:

$$\theta \text{ (radians)} = \theta \text{ (revs)} \times 2\pi = \theta^\circ \times \frac{\pi}{180}$$

Angular velocity (ω)

The angular velocity may be measured in rad/s or revolutions per second (rps) or revolutions per minute (rpm). Mechanics formulas use rad/s. If the rotational velocity is given in rpm, the conversion is:

$$\omega = \frac{2\pi N}{60} = \frac{\pi N}{30}$$

where N = rotational velocity in rpm.

Angular acceleration (α)

The angular acceleration is measured in rad/s^2.

Kinematics of rotation

All the formulas for linear motion may be converted to rotational motion formulas simply by substituting the correct rotational terms. For example, the linear formula:

$$v_2 = v_1 + at$$

for rotational motion becomes:

$$\omega_2 = \omega_1 + \alpha t$$

Note that the time (t) in seconds is the same in both cases.

The following relationships between linear and rotational motion are also most useful:

$$v = r\omega$$ **(7.10) linear and rotational velocity**

$$a = r\alpha$$ **(7.11) linear and rotational acceleration**

where r = radius about the centre of rotation (centre of curvature).

Notes

1. Motion diagrams or formulas may be used to solve problems involving rotational motion just as for linear motion.

2. If a wheel or disc rotates with constant angular velocity (ω) the linear velocity is not constant but varies with the radius, reaching a maximum value at the outside of the rim. Similarly if the wheel or disc accelerates with constant angular acceleration (α), the linear acceleration varies with radius and reaches a maximum on the outer periphery.

Example 7.9

The tyres of a car have an outside diameter of 500 mm. The car accelerates from rest at a constant rate of 3 m/s^2 for a distance of 120 m.

Assuming no slip on the road surface, determine:

(a) angular acceleration of the wheels
(b) maximum angular velocity of the wheels in rad/s and rpm.

Solution

(a) Using the conversion $a = r\alpha$:

$$\alpha = \frac{a}{r} = \frac{3}{0.25} = \mathbf{12\ rad/s^2}$$

(b) In order to determine the final velocity of the car, apply the formula:

$$v_2^2 = v_1^2 + 2as$$

$$v_2^2 = 0 + 2 \times 3 \times 120$$

$$\therefore\ v_2 = 26.83\ \text{m/s}$$

Using the conversion $v = r\omega$:

$$\therefore\quad \omega = \frac{v}{r} = \frac{26.83}{0.25} = \mathbf{107.3\ rad/s}$$

$$\text{Since } \omega = \frac{\pi N}{30} \quad \therefore\ N = \frac{30 \times 107.3}{\pi} = \mathbf{1025\ rpm}$$

Example 7.10

A shaft rotates at 5200 rpm and is brought to rest with uniform deceleration in 30 s.

Determine:

(a) angular deceleration
(b) number of revolutions during the deceleration period.

Solution

(a) $\omega_1 = 5200\ \text{rpm} = 5200 \times \frac{\pi}{30} = 544.54\ \text{rad/s}$

$\omega_2 = 0 \qquad t = 30\ \text{s}$

Using the formula: $\omega_2 = \omega_1 + \alpha t \quad (v_2 = v_1 + at)$

$$\therefore\quad \alpha = \frac{\omega_2 - \omega_1}{t}$$

$$\therefore\quad \alpha = -\frac{544.54}{30} = \mathbf{-18.15\ rad/s^2}$$ (negative sign indicates deceleration)

(b) Using the formula:

$$\theta = \omega_1 t + \tfrac{1}{2}\alpha t^2 \quad (s = v_1 t + \tfrac{1}{2}at^2)$$

$$= 544.54 \times 30 - \tfrac{1}{2} \times 18.15 \times 30^2$$

$$= 8169\ \text{rads} = \mathbf{1300\ revs}$$

Example 7.11

A motor shaft increases speed with uniform acceleration from rest to 2400 rpm in 10 s. The shaft rotates at constant speed for 480 revolutions and then decelerates at a constant rate for 160 revolutions until it comes to rest again.

Draw motion diagrams (in terms of revolutions and seconds) and hence determine the total number of revolutions of the shaft and the elapsed time.

Solution

Calling the key points A, B, C and D, the motion diagrams may be drawn as shown in Figure 7.15. Calculated values are shown to the right of the diagram.

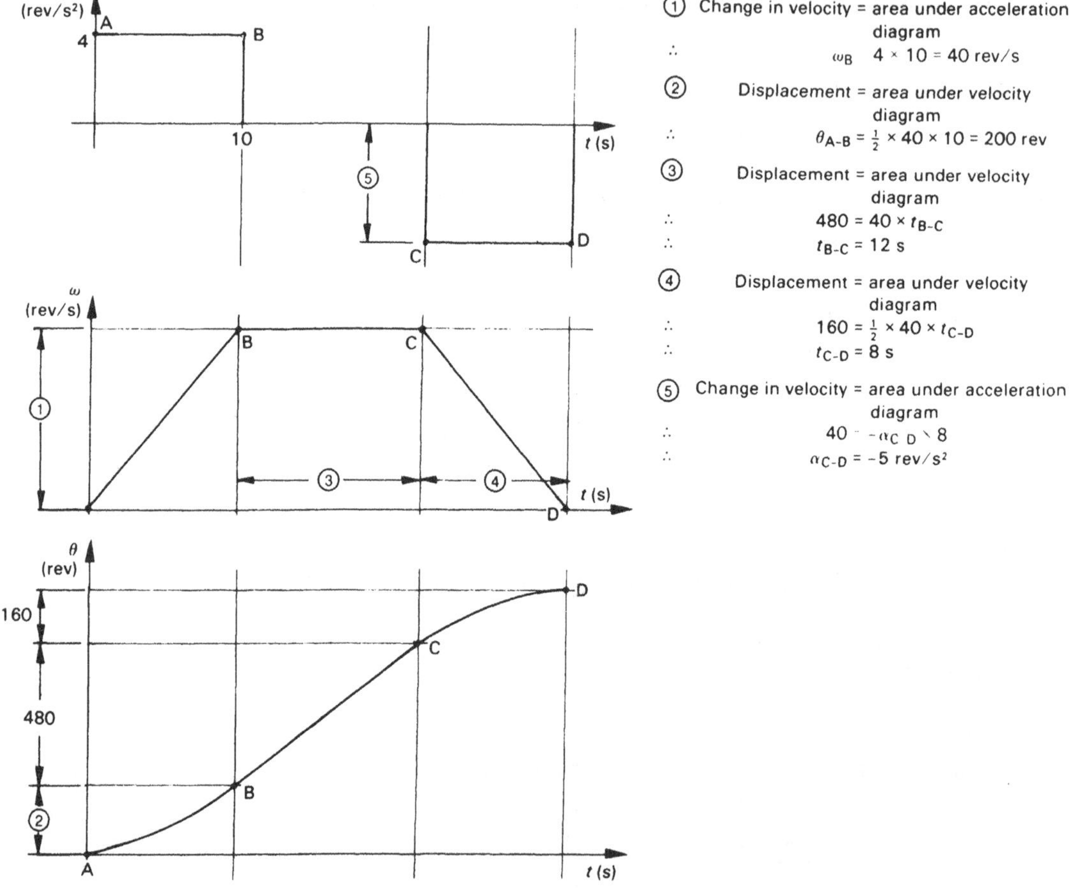

Fig. 7.15

Hence total number of revolutions = 200 + 480 + 160 = **840 revs**

and elapsed time = 10 + 12 + 8 = **30 s.**

Note: This problem could also have been solved by use of the formulas in each of the three phases of motion. This is a suitable student exercise.

Computer solution

The program MOTION 1 discussed previously for linear motion may also be used for rotational motion. Velocities must be input in rad/s (the conversion from rpm being to divide by 9.5493 $(30/\pi)$).

Displacement must be input as radians (the conversion from revolutions being to multiply by 6.2832 (2π)).

Output velocities and displacements are given in both radians and revolutions (rounded off to two decimal places).

The solution to example 7.10 is given below the program listing in Appendix 3.8.

7.10 Centripetal acceleration

A body moving in a curved path with constant speed is subject to an acceleration. This follows because a body moving without any acceleration will move only in a straight line. Hence if the body follows a curved path there must be an acceleration directed toward the centre of curvature in order to cause the change in direction.

Of particular interest is the case of a body moving in a circular path. In this case the centre of curvature remains a fixed point and there is an acceleration directed toward the centre known as the **centripetal acceleration**.

An expression for the centripetal acceleration may be derived as follows:

Consider a particle moving in a circular path of radius r with constant angular velocity ω and constant linear speed v. Two positions of the particle at two points separated by a short time interval is as shown in Figure 7.16.

Fig. 7.16 *Circular motion*

From the vector diagram it is clear that there is a change in velocity Δv and this change in velocity is directed toward the centre 0.

If the angle $\Delta\theta$ is small:

$$\frac{\Delta v}{v} = \Delta\theta \text{ (rads)}$$

but $\Delta\theta = \omega\Delta t$

$$\therefore \quad \frac{\Delta v}{v} = \omega\Delta t$$

but $v = r\omega$

$$\therefore \quad \frac{\Delta v}{\Delta t} = r\omega^2$$

In the limit as $\Delta t \rightarrow 0, \frac{\Delta v}{\Delta t} \rightarrow \frac{dv}{dt} = a$

$$\therefore \quad \boxed{a_c = r\omega^2} \qquad \textbf{(7.12) centripetal acceleration}$$

Since $v = r\omega$ or $\omega = \frac{v}{r}$, this equation may also be written:

$$a_c = r\frac{v^2}{r^2}$$

$$\boxed{a_c = \frac{v^2}{r}} \qquad \textbf{(7.13) centripetal acceleration}$$

Example 7.12

A car travelling at 72 km/h turns a bend of radius 80 m.

Determine the rotational velocity of the car and the centripetal acceleration.

Solution

The linear velocity in m/s is $\frac{72}{3.6} = 20$ m/s

Now $v = r\omega \quad \therefore \omega = \frac{v}{r} = \frac{20}{80} = \mathbf{0.25\ rad/s}$

From equation 7.12,

$$a_c = r\omega^2 = 80 \times 0.25^2 = \mathbf{5\ m/s^2}$$

From equation 7.13,

$$a_c = \frac{v^2}{r} = \frac{20^2}{80} = \mathbf{5\ m/s^2}$$

7.11 Coriolis acceleration

When a body which is rotating is also moving radially at the same time, there is another acceleration present known as the Coriolis acceleration.

Consider a block sliding outward with radial velocity of constant magnitude v on an arm rotating with constant angular velocity ω as shown in Figure 7.17.

At a given instant, the block is at a radius r_1, a short time later, Δt, it is at radius r_2. The magnitude of the radial velocity v_R remains constant but the magnitude of the tangential velocity v_T changes, (increasing if the block is sliding outward and decreasing if the block is sliding inward). This change in tangential velocity implies an acceleration which is in fact the Coriolis acceleration.

The Coriolis acceleration has two components:

1. the component due to the change in *magnitude* of the tangential velocity, and
2. the component due to the change in *direction* of the radial velocity.

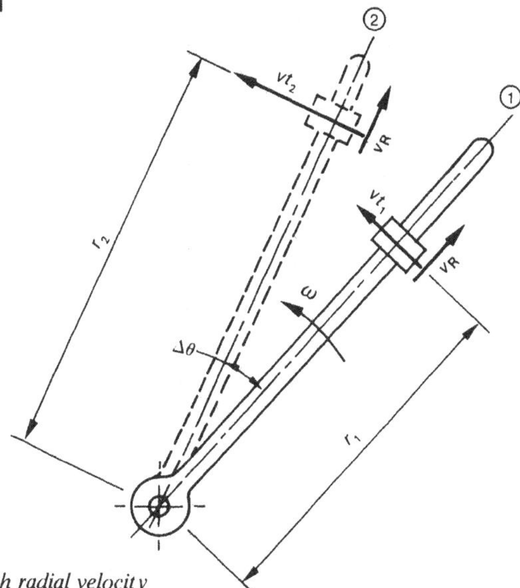

Fig. 7.17 *Rotating block with radial velocity*

Note: The component due to the change in direction of the tangential velocity is not considered in the Coriolis acceleration because in fact it gives rise to the centripetal acceleration. This is always treated separately since it acts radially whereas the Coriolis acceleration acts tangentially.

Component 1 is given by:

$$\Delta v = r_2\omega_2 - r_1\omega_1 = (r_2 - r_1)\,\omega = \Delta r\omega$$

Dividing by Δt:

$$\frac{\Delta v}{\Delta t} = \frac{\Delta r}{\Delta t}\,\omega$$

But in the limit as $\Delta\theta \to 0$,

$$\frac{dv}{dt} = \frac{dr}{dt}\,\omega$$

$$\therefore a = v_R\omega \quad \text{because } \frac{dr}{dt} = v_R$$

Component 2 is given by:

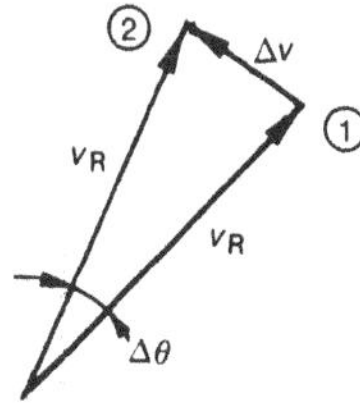

Fig. 7.18 *Vector diagram for radial velocity change*

$$\Delta v = v_R\,\Delta\theta$$

Dividing by Δt:

$$\frac{\Delta v}{\Delta t} = v_R\,\frac{\Delta\theta}{\Delta t}$$

In the limit as $\Delta\theta \rightarrow 0$,

$$\frac{dv}{dt} = v_R\omega \quad \text{because} \quad \frac{d\theta}{dt} = \omega$$

$$\therefore a = v_R\omega$$

Thus it will be seen that the two components are of equal magnitude and are equal to $v_R\omega$. Thus the total tangential acceleration (the Coriolis acceleration), is:

$$a_{COR} = 2\, v_R\omega \qquad \textbf{(7.14) Coriolis acceleration}$$

Note that the direction of the Coriolis acceleration is in the direction of the tangential velocity when the body moves outward and in the opposite sense when the body moves inward.

Example 7.13

A block moves outward with velocity 2.5 m/s on an arm rotating anticlockwise at 200 rpm.

Determine the magnitude and direction of the acceleration of the block when the arm is horizontal and the block is 300 mm from the centre of rotation.

Solution

The Coriolis acceleration is given by:

$$a_{COR} = 2\, v_R\omega \qquad \omega = \frac{\pi \times 200}{30} = 20.94 \text{ rad/s}$$

$$= 2 \times 2.5 \times 20.94$$

$$= 105 \text{ m/s}^2 \text{ acting tangentially in the direction of rotation.}$$

The centripetal acceleration is given by:

$$a_C = r\omega^2$$

$$= 0.3 \times 20.94^2$$

$$= 132 \text{ m/s}^2 \text{ acting toward the centre of rotation.}$$

The resultant acceleration is the vector sum of these two accelerations (Fig. 7.19).

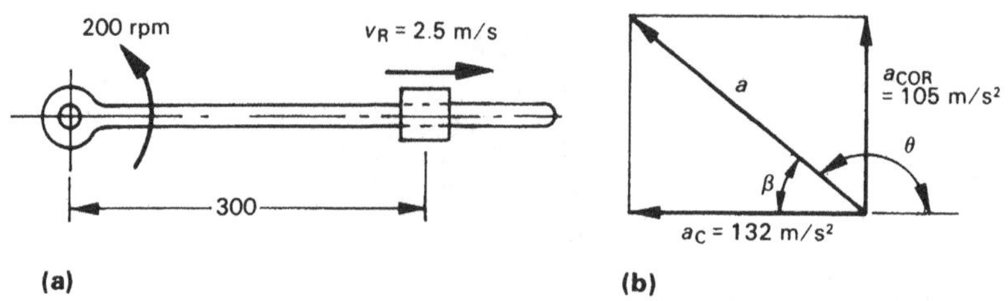

Fig. 7.19 *(a) Configuration diagram. (b) Vector acceleration diagram*

Hence the resultant acceleration:

a = **169 m/s² @ 141.5°** (β = 38.5°)

Example 7.14

A flat disc is accelerated uniformly from rest clockwise to a speed of 60 rpm in 1 s. At the instant when the disc starts to rotate, a small mass at the centre of the disc moves radially outward with constant acceleration to a position 400 mm from the centre in time 0.7 s.

Determine the magnitude and direction of the acceleration of the mass at time $t = 0.7$ s if at this time it is in a horizontal position to the right of the centre.

Solution

The mass has four acceleration components: (a) tangential, (b) radial, (c) Coriolis and (d) centripetal. Each component must be determined in turn and then combined to determine the resultant acceleration.

(a) Tangential acceleration

$$\omega_2 = \omega_1 + \alpha t \qquad \omega_2 = \frac{\pi \times 60}{30} = 6.283 \text{ rad/s}$$

$$\therefore\ 6.283 = 0 + \alpha \times 1$$

$$\therefore\ \alpha = 6.283 \text{ rads/s}^2$$

$$a = r\alpha = 0.4 \times 6.283 = 2.51 \text{ m/s}^2 \downarrow$$

(b) Radial acceleration

$$s = v_1 t + \tfrac{1}{2} at^2$$

$$\therefore\ 0.4 = 0 + \tfrac{1}{2} \times a \times 0.7^2$$

$$\therefore\ a = 1.633 \text{ m/s}^2 \rightarrow$$

(c) Coriolis acceleration

$$a_{COR} = 2\, v_R \omega$$

The radial velocity is given by:

$$v_R = v_1 + at$$

$$= 0 + 1.633 \times 0.7$$

$$= 1.143 \text{ m/s}$$

$$\therefore\ a_{COR} = 2 \times 1.143 \times 6.283$$

$$= 14.36 \text{ m/s}^2 \downarrow$$

(d) Centripetal acceleration

$$a_C = r\omega^2$$

$$= 0.4 \times 6.283^2$$

$$= 15.8 \text{ m/s}^2 \leftarrow$$

Hence the resultant vertical acceleration is:

$$2.51 \downarrow + 14.36 \downarrow = 16.87 \text{ m/s}^2 \downarrow$$

The resultant horizontal acceleration is:

$$1.63 \rightarrow - 15.8 \leftarrow = 14.17 \text{ m/s}^2 \leftarrow$$

The horizontal and vertical components may now be combined to determine the resultant as shown in Figure 7.20.

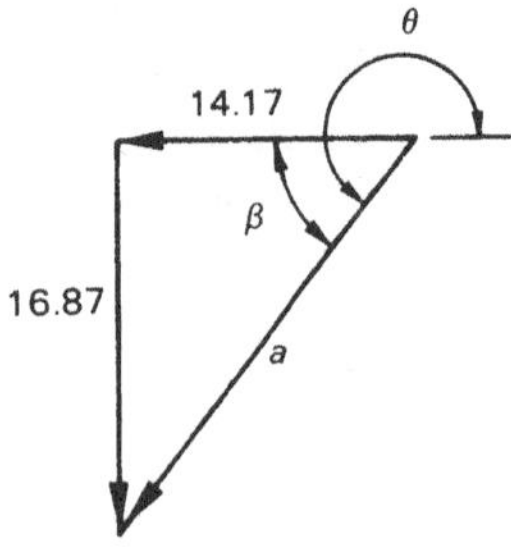

Fig. 7.20

From which $\boldsymbol{a} = \mathbf{22\ m/s^2}$ **@ 230°** ($\beta = 50°$)

Problems

7.1 The displacement of a vehicle travelling in a straight line is given by:

$$s = 3t^2 - 4t + 3$$

where s is in m and t is in seconds.

Draw motion diagrams to scale between time $t = 0$ and $t = 5$ s (marking the scales on the x and y axes). Hence determine:

(a) velocity at time $t = 2$ s
(b) displacement between time $t = 1$ s and 5 s
(c) acceleration at time $t = 3$ s
(d) change in velocity between time $t = 1$ and $t = 5$ s.

(a) 8 m/s (b) 56 m (c) 6 m/s² (d) 24 m/s

7.2 Repeat problem 7.1 for a vehicle whose displacement is given by:

$$s = \frac{t^2}{4} - t + 3$$

(a) 0 (b) 2 m (c) 0.5 m/s² (d) 2 m/s

7.3 A train starts from rest at station A and has a constant acceleration for 30 s to velocity 90 km/h which is maintained for 2 min. The brakes are then applied and the train decelerates at a constant rate for 20 s to come to rest again at station B.

Draw motion diagrams to scale for the journey (marking the scales on the x and y axes). Hence determine:

(a) maximum acceleration
(b) maximum deceleration
(c) distance between stations A and B
(d) total elapsed time for the journey.

(a) 0.833 m/s² (b) − 1.25 m/s² (c) 3.625 km (d) 170 s

7.4 Solve problem 7.3 using the analytical method (equations).

7.5 A valve with total lift 6 mm is opened and closed by a cam such that opening and closing take place at constant velocity. The valve opens in 1 ms and remains open for 0.5 ms then drops to the closed position in 0.75 ms and remains closed for 0.75 ms before the cycle is repeated.

Draw motion diagrams to scale for the complete cycle of the cam (marking the scales on the x and y axes).

Comment on the acceleration diagram.

7.6 Repeat problem 7.5 if a different cam is used such that opening and closing take place with constant acceleration and deceleration for one half of the opening and closing period respectively.

7.7 A vehicle starts from rest at A and is given an acceleration which increases uniformly from 3 m/s^2 at A to 9 m/s^2 at B in time 4 s. The vehicle then travels at constant velocity for 84 m to C at which point it decelerates at constant rate 6 m/s^2 until it comes to rest at D.

Draw motion diagrams to scale for the journey (marking scales on the x and y axes). Mark points A, B, C and D clearly on each diagram. Hence determine:

(a) maximum acceleration
(b) maximum velocity
(c) displacement between A and B, B and C, and C and D
(d) distance between A and D
(e) total elapsed time.

(a) 9 m/s^2 (b) 24 m/s (c) 32 m, 84 m, 48 m (d) 164 m (e) 11.5 s

7.8 A vehicle travelling at 18 km/h, accelerates at a constant rate to 60 km/h in time 5 s.

Write equations for the displacement and velocity of the vehicle (in terms of metres and seconds) and hence determine the velocity and displacement at time t = 3 s.

12 m/s, 25.5 m

7.9 A torpedo is launched at 50 km/h and angle 30° horizontally and to the right of the centre line of a vessel travelling at 20 km/h in direction NW.

Determine the magnitude and direction of the velocity of the torpedo at launch relative to the ground.

68.1 km/h @ 113.45°

7.10 A stone is dropped from a building 20 m in height. A cross wind carries the stone at 30 km/h horizontally during its descent.

If the downward acceleration of the stone = g, determine the time taken for the stone to hit the ground and the velocity of the stone (magnitude and direction) relative to the ground just before impact.

2.02 s, 77.4 km/h @ 67.2°

7.11 Determine the maximum belt speed (in km/h) of the coal conveyor shown in Figure P7.11 so that all of the coal will enter the hopper.

7.45 km/h

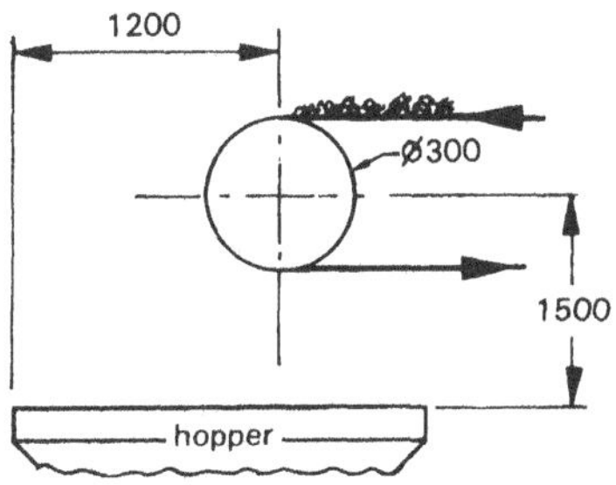

Fig. P7.11

7.12 Rain falls vertically with terminal velocity 90 km/h.

Determine the apparent velocity (magnitude and direction) of the rain relative to a vehicle which travels at 60 km/h:

(a) horizontally
(b) upward on a hill inclined at 10° to the horizontal
(c) downward on a hill inclined at 10° to the horizontal.

(a) 108 km/h @ 56.3° (b) 116.5 km/h @ 49.5° (c) 99.1 km/h @ 63.4°

7.13 A yacht sails at 10 km/h in a westerly direction. The wind is blowing at 20 km/h from the north-east.

Determine the velocity of the wind (magnitude and direction) relative to the yacht.

14.7 km/h @ 73.7° N of E

7.14 A yacht moving at 12 km/h in a north-easterly direction is fitted with a wind speed indicator which shows a wind of 18 km/h blowing from due east relative to the yacht.

Determine the true wind speed and direction.

12.75 km/h @ 41.73° S of E

7.15 A rotor of diameter 800 mm accelerates from rest at a constant rate of 5 rad/s^2. After 6 s a balance mass attached to the edge of the rotor dislodges.

Determine:

(a) velocity of the mass as it leaves the rotor
(b) number of revolutions made by the rotor when the balance mass dislodges
(c) maximum centripetal acceleration of the balance mass.

(a) 12 m/s (b) 14.3 revs (c) 360 m/s^2

7.16 A conveyor belt is driven by sprockets, pitch diameter 420 mm, through a reduction gearbox, ratio 50:1, and an electric motor. When the motor is switched on it takes 2.5 s to reach working speed of 1440 rpm.

Determine:

(a) angular acceleration of the motor
(b) angular acceleration of the sprockets
(c) linear acceleration of the belt
(d) maximum belt speed in km/h.

(a) 60.32 rad/s^2 (b) 1.206 rad/s^2 (c) 0.253 m/s^2 (d) 2.28 km/h

7.17 A train with wheels of rim diameter 700 mm accelerates from rest to 110 km/h in a distance of 450 m.

Assuming no slip of the wheels determine:

(a) linear acceleration of the train
(b) angular acceleration of the wheels
(c) maximum rotational speed of the wheels (in rpm)
(d) centripetal acceleration of a point on the rim of the wheels at speed
(e) velocity (relative to the rails) of a point on the rim of the wheels when this point is:
  (i) in contact with the rails
  (ii) vertically above the contact point with the rails
  (iii) at 90° to the contact point with the rails.

(a) 1.04 m/s^2 (b) 2.96 rad/s^2 (c) 834 rpm (d) 2668 m/s^2
(e)(i) 0 (ii) 61.1 m/s (iii) 43.2 m/s

7.18 The train given in problem 7.17 travels at maximum speed for 1.6 km before the brakes are applied and it decelerates to rest at constant rate in 6.4 s.

Draw motion diagrams in terms of radians and seconds for the wheels for the entire journey and hence determine the number of revolutions made by the wheels and the elapsed time.

977 revs, 88.2 s

7.19 A block slides outward at 8 m/s on an arm rotating anticlockwise at 125 rpm. When the arm is in the vertical position the block is 800 mm from the centre of rotation.

Determine for this position of the block:

(a) resultant velocity relative to the frame of the machine (magnitude and direction)
(b) centripetal acceleration
(c) Coriolis acceleration
(d) resultant acceleration relative to the frame of the machine (magnitude and direction).

(a) 13.2 m/s @ 142.6° (b) 137 m/s^2 (c) 209.4 m/s^2 (d) 250 m/s^2 @ 213.2°

7.20 The arm of a machine is fitted with a sliding block. The arm accelerates uniformly from rest to a speed of 150 rpm in 2 s. At the instant when the block is vertically above the centre of rotation and 300 mm from it, the arm is rotating anticlockwise at 120 rpm and the block is moving outward with velocity 5 m/s and acceleration 2.6 m/s^2.

Determine for this position of the block:

(a) tangential velocity
(b) tangential acceleration
(c) radial acceleration
(d) Coriolis acceleration
(e) centripetal acceleration
(f) resultant acceleration.

(a) 3.77 m/s (b) 2.36 m/s^2 (c) 2.6 m/s^2 (d) 125.7 m/s^2 (e) 47.4 m/s^2 (f) 135.7 m/s^2 @ 199.3°

8

Dynamics of motion

In Chapter 7 the kinematics of motion was treated, that is the motion of bodies was considered without regard to the forces involved. In this chapter, the forces are considered and this is known as **dynamics**. The main distinction between dynamics and statics (which considers only static bodies or structures) is that in dynamics there is an additional force which arises due to the acceleration or deceleration of a mass. This force is described by Newton's first law of motion which states that a body will remain at rest or continue to move with constant velocity in a straight line unless acted upon by an *unbalanced* force.

Newton's law produces some apparently paradoxical conclusions. For example, when a car moves along a level road with constant velocity, the law implies that there is no unbalanced force acting upon it. Yet it is common knowledge that considerable force is required to keep a car moving along level ground and that this force is provided by the engine. The key to this apparent paradox lies in the word "unbalanced". The law does not state that there are *no* forces acting upon the car in uniform motion but that the forces acting upon it are balanced. That is, there is a driving force provided by the engine and this is balanced by the resistance forces due to mechanical and fluid friction. If the car moves up an incline, there is an additional force due to the component of the weight of the car which also acts to oppose the motion and hence an additional force must be provided by the engine.

The first law also implies that a mass moving with constant velocity in a straight line is in fact equivalent to a mass at rest, as far as the forces are concerned, and may be treated by the normal methods of statics. For example, travellers in a soundproof, windowless, carriage moving in a straight line along a level track would have no way of distinguishing whether they were at rest or moving and, indeed, if they were moving, they would not be able to infer the direction of motion. Furthermore, they could conduct no experiment within the carriage which could provide the answer. However, if the carriage were to slow down, speed up or round a bend, the motion would immediately be obvious to them because of the force which results due to the change in velocity (acceleration).

8.1 Accelerating force

Consider the case of a mass acted upon by forces which are not balanced. Newton's second law of motion states that the unbalanced force will produce an acceleration in the direction of the resultant unbalanced force and of magnitude given by:

$$F = ma \qquad (8.1)$$

(8.1) accelerating force

where F = resultant unbalanced force (N)
m = mass of the body (kg)
a = resultant acceleration of the body (m/s^2)

Note that if there is no rotational acceleration of the mass, the resultant force passes through the centre of mass (also known as the centre of gravity CG) which is the point at which the mass could be considered to be concentrated (point mass).

Example 8.1

A mass of 5 kg rests on a horizontal surface. The coefficient of friction is 0.3. A force of 50 N is then applied at an angle of 30° as shown in Figure 8.1.

Determine the magnitude and direction of the acceleration of the mass.

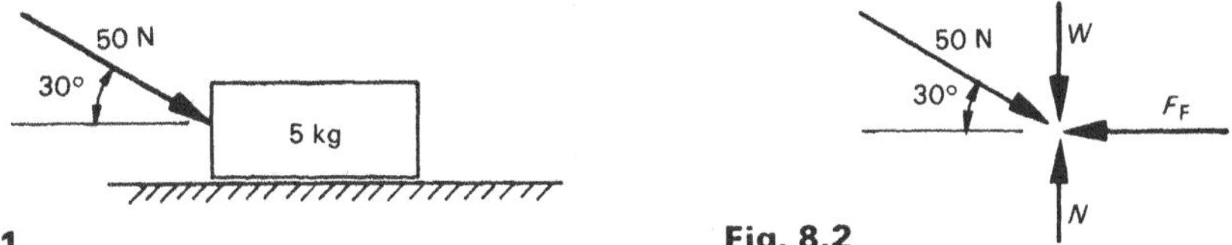

Fig. 8.1 **Fig. 8.2**

Solution

Draw a free body diagram of the forces acting on the mass as shown in Figure 8.2.

Since there will be no motion in the vertical direction, it follows that the vertical forces are balanced.

$$\Sigma F_V = 0$$

$$\therefore -W - 50 \sin 30° + N = 0$$

$$\therefore \quad N = 50 \sin 30° + 5 \times 9.81$$

$$= 74.05 \text{ N}$$

In the horizontal direction there are two forces acting, the component of the 50 N force (positive) and the friction force F_F (negative).

The resultant force F in the horizontal direction is therefore given by:

$$F = 50 \cos 30° - F_F$$

$$\text{But } F_F = \mu N$$

$$= 0.3 \times 74.05 = 22.215 \text{ N}$$

$$\therefore \quad F = 50 \cos 30° - 22.215 = 21.09 \text{ N}$$

From equation 8.1:

$$F = ma$$

$$\therefore \quad a = \frac{F}{m} = \frac{21.09}{5} = \mathbf{4.22 \ m/s^2}$$

Since the resultant force is positive (acts to the right), the acceleration will therefore also be horizontal and to the right.

8.2 Inertia force

An alternative method of dealing with accelerating masses is to consider that the mass has inertia, that is a resistance to acceleration. This implies that there is an inertia force acting in an opposite direction to the acceleration. If the inertia force is included, the forces acting on the mass will then be in equilibrium and may be treated by the methods of statics. This principle is referred to as **d'Alembert's principle**.

Using F_i to denote the inertia force, it is evident that:

$F_i = ma$

and the direction of F_i is opposite to the direction of the acceleration.

Example 8.2
Solve example 8.1 using inertia force.

Solution
The free body diagram of the forces acting upon the mass are shown in Figure 8.3.

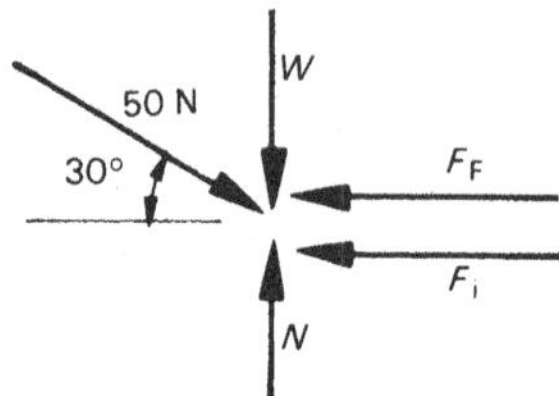

Fig. 8.3

It is clear that the direction of the acceleration is to the right so the inertia force acts to the left (negative).

In the vertical direction there is no difference to the forces and $N = 74.05$ N (as previously calculated).

In the horizontal direction:

$50 \cos 30° - F_F - F_i = 0$

and $F_F = 22.215$ N (as previously calculated).

$\therefore F_i = 50 \cos 30° - 22.215 = 21.09$ N

Since $F_i = ma$,

$$a = \frac{F_i}{m} = \frac{21.09}{5} = \mathbf{4.22\ m/s^2} \quad \text{(as before)}$$

Choice of method

It is evident therefore, that there are two methods of dealing with accelerating masses:

1. Use the principles of statics to obtain the resultant of the applied unbalanced forces acting on the mass (in magnitude and direction). Hence determine the resultant acceleration.
2. Assume the forces are balanced by including the inertia force acting in the opposite direction to the assumed or known acceleration direction. Use the principles of statics to determine the unknown inertia force. Hence determine the resultant acceleration of the mass.

The choice of method is a matter of preference since both require the same amount of calculation and give the same result. However, in more complex problems, the inertia force method is less likely to lead to error since it gives a standard method of attack.

Example 8.3
A mass of 1 t is lifted vertically off the ground by a winch cable to a height of 21 m when it comes to rest again. The movement consists of three phases:

1. constant (uniform) acceleration for 3 s to a height of 6 m

2. constant velocity motion for 3 s
3. constant deceleration to rest.

Draw motion diagrams marking all key points and determine the tension in the cable during each of the three phases of the motion.

Solution

Phase 1. $v_1 = 0, \quad s = 6\text{ m}, \quad t = 3\text{ s}$

$$s = \tfrac{1}{2}(v_1 + v_2)t$$
$$\therefore\ 6 = \tfrac{1}{2}(0 + v_2)3$$
$$\therefore\ v_2 = 4\text{ m/s}$$
$$v_2 = v_1 + at$$
$$4 = 0 + a \times 3$$
$$\therefore\ a = 1.33\text{ m/s}^2$$

Phase 2. $v_1 = v_2 = 4\text{ m/s}, \quad a = 0, \quad t = 3\text{ s}$

$$s = vt$$
$$= 4 \times 3 = 12\text{ m}$$

Phase 3. $v_1 = 4\text{ m/s}, \quad v_2 = 0, \quad s = 3\text{ m} \quad (21 - 6 - 12)$

$$s = \tfrac{1}{2}(v_1 + v_2)t$$
$$\therefore\ 3 = \tfrac{1}{2}(4 + 0)t$$
$$\therefore\ t = 1.5\text{ s}$$
$$v_2 = v_1 + at$$
$$\therefore\ 0 = 4 + a \times 1.5$$
$$\therefore\ a = -2.67\text{ m/s}^2 \quad (-\text{ indicating deceleration})$$

The motion diagrams may now be drawn as shown in Figure 8.4.

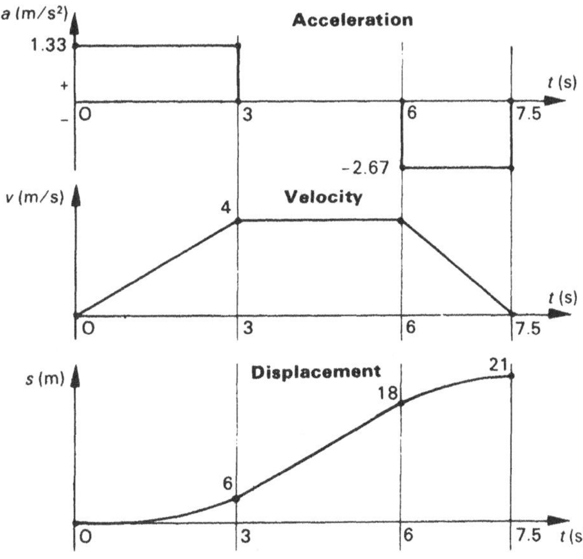

Fig. 8.4

In order to determine the tension in the cable during each phase of motion, it is desirable to draw free body diagrams first. Using the inertia force method, the diagrams are as shown in Figure 8.5.

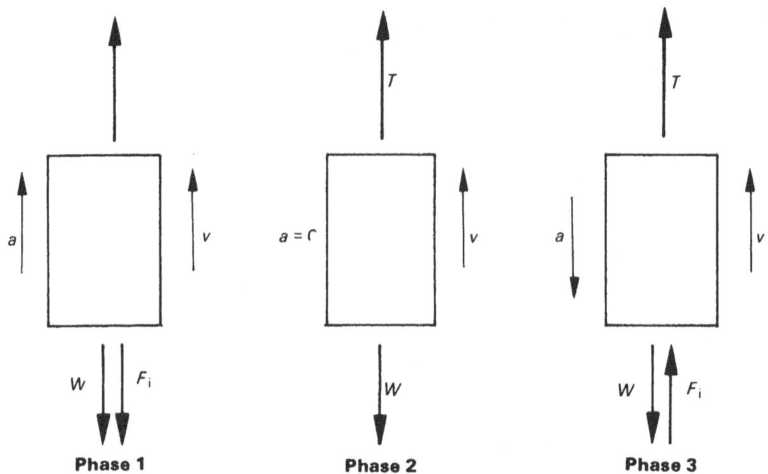

Fig. 8.5

Note: On the free body diagrams it is useful to show the direction of the velocity and acceleration.

The equations for the force equilibrium in each case may now be written.

Phase 1. $T - W - F_i = 0$

$F_i = ma = 1000 \times 1.33$

$\therefore T - 1000 \times 9.81 - 1000 \times 1.33 = 0$

$\therefore \boldsymbol{T} = \mathbf{11.14\ kN}$

Phase 2. $T - W = 0$

$\therefore T - 1000 \times 9.81 = 0$

$\therefore \boldsymbol{T} = \mathbf{9.81\ kN}$

Phase 3. $T - W + F_i = 0$

$F_i = ma = 1000 \times 2.67$

$\therefore T - 1000 \times 9.81 + 1000 \times 2.67 = 0$

$\therefore \boldsymbol{T} = \mathbf{7.14\ kN}$

Notes

1. It is not necessary to include the minus sign for the acceleration in the above force calculations because the direction of the inertia force (opposite to the acceleration) is taken into account when writing the equations for force equilibrium.
2. As expected, the maximum tension occurs during phase 1. During phase 2 the tension is the same as the suspended weight at rest. Minimum tension occurs during phase 3 because the inertia force acts opposite to the weight. In fact, if the deceleration had been 9.81 m/s^2, the tension would have been zero. This is a condition of zero gravity (weightlessness).

Example 8.4

A forklift truck as shown in Figure 8.6 has an unladen mass of 1.2 t and a payload of 1 t.

Determine:

(a) maximum upward acceleration which can be applied to the payload before the rear wheels will lift off the ground
(b) reaction at the front wheels when this occurs
(c) reaction at the front and rear wheels when the upward acceleration of the payload is 2 m/s^2.

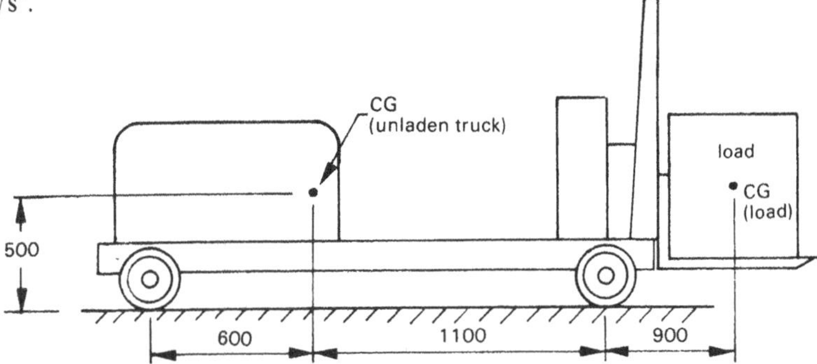

Fig. 8.6

Solution

A free body diagram showing the forces acting on the forklift when the rear wheels are in contact with the ground and the load is being given an upward acceleration is shown in Figure 8.7. Note that the force lifting the load has not been included since this force is balanced by an equal and opposite reaction force within the frame of the truck. This would, of course, not be the case if the lifting force were applied externally, e.g. by a cable not attached to the truck.

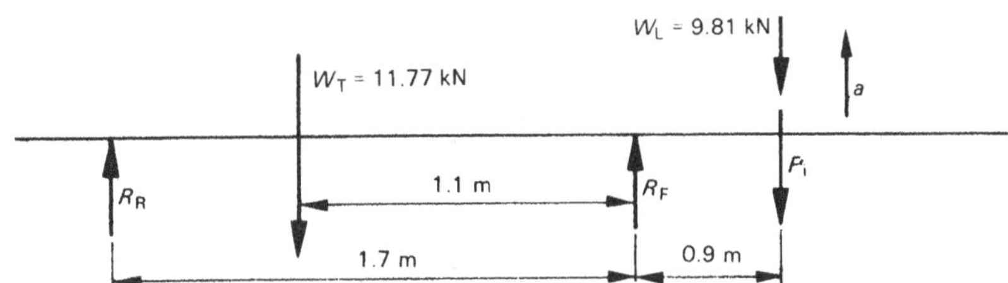

Fig. 8.7

(a) The rear wheels will lift off the ground when the clockwise moment of the forces about the front wheels is greater than the anticlockwise moment. When this occurs the reaction at the rear wheels is zero.

When $R_R = 0$, taking moments about R_F:

$$-W_T \times 1.1 + (F_i + W_L) \times 0.9 = 0$$

$F_i = m_L a = 10^3 \times a$

$\therefore -11.77 \times 10^3 \times 1.1 + (10^3 \times a + 10^3 \times 9.81) \times 0.9 = 0$

$\therefore \boldsymbol{a = 4.58 \text{ m/s}^2}$

$$\text{Hence, } F_i = 10^3 \times 4.58 = 4.58 \text{ kN}$$

(b) $\Sigma F_V = 0$,

$\therefore -W_T + R_F - W_L - F_i = 0$

$$\therefore R_F = W_T + W_L + F_i$$
$$= 11.77 + 9.81 + 4.58$$
$$= \mathbf{26.2\ kN}$$

(c) When the upward acceleration of the load is $2\ m/s^2$, $F_i = 1 \times 10^3 \times 2 = 2$ kN. Taking moments about R_F:

$$R_R \times 1.7 - W_T \times 1.1 + (W_L + F_i) \times 0.9 = 0$$

$$\therefore R_R \times 1.7 - 11.77 \times 1.1 + (1 \times 9.81 + 1 \times 2) \times 0.9 = 0$$

$$\therefore \boldsymbol{R_R} = \mathbf{1.36\ kN}$$

Similarly taking moments about R_R:

$$W_T \times 0.6 - R_F \times 1.7 + (W_L + F_i) \times 2.6 = 0$$

$$\therefore 11.77 \times 0.6 - R_F \times 1.7 + (9.81 + 2) \times 2.6 = 0$$

$$\therefore \boldsymbol{R_F} = \mathbf{22.2\ kN}$$

Check: $\Sigma F_V = 0$

$$1.36 - 11.77 + 22.2 - 9.81 - 2 = 0$$

which checks.

8.3 Acceleration on the inclined plane

In Chapter 6 equations were developed for the motion of non-accelerated masses on the inclined plane, that is masses which were at rest on the plane or moving with constant velocity up or down the plane. Also a computer program INPLANE 1 was developed for determining the force necessary for equilibrium.

Now the case of accelerated masses on the inclined plane will be considered, the typical situation being that the applied force is given and is different to the force necessary for equilibrium so that the mass accelerates up or down the plane.

Consider the general case of accelerated motion up an inclined plane as shown in Figure 8.8.

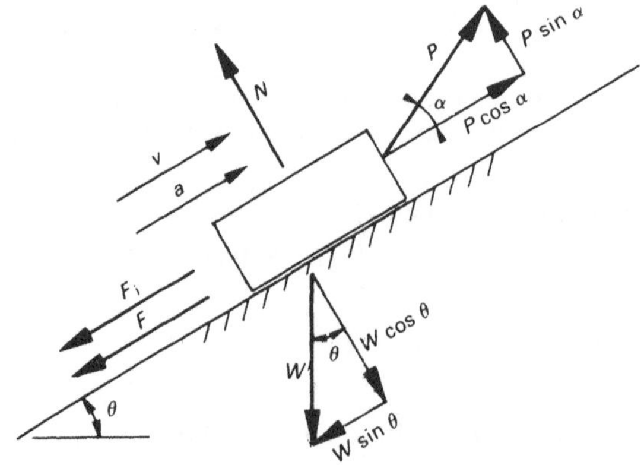

Fig. 8.8 *General case of accelerated motion up an inclined plane*

The forces are the same as for the case of non-accelerated motion up the plane except for the inclusion of the inertia force which acts down the plane (opposite to the direction of the acceleration).

Resolving perpendicular to the plane:

$$N = W\cos\theta - P\sin\alpha$$

Resolving parallel to the plane:

$$P\cos\alpha - F - W\sin\theta - F_i = 0$$

$$\therefore P\cos\alpha - \mu(W\cos\theta - P\sin\alpha) - W\sin\theta - ma = 0$$

Hence:

$$a = \frac{P}{m}(\cos\alpha + \mu\sin\alpha) - g(\sin\theta + \mu\cos\theta) \qquad \textbf{(8.2)}$$

accelerated motion up plane

A similar equation may be derived for accelerated motion down the plane. With angle α measured anticlockwise to the plane (with P pointing away from the block), only one additional equation is necessary:

$$a = \frac{P}{m}(\mu\sin\alpha - \cos\alpha) + g(\sin\theta - \mu\cos\theta) \qquad \textbf{(8.3)}$$

accelerated motion down plane

Notes

1. It is not strictly necessary to use another equation for motion down the plane and equation 8.2 may be used for motion down the plane as well as up the plane if, when the motion is down the plane, a is negative and μ is negative.
2. If the direction of motion is assumed and a negative acceleration is obtained, then this indicates the wrong direction was assumed. The answer obtained will not have the correct numerical value and the correct value may be determined by resubstitution with motion in the opposite direction.
3. Equation 8.2 may be used for a horizontal surface by substituting $\theta = 0$ so that $\sin\theta = 0$ and $\cos\theta = 1$. Also the equation is valid for a free block by substituting $P = 0$.

Computer solution

The computer program INPLANE 1 was modified to include the case of accelerated motion and the new program named INPLANE 2. This program is suitable for both accelerated and constant velocity motion on the inclined plane and is listed in Appendix 3.10. The solution to example 8.1 and example 8.5 (which follows) is given below the program listing and should be self-explanatory.

Example 8.5

Determine the acceleration of the 50 kg block shown in Figure 8.9 if the coefficient of friction between the block and the plane is 0.25 and:

(a) applied force P is 180 N
(b) applied force P is zero.

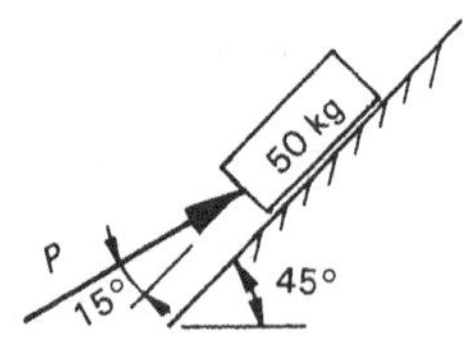

Fig. 8.9

Solution

(a) Assuming motion up the plane, equation 8.2 may be used or the computer program INPLANE 2.

Here $P = 180$ N

$m = 50$ kg

$\alpha = -15°$ (the minus sign is due to the convention in specifying this angle)

$\theta = 45°$

$\mu = 0.25$

Substituting: $a = -5.43 \text{ m/s}^2$

The minus sign indicates the wrong assumption was made about the direction of motion. The data must now be input again with motion down the plane (or in equation 8.3).

Substituting: $\boldsymbol{a} = \mathbf{1.49 \text{ m/s}^2}$ (down the plane).

(b) If $P = 0$, motion must be down the plane.

Using the computer program or substituting in equation 8.3:

$$\begin{aligned} a &= g(\sin\theta - \mu\cos\theta) \\ &= 9.81(\sin 45° - 0.25\cos 45°) \\ &= \mathbf{5.2 \text{ m/s}^2} \end{aligned}$$

8.4 Centrifugal force (F_c)

In Chapter 7 it was seen that a rotating particle is subject to a centripetal acceleration. This means that a rotating mass is subject to an accelerating force known as the centripetal force. For example, consider the mass m attached to a cord and rotating about centre O as shown in Figure 8.10.

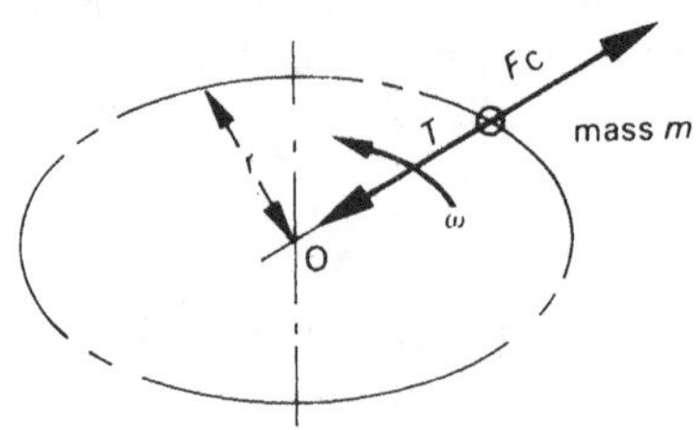

Fig. 8.10 *Centripetal force (T) and centrifugal force (F_c)*

It is evident that the centripetal force (T) is transmitted by the tension in the cord for, if the cord were suddenly cut, the mass would stop its circular motion and revert to linear motion.

Because of the inward centripetal acceleration there is an outward inertia force (F_C) called the centrifugal force. D'Alembert's principle may be applied and the system considered to be in force equilibrium and treated by the methods of statics.

From equation 7.12:

$$a_C = r\omega^2$$

and since $F = ma$

$$\boxed{F_C = mr\omega^2} \qquad \textbf{(8.4) centrifugal force}$$

Similarly from equation 7.13:

$$a = \frac{v^2}{r}$$

$$\boxed{F_C = m\frac{v^2}{r}} \qquad \textbf{(8.5) centrifugal force}$$

Example 8.6

In an amusement park, a car travels in a vertical circular path of diameter 30 m as shown in Figure 8.11.

Determine the minimum speed of the car so that there is no tendency for it to fall off the track.

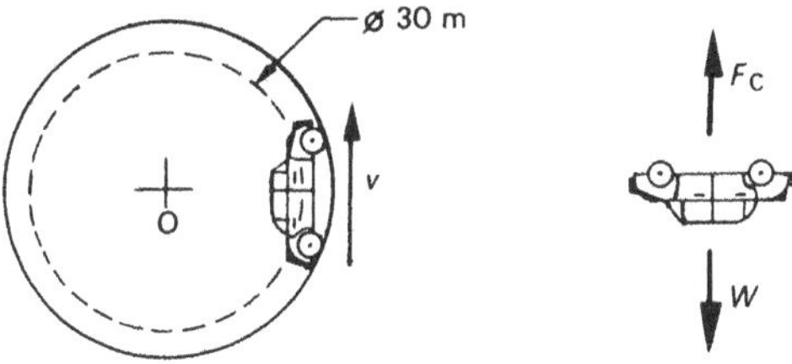

Fig. 8.11

Solution

The car will not fall off the track when $F_C > W$. The minimum velocity therefore occurs when:

$$m\frac{v^2}{r} = mg$$

or
$$v = \sqrt{gr}$$
$$= \sqrt{9.81 \times 15}$$
$$= \mathbf{12.1\ m/s} \quad \textbf{or} \quad \mathbf{43.7\ km/h}$$

Example 8.7

A rail car is shown in cross-section in Figure 8.12. It has a top speed of 300 km/h.

Determine the maximum radius flat curve to avoid overturning (rolling).

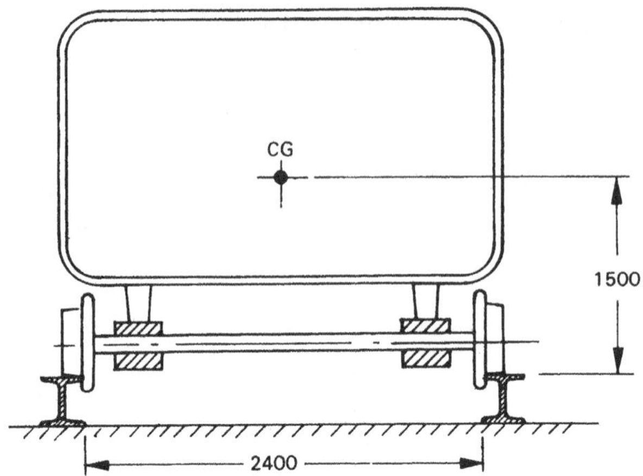

Fig. 8.12

Solution

Rolling occurs when the moment of the centrifugal force acting about the outer wheel (rail) is greater than the moment due to the weight.

The forces are shown in Figure 8.13.

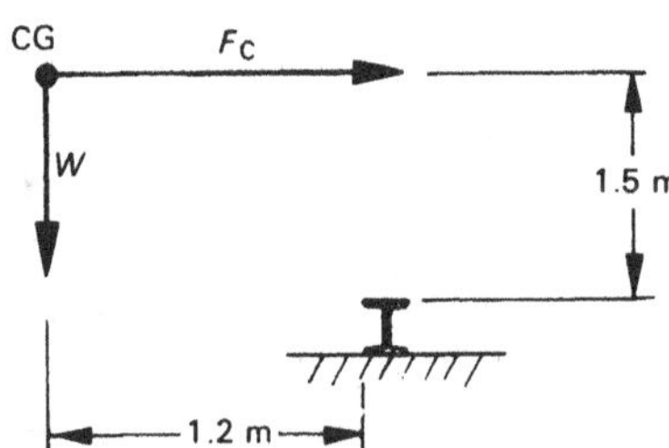

Fig. 8.13

Taking moments about the rail when overturning occurs:

$$F_C \times 1.5 = W \times 1.2$$

$$m\frac{v^2}{r} \times 1.5 = mg \times 1.2$$

or

$$r = \frac{v^2 \times 1.5}{9.81 \times 1.2}$$

Now $v = 300 \text{ km/h} = 83.3 \text{ m/s}$

$$\therefore r = \frac{83.3^2 \times 1.5}{9.81 \times 1.2} = \mathbf{885\ m}$$

8.5 Rotation on the inclined plane

This situation occurs when bodies are on a rotating inclined plane or when a vehicle travels around a banked curve. Figure 8.14 shows the forces involved when motion is impending up the plane.

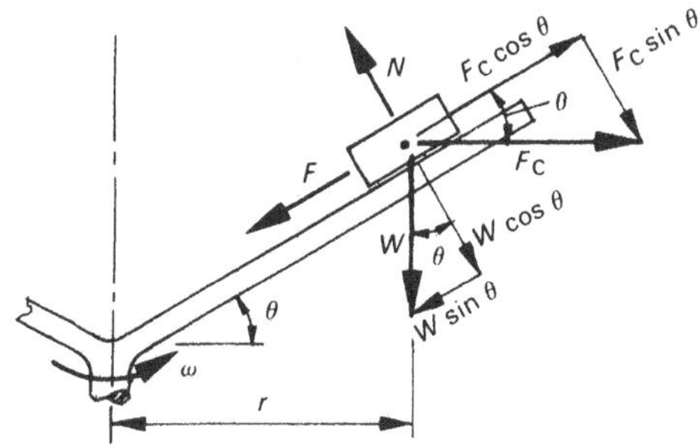

Fig. 8.14 *Rotation on an inclined plane*

Resolving the forces perpendicular to the plane:

$$N = F_C \sin\theta + W\cos\theta$$

Parallel to the plane:

$$\begin{aligned} F_C\cos\theta &= W\sin\theta + F \\ &= W\sin\theta + \mu N \\ &= W\sin\theta + \mu(F_C\sin\theta + W\cos\theta) \\ &= W\sin\theta + \mu F_C\sin\theta + \mu W\cos\theta \end{aligned}$$

Dividing both sides by $\cos\theta$:

$$\begin{aligned} F_C - \mu F_C\tan\theta &= W\tan\theta + \mu W \\ \therefore F_C(1 - \mu\tan\theta) &= W(\mu + \tan\theta) \\ \therefore \frac{mv^2}{r}(1 - \mu\tan\theta) &= mg(\mu + \tan\theta) \end{aligned}$$

$$\therefore v = \sqrt{\frac{rg(\mu + \tan\theta)}{1 - \mu\tan\theta}}$$

This equation gives the velocity at which sliding up the plane will just commence. An alternative form of this equation may be derived by use of the angle of friction ϕ. Refer to Figure 8.15.

Resolve vertically:

$$W = R\cos(\theta + \phi)$$

$$\therefore R = \frac{W}{\cos(\theta + \phi)}$$

Resolving horizontally:

$$F_C = R\sin(\theta + \phi)$$

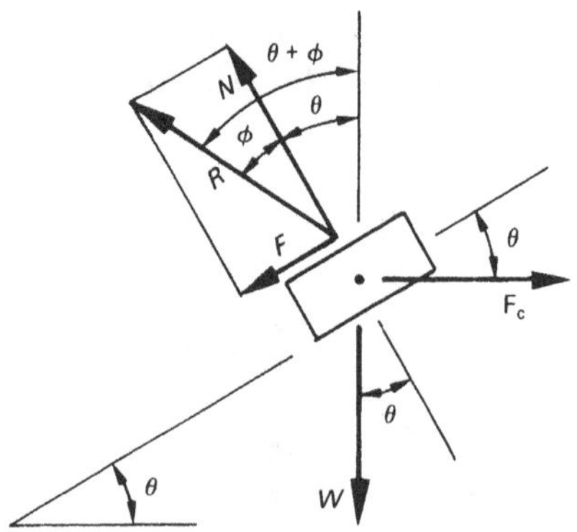

Fig. 8.15 *Rotation on an inclined plane using angle of friction (φ)*

$$\therefore F_C = \frac{W \sin(\theta + \phi)}{\cos(\theta + \phi)}$$

$$\therefore m\frac{v^2}{r} = \frac{mg \sin(\theta + \phi)}{\cos(\theta + \phi)}$$

$$\boxed{v = \sqrt{rg \tan(\theta + \phi)}}$$ **(8.6) motion up a rotating inclined plane**

Notes

1. When motion is impending down the plane, the equation is:

$$\boxed{v = \sqrt{rg \tan(\theta - \phi)}}$$ **(8.7) motion down a rotating inclined plane**

2. There is no tendency for sliding inward or outward when the component of the weight acting down the plane balances the component of the centrifugal force acting up the plane. That is when:

$$W \sin\theta = F_C \cos\theta$$

$$\text{or } mg \tan\theta = m\frac{v^2}{r}$$

$$\boxed{v = \sqrt{rg \tan\theta}}$$ **(8.8) balanced forces on a rotating inclined plane**

Example 8.8

For the block illustrated in Figure 8.16, determine the rotational speed in rpm at which:

(a) the block will just slide inward
(b) the block will just slide outward
(c) the forces acting on the block parallel to the plane are balanced so there is no tendency for sliding inward or outward.

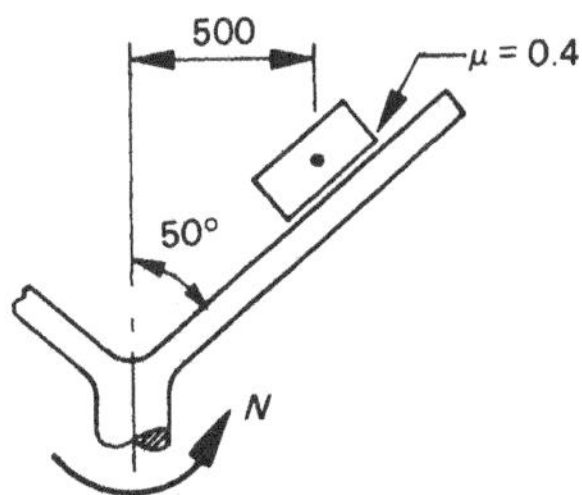

Fig. 8.16

Solution

$\theta = 40°$, $\phi = \tan^{-1} 0.4 = 21.8°$.

(a) Using equation 8.7, the speed at which the block will just slide inward is given by:

$$v = \sqrt{rg \tan(\theta - \phi)}$$

$$= \sqrt{0.5 \times 9.81 \times \tan(40° - 21.8°)}$$

$$= 1.27 \text{ m/s}$$

Since $v = r\omega$

$$\omega = \frac{1.27}{0.5} = 2.54 \text{ rad/s}$$

$\therefore$ **N = 24.2 rpm**

(b) Using equation 8.6, the speed at which the block will just slide outward is given by:

$$v = \sqrt{rg \tan(\theta + \phi)}$$

$$= \sqrt{0.5 \times 9.81 \times \tan(40° + 21.8°)}$$

$$= 3.025 \text{ m/s}$$

$$\therefore \omega = \frac{3.025}{0.5} = 6.045 \text{ rad/s}$$

$\therefore$ **N = 57.8 rpm**

(c) Balance occurs when:

$$v = \sqrt{rg \tan \theta}$$

$$\therefore v = \sqrt{0.5 \times 9.81 \times \tan 40°}$$

$$= 2.03 \text{ m/s}$$

$$\therefore \omega = \frac{2.03}{0.5} = 4.06 \text{ rad/s}$$

$\therefore$ **N = 38.7 rpm**

Example 8.9

A road curve of radius 120 m is sloped at 1 in 20 (tan).

Determine the maximum speed in km/h at which a vehicle can round the curve without skidding if the minimum coefficient of friction between the tyres and the road is 0.2.

Also determine the most comfortable road speed for the occupants of the vehicle.

Solution

$$\tan\theta = \frac{1}{20} \quad \therefore \theta = 2.86°$$

$$\tan\phi = 0.2 \quad \therefore \phi = 11.31°$$

Maximum speed without skidding is given by equation 8.6:

$$\begin{aligned} v &= \sqrt{rg\tan(\theta + \phi)} \\ &= \sqrt{120 \times 9.81 \times \tan(2.86° + 11.31°)} \\ &= 17.24 \text{ m/s} = \mathbf{62.1\ km/h} \end{aligned}$$

The most comfortable road speed for the occupants is when there is no tendency for them to be thrown outward or inward, that is from equation 8.8 when:

$$\begin{aligned} v &= \sqrt{rg\tan\theta} \\ &= \sqrt{120 \times 9.81 \times \tan 2.86°} \\ &= 7.67 \text{ m/s} = \mathbf{27.6\ km/h} \end{aligned}$$

8.6 Accelerating torque and inertia torque: Point mass

When unbalanced or resultant torque acts on a mass which is free to rotate about a centre, an angular acceleration occurs. Consider torque τ applied to the *point* mass m as shown in Figure 8.17.

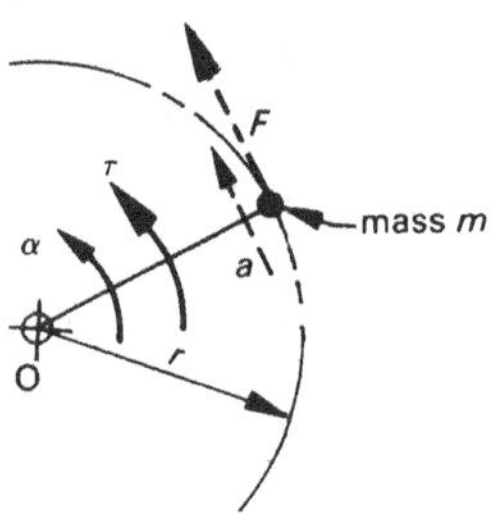

Fig. 8.17 *Torque applied to a point mass*

The torque τ causes an equivalent force F at radius r and an angular acceleration α about the centre O.

Since torque $\tau = Fr$

$$\therefore F = \frac{\tau}{r} \qquad (1)$$

Force $F = ma$, but $a = r\alpha$

$$\therefore F = mr\alpha \qquad (2)$$

Substituting (2) into (1):

$$mr\alpha = \frac{\tau}{r}$$

or $$\boxed{\tau = mr^2\alpha}$$ **(8.9) torque applied to a point mass**

where τ = resultant torque (Nm)
m = mass (kg)
r = radius (m)
α = angular acceleration (rad/s^2)

The applied torque therefore produces an angular acceleration in the same way that an applied or unbalanced force produces a linear acceleration.

D'Alembert's principle may be used if an inertia torque is included acting to oppose the applied torque. The system may then be considered as if it were in static equilibrium as shown in Figure 8.18.

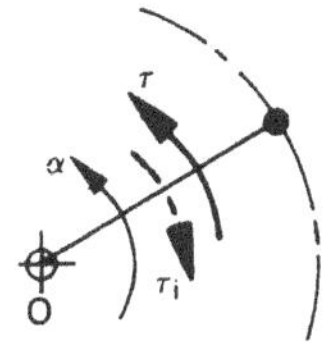

Fig. 8.18 *Applied torque (τ) and inertia torque (τ$_i$)*

Example 8.10

A 0.5 kg mass is attached to a horizontal spoke and is free to rotate about a vertical axis located at a distance of 750 mm from it. A torque of 4 Nm is applied to the system for 6 s.

Determine the final rotational speed in rpm and the number of revolutions made in reaching this speed. The mass may be considered as a point mass.

Solution

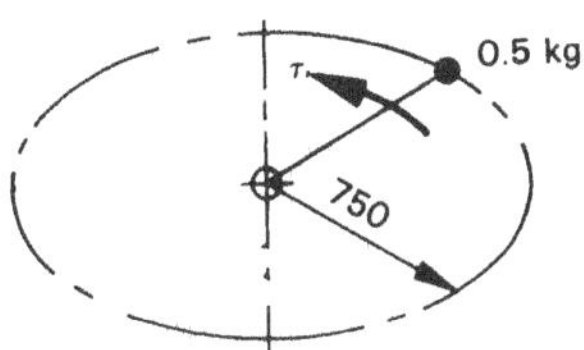

Fig. 8.19

Using equation 8.9:

$$\tau = mr^2\alpha$$
$$\therefore 4 = 0.5 \times 0.75^2 \times \alpha$$
$$\therefore \alpha = 14.22 \text{ rad/s}^2$$

Now $\omega_2 = \omega_1 + \alpha t$

$\therefore \omega_2 = 0 + 14.22 \times 6 \quad = 85.33$ rad/s = **815 rpm**

Similarly $\theta = \omega_1 t + \frac{1}{2}\alpha t^2$

$\therefore \theta = 0 + \frac{1}{2} \times 14.22 \times 6^2 \quad = 256$ rad = **40.7 revs**

8.7 Rotating non-point mass and mass moment of inertia

When a mass has substantial dimensions relative to the centre of rotation it may not be considered as a point mass. Consider such a mass as shown in Figure 8.20.

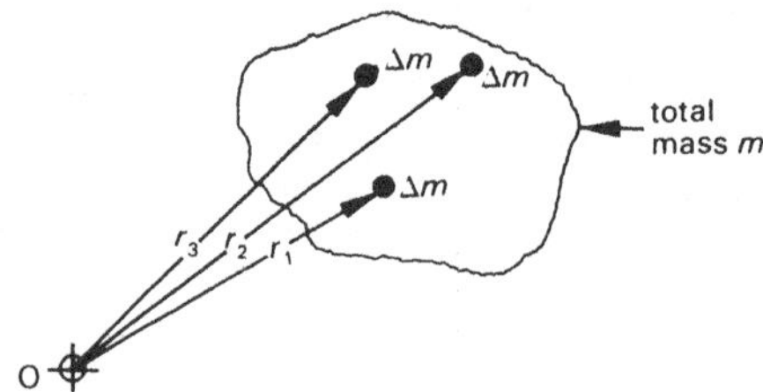

Fig. 8.20 *Rotating non-point mass*

The total mass m may be considered to consist of a number of small (point) masses, Δm, at radius r_1, r_2, r_3, etc. Equation 8.6 may be written:

$$\tau = (\Delta m r_1^2 + \Delta m r_2^2 + \Delta m r_3^2 + \ldots)\alpha$$
$$= \Sigma(\Delta m r^2)\alpha$$

The expression $\Sigma \Delta m r^2$ is usually called the mass moment of inertia or the second moment of mass because the first moment of mass is $\Sigma \Delta m r$ and the second moment of mass is $(\Sigma \Delta m r) \times r = \Sigma \Delta m r^2$. It is given the symbol I and has units kg m × m = kgm^2.

$$I = \Sigma \Delta m r^2$$

(8.10) second moment of mass or mass moment of inertia

Hence equation 8.9 for a non-point mass becomes:

$$\tau = I\alpha$$

(8.11) torque applied to a non-point mass

Comparing equation 8.11 with equation 8.1, $F = ma$, it is evident that dynamic linear formulas have rotational equivalent formulas in the same way that kinematic linear formulas have rotational equivalents. Dynamic equivalents are given in Table 8.1.

Table 8.1 Dynamic linear and rotational equivalents

Linear	*Rotational*
Force F (N)	Torque τ (Nm)
Acceleration a (m/s^2)	Angular acceleration α (rad/s^2)
Mass m (kg)	Second moment of mass or mass moment of inertia I (kgm^2)
Applied force $F = ma$	Applied torque $\tau = I\alpha$
Inertia force $F_i = ma$	Inertia torque $\tau_i = I\alpha$

Example 8.11

A rotor supported by plain bearings has mass moment of inertia 2 kgm^2. The rotor is accelerated uniformly from rest to 2900 rpm by a torque of 35 Nm which is applied for 20 s.

Determine the frictional torque.

The rotor is then allowed to coast to rest. Assuming the frictional torque is constant, determine the number of revolutions made by the rotor during this period.

Solution

Calculate the angular acceleration α.

$$\omega_2 = \omega_1 + \alpha t$$

$$\therefore 303.7 = 0 + \alpha \times 20$$

$$\therefore \alpha = 15.18 \text{ rad/s}^2$$

$$\omega_2 = \pi \times \frac{2900}{30} = 303.7 \text{ rad/s}$$

Denoting the frictional torque τ_F and the inertia torque τ_i:

$$\tau - \tau_F - \tau_i = 0$$

$$\therefore \tau - \tau_F - I\alpha = 0$$

$$\therefore 35 - \tau_F - 2 \times 15.18 = 0$$

$$\therefore \boldsymbol{\tau_F = 4.63 \text{ Nm}}$$

In the period of deceleration, there is no applied torque. Hence:

$$I\alpha - \tau_F = 0$$

$$\therefore 2\alpha - 4.63 = 0$$

$$\therefore \alpha = 2.315 \text{ rad/s}^2$$

Using $\omega_2^2 = \omega_1^2 + 2\alpha\theta$

With $\alpha = -2.315 \text{ rad/s}^2$ (deceleration)

$$\therefore 0 = 303.7^2 - 2 \times 2.315 \times \theta$$

$$\therefore \theta = 19\,920 \text{ rad}$$

$$= \mathbf{3170 \text{ revs}}$$

8.8 Calculation of the mass moment of inertia

The mass moment of inertia of a mass about a centre of rotation may be determined experimentally or by calculation. For complex shapes, the calculation may be lengthy but for common shapes involves only a few formulas.

A common simple shape is the solid disc rotating about its centre. Consider a small section of such a disc at radius r as shown in Figure 8.21.

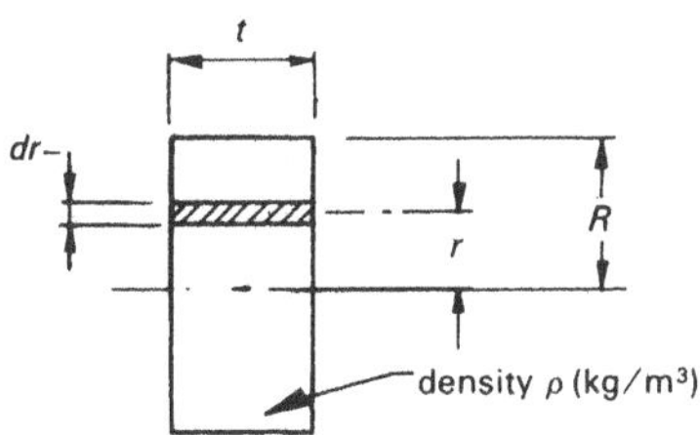

Fig. 8.21 *Section of a solid disc*

Using the methods of calculus:

$$dI = \int_0^R r^2\, dm$$

Now $dm = \rho dV$

$$= \rho \times 2\pi \times r \times dr \times t$$

$$= 2\pi \rho t r dr$$

$$\therefore dI = 2\pi \rho t \int_0^R r^3\, dr$$

$$\therefore I = \frac{\pi \rho t}{2} R^4$$

But the total mass m is $\pi R^2 t \rho$.

$$\therefore \boxed{I = \tfrac{1}{2} m R^2}$$ **(8.12) mass moment of inertia of a solid disc**

where I = mass moment of inertia (kgm^2)
m = mass of disc (kg)
R = radius of the disc ($\frac{1}{2}$ the diameter) (m)

Parallel axis theorem

If the mass moment of inertia about an axis passing through the centre of mass is known, the mass moment of inertia about any other parallel axis located a distance x from it may be calculated by use of the parallel axis theorem.

$$\boxed{I_{XX} = I_{CG} + m x^2}$$ **(8.13) parallel axis theorem**

where I_{XX} = moment of inertia about axis x-x (kgm^2)
I_{CG} = moment of inertia about the centre of mass axis (kgm^2)
m = mass (kg)
x = distance between axis x-x and the centre of mass axis (m)

By use of equations 8.12 and 8.13, the mass moment of inertia of common sections may be determined.

Note that for any composite body, the total mass moment of inertia is the sum of the moments of inertia of each part considered separately. Similarly for a body with one or more voids or hollows, the net mass moment of inertia is the moment of inertia of the body considered as a solid, less the moment of inertia of the voids or hollows considered to be made of solid material.

Radius of gyration (k)

The radius of gyration is that radius at which the mass if concentrated (point mass) would have the same mass moment of inertia as the original non-point mass.

Hence for any mass $I = mk^2$

or $$\boxed{k = \sqrt{\frac{I}{m}}}$$ **(8.14) radius of gyration**

Note that the radius of gyration is not located at the mean radius nor is it equal to the distance between the centre of rotation and the centre of mass.

Example 8.12

Calculate the mass moment of inertia and the radius of gyration of the steel tube section shown in Figure 8.22 about the centreline axis. The density of steel is 7800 kg/m^3.

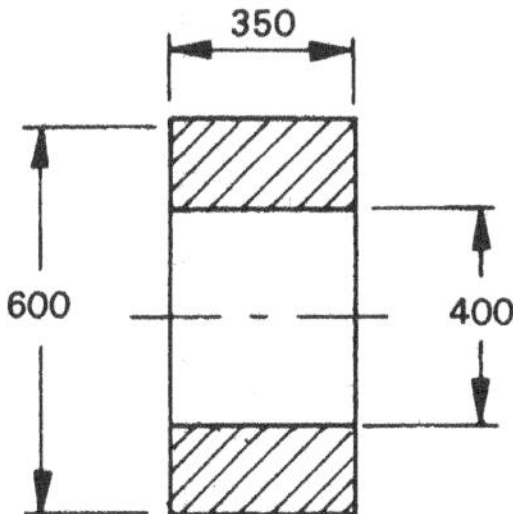

Fig. 8.22

Solution

This problem may be solved by first calculating the mass moment of inertia of the solid disc and then subtracting the mass moment of inertia of the hollow disc.

For the solid disc:

$$m = \rho V$$
$$= 7800 \times \pi \times 0.3^2 \times 0.35$$
$$= 772 \text{ kg}$$

Using equation 8.12:

$$I = \tfrac{1}{2} m R^2$$
$$= \tfrac{1}{2} \times 772 \times 0.3^2$$
$$= 34.74 \text{ kgm}^2$$

Similarly for the hollow disc:

$$m = 7800 \times \pi \times 0.2^2 \times 0.35$$
$$= 343 \text{ kg}$$
$$I = \tfrac{1}{2} \times 343 \times 0.2^2$$
$$= 6.86 \text{ kgm}^2$$

$$\therefore I \text{ net} = 34.74 - 6.86 = \mathbf{27.9 \text{ kgm}^2}$$

Using equation 8.14:

$$k = \sqrt{\frac{I}{m}}$$

and $m = 772 - 343 = 429$ kg

$$\therefore k = \sqrt{\frac{27.9}{429}} \quad \text{(m)}$$
$$= \mathbf{255 \text{ mm}}$$

Note that the mean radius is 250 mm, so the radius of gyration is located further out than the mean radius. This occurs because the moment of inertia increases with the *square* of the radius.

Example 8.13

Determine the mass moment of inertia and the radius of gyration of the flywheel illustrated in Figure 8.23. The flywheel is manufactured from cast iron (density 7200 kg/m^3). The inertia of the centre boss is negligible and need not be calculated.

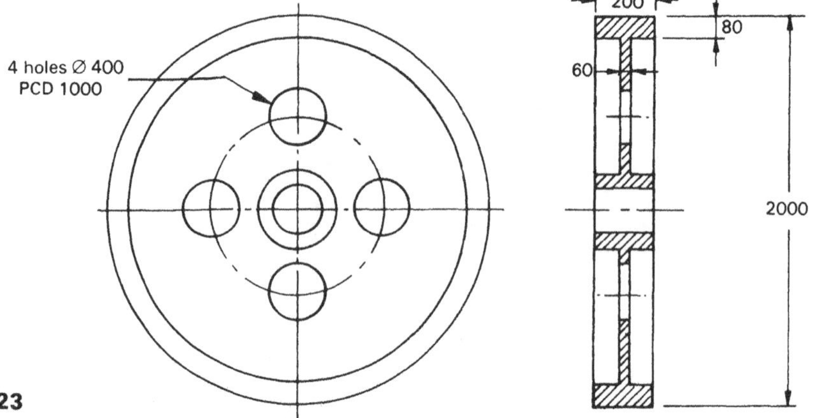

Fig. 8.23

Solution

Outer disc $m = \rho \frac{\pi}{4} d^2 t$

$= 7200 \times \frac{\pi}{4} \times 2^2 \times 0.2$

$= 4524$ kg

$I = \frac{1}{2} m R^2$

$= \frac{1}{2} \times 4524 \times 1^2$

$= 2262$ kgm^2

Inner disc $m = 7200 \times \frac{\pi}{4} \times 1.84^2 \times 0.2$

$= 3829$ kg

$I = \frac{1}{2} \times 3829 \times 0.92^2$

$= 1620$ kgm^2

Rim $m = 4524 - 3829 = 695$ kg

$I = 2262 - 1620 = 642$ kgm^2

Solid web $m = 7200 \times \frac{\pi}{4} \times 1.84^2 \times 0.06$

$= 1149$ kg

$I = \frac{1}{2} \times 1149 \times 0.92^2$

$= 486$ kgm^2

Holes $m = 7200 \times \frac{\pi}{4} \times 0.4^2 \times 0.06$

$= 54.3$ kg (for each hole)

In order to determine I for the holes, the parallel axis theorem may be used because the centre of the holes is displaced 500 mm from the centre of rotation.

$$\begin{aligned} I_{xx} &= I_{CG} + mx^2 \\ &= \tfrac{1}{2}mR^2 + mx^2 \\ &= m(\tfrac{1}{2}R^2 + x^2) \\ &= 54.3(\tfrac{1}{2} \times 0.2^2 + 0.5^2) \\ &= 14.66\ \text{kgm}^2 \end{aligned}$$

Therefore for 4 holes:

$$m = 217\ \text{kg},\ I = 59\ \text{kgm}^2$$

Therefore web with 4 holes:

$$\begin{aligned} m &= 1149 - 217 = 932\ \text{kg} \\ I &= 486 - 59 = 427\ \text{kgm}^2 \end{aligned}$$

Therefore for the total flywheel:

$$\begin{aligned} m &= 695 + 932 = 1627\ \text{kg} \\ I &= 642 + 427 = \mathbf{1069\ kgm^2} \end{aligned}$$

The radius of gyration may be determined from:

$$\begin{aligned} k &= \sqrt{\frac{I}{m}} \\ &= \sqrt{\frac{1069}{1627}} \quad \text{(m)} \\ &= \mathbf{811\ mm} \end{aligned}$$

That is, the mass moment of inertia is the same as that of a thin ring of mass 1627 kg and radius 811 mm.

Computer solution

The calculation of mass moment of inertia for a flywheel may readily be performed using a computer. The program FLYINT listed in Appendix 3.11 calculates the mass moment of inertia and radius of gyration of a flywheel shape as shown in Figure 8.23. The program inputs are:

1. material from which the flywheel is made
2. outside diameter of the rim
3. inside diameter of the rim
4. width of the rim
5. thickness of the web
6. number of holes
7. hole diameter and PCD.

The program outputs are:

1. mass of the rim
2. mass moment of inertia of the rim
3. mass of the web

4. mass moment of inertia of the web
5. mass of the flywheel
6. moment of inertia of the flywheel
7. radius of gyration of the flywheel.

Notes

1. The inertia of the boss is neglected.
2. The material choice is from cast iron (density 7200 kg/m^3), steel (density 7800 kg/m^3) or any other material whose density may be input by the user.
3. If the web is solid, the program will run correctly if the user inputs 0 for the number of holes. If there is no web (i.e. a ring) the program will also run if the user inputs 0 for the web thickness.

The solution to examples 8.12 and 8.13 is given below the program listing in Appendix 3.11 and should be self-explanatory.

8.9 Systems of connected masses: Translation

So far this chapter has dealt with single masses having linear or rotational acceleration. However it is common in engineering that a system may consist of masses connected in such a way that their motion is interdependent. In some cases, the connection may be such that the masses have a common motion and may effectively be considered as a single mass, for example in the case of a number of rail cars coupled together. However, in other cases the motion or the forces acting on each mass are different, in which case a separate analysis of each mass is usually required (taking into account the common motion or forces). If the connection between the masses is by means of pinned joints, the system becomes a mechanism and analysis of the motion and forces is involved and becomes part of specialist subjects such as "mechanics of machines". Where the connection between the masses is by means of a flexible cord or rope with negligible mass and where any pulleys included in the system have negligible friction and mass (inertia), the analysis is relevant to mechanics. There is no general formula or method which may be applied in all cases; problems are solved by application of the general principles of force and moment equilibrium in conjunction with free body diagrams and recognition of the common forces or motion. Typical problems and their solutions will now be illustrated with examples.

Example 8.14

For the blocks illustrated in Figure 8.24, determine:

(a) tension in the connecting cord when block A is held
(b) acceleration of the blocks when block A is released
(c) tension in the connecting cord when the blocks are moving.

The coefficient of friction is 0.3.

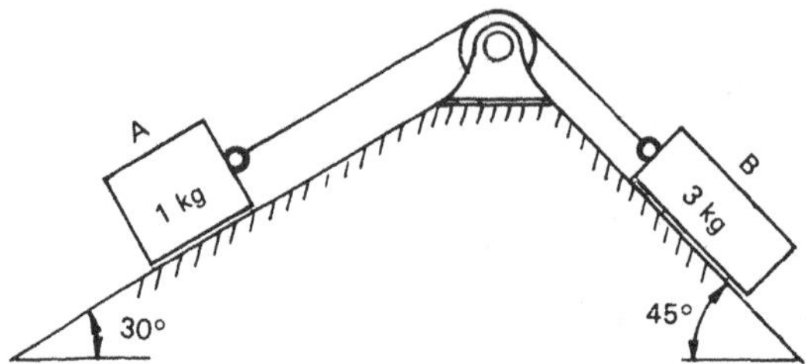

Fig. 8.24

Solution

(a) When block A is held, block B is also stationary and the tension in the cord may be determined by a free body diagram drawn for block B. This is shown in Figure 8.25.

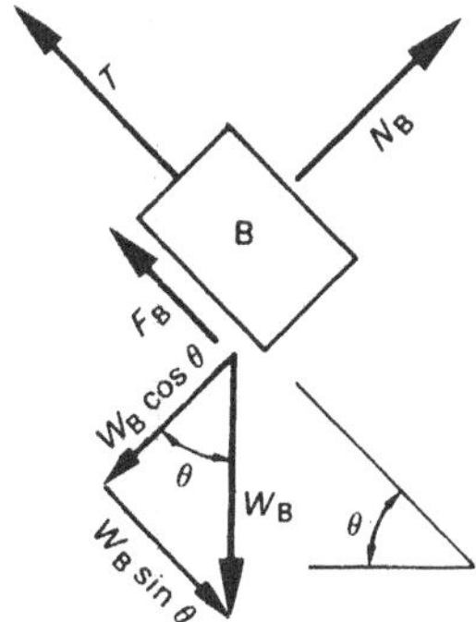

Fig. 8.25

Resolving perpendicular to the plane:

$$N_B = W_B \cos\theta$$

Resolving parallel to the plane:

$$T + F_B = W_B \sin\theta$$

But $F_B = \mu N_B$,

$$\therefore T + \mu W_B \cos\theta = W_B \sin\theta$$

$$\therefore T = W_B(\sin\theta - \mu\cos\theta)$$

$$= 3 \times 9.81 \times (\sin 45^\circ - 0.3 \cos 45^\circ)$$

$$\mathbf{T = 14.57\ N}$$

(b) By inspection of the plane angles and masses it is evident that when block A is released, block B will move down the plane and block A up the plane. Denoting the acceleration as a (common to both blocks) and the tension in the cord T (also common if there is negligible pulley friction and inertia), the free body diagrams for both masses are shown in Figure 8.26.

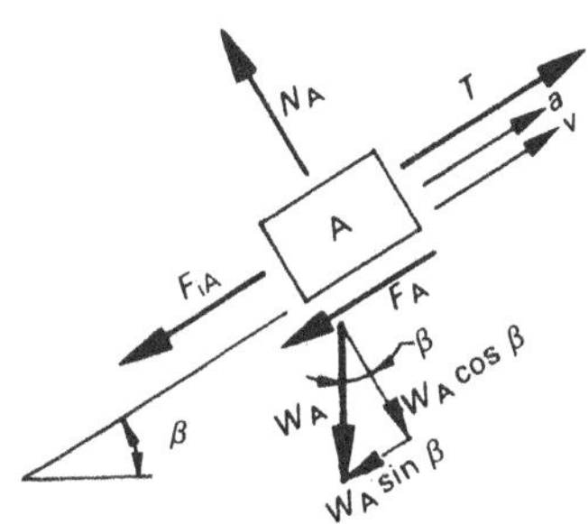

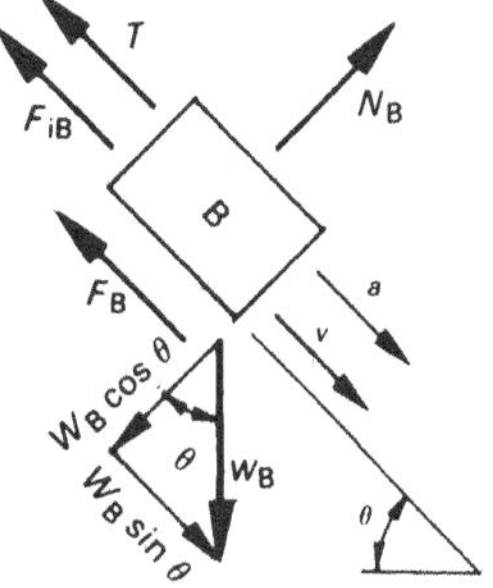

Fig. 8.26

Block B

The free body diagram is the same as in Figure 8.25 except that the inertia force acting in the opposite direction to the acceleration has been included.

Perpendicular to the plane:

$N_B = W_B \cos\theta$

Parallel to the plane:

$$T + F_B + F_{iB} = W_B \sin\theta$$

$\therefore T + \mu W_B \cos\theta + m_B a = W_B \sin\theta$

$\therefore T = W_B(\sin\theta - \mu\cos\theta) - m_B a$

$= 3 \times 9.81 \times (\sin 45° - 0.3\cos 45°) - 3a$

$\therefore T = 14.57 - 3a$ **(1)**

Block A

Perpendicular to the plane:

$N_A = W_A \cos\beta$

Parallel to the plane:

$T = F_A + F_{iA} + W_A \sin\beta$

$= \mu W_A \cos\beta + m_A a + W_A \sin\beta$

$= W_A(\sin\beta + \mu\cos\beta) + m_A a$

$= 1 \times 9.81 \times (\sin 30° + 0.3\cos 30°) + 1a$

$= 7.45 + a$ **(2)**

Equating (1) and (2):

$14.57 - 3a = 7.45 + a$

$\therefore 4a = 7.12$

$\mathbf{a = 1.78\ m/s^2}$

(c) The cord tension may be determined by substitution in either (1) or (2):

Substituting in (1):

$T = 14.57 - 3 \times 1.78$

$\mathbf{T = 9.23\ N}$

Example 8.15

For the system of connected masses shown in Figure 8.27, determine:

(a) acceleration of block A up the plane
(b) tension in the common cord
(c) tension in the cord holding block B.

The coefficient of friction is 0.35.

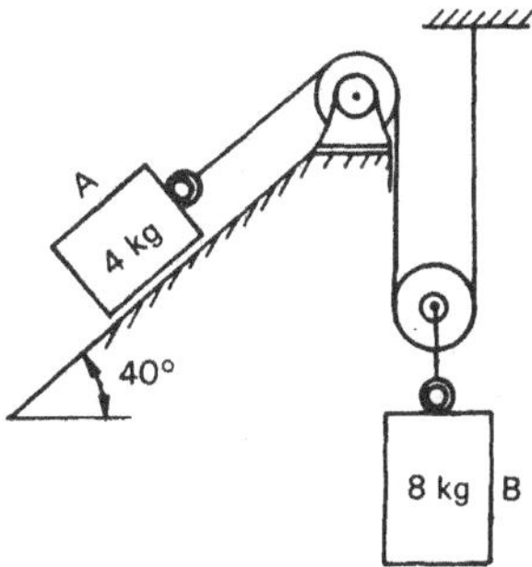

Fig. 8.27

Solution

Denoting the acceleration of block A as a, then the acceleration of block B will be $\frac{a}{2}$ because by inspection it is clear that block B moves one-half the distance that block A moves in a given time. Also denoting the tension in the common cord as T, then the tension in the single cord holding block B is $2T$. The free body diagrams are shown in Figure 8.28.

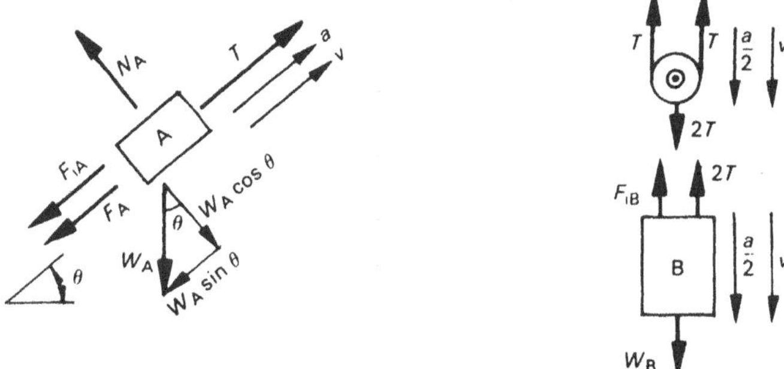

Fig. 8.28

(a) *Block A*

The free body diagram for block A is the same as that drawn for block A in example 8.14. Hence:

$$T = W_A(\sin\theta + \mu\cos\theta) + m_A a$$

$$= 4 \times 9.81 \times (\sin 40° + 0.35\cos 40°) + 4a$$

$$\therefore T = 35.74 + 4a \qquad \textbf{(1)}$$

Block B

$$2T + F_{iB} = W_B$$

$$\therefore 2T = 8 \times 9.81 - 8 \times \frac{a}{2}$$

$$\therefore T = 39.24 - 2a \qquad \textbf{(2)}$$

Equating (1) and (2):

$35.74 + 4a = 39.24 - 2a$ $\qquad \therefore a = \mathbf{0.583\ m/s^2}$

(b) Substituting in (1)

$T = 35.74 + 4 \times 0.583$

$\mathbf{T = 38\ N}$ (tension in the common cord)

(c) The tension in the cord holding block B is therefore:

$2 \times 38 = \mathbf{76\ N}$

8.10 Systems of connected masses: Translation and rotation

It is also common in engineering that systems of connected masses may experience simultaneous translation and rotation. For example, this would occur in the previous problems if the moment of inertia of the rotating parts could not be neglected as would be the case with substantial pulleys, shafts and flywheels.

Several examples of the approach to the solution of such problems are now presented.

Example 8.16

A winch drum is lowering a load of 50 kg down an inclined plane as shown in Figure 8.29.

If the drum is released, determine the acceleration of the load down the plane and the tension in the cable. The coefficient of friction between the load and the plane is 0.25 and bearing friction in the winch drum is negligible.

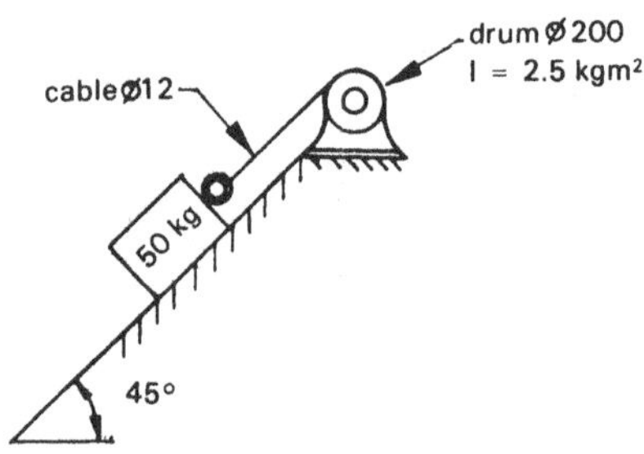

Fig. 8.29

Solution

The free body diagrams are shown in Figure 8.30.

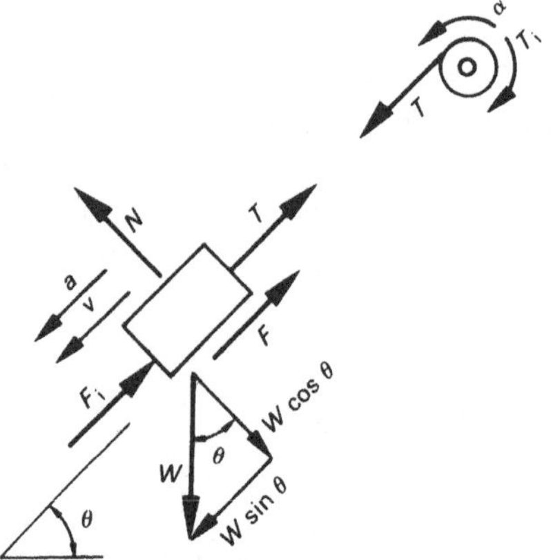

Fig. 8.30

Resolving perpendicular to the plane:

$$N = W\cos\theta$$

Parallel to the plane:

$$T + F_i + F = W\sin\theta$$

$$\therefore T = W\sin\theta - \mu W\cos\theta - ma$$

$$= 50 \times 9.81 \times (\sin 45^\circ - 0.25\cos 45^\circ) - 50a$$

$$= 260 - 50a \qquad (1)$$

For the equilibrium of the drum:

$$\tau = \tau_i \qquad (2)$$

where $\tau_i = I\alpha$ (inertia torque)

$\tau = Fr$ (applied torque)

$= T \times 0.106$ (r = mean radius to the centre of the cable which is 100 + 6 mm = 0.106 m)

Also $a = r\alpha \quad \therefore \alpha = \dfrac{a}{r} = \dfrac{a}{0.106}$

$I = 2.5\ \text{kgm}^2$ (given)

Hence (2) becomes:

$$T \times 0.106 = 2.5 \times \frac{a}{0.106}$$

$$\therefore a = 0.0045\,T$$

Substituting in (1):

$$T = 260 - 50 \times 0.0045\,T$$

$$\therefore 1.225\,T = 260$$

$$\therefore \mathbf{T = 212\ N}$$

Substituting in (1):

$$212 = 260 - 50a$$

$$\therefore \mathbf{a = 0.954\ m/s^2}$$

Example 8.17

A flywheel and shaft, combined mass 23 kg, are mounted in horizontal bearings. The shaft diameter is 72 mm and a light cord, diameter 2 mm, is wrapped around it, one end being secured to the shaft and the other end attached to a mass of 4 kg. In an experiment, the mass is allowed to fall from rest a distance of 1 m and the time measured and found to be 12 s.

Determine:

(a) tension in the cord when the mass is falling
(b) mass moment of inertia of the flywheel and shaft
(c) radius of gyration of the flywheel and shaft.

Bearing friction and air resistance are negligible.

Solution

The free body diagrams are shown in Figure 8.31.

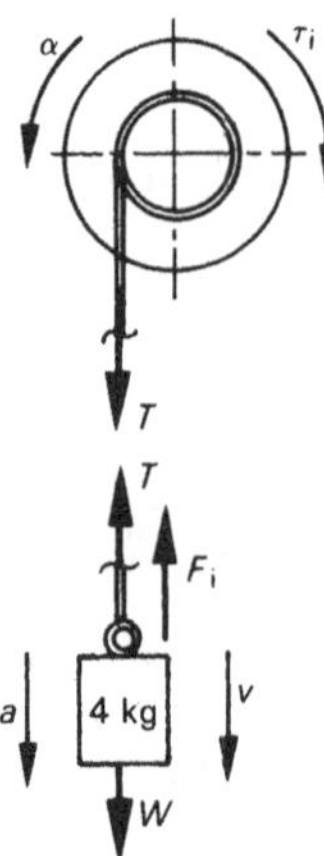

Fig. 8.31

Calculate the acceleration a:

$$s = v_1 t + \tfrac{1}{2}at^2 \quad \text{(from equation 7.8)}$$

$$\therefore 1 = 0 + \tfrac{1}{2} \times a \times 12^2$$

$$\therefore a = 0.0139 \text{ m/s}^2$$

(a) From the equilibrium of the weight:

$$W = T + F_i$$

$$T = mg - ma = m(g - a)$$

$$= 4(9.81 - 0.0139)$$

$$\therefore \boldsymbol{T = 39.18 \text{ N}}$$

(b) From the equilibrium of the flywheel:

$$\tau = \tau_i$$

where τ = applied torque = Tr

τ_i = inertia torque = $I\alpha$

$$\therefore Tr = I\alpha$$

Now $$r = \frac{72}{2} + \frac{2}{2} = 37 \text{ mm} = 0.037 \text{ m} \quad \text{(mean radius)}$$

$$\alpha = \frac{a}{r} = \frac{0.0139}{0.037} = 0.375 \text{ rad/s}^2$$

$$T = 39.18 \text{ N} \quad \text{(determined in (a))}$$

$$\therefore 39.18 \times 0.037 = I \times 0.375$$

$$\therefore \boldsymbol{I = 3.87 \text{ kgm}^2}$$

(c) From equation 8.14:

$$k = \sqrt{\frac{I}{m}} = \sqrt{\frac{3.87}{23}} = 0.410 \text{ m}$$

$$\therefore \boldsymbol{k = 410 \text{ mm}}$$

Problems

8.1 A mass of 6 kg rests on a horizontal surface. The coefficient of friction is 0.25. A force of 40 N is applied downward at an angle of 35° to the mass.
Determine the acceleration of the mass.
2.05 m/s^2

8.2 A mass rests upon an inclined plane which makes an angle of 30° to the horizontal. The mass is allowed to slide down the plane from rest a distance of 4 m which is done in time 1.85 s.
Determine the coefficient of friction between the mass and the plane.
0.302

8.3 A mass of 25 kg rests on a plane inclined at 30° to the horizontal. A horizontal force of 350 N is applied so that the mass accelerates up the plane.
Determine the acceleration of the mass if the coefficient of friction is 0.2.
4.12 m/s^2

8.4 A mass of 20 kg rests on a plane inclined at 15° to the horizontal.
Determine the magnitude of a horizontal force necessary to move the mass down the plane,
(a) with constant velocity
(b) with uniform acceleration 2.5 m/s^2.
The coefficient of friction is 0.35.
(a) 14.7 N (b) 62 N

8.5 A mass of 500 kg is lifted vertically from the ground by a winch cable to a height of 48 m when it is at rest again. The movement consists of three phases:
(a) constant (uniform) acceleration for 4 s to a height of 8m
(b) constant velocity motion for 5 s
(c) constant deceleration to rest.
Draw motion diagrams marking all key points and determine the tension in the cable during each phase of motion.
(a) 5.41 kN (b) 4.91 kN (c) 4.71 kN

8.6 A mass of 200 kg hangs on the end of a winch cable. The winch accelerates horizontally at 2 m/s^2.
Determine:
(a) angle made by the winch cable with the vertical
(b) tension in the winch cable.
Neglect wind resistance.
(a) 11.5° (b) 2 kN

8.7 The forklift truck shown in Figure P8.7 has an unladen mass of 1 t.
Determine:
(a) maximum upward acceleration of the 900 kg load in order that the normal reaction on *each* of the two rear wheels is not less than 450 N
(b) normal reaction on each of the two front wheels under condition (a).
(a) 2.72 m/s^2 (b) 10.1 kN

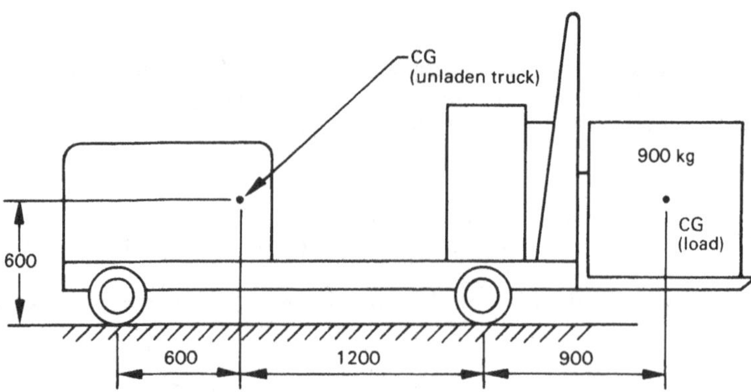

Fig. P8.7

8.8 The load shown in Figure P8.7 is raised until its centre of mass is 4.5 m above ground level.

Determine:

(a) maximum rearward acceleration of the forklift in order to maintain a minimum normal reaction of 450 N on each of the two rear wheels

(b) normal reaction on each of the two front wheels under condition (a).

(a) 0.474 m/s^2 (b) 8.87 kN

8.9 The aircraft shown in Figure P8.9 has a mass of 1 t and stops after landing by a constant deceleration force applied to each of the two rear wheels. The plane lands at 140 km/h and comes to rest in 200 m.

Determine:

(a) normal reaction on the front wheel just before the plane comes to rest (when the lift from the wings is zero)

(b) static reaction on the front wheel.

(a) 3.5 kN (b) 981 N

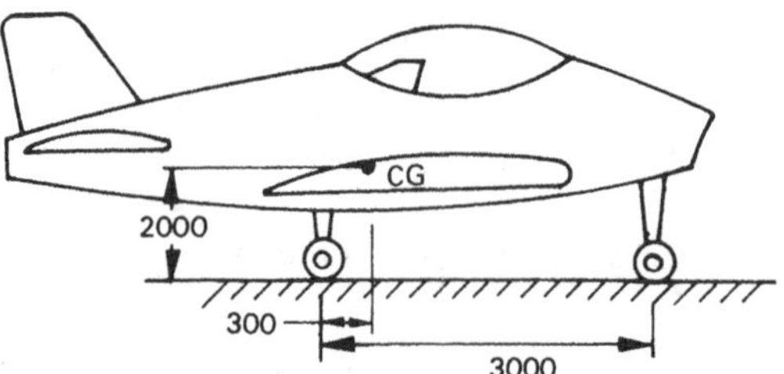

Fig. P8.9

8.10 The loaded four-wheel trailer shown in Figure P8.10 has a total mass of 5 t and is pulled along a level road. Average resistance to motion (friction and windage) is 2 per cent of the total weight and acts at the same horizontal level as the applied force *P*.

Determine the normal reaction at each of the front and rear wheels when the trailer moves:

(a) at constant velocity

(b) with forward acceleration 1.5 m/s^2.

(a) 14.7 kN, 9.81 kN (b) 14.0 kN, 10.6 kN

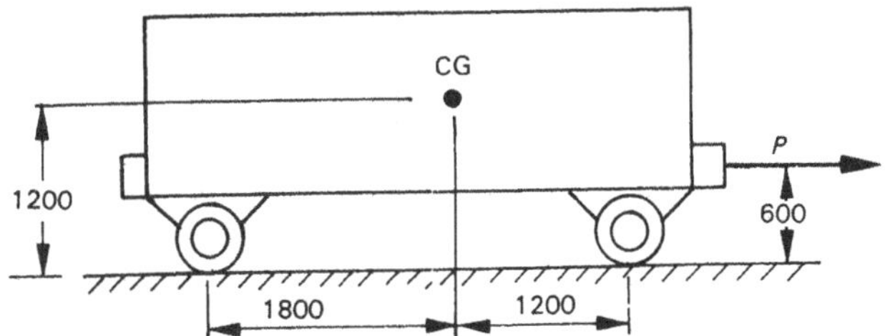

Fig. P8.10

8.11 The garage door shown in Figure P8.11 has a mass of 100 kg. It moves on rollers which run on an overhead track. The frictional resistance acting at a roller is 10 per cent of the normal reaction.

Determine:

(a) acceleration of the door caused by force $P = 200$ N
(b) normal reactions N_1 and N_2 at each roller
(c) friction forces F_1 and F_2 at each roller.

(a) 1.02 m/s² (b) 579 N, 402 N (c) 57.9 N, 40.2 N

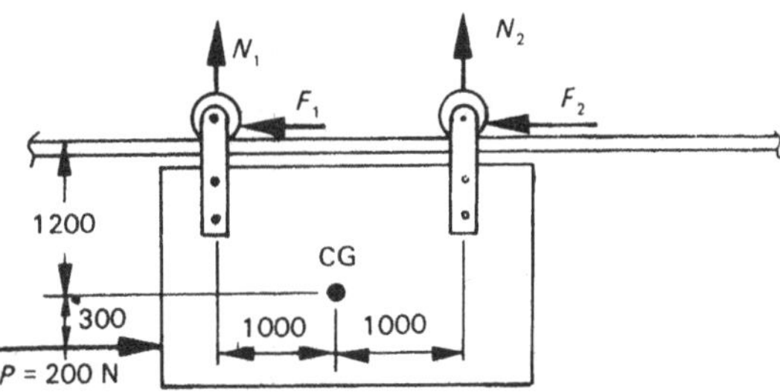

Fig. P8.11

8.12 A mass of 0.5 kg is suspended from a 1.5 m cord and rotated about a vertical axis at a speed of 50 rpm.

(a) Determine the angle made by the cord with the vertical axis and the tension in the cord.
(b) If the mass were changed to 1 kg, what would the angle and tension now be?

(a) 76.2°, 20.6 N (b) 76.2°, 41.2 N

8.13 The rotor in an amusement park is illustrated in Figure P8.13.

Determine the rotational speed in rpm at which the riders will remain suspended to the walls when the floor retracts if the coefficient of friction at the sides of the rotor is 0.2.

56.5 rpm

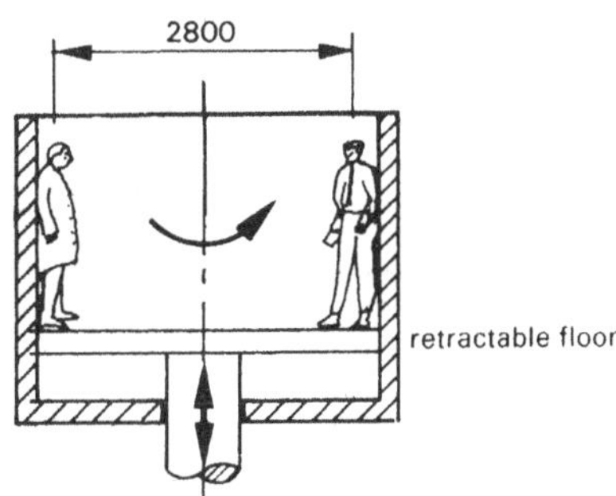

Fig. P8.13

8.14 The cross-section of a caravan is shown in Figure P8.14.

Determine:

(a) maximum speed at which the caravan can travel around a 100 m radius level curve without skidding on a wet day if the coefficient of friction is 0.2

(b) maximum speed and corresponding minimum coefficient of friction at which overturning rather than skidding will occur.

(a) 50.4 km/h (b) 97.6 km/h, 0.75

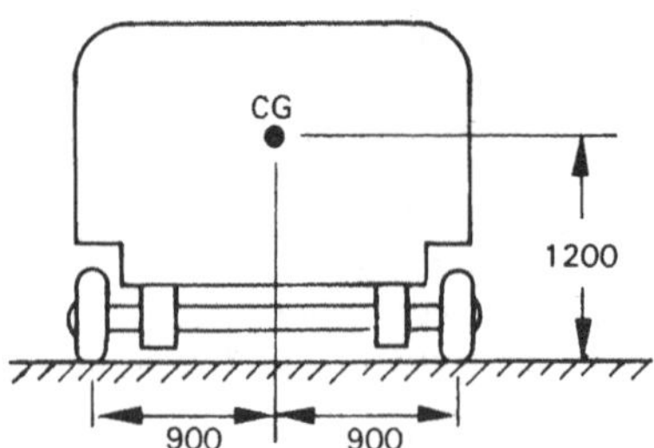

Fig. P8.14

8.15 For the plate shown in Figure P8.15, determine the maximum rotational speed in rpm at which the mass will slide off if the coefficient of friction is 0.25.

37.1 rpm

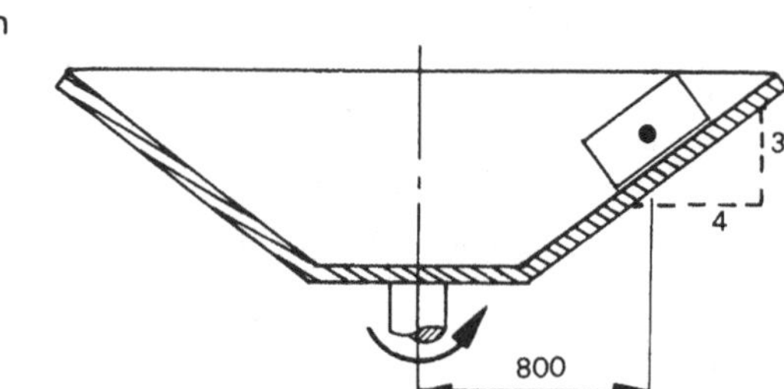

Fig. P8.15

8.16 If a mass of different material is placed in the same position on the plate given in problem 8.15 and sliding just takes place at 42 rpm, determine the coefficient of friction.

0.38

8.17 A 300 m radius road curve is to be elevated so that outward skidding will not occur at speeds below 100 km/h.

If the minimum coefficient of friction is 0.2 and the road is 8 m wide, determine the required elevation. Also determine the most comfortable road speed for passengers on this curve with elevation as determined.

472 mm, 47.5 km/h

8.18 A motor vehicle mass 2 t has the front wheels spaced 1500 mm apart. The centre of mass is 670 mm above the road and located so that 60 per cent of the vehicle weight is supported by the front wheels. The vehicle rounds a level curve 240 m radius at a speed of 100 km/h. Determine for the inner and outer front wheels:

(a) normal reaction

(b) friction force.

(a) 4.16 kN, 7.61 kN (b) 1.36 kN, 2.49 kN

8.19 Determine the angular acceleration of the mass shown in Figure P8.19 by:
(a) use of rotational formula $\tau = I\alpha$
(b) use of the linear formula $F = ma$ and then converting the linear acceleration to angular acceleration.
The inertia of the slender rod may be neglected and the mass treated as a point mass.
2.4 rad/s^2

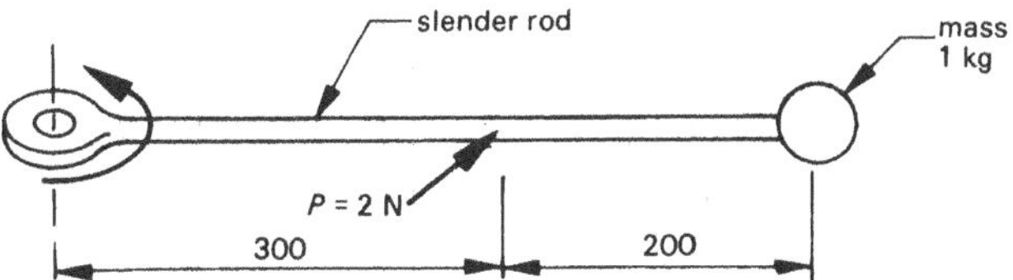

Fig. P8.19

8.20 A circular grindstone (relative density 2.35) has a diameter of 1 m and a uniform thickness of 100 mm.
Determine:
(a) mass moment of inertia
(b) radius of gyration
(c) time taken for the grindstone to come to rest from a speed of 150 rpm by a frictional torque of 3 Nm.
(a) 23.1 kg m^2 (b) 354 mm (c) 2 min

8.21 A cast-iron ring (relative density 7.2) is 200 mm wide and 40 mm thick with a mean diameter of 600 mm.
Determine:
(a) mass moment of inertia considering the ring as a thin ring (point mass)
(b) exact mass moment of inertia
(c) torque necessary to accelerate the ring from 0 to 300 rpm in 5 s.
(a) 9.77 kgm^2 (b) 9.82 kgm^2 (c) 61.7 Nm

8.22 The rim of a large cast-iron flywheel (relative density 7.2) is 600 mm wide and the inner and outer diameters are 3 m and 3.3 m.
Determine the mass moment of inertia and radius of gyration of the rim about its rotational axis.
15.9×10^3 kgm^2, 1.58 m

8.23 A steel ring 500 mm outside diameter, 200 mm inside diameter and 900 mm wide, is rotated with angular acceleration 0.5 rad/s^2.
Determine the torque required. The density of steel is 7800 kg/m^3.
21 Nm

8.24 The winding drum and rotor of a large earth-moving plant have a combined mass of 500 t and radius of gyration 3 m. The motor output torque is 1 MNm and friction torque is 20 KNm.
Determine the time taken and number of revolutions made by the drum in achieving working speed of 120 rpm from rest.
57.7 s, 57.7 revolutions

8.25 The brake shown in Figure P8.25 is applied to the drum which rotates at 180 rpm and has a mass moment of inertia 50 kgm^2.
Determine the force P necessary to bring the drum to rest in 6 s if the coefficient of friction between the brake and drum is 0.5.
62.8 N

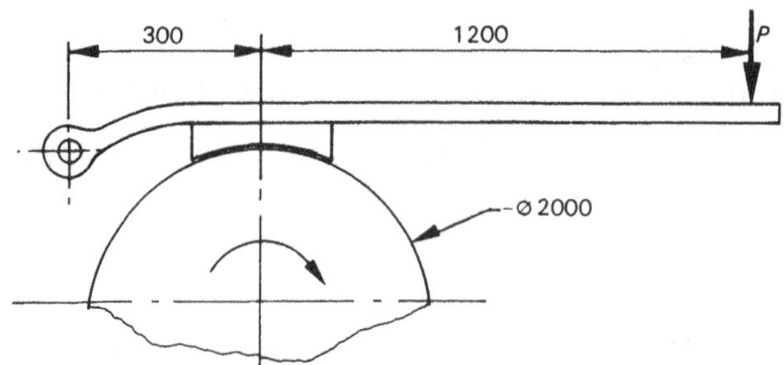

Fig. P8.25

8.26 The flywheel shown sectioned in Figure P8.26 is manufactured from steel (RD 7.8).

Determine:

(a) mass of flywheel
(b) mass moment of inertia
(c) radius of gyration.

The boss may be neglected and the following dimensions in mm are given: A 1200, B 60, C 100, D 30, E 200, F 600.

(a) 338 kg (b) 81.7 kgm^2 (c) 492 mm

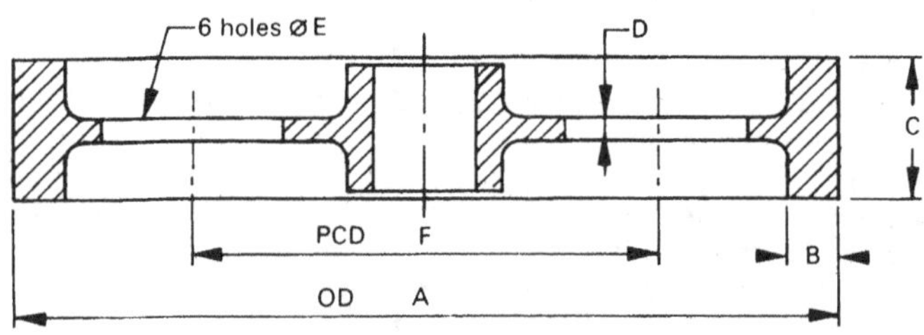

Fig. P8.26

8.27 Repeat the calculations for problem 8.26 given a cast-iron flywheel (RD 7.2) of the following dimensions: A 1500, B 70, C 150, D 50, E 250, F 800.

(a) 757 kg (b) 277 kgm^2 (c) 605 mm

8.28 For the system of masses shown in Figure P8.28, determine after release:

(a) acceleration of the masses
(b) tension in the common cord
(c) tension in upper cord
(d) velocity of the masses after moving 3 m from rest.

(a) 0.892 m/s^2 (b) 53.5 N (c) 108 N (d) 2.31 m/s

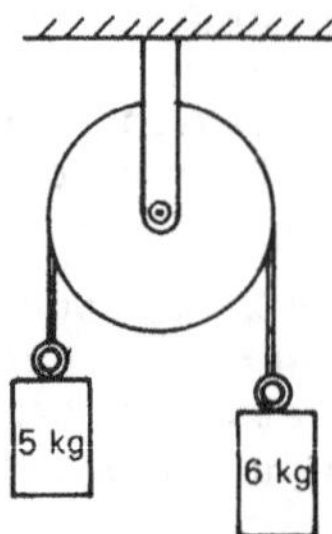

Fig. P8.28

8.29 In the system of masses shown in Figure P8.29, block A is 25 kg and block B 40 kg.

Determine after release:

(a) acceleration of block A
(b) acceleration of block B
(c) common cord tension
(d) tension in cord supporting block B.

(a) 1.4 m/s^2 (b) 0.7 m/s^2 (c) 210 N (d) 420 N

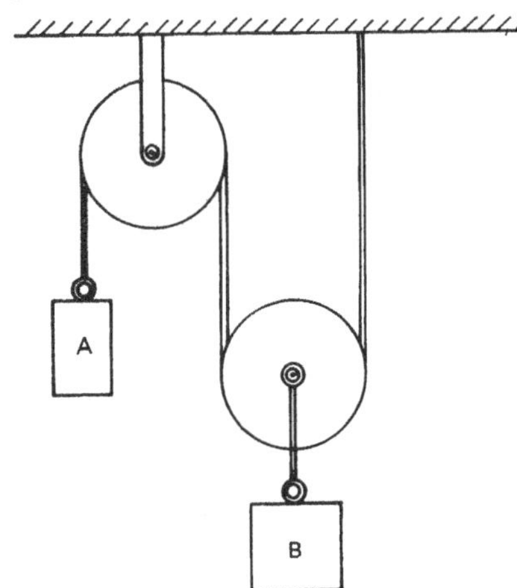

Fig. P8.29

8.30 For the system shown in Figure P8.30 determine the velocity of the system 3 s after release if block A is 20 kg and block B 10 kg and the coefficient of friction is 0.3. Also determine the tension in the cord.

3.92 m/s, 85 N

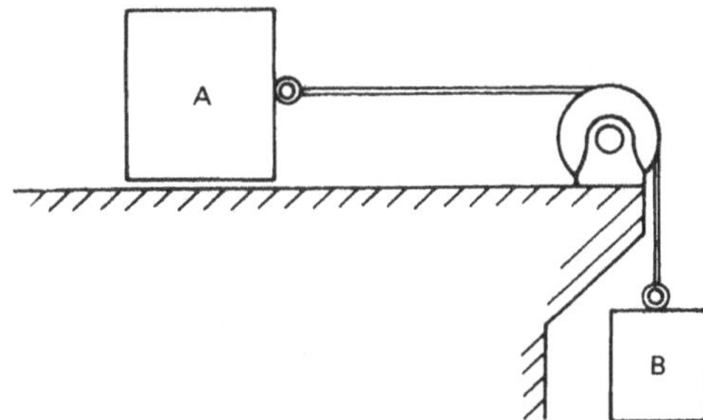

Fig. P8.30

8.31 For the masses shown in Figure P8.30, A = 40 kg and the coefficient of friction is 0.25.

Determine mass B in order that the system move from rest a distance of 3 m in 2 s.

19 kg

8.32 For the system shown in Figure P8.32, determine the acceleration of the system and the tension in the cord. Mass A is 30 kg, mass B is 20 kg, coefficient of friction is 0.25.

0.528 m/s^2, 186 N

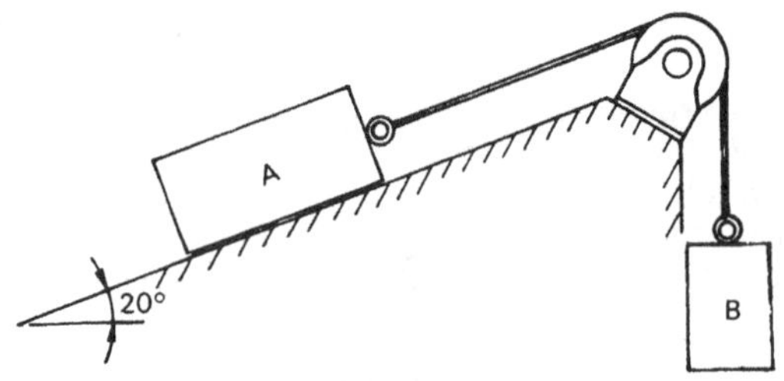

Fig. P8.32

8.33 Repeat problem 8.32 with mass A 60 kg, mass B 5 kg.

0.215 m/s^2, 50.1 N

8.34 For the system shown in Figure P8.34, determine the velocity of block B after it has moved 1.5 m from rest. Also determine the tension in the common cord. Block A is 25 kg, block B is 45 kg, the coefficient of friction is 0.3.

0.78 m/s, 216 N

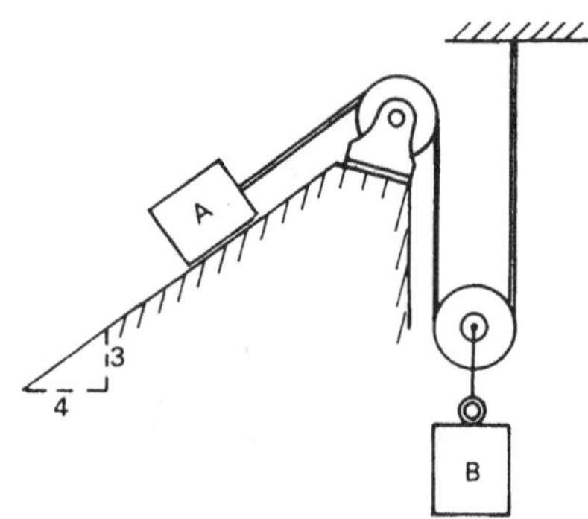

Fig. P8.34

8.35 For the system of masses shown in Figure P8.35, A is 50 kg, B is 130 kg and the coefficient of friction is 0.2.

Determine:

(a) acceleration of system after release
(b) tension in cord after release
(c) velocity of masses at the instant B strikes the stop at C
(d) distance A will continue moving after B stops.

(a) 0.872 m/s^2 (b) 142 N (c) 2.29 m/s (d) 1.33 m

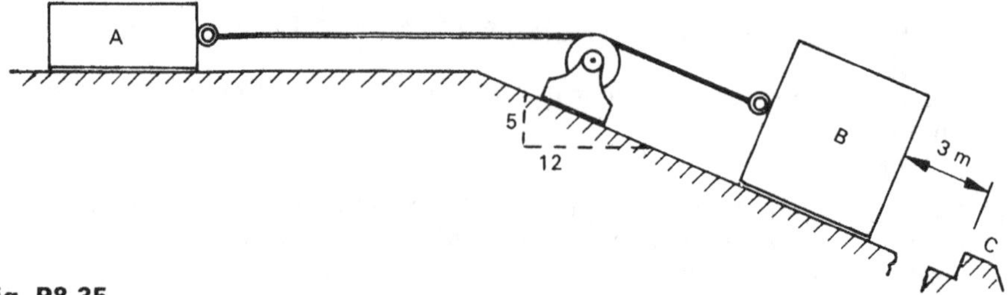

Fig. P8.35

8.36 For the system of masses shown in Figure P8.36, A is 50 kg, B is 100 kg and the coefficient of friction is 0.4.

Determine the distance moved by the masses 3 s after release and the tension in the cord.

5.04 m, 541 N

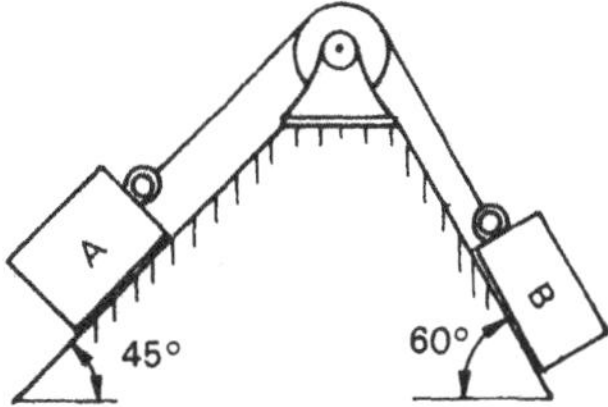

Fig. P8.36

8.37 The flywheel shown in Figure P8.37 has a diameter of 300 mm. A cord is pegged to the flywheel and then wrapped around it several times. Mass *m* of 2 kg is attached to the end of the cord.

If the mass moment of inertia of the flywheel and shaft is 1.5 kgm^2, determine how long it will take for the mass to descend a distance of 3 m after release. The diameter of the cord and bearing friction may be considered negligible.

4.58 s

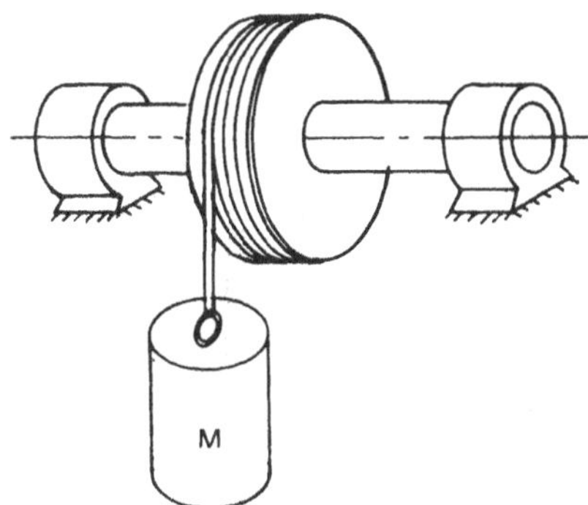

Fig. P8.37

8.38 For a flywheel system similar to that shown in Figure P8.37, a mass of 1.5 kg descends a distance of 1 m in 3.5 s after release. The diameter of the flywheel is 420 mm and the mass of the flywheel and shaft is 180 kg.

Determine the mass moment of inertia and radius of gyration if the diameter of the cord and bearing friction may be considered negligible.

3.91 kgm^2, 147 mm

8.39 The flywheel shown in Figure P8.39 has a diameter of 260 mm and the shaft has a diameter of 22 mm. A cord is pegged to the shaft and then wrapped around it several times. A mass *m* of 80 g attached to the end of the cord lowers at constant velocity after release. An additional mass of 1.5 kg attached to the first mass lowers 1.2 m in 12.6 s before the cord and mass drop off the shaft. The diameter of the cord is 2.5 mm and the mass of the flywheel and shaft is 5.2 kg.

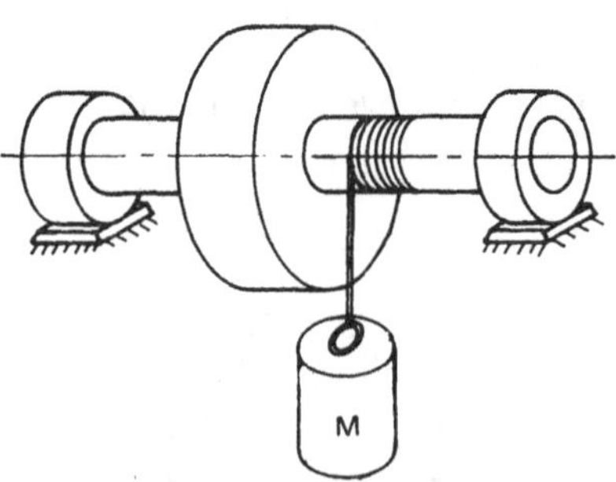

Fig. P8.39

Determine:
(a) bearing frictional torque
(b) tension in the cord as the combined mass is lowering
(c) mass moment of inertia of the flywheel and shaft
(d) radius of gyration of the flywheel and shaft
(e) number of revolutions made by the flywheel after the combined mass and cord drop off the shaft.

(a) 0.0096 Nm (b) 15.5 N (c) 0.146 kgm^2 (d) 168 mm (e) 292 revs

8.40 For the system shown in Figure P8.40, determine the tension in the cord and the pull P necessary to give the wheel a clockwise angular acceleration of 4 rad/s^2. The wheel has a mass moment of inertia 80 kg m^2. The coefficient of friction on the inclined plane is 0.25 and bearing friction is negligible.

512 N, 697 N

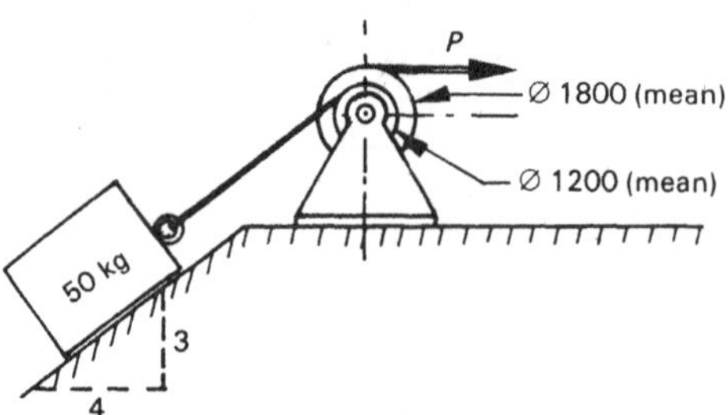

Fig. P8.40

9

Momentum, energy and power

In previous chapters, mechanics problems have been solved by use of the principles of force and moment equilibrium: that is, for any system in equilibrium, the sum of the forces in any direction and the sum of the moments about any point is zero.

The principles of force and moment equilibrium may be applied to masses at rest or in uniform motion and may be extended to apply to accelerating masses by use of the inertia force and d'Alembert's principle.

For systems of masses in motion (either uniform or non-uniform) there are two other principles which may be applied to give an alternative solution. These are the principles of conservation of momentum and conservation of energy. The use of the word "principle" is common but more correctly they should be called "laws" because of their fundamental nature and general applicability to all branches of science and technology.

When dealing with energy, it is also important to consider the rate at which energy is converted. The rate of conversion of energy (with time) is known as power. In practice, energy conversion inevitably involves expenditure of energy into some non-useful forms (e.g. friction). The non-useful energy is also known as "losses". Energy conversion devices are known as machines and the ability of a machine to convert energy with a minimum of loss is a most important consideration and is the measure of the efficiency of the machine.

9.1 Momentum

Momentum is a property possessed by a body which has both mass and velocity (relative to another body or frame of reference). Momentum may be linear (translational) or angular (rotational) depending upon the motion of the body.

Linear momentum $= mv$ (kg m/s)

Angular momentum $= I\omega$ (kg m^2/s)

where m = mass (kg)
v = linear velocity (m/s)
I = mass moment of inertia (kg m^2)
ω = angular (rotational) velocity (rad/s)

Momentum is an important property in mechanics because of Newton's second law of motion which states that:

force = time rate of change in momentum

i.e. $$F = \frac{\Delta(mv)}{\Delta t} \quad \text{(linear)}$$

or alternatively,

impulse = change in momentum

i.e. $$F\,\Delta t = \Delta(mv) \quad \text{(linear)}$$

both being equivalent statements.

In the limit, since $\frac{\Delta v}{\Delta t} \rightarrow \frac{dv}{dt} \rightarrow a$

$$F = ma$$

which is the familiar equation relating force and acceleration of a mass.

Note: Since velocity is a vector quantity, momentum is also a vector quantity and the direction of the force and the change in momentum must be the same.

Example 9.1

The horizontal coal conveyor shown in Figure 9.1 conveys 270 t/h of coal at a speed of 5.4 km/h.

Determine the torque at the drive shaft if 30 per cent of the torque applied at the drive shaft is used to overcome friction.

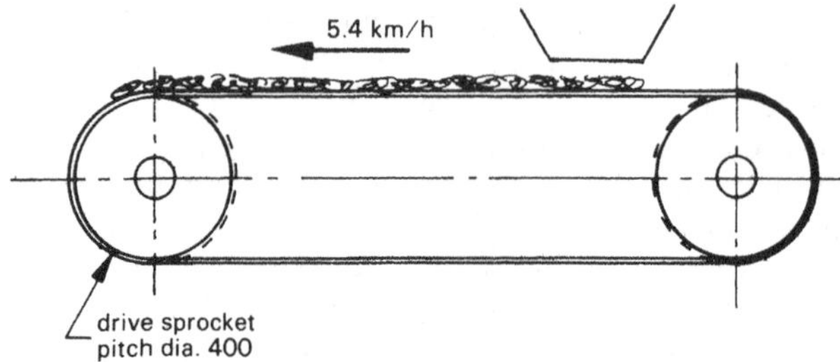

Fig. 9.1

Solution

In the horizontal direction, the change in velocity of the coal is 5.4 km/h = 1.5 m/s. Using a time base of one second, the mass of coal conveyed is:

$$\frac{270 \times 10^3}{3.6 \times 10^3} = 75 \text{ kg/s}$$

Therefore the momentum change per second in the horizontal direction is:

$$\Delta(mv) = 75 \times 1.5 = 112.5 \text{ kg m/s}$$

Using Newton's second law:

$$\begin{aligned} \text{force } F &= \text{momentum change per second} \\ &= 112.5 \text{ N} \end{aligned}$$

If there were no friction, the ideal drive shaft torque would be:

$$\begin{aligned} \tau &= Fr \\ &= 112.5 \times \frac{0.4}{2} \\ &= 22.5 \text{ Nm} \end{aligned}$$

Since friction is 30 per cent of the applied torque τ_A, the frictional torque is 0.3 τ_A.

$$\therefore \quad \tau_A = 22.5 + 0.3\,\tau_A$$

$$\therefore \quad 0.7\,\tau_A = 22.5$$

$$\therefore \quad \tau_A = \mathbf{32.1\ Nm}$$

9.2 Principle of momentum conservation

This principle (law) follows as a result of Newton's second law of motion. Since:

force = time rate of change in momentum

it follows that if there is no force, there cannot be a change in momentum.

The principle of momentum conservation may thus be stated:

In any system, if no forces or moments are transferred into or out of the system, the total momentum within the system remains constant.

Notes

1. Momentum is a vector quantity and therefore the principle applies to linear momentum in any chosen direction or to angular momentum about any chosen point. Momentum will be negative if it has the opposite direction to that specified as positive.
2. System was defined in Chapter 2 and is any region of space or any group of masses or forces being considered. The system is separated from the surroundings by a boundary which does not have to be a real physical one and indeed in mechanics it is usually conceptual.
3. Forces or moments which may be transferred across a system boundary are also known as **external** forces or moments to distinguish them from the **internal** forces or moments which may occur as a result of momentum transfer within the system.
4. Where external forces or moments are transferred across a system boundary, the principle may still be applied if the system boundary is extended to include the body upon which the external forces or moments react so that they become internal relative to the new system boundary. This is explained more fully in example 9.2.
5. In mechanics, the principle of momentum conservation is most often applied to problems involving collision or departure of masses.

Example 9.2

A projectile of mass 5 kg is fired horizontally from a cannon of mass 1.8 t which is mounted on rails as shown in Figure 9.2. The velocity of the projectile leaving the barrel is 1.5 km/s.

(a) Determine the recoil velocity of the cannon at the instant when the projectile leaves the barrel if friction between the cannon and the ground is negligible.

(b) Recalculate the recoil velocity of the cannon if the firing angle is changed to 50° above the horizontal.

(c) Explain how the principle of momentum conservation applies to the vertical direction in (b).

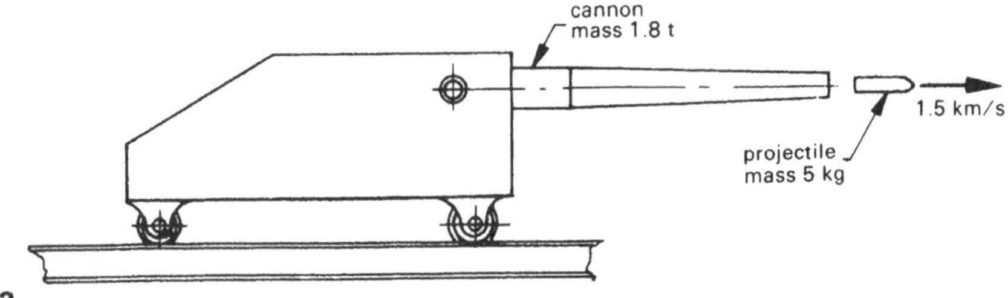

Fig. 9.2

Solution

(a) Define the system by drawing the boundary as shown in Figure 9.3.

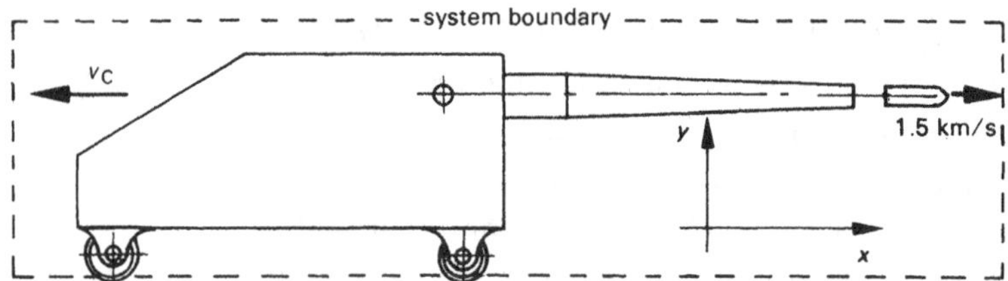

Fig. 9.3

Before firing, the total momentum within the system is zero. Since friction is negligible there is no transfer of force across the system boundary on firing and the momentum in the system must be zero during firing. This can only be true if the cannon recoils backward during firing (negative momentum in the x direction) because the projectile has positive momentum in the x direction during firing.

Denoting v_C as the velocity of the cannon after firing and applying the principle of momentum conservation to the x direction:

$$0 = 5 \times 1500 - 1800 \times v_C$$

$$\therefore \quad v_C = \mathbf{4.17\ m/s} \leftarrow$$

(b) Again considering the conservation of momentum in the x direction:

$$0 = 5 \times 1500 \cos 50° - 1800 \times v_C$$

$$\therefore \quad v_C = \mathbf{2.68\ m/s} \leftarrow$$

(c) If the principle of momentum conservation is applied in (b) to the y direction, it appears to be violated because the cannon cannot move downward through the rails! The reason why the principle does not apply to the specified system in the y direction is because in this direction there is a transfer of force across the system boundary which results in a reaction force at the ground. Note that this reaction force does not occur in the horizontal direction because friction is negligible.

In order to apply the principle of momentum conservation in the y direction, it is necessary to extend the system boundary to include the body upon which the transferred force reacts, in this case the earth. Hence the conclusion is that the earth moves downward as a result of the firing of the cannon at an angle upward (see Fig. 9.4)!

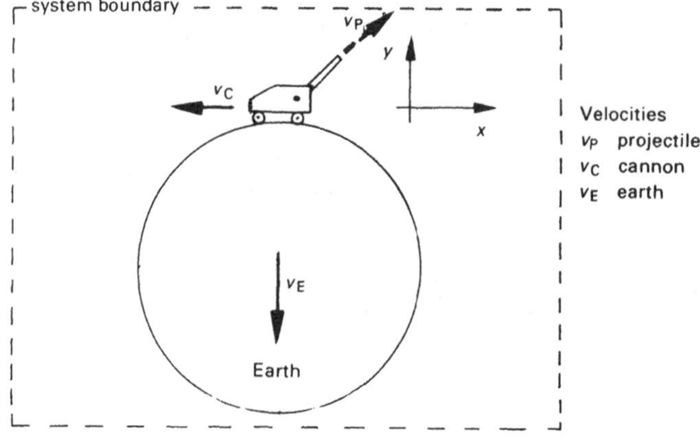

Fig. 9.4

Of course the movement is extremely small because the earth has such a huge mass compared to the projectile.

Example 9.3

A vehicle of mass 1.4 t travelling north-east at 60 km/h collides with another vehicle of mass 2.2 t travelling due west at 40 km/h.

Determine the resultant motion of the combined vehicle mass immediately after impact if the effect of friction on the road surface is neglected.

Solution

Choose a boundary to include both vehicles as shown in Figure 9.5.

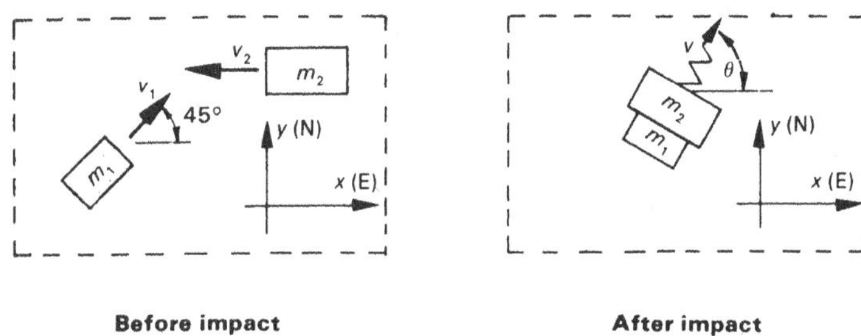

Fig. 9.5

Because friction on the road surface is neglected, the momentum within the system in any direction will be the same before and after impact.

Choosing the x direction:

Before impact *After impact*

$$m_1 v_1 \cos 45° - m_2 v_2 = (m_1 + m_2) v_x$$

$$\therefore \; 1.4 \times 60 \times \cos 45° - 2.2 \times 40 = 3.6 \times v_x$$

$$\therefore \; v_x = -7.945 \text{ km/h}$$

Similarly in the y direction:

$$m_1 v_1 \sin 45° = (m_1 + m_2) v_y$$

$$\therefore \; 1.4 \times 60 \times \sin 45° = 3.6 \times v_y$$

$$\therefore \; v_y = 16.5 \text{ km/h}$$

Hence the resultant velocity may be determined:

v = **18.3 km/h**

θ = **115.7°**

Note: It is not necessary to convert the mass to kg or the velocity to m/s when applying the principle of momentum conservation, provided that consistency is maintained.

Example 9.4

A shaft and flywheel, mass moment of inertia 1.2 kgm^2, rotates freely at 1440 rpm before being clutched to a stationary shaft which has a mass moment of inertia 0.4 kgm^2. After the clutch is released both shafts rotate at the same speed.

Determine this speed if friction is negligible.

Solution

The system boundary may be defined to include both shafts. Since bearing friction is negligible, the angular momentum within this system is the same before and after clutching. Hence:

Before clutching   *After clutching*

$$I_1 \omega_1 = (I_1 + I_2)\,\omega$$

$$\therefore\ 1.2 \times 1440 = (1.2 + 0.4)\,\omega$$

$$\therefore\ \omega = \mathbf{1080\ rpm}$$

Note: Again it is not necessary to convert angular velocity to rad/s, provided consistency is maintained in the momentum equation.

9.3 Mechanical energy

Although there are many forms of energy, in mechanics the only forms normally considered are potential energy and kinetic energy.

Potential energy

This is the energy possessed by a mass by virtue of its position in a gravitational field (relative to another mass). According to Newton's law of universal gravitation, every mass exerts a force of attraction on every other mass. Therefore when one mass is moved away from another, work is done in overcoming the gravitational force and there is a corresponding energy increase which is an increase in potential energy. The gravitational force is very small unless one or both of the masses is very large or the distance between the centres of mass is very small. The earth is a very large mass and therefore exerts a significant gravitational force (weight) and hence the corresponding potential energy is significant.

Potential energy is given by:

$$\boxed{PE = mgh} \qquad \textbf{(9.1) potential energy}$$

where PE = potential energy (J)
m = mass of the body (kg)
g = earth's gravitational constant (N/kg), also
= acceleration due to gravity (m/s^2)
h = height above the earth (m)
(measured from any convenient datum)

Notes

1. Although the gravitational constant g varies with distance above the earth, for most mechanics problems, the variation is slight and a mean value of 9.81 may be used.
2. The height h is measured vertically above the earth to the centre of mass of the body (CG).

In all problems, changes in potential energy need only be calculated so that any convenient height may be used for datum and the potential energy calculated relative to this datum.

3. If a body moves horizontally, there is no change in potential energy.

Kinetic energy

Kinetic energy is the energy possessed by a mass by virtue of its velocity (relative to another mass). Because an increase in velocity requires an acceleration, there must also be a force which, in moving through a distance, does work. The corresponding energy increase is known as kinetic energy. Linear kinetic energy is given by:

$$KE = \tfrac{1}{2}mv^2$$

(9.2) linear kinetic energy

where KE = linear kinetic energy (J)
m = mass of the body (kg)
v = velocity of the body (m/s)

Rotational kinetic energy is given by:

$$KE = \tfrac{1}{2}I\omega^2$$

(9.3) rotational kinetic energy

where KE = rotational kinetic energy (J)
I = mass moment of inertia of the body (kgm^2)
ω = angular velocity of the body (rad/s)

Notes

1. Since velocity of any body can only be measured relative to another body (or frame of reference), kinetic energy can only be specified with relation to the other body (or frame of reference). The usual frame of reference is the earth.
2. A body can have linear and rotational energy simultaneously and indeed this is often the case. For example, the wheels of a vehicle in contact with the ground while the vehicle moves forward have linear motion at their centre as well as rotational motion about the centre. Also in most ball games, such as golf, cricket, snooker, etc., the ball rotates (spins) as it moves.

9.4 Work and power

Kinetic energy and potential energy are forms of mechanical energy which are stored in a system, that is these energy levels do not change unless there is a transfer of energy into or out of the system. In mechanics, the transfer of energy into or out of a system occurs as work. If the work is transferred into the system there is a corresponding increase in energy, for example when a motor is used to drive a machine. If the work is transferred out of the system, there is a corresponding decrease in energy, for example through the mechanism of friction.

Work measures the total amount of energy transferred when a force is displaced but does not measure the rate of transfer of the energy. The rate of transfer (with respect to time) is known as power.

Linear work and power are given by:

$$W = Fs \quad \textbf{(9.4) linear work}$$

$$P = Fv \quad \textbf{(9.5) linear power}$$

where W = linear work (J)
P = linear power (W)
F = force (N)
s = linear distance moved by the force (along its line of action) (m)

Rotational work and power are given by:

$$W = \tau\theta \quad \textbf{(9.6) rotational work}$$

$$P = \tau\omega \quad \textbf{(9.7) rotational power}$$

where W = rotational work (J)
P = rotational power (W)
τ = torque (Nm)
θ = angular distance moved by the torque (rad)

Example 9.5

A winch has a drum of pitch diameter 250 mm and is used to raise a load of 1.2 t vertically. The load is accelerated from rest to the working velocity of 0.6 m/s in 2 m.

If frictional losses are neglected, determine:

(a) maximum power required to raise the load during the accelerating period
(b) power required to raise the load at working velocity
(c) change in potential energy during the accelerating period
(d) change in kinetic energy during the accelerating period
(e) rotational speed of the winch drum at working velocity
(f) maximum torque at the winch drum
(g) maximum power at the winch drum.

Solution

During the period of acceleration, the acceleration a may be determined from:

$$v_2^2 = v_1^2 + 2as$$

$$\therefore 0.6^2 = 0^2 + 2 \times a \times 2$$

$$a = 0.09 \text{ m/s}^2$$

(a) Force required to raise the load during the accelerating period (which is also the tension in the cable) is:

$$F_{max} = m(g + a)$$

$$= 1.2 \times 10^3 \times (9.81 + 0.09) \quad \text{(N)}$$

$$= 11.88 \text{ kN}$$

Since F is constant during the accelerating period, the maximum power occurs at the point of maximum velocity, that is when $v = 0.6$ m/s (just before the motion changes to one of constant velocity).

$$P_{max} = F_{max}\,v$$
$$= 11.88 \times 0.6$$
$$= \mathbf{7.128\ kW}$$

(b) At working velocity:

$$F = mg$$
$$= 1.2 \times 10^3 \times 9.81 \quad \text{(N)}$$
$$= 11.772\ \text{kN}$$
$$P = Fv$$
$$= 11.772 \times 0.6$$
$$= \mathbf{7.06\ kW}$$

(c)
$$PE = mgh$$
$$= 1.2 \times 10^3 \times 9.81 \times 2 \quad \text{(J)}$$
$$= \mathbf{23.54\ kJ}$$

(d)
$$KE = \tfrac{1}{2}\,mv^2$$
$$= \tfrac{1}{2} \times 1.2 \times 10^3 \times 0.6^2 \quad \text{(J)}$$
$$= \mathbf{216\ J}$$

(e)
$$v = r\omega$$
$$\therefore\ \omega = \frac{v}{r} = \frac{0.6}{0.125} = 4.8\ \text{rad/s}$$
$$= \mathbf{45.8\ rpm}$$

(f)
$$\tau_{max} = F_{max}\,r$$
$$= 11.88 \times 0.125$$
$$= \mathbf{1.485\ kNm}$$

(g) Maximum winch drum power:

$$P_{max} = \tau_{max}\,\omega$$
$$= 1.485 \times 4.8$$
$$= \mathbf{7.128\ kW}$$

which is the same as the maximum power required to raise the load since frictional losses have been neglected.

Example 9.6

The flywheel of a punching press has a mass of 52.5 kg and a radius of gyration 350 mm. The wheel is accelerated from rest to operating speed 850 rpm by a constant torque of 100 Nm. The punching operation causes 10 per cent reduction in the speed of the flywheel over one-half a revolution.

If frictional losses are negligible, determine:

(a) maximum power input required to bring the flywheel to operating speed

(b) energy stored in the flywheel at operating speed
(c) number of revolutions made in reaching operating speed
(d) time taken to reach operating speed
(e) work done by the flywheel during punching
(f) average power output by the flywheel during punching.

Solution

(a) Since the torque is constant, the maximum power input occurs just as the flywheel reaches operating speed (maximum speed).

$$\text{Now } \omega_{max} = \pi \times \frac{850}{30} = 89 \text{ rad/s}$$

$$P_{max} = \tau\,\omega_{max} = 100 \times 89 \quad \text{(W)}$$

$$= \mathbf{8.9\ kW}$$

(b) The flywheel stores energy in the form of rotational kinetic energy.

$$\text{Now } KE = \tfrac{1}{2} I \omega^2$$

$$\text{and } \quad I = mk^2$$

$$= 52.5 \times 0.35^2$$

$$= 6.431 \text{ kg m}^2$$

$$\therefore \quad KE = \tfrac{1}{2} \times 6.431 \times 89^2$$

$$= 25\,470 \text{ J}$$

$$= \mathbf{25.5\ kJ}$$

(c) Since frictional losses are negligible, the energy equation is simply:

$$W = KE$$

$$\therefore \quad \tau\theta = KE$$

$$\therefore \quad 100 \times \theta = 25\,470$$

$$\therefore \quad \theta = 254.7 \text{ rad}$$

$$= \mathbf{40.5\ revs}$$

Note: This could also be solved by determining the angular acceleration of the flywheel.

$$\tau = I\alpha$$

$$\therefore \quad 100 = 6.431 \times \alpha$$

$$\therefore \quad \alpha = 15.55 \text{ rad/s}^2$$

$$\text{and } \omega_2^{\,2} = \omega_1^{\,2} + 2\,\alpha\,\theta$$

$$\therefore \quad 89^2 = 0^2 + 2 \times 15.55 \times \theta$$

$$\therefore \quad \theta = 254.7 \text{ rad (as before)}$$

(d) The average power supplied to the flywheel is:

$$\frac{8.9}{2} = 4.45 \text{ kW}$$

$$\text{Now } P = \frac{\text{energy change}}{\text{time}} \qquad \therefore\ t = \frac{KE}{P}$$

$\therefore \quad t = \frac{25.5}{4.45} = \mathbf{5.73\ s}$

Note: This could also be solved by:

$$\theta = \tfrac{1}{2}(\omega_1 + \omega_2)\,t$$

$$\therefore\ 254.7 = \tfrac{1}{2}(0 + 89)\,t$$

$$\therefore\quad t = 5.73\text{ s (as before)}$$

(e) The work done during punching equals the difference in kinetic energy before and after punching.

Since the punching operation causes a 10 per cent reduction in speed, the speed after punching is $0.9 \times 89 = 80.1$ rad/s.

$$W = KE_1 - KE_2$$

$$= \tfrac{1}{2}\,I(\omega_1^2 - \omega_2^2)$$

$$= \tfrac{1}{2} \times 6.431 \times (89^2 - 80.1^2) \quad \text{(J)}$$

$$= \mathbf{4.84\ kJ}$$

(f) In order to determine the average power output during punching, it is first necessary to calculate the time taken. This may be determined from:

$$\theta = \tfrac{1}{2}(\omega_1 + \omega_2)\,t$$

$$\text{Now } \theta = \tfrac{1}{2}\text{ rev} = \pi \text{ rad}$$

$$\therefore\quad \pi = \tfrac{1}{2} \times (80.1 + 89) \times t$$

$$\therefore\quad t = 0.0372\text{ s}$$

$$\text{and } P = \frac{W}{t}$$

$$= \frac{4.84}{0.0372}$$

$$= \mathbf{130\ kW}$$

9.5 Principle of energy conservation

This principle (law) may be stated thus:

In any system, if no energy is transferred into or out of the system, the total energy within the system remains constant.

This all-embracing principle is applicable to all forms of energy and energy transfer but mechanics deals only with mechanical energy (kinetic and potential) and energy transfer in the form of work.

Therefore for mechanical systems, the following equation applies:

$$\boxed{PE_1 + KE_1 \pm W = PE_2 + KE_2}$$

(9.8) energy equation (mechanical systems)

Notes

1. Energy and work are scalar quantities. The sign of the work transfer will be taken as positive if the work is into the system (increases the system energy) and negative if the

work is out of the system (decreases the system energy). Hence friction always causes negative work.

2. Equation 9.8 provides an alternative method of solution to many problems in mechanics and does not require the calculations of the acceleration and inertia force. This method is also known as the **work-energy method**.
3. If the motion is horizontal, the *PE* terms are the same on both sides of the equation and hence disappear.
4. The work-energy method *cannot* be used for collision or departure problems because mechanical energy is not conserved due to the heat generated by the deformation which occurs.

Example 9.7

A mass of 20 kg rests on an inclined plane as shown in Figure 9.6 and is acted upon by an applied force $P = 300$ N which moves the mass a distance of 15 m along the plane upward. The coefficient of friction is 0.4.

Using the initial position of the mass as datum, determine:

(a) final potential energy
(b) final velocity and kinetic energy using the force-acceleration method
(c) final velocity and kinetic energy using the work-energy method, i.e. without determination of the acceleration of the mass.

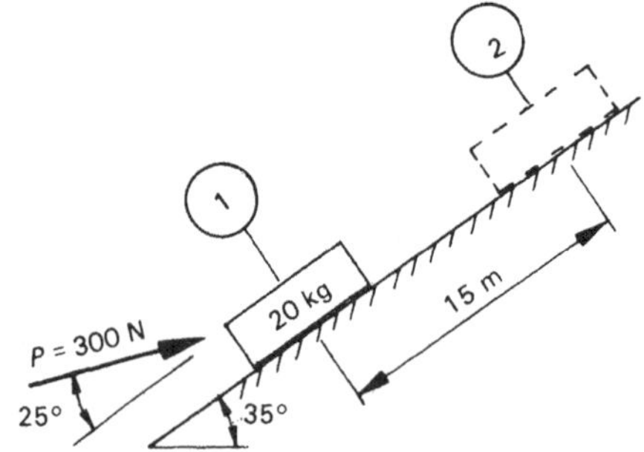

Fig. 9.6

Solution

(a) The vertical distance h moved by the mass is:

$$h = 15 \sin 35° = 8.6 \text{ m}$$

Therefore the potential energy in position ② is:

$$PE_2 = mgh = 20 \times 9.81 \times 8.6 \quad \text{(J)}$$
$$= \mathbf{1.69 \text{ kJ}}$$

(b) A free body diagram for the mass is shown in Figure 9.7.

The acceleration may be determined from (1) first principles, (2) equation 8.2 or (3) computer program INPLANE 2. Using the computer program with:

$$P = 300 \text{ N}, m = 20 \text{ kg}, \alpha = -25°, \theta = 35°, \mu = 0.4$$

the output is: $a = 2.218 \text{ m/s}^2$ (up the plane).

Since $v_2^2 = v_1^2 + 2as$

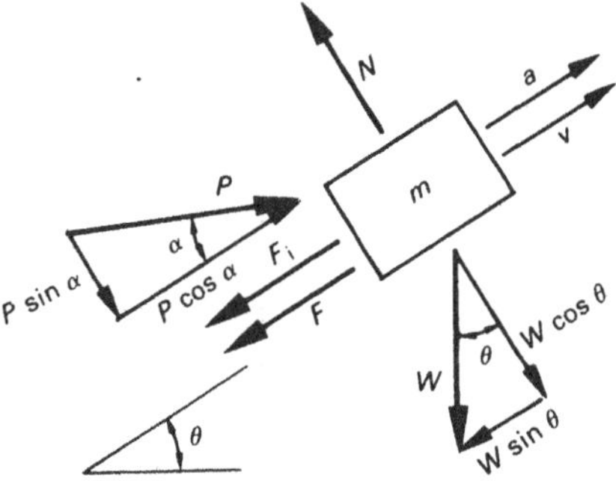

Fig. 9.7

$$\therefore\ v_2^2 = 0^2 + 2 \times 2.218 \times 15$$
$$\therefore\ v_2 = 8.15 \text{ m/s}$$

The kinetic energy in position ② is:

$$KE_2 = \tfrac{1}{2}\, mv_2^2 = \tfrac{1}{2} \times 20 \times 8.15^2$$
$$= \mathbf{665\ J}$$

(c) Define the system as shown in Figure 9.8.

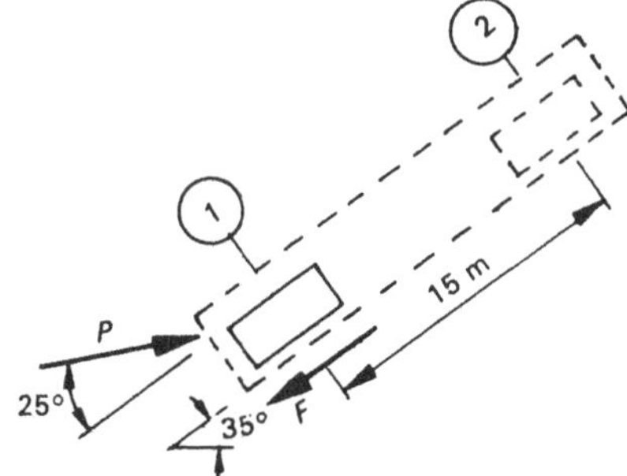

Fig. 9.8

The mechanical energy equation is:

$$PE_1 + KE_1 \pm W = PE_2 + KE_2$$

But PE_1 and KE_1 are both zero.

Hence:

$$W = PE_2 + KE_2$$

The positive work input to the system is done by the component of force P acting parallel to the plane. Note that the component of this force acting perpendicular to the plane does no work since it is not displaced along its line of action.

Now $W = Fs$

$$\therefore\ +W = P\cos 25° \times 15 \quad \text{(J)}$$
$$= 4.08 \text{ kJ}$$

The negative work which decreases the energy of the system is the work done by the friction force.

Now $F = \mu N$

and $N = W \cos \theta + P \sin \alpha$

$\therefore \quad F = 0.4\,(20 \times 9.81 \times \cos 35° + 300 \times \sin 25°)$

$= 115 \text{ N}$

$\therefore \; -W = 115 \times 15 \quad (\text{J})$

$= 1.725 \text{ kJ}$

Substituting in the mechanical energy equation (with $PE_2 = 1.69$ kJ as calculated previously)

$4.08 - 1.725 = 1.69 + KE_2$

$\therefore \; KE_2 = 0.665$ kJ (which is the same as the value 665 J calculated in (b)).

Hence v_2 may be calculated from

$KE_2 = \frac{1}{2} m v_2^2$

$\therefore \; 665 = \frac{1}{2} \times 20 \times v_2^2$

$\therefore \; \mathbf{v_2 = 8.15 \text{ m/s}}$ (as before)

Example 9.8

Solve example 8.17 using the energy method. This example is reproduced below:

A flywheel and shaft, combined mass 23 kg, are mounted in horizontal bearings. The shaft diameter is 72 mm and a light cord, diameter 2 mm, is wrapped around it, one end being secured to the shaft and the other end attached to a mass of 4 kg.

In an experiment, the mass is allowed to fall from rest a distance of 1 m and the time measured and found to be 12 s.

Determine:

(a) tension in the cord when the mass is falling
(b) mass moment of inertia of the flywheel and shaft
(c) radius of gyration of the flywheel and shaft.

Bearing friction and air resistance are negligible.

Solution

(a) The velocity v_2 may be determined from:

$h = \frac{1}{2}(v_1 + v_2)\,t$

$\therefore \quad 1 = \frac{1}{2} \times (0 + v_2) \times 12$

$\therefore \quad v_2 = 0.167 \text{ m/s}$

But $v_2 = r\omega_2 \quad \therefore \; \omega_2 = \dfrac{v_2}{r}$

and the mean radius $r = 37$ mm.

Hence $\omega_2 = \dfrac{0.167}{0.037} = 4.5 \text{ rad/s}$

In order to determine the tension in the cable, it is necessary to define the system boundary around the mass itself as shown in Figure 9.9.

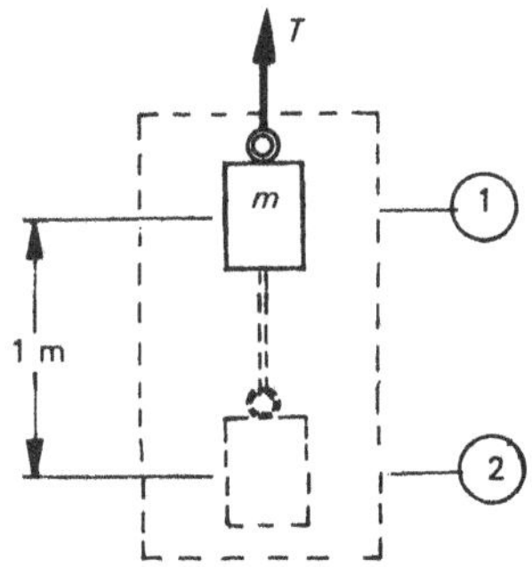

Fig. 9.9

Considering the energy balance of this system:

$$PE_1 + KE_1 \pm W = PE_2 + KE_2$$
$$\therefore PE_1 + 0 - T \times h = 0 + KE_2$$
$$\therefore PE_1 - KE_2 = T \times h$$
$$\therefore mgh - \tfrac{1}{2} mv_2^2 = T \times h$$

Substituting with $m = 4$ kg, $v_2 \Rightarrow 0.167$ m/s, $h = 1$ m:

$$4 \times 9.81 \times 1 - \tfrac{1}{2} \times 4 \times 0.167^2 = T \times 1$$

$\boldsymbol{T} = \mathbf{39.18\ N}$ (as before)

(b) Now define the system boundary to include both the mass and the flywheel as shown in Figure 9.10.

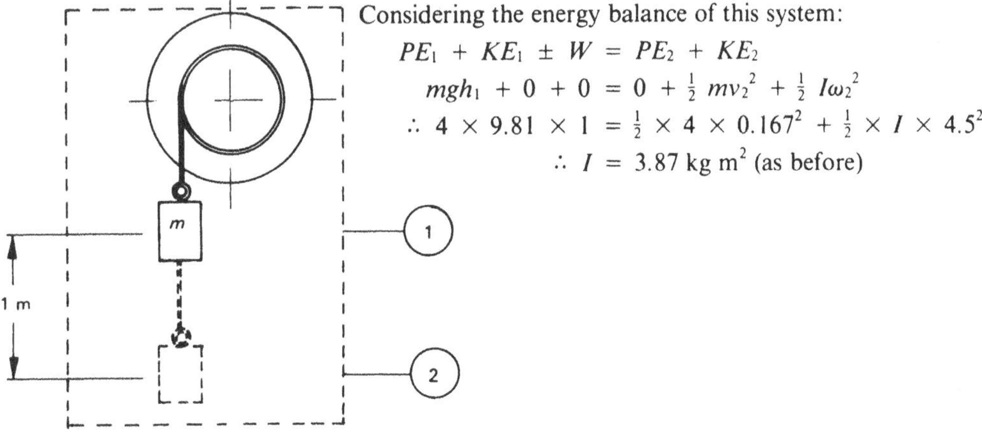

Fig. 9.10

Considering the energy balance of this system:

$$PE_1 + KE_1 \pm W = PE_2 + KE_2$$
$$mgh_1 + 0 + 0 = 0 + \tfrac{1}{2} mv_2^2 + \tfrac{1}{2} I\omega_2^2$$
$$\therefore 4 \times 9.81 \times 1 = \tfrac{1}{2} \times 4 \times 0.167^2 + \tfrac{1}{2} \times I \times 4.5^2$$
$$\therefore I = 3.87 \text{ kg m}^2 \text{ (as before)}$$

(c) $k = \sqrt{\dfrac{I}{m}} = \sqrt{\dfrac{3.87}{23}}$ (m)

$\boldsymbol{k} = \mathbf{410\ mm}$ (as before)

Example 9.9

It has been proposed that a water package fired from a cannon may be used to put out oil-well fires. An experiment set up to measure the velocity of the water package is shown in Figure 9.11.

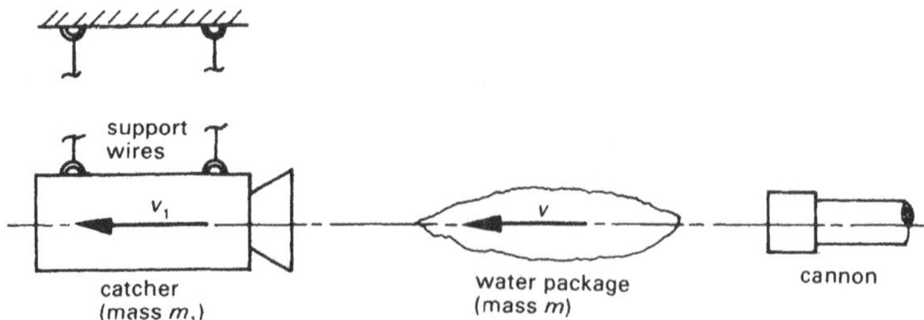

Fig. 9.11

Experimental data from a test run are:

Mass of water package fired (m) = 0.3 kg

Mass of catcher (m_1) = 14.3 kg

Vertical displacement of catcher after firing (h) = 316 mm

Use the data of the experiment to determine:

(a) velocity of the water package (v)
(b) loss in mechanical energy caused by the impact of the water package in the catcher.

Solution

(a) Define the system boundary around the catcher and the water package. Air resistance and friction in the catcher are obviously negligible. Applying the conservation of mechanical energy equation after impact:

$$PE_1 + KE_1 \pm W = PE_2 + KE_2$$

But $PE_1 = 0$, $W = 0$, $KE_2 = 0$

$$\therefore \tfrac{1}{2}(m + m_1) v_1^2 = (m + m_1) gh$$

$$\therefore v_1^2 = 2 \times 9.81 \times 0.316$$

$$\therefore v_1 = 2.49 \text{ m/s}$$

which is the velocity of the catcher and water package immediately after impact.

Applying the conservation of momentum principle to the impact itself:

$$(m + m_1) v_1 = mv$$

$$\therefore (0.3 + 14.3) \times 2.49 = 0.3 \times v$$

$$\therefore v = \mathbf{121 \text{ m/s}} \quad \mathbf{(436 \text{ km/h})}$$

which is the velocity of the water package just before impact.

(b) The mechanical energy before impact is the kinetic energy of the water package.

$$KE = \tfrac{1}{2} mv^2$$
$$= \tfrac{1}{2} \times 0.3 \times 121^2$$
$$= 2196 \text{ J}$$

The mechanical energy after impact is the combined kinetic energy of the catcher and water package.

$$KE = \frac{1}{2}(m + m_1)v_1^2$$
$$= \frac{1}{2} \times 14.6 \times 2.49^2$$
$$= 45 \text{ J}$$

The loss in mechanical energy caused by the impact is:

$$2196 - 45 = 2151 \text{ J}$$
$$= \mathbf{2.15\ kJ}$$

Therefore it is evident that almost all the mechanical energy is "lost". Actually what occurs is that most of the kinetic energy is converted into heat energy.

Example 9.10

A pile driver of mass 1.2 t falls vertically through a height of 3 m on to a pile of mass 600 kg. The average soil resistance is 150 kN.

Neglecting air resistance and assuming momentum conservation on impact with no rebound, determine:

(a) pile driver velocity just before impact
(b) velocity of the combined driver and pile just after impact
(c) mechanical energy lost by the impact
(d) penetration of the pile into the ground.

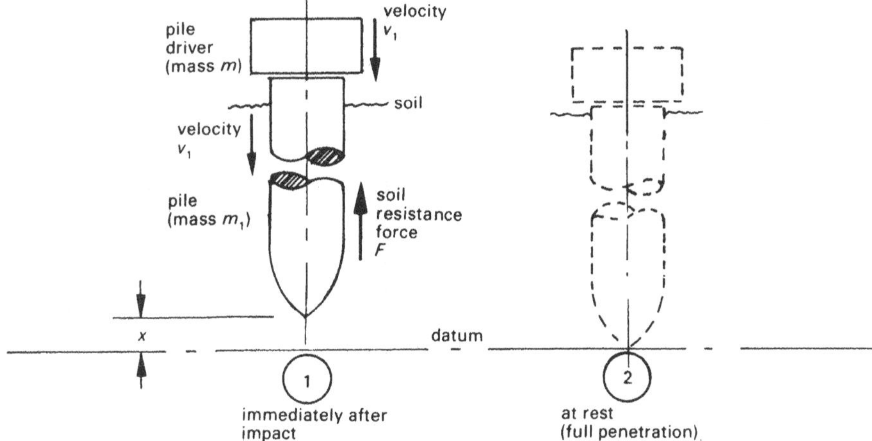

Fig. 9.12

Solution

(a) For a freely falling body:

$$v = \sqrt{2gh}$$
$$\therefore v = \sqrt{2 \times 9.81 \times 3}$$
$$= \mathbf{7.672\ m/s}$$

which is the velocity of the driver just before impact.

(b) Applying the conservation of momentum principle to the impact:

$$m \times v = (m + m_1)\, v_1$$

$$\therefore\ 1.2 \times 7.672 = (1.2 + 0.6) \times v_1$$

$$\therefore\ \mathbf{v_1 = 5.115\ m/s}$$

which is the velocity of the combined driver and pile just after impact.

(c) Before impact, the mechanical energy is the energy of the pile driver which is:

$m \times g \times h$ or $\frac{1}{2}\, m \times v^2$

$= 1.2 \times 10^3 \times 9.81 \times 3 \quad (J)$ $= \frac{1}{2} \times 1.2 \times 10^3 \times 7.672^2 \quad (J)$

$= 35.3\ kJ$ $= 35.3\ kJ$

After impact, the mechanical energy is the kinetic energy of the combined driver and pile. This is:

$$\tfrac{1}{2}\,(m + m_1)\, v_1^2 = \tfrac{1}{2} \times 1.8 \times 10^3 \times 5.115^2\ (J)$$

$$= 23.5\ kJ$$

Therefore the mechanical energy lost on impact is:

$$35.3 - 23.5 = \mathbf{11.8\ kJ}$$

9.6 Mechanical machines and efficiency

A device which uses a source of mechanical energy (mechanical power input) to produce mechanical work (mechanical power output) is known as a mechanical machine.

There are many other types of machine which use other types of power, such as electrical machines and heat machines, but these are not considered in mechanics.

For a mechanical machine the efficiency is defined as:

$$\eta = \frac{\text{useful power output}}{\text{power input}} \qquad \textbf{(9.9) efficiency}$$

The conservation of energy principle implies that there is no loss in energy in total so that the total energy output is equal to the total energy input. However for a mechanical machine, output of energy other than mechanical energy is not useful and is considered a loss. Typical losses in a mechanical machine are due to mechanical friction and heat loss, fluid friction and pumping loss, and sound and vibrational loss.

A mechanical machine with no mechanical losses would have an efficiency of 100 per cent and is known as an ideal machine. Such a machine is impossible on the earth because of atmospheric air friction which can never completely be excluded. The efficiency of most practical mechanical machines, such as gear and chain drives, lever devices, pulley systems, lifting hoists and jacks, may vary from as low as 5 per cent to as high as 98 per cent depending upon the type of machine and the conditions under which it is operating. In general, mechanical machines operate most efficiently under high load conditions, i.e. close to the maximum power condition.

The input force or torque to a mechanical machine is known as the **effort** whereas the output force or torque is known as the **load**. The ratio between the load and the effort is known as the **mechanical advantage** (*MA*).

In most mechanical machines, the ratio between the distance moved by the load and the distance moved by the effort is in a fixed ratio, known as the **velocity ratio** (VR).

For a machine with 100 per cent efficiency:

$MA = VR$ (ideal machine)

However, because a practical machine is less than 100 per cent efficient, the mechanical advantage is less than the velocity ratio. This leads to an alternative definition of efficiency,

$$\eta = \frac{MA}{VR}$$ **(9.10) efficiency**

The velocity ratio is determined by the design of the machine and if there is no slip:

MA (ideal) $= VR$

However the actual mechanical advantage is always less than the ideal by the ratio of the efficiency. Hence there is a third way of defining the efficiency of a mechanical machine:

$$\eta = \frac{MA \text{ (actual)}}{MA \text{ (ideal)}}$$ **(9.11) efficiency**

Machines in series

In many practical situations, several machines may be linked in series to form a more complex unit. For example, in a motor car transmission the motor is coupled to a gearbox which is in turn coupled to a drive shaft and then to a differential unit. For machines in series, the overall efficiency, velocity ratio and mechanical advantage are the product of the efficiency, velocity ratio and mechanical advantage of each machine considered as a separate unit.

Example 9.11

In the hoist system illustrated in Figure 9.13, the electric motor rotates at 1440 rpm and the gearbox has a reduction ratio of 30:1 and an efficiency under load of 82 per cent. The winch drum has a pitch diameter of 200 mm and a maximum load capacity of 1 tonne. Frictional losses in the winch drum itself are negligible.

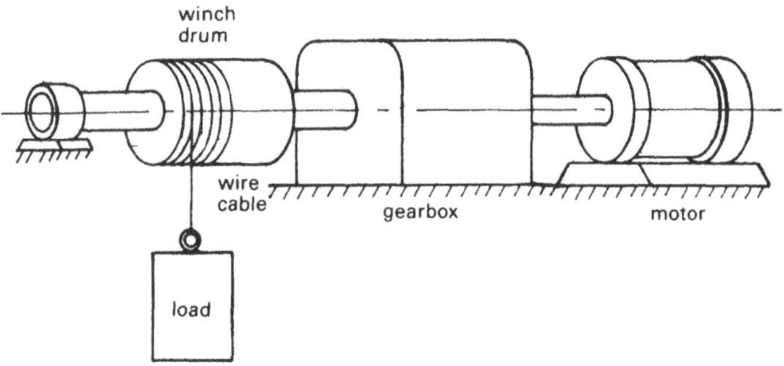

Fig. 9.13

If the maximum load is lifted at constant velocity, determine:

(a) rotational speed of the winch drum
(b) speed at which the load will be lifted

(c) output torque from the gearbox
(d) output power from the gearbox
(e) motor torque
(f) motor power.

Solution

(a) The reduction ratio of the gearbox (velocity ratio) is 30:1. Therefore the rotational speed of the winch drum is:

$$1440/30 = \mathbf{48\ rpm}$$

(b) Now $v = rw$

therefore the linear velocity of the load is:

$$v = 0.1 \times \pi \times \frac{48}{30}$$
$$= \mathbf{0.503\ m/s}$$

(c) The output torque from the gearbox is:

$$\tau = Fr$$
$$= 1 \times 10^3 \times 9.81 \times 0.1$$
$$= \mathbf{981\ Nm}$$

(d) The gearbox output power is:

$$P = Fv$$
$$= 1 \times 10^3 \times 9.81 \times 0.503 \quad \text{(W)}$$
$$= \mathbf{4.93\ kW}$$

or

$$P = \tau\omega$$
$$= 981 \times \pi \times \frac{48}{30} \quad \text{(W)}$$
$$= \mathbf{4.93\ kW}$$

(e) The velocity ratio of the gearbox is 30 and since $\eta = \frac{MA}{VR}$,

$$MA = VR \times \eta$$
$$= 30 \times 0.82 = 24.6$$

Also $MA = \frac{\text{output torque}}{\text{input torque}}$ and the input torque to the gearbox = motor torque.

$$\therefore \text{ motor torque} = \frac{\text{output torque}}{MA}$$
$$= \frac{981}{24.6}$$
$$= \mathbf{39.9\ Nm}$$

(f) The motor power is given by:

$$P = \tau\omega$$
$$= 39.9 \times \pi \times \frac{1440}{30} \quad \text{(W)}$$
$$= \mathbf{6.01\ kW} \quad \text{(say 6 kW)}$$

Alternatively:

$$\eta \text{ (gearbox)} = \frac{\text{output power}}{\text{input power}} = \frac{\text{output power}}{\text{motor power}}$$

$$\therefore \text{ motor power} = \frac{\text{gearbox output power}}{\eta}$$

$$= \frac{4.93}{0.82}$$

$$= \mathbf{6.01\ kW} \quad \text{(as before)}$$

Example 9.12

The preliminary design of an escalator is illustrated in Figure 9.14. At peak capacity the escalator will transport 10 000 persons, of average mass 65 kg each, per hour. Power to the drive sprocket will be provided by an electric motor through a 30:1 reduction worm gearbox. Frictional losses in the escalator are estimated at 40 per cent of the power supplied to the drive sprocket and the gearbox has an efficiency under load of 88 per cent.

Assuming the persons step on to the escalator at speed (smoothly), determine for peak load conditions:

(a) rotational speed of the drive sprocket
(b) power at the drive sprocket
(c) torque at the drive sprocket
(d) motor speed
(e) motor power
(f) motor torque.

Fig. 9.14

Solution

(a) Linear velocity = 2.7 km/h = 0.75 m/s

$$\omega = \frac{v}{r} = \frac{0.75}{0.3} = 2.5 \text{ rad/s} = \mathbf{23.9\ rpm}$$

(b) The potential energy change per second is:

$$\dot{P}E = \dot{m}gh \quad (\dot{m} = \text{mass raised per second})$$

$$= \frac{10\,000}{3600} \times 65 \times 9.81 \times 6 \quad \text{(W)}$$

$$= 10.63 \text{ kW}$$

Since there is no change in kinetic energy, this is the ideal power at the drive sprocket. However since frictional losses are 40 per cent, the actual power which must be provided at the drive sprocket is:

$$P = \frac{10.63}{0.6} = \mathbf{17.7\ kW}$$

(c) $P = \tau\omega \therefore \tau = \frac{P}{\omega} = \frac{17.7}{2.5} = \mathbf{7.08\ kNm}$

(d) Motor speed $=$ sprocket speed $\times$ VR (gearbox)

$= 23.87 \times 30$

$= \mathbf{716\ rpm}$

(e) Since the efficiency of the gearbox is given by:

$$\eta = \frac{\text{sprocket power}}{\text{motor power}}$$

$$\therefore \text{motor power} = \frac{\text{sprocket power}}{\eta} = \frac{17.7}{0.88} = \mathbf{20.1\ kW}$$

(f) $\tau = \frac{P}{\omega} = \frac{20.1 \times 10^3}{\pi \times \frac{716}{30}} = \mathbf{268\ Nm}$

Note: In fact a standard 20 kW 720 rpm motor would be suitable for this application.

Problems

9.1 A loaded rail car, mass 15 t, travelling at 20 km/h collides with and locks on to another rail car, mass 18 t, travelling in the same direction at 16 km/h.

Determine the velocity of the combined cars after impact and the loss in kinetic energy resulting from the impact. Neglect friction.

17.8 km/h, 5.05 kJ

9.2 A man, mass 75 kg, jumps upward at an angle of 45° from the stern of a boat, mass 125 kg, free in the water.

If the velocity of the man as he leaves the boat is 6 km/h, determine the forward speed of the boat at this instant if friction is negligible.

2.55 km/h

9.3 A vehicle, mass 1.6 t, travelling at 80 km/h due W collides with another vehicle, mass 2 t, travelling NE at 60 km/h.

Determine:

(a) velocity of the second vehicle relative to the first before impact

(b) resultant motion of the combined vehicle mass immediately after impact if friction on the road surface is neglected.

(a) 130 km/h @ 19.1° (b) 26.4 km/h @ 117°

9.4 A projectile, mass 1.5 kg, is fired horizontally from a cannon, mass 2.5 t, mounted on rails. The barrel length is 2200 mm and the velocity of the projectile leaving the cannon is 650 m/s.

If friction is negligible, determine:

(a) recoil velocity of the cannon as the projectile leaves the barrel

(b) average acceleration of the projectile (relative to the barrel)

(c) average recoil acceleration of the cannon

(d) average force on projectile

(e) average force on the cannon.

(a) 0.39 m/s (b) 96 km/s^2 (c) 57.6 m/s^2 (d) 144 kN (e) 144 kN

9.5 Repeat problem 9.4 if the barrel of the cannon is angled upward at 30° to the horizontal.

(a) 0.338 m/s (b) 96 km/s^2 (c) 49.9 m/s^2 (d) 144 kN
(e) 125 kN (horizontal)

9.6 A motor vehicle, mass 1.5 t, is accelerated from 10 to 20 km/h in 4 s along a horizontal road surface. It is then allowed to coast to rest.

If the average friction and air resistance is 5 per cent of the vehicle weight, determine:
(a) work done by the accelerating force
(b) distance travelled during the acceleration period
(c) total distance travelled.
Solve this problem using both the force method and the energy method.

(a) 29.6 kJ (b) 16.67 m (c) 48.1 m

9.7 A child of mass 55 kg swings back and forth through a total angle of 60° on a swing. The centre of mass of the child is 4 m from the pivot of the swing.

If friction may be neglected, determine:
(a) maximum velocity of the child
(b) tension in each of the two supporting ropes when they are vertical.

(a) 3.24 m/s (b) 342 N

9.8 In an Izod test, a block of mass 2.5 kg is swung through an arc of 90° to impact a test specimen when the mass is directly below the pivot point. The mass then swings through a further arc of 78° after impact. The centre of mass is located 1.2 m from the pivot point and friction is negligible.

Determine:
(a) maximum kinetic energy of the mass
(b) energy required to fracture the test specimen.

(a) 29.4 J (b) 6.12 J

9.9 In a test of the collision safety of a motor vehicle, a remote-controlled motor vehicle, mass 1.5 t, is accelerated to 60 km/h before being driven into a solid wall. A film sequence of the event with a high speed camera shows that the total impact takes place in 80 frames.

If the film speed is 250 frames/second, determine the average force on the wall during impact.

78.1 kN

9.10 A horizontal conveyor, travelling at 6.5 km/h, conveys 150 t/h of pulverised material which drops vertically on to the conveyor.

Determine:
(a) ideal power required to drive the conveyor
(b) input power if frictional losses are 40 per cent of the input power.

(a) 135.8 W (b) 226 W

9.11 The velocity of gas leaving the nozzles of a vertical rocket, mass 2.6 t, is 650 m/s.

Determine the mass of gas which must be generated per second to:
(a) just lift the rocket off the ground
(b) give the rocket an upward acceleration of 2.5 *g*.

(a) 39.2 kg/s (b) 137 kg/s

9.12 A block of material has a square cross-section. The block may be moved horizontally by sliding along the ground or by rolling edge over edge. When the coefficient of friction on the ground is low, the sliding method consumes less energy; however as the coefficient of friction increases, the rolling method is better.

Determine the coefficient of friction when this occurs. Assume the block is moved slowly so that the kinetic energy change is negligible.

0.207

9.13 A freely rotating shaft and flywheel, mass moment of inertia 2.6 kg m^2, rotating at 1260 rpm, is clutched to a stationary shaft, mass moment of inertia 1.4 kg m^2, so that both rotate at the same speed.

Determine this speed if bearing friction is negligible. Also determine the loss of kinetic energy caused by the clutching.

819 rpm, 7.92 kJ

9.14 A shaft rotating at 1200 rpm has four movable masses of 250 g each located at a radius of 120 mm from the centre of rotation.

If the shaft speed was changed to 1450 rpm, determine the radius at which the masses would now be located in order to conserve angular momentum.

109 mm

9.15 An ice skater, mass 70 kg, spins around with arms outstretched at 140 rpm. In this position the radius of gyration of her body is 100 mm and of her arms is 360 mm.

If she brings her arms in to her side so that their radius of gyration is 200 mm, determine the new speed of rotation. Assume the arms are 8 per cent of the body weight and friction on the ice surface is negligible.

221 rpm

9.16 A projectile mass 350 g is fired vertically upward at 640 km/h. It attains a maximum height of 400 m.

Determine the work done in overcoming air resistance as a percentage of the initial energy of the projectile.

75.2 per cent

9.17 A mass of 20 kg is accelerated to 15 km/h by an applied force which is then removed and the mass allowed to coast to rest along a level surface.

If the coefficient of friction is 0.3, determine using the energy method the distance the mass will move along the surface before coming to rest after the force is removed.

2.95 m

9.18 Solve problem 9.17 if the surface is inclined, the angle being:

(a) 10° upward
(b) 10° downward.

(a) 1.89 m (b) 7.27 m

9.19 A mass of 15 kg rests on an inclined plane inclined at 30° to the horizontal. An applied force of 200 N acting parallel to the plane moves the mass a distance of 12 m along the plane upward.

Using the initial position of the block as a datum and a coefficient of friction of 0.45, determine:

(a) final potential energy
(b) work done by the applied force

(c) work done in overcoming friction
(d) final kinetic energy
(e) final velocity.

(a) 883 J (b) 2.4 kJ (c) 688 J (d) 829 J (e) 10.5 m/s

9.20 Solve problem 9.19 if the applied force P acts upward at an angle of 20° to the plane, all other conditions being the same.

(a) 883 J (b) 2.26 kJ (c) 319 J (d) 1054 J (e) 11.85 m/s

9.21 A wheel, mass 30 kg, radius of gyration 420 mm, spins freely at 1600 rpm. Frictional torque is 2.3 Nm.

Using both force and energy methods, determine the number of revolutions made by the wheel in coming to rest. Also determine the time taken (in minutes).

5140 revs, 6.43 min

9.22 A flywheel, mass 60 kg and radius of gyration 400 mm, is accelerated from rest to a speed of 1470 rpm by a constant torque of 120 Nm.

If friction is negligible, determine:
(a) energy stored in flywheel
(b) revolutions made by the flywheel in reaching speed
(c) time taken to reach speed
(d) energy available from the flywheel by reducing its speed by 2 per cent

(a) 114 kJ (b) 151 rev (c) 12.3 s (d) 4.5 kJ

9.23 The flywheel of a stamping press has a mass moment of inertia of 120 kg m^2 and rotates at 250 rpm. It is driven by a 10 kW electric motor through a belt drive. The stamping operation requires 20 kJ of energy and takes 0.75 s to complete.

If frictional losses are neglected, determine:
(a) speed of the flywheel (in rpm) immediately on completion of stamping
(b) time taken to regain speed after stamping
(c) maximum number of stamping operations which can be performed per minute
(d) maximum torque delivered by the motor to the flywheel.

(a) 209 rpm (b) 1.25 s (c) 30 (d) 458 Nm

9.24 In an experiment to determine its velocity, a projectile, mass 120 g, is fired into a block of wood, mass 3.5 kg, and becomes embedded in it. The block is suspended from wires such that its centre of mass is 4.5 m below the point of suspension. After firing, the maximum angle made by the wires to the vertical is 26°.

If friction is negligible, determine:
(a) velocity of the projectile
(b) energy absorbed by the impact as a percentage of the initial energy of the projectile.

(a) 90.2 m/s (b) 96.7 per cent

9.25 A pile driver of mass 2 t falls vertically through a height of 4 m before striking a pile of mass 1 t. The average soil resistance is 800 kN.

Neglecting air resistance and assuming conservation of momentum on impact with no rebound, determine:
(a) pile driver velocity just before impact
(b) velocity of driver and pile just after impact
(c) mechanical energy lost by the impact
(d) penetration of the pile into the ground.

(a) 8.86 m/s (b) 5.91 m/s (c) 26.2 kJ (d) 67.9 mm

9.26 Repeat problem 9.25 if the pile driver has a mass of 135 kg and falls through 2 m, the pile has a mass of 120 kg and soil resistance of 45 kN.

(a) 6.26 m/s (b) 3.32 m/s (c) 1.25 kJ (d) 33 mm

9.27 A flywheel mass 90 kg mounted in horizontal bearings has outside diameter 250 mm. A light cord is pegged to the outside of the flywheel then wrapped around it several times. A mass of 1.8 kg is attached to the end of the cord and descends a distance of 3 m after release in 4.26 s.

Using the energy method, determine:

(a) tension in the cord
(b) mass moment of inertia of flywheel and shaft
(c) radius of gyration of flywheel and shaft.

Friction and the diameter of the cord are negligible.

(a) 17.1 N (b) 0.806 kg m^2 (c) 94.6 mm

9.28 For the system illustrated in Figure P9.28, determine the velocity of the mass and the tension in the cord after the mass has descended a distance of 5 m from rest along the inclined plane.

Use the energy method and neglect friction and the diameter of the cord.

3.83 m/s, 219 N

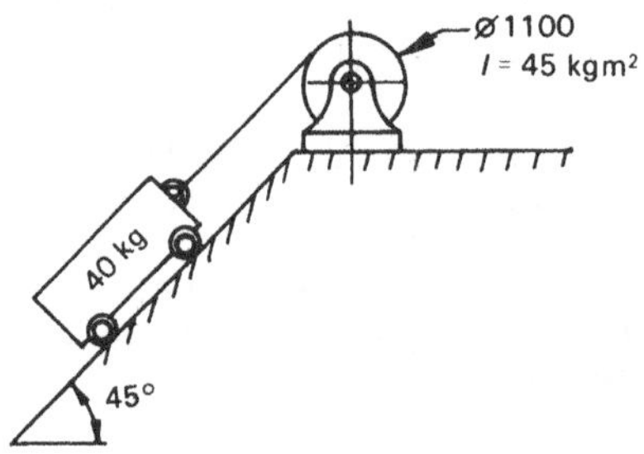

Fig. P9.28

9.29 Solve problem 9.28 if the coefficient of friction on the inclined plane is 0.2, all other data being the same.

3.43 m/s, 175 N

9.30 A motorised gearbox has a reduction ratio of 50:1. The power input to the gearbox is 5 kW at 1440 rpm.

If the gearbox efficiency is 85 per cent under maximum load conditions, determine:

(a) input torque
(b) output speed
(c) output torque
(d) output power.

(a) 33.16 Nm (b) 28.8 rpm (c) 1409 Nm (d) 4.25 kW

9.31 The output shaft of the gearbox given in problem 9.30 is attached to a winch drum of pitch diameter 220 mm and is used to hoist a load.

Determine the maximum speed at which a load may be lifted and the maximum load capacity (in tonnes) of the winch. Frictional losses in the winch drum itself may be neglected.

0.332 m/s, 1.3 t

9.32 A conveyor is illustrated in Figure P9.32. Power input to the conveyor is through four motorised rollers providing 1.2 kW (maximum) each. Frictional losses are 35 per cent of the input energy to the conveyor.

Determine the maximum capacity of the conveyor in t/h and the required speed and torque of each motorised roller.

278 t/h, 67.2 rpm, 171 Nm

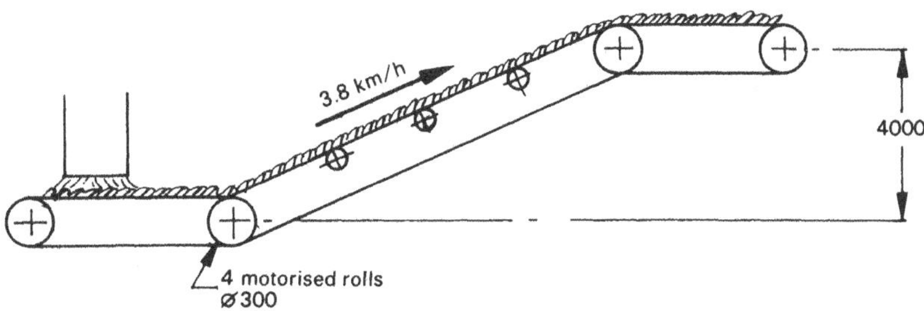

Fig. P9.32

9.33 A hoist system is to be designed to transport 1500 persons, of average mass 65 kg, per hour to the top of a mountain 800 m high at a linear speed of 3.6 km/h. The system will be driven by an electric motor rotating at 720 rpm through a gearbox to a drive sprocket of pitch diameter 1200 mm. Frictional losses in the hoist system are estimated to be 35 per cent of the power at the drive sprocket and the gearbox has an efficiency of 85 per cent.

Determine:

(a) rotational speed of the drive sprocket
(b) theoretical reduction ratio of the gearbox
(c) output power of the gearbox
(d) motor power.

(a) 15.9 rpm (b) 45.2 (c) 327 kW (d) 385 kW

10

Mechanical vibration

Mechanical vibration occurs when a system having both mass and elasticity undergoes a cyclic motion, that is an oscillating or reciprocating movement where there is a periodic reversal in the direction of the displacement (and velocity) so that the motion repeats itself after a certain time interval. Examples of this are a bouncing ball, an oscillating spring, or a vibrating reed valve.

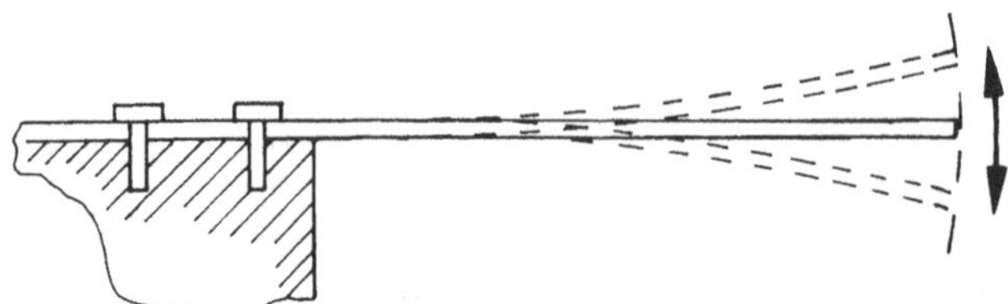

Fig. 10.1 *Mechanical vibration of a reed valve*

When the vibration is caused by a periodic disturbing force applied to the system, it is known as a **forced vibration**. It is also possible that the vibration occurs without the continuous presence of a periodic disturbing force, for example when a piano key is struck. Here a force is applied initially to disturb the string, the force is then removed and the string vibrates of its own accord. Such a vibration is known as a **free vibration** and it will attenuate with time, that is die down and eventually stop due to the natural damping forces such as air resistance and friction which are present. A displacement–time diagram for a damped free vibration is as shown in Figure 10.2.

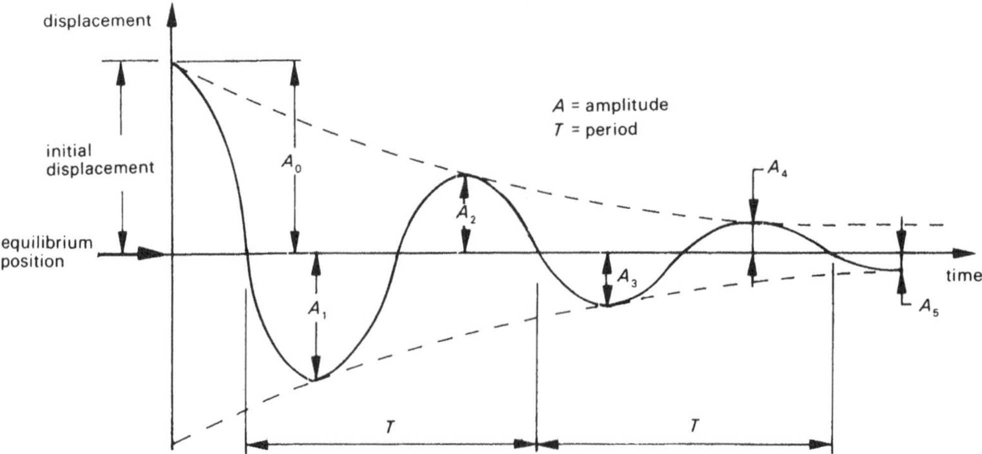

Fig. 10.2 *Displacement–time diagram for a free vibration (damped)*

10.1 Definition of terms

Amplitude (A)

The amplitude is the maximum distance moved from the equilibrium position. In the case of a damped free vibration, the amplitude decreases with time. In the case of a forced vibration, the amplitude will generally become constant as a steady state condition is achieved (after an initial transient period).

Frequency (f)

The frequency is the number of cycles of the vibration completed per second. In the case of a free vibration, the frequency is known as the **natural frequency** because it is the frequency the system will adopt of its own accord. In the case of audible frequencies, the natural frequency is also known as the **pitch** and the greater the frequency, the higher the pitch. For a forced vibration, the steady state frequency is equal to the frequency of the disturbing force as the system will adopt this frequency regardless of the natural frequency.

Frequency has units of cycles per second, also known as hertz (Hz).

Period (T)

The period is the time in seconds taken to complete one cycle of the vibration. It is evident, therefore, that period is the reciprocal of the frequency, that is:

$$T = \frac{1}{f}$$ **(10.1) period of vibration**

10.2 Simple harmonic motion

When the displacement–time diagram is a sine curve (or a cosine curve), the motion is said to be **simple harmonic**. In an ordinary piston and crank mechanism, the motion of the piston is not exact simple harmonic motion but approximates it when the connecting rod is long compared to the stroke. However exact simple harmonic motion may be obtained with the scotch yoke mechanism illustrated in Figure 10.3.

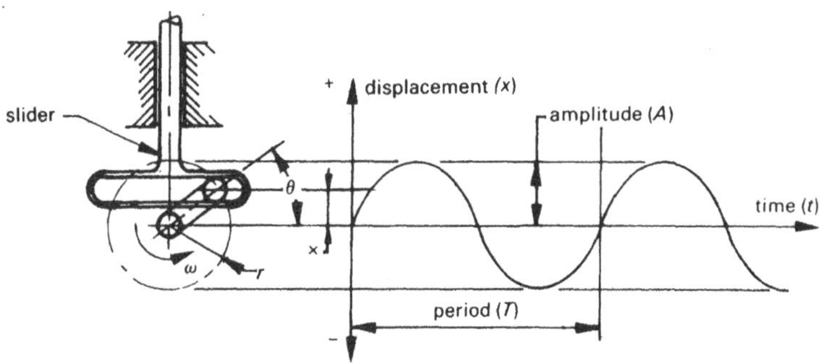

Fig. 10.3 *Scotch yoke mechanism*

The slider is constrained to move vertically while the crank rotates with uniform angular velocity (ω). The slider then moves with simple harmonic motion because the vertical displacement is $r \sin \theta$, where r is the crank radius (one-half the stroke).

Amplitude (A)

It is evident that the amplitude is the radius of the crank arm since this is the maximum displacement of the follower from the central or equilibrium position, that is $A = r$

Displacement (x)

The displacement of the slider is:

$$x = r \sin \theta$$

But $\theta = \omega t$ and $A = r$

$$\therefore \quad \boxed{x = A \sin (\omega t)} \qquad \textbf{(10.2) displacement in SHM}$$

Since ω is constant, the displacement will vary sinusoidally with time; the motion is therefore simple harmonic motion.

Note: The maximum displacement occurs when $\sin(\omega t) = 1$, i.e.

$$x_{(\max)} = A = r.$$

Velocity (v)

Since $v = \dfrac{dx}{dt}$

$$v = \frac{d}{dt} A \sin (\omega t)$$

$$\therefore \quad \boxed{v = A \omega \cos (\omega t)} \qquad \textbf{(10.3) velocity in SHM}$$

Notes

1. The maximum velocity occurs when $\cos(\omega t) = 1$, i.e.

   $$v_{(\max)} = A \omega.$$

2. The cos curve has the same shape as the sine curve but displaced 90° relative to it, so that when the displacement is maximum (top and bottom of the stroke), the velocity is zero and when the velocity is a maximum (at the centre of the stroke), the displacement is zero.
3. Equation 10.3 may also be written in an alternative form. Since

   $$x = A \sin \theta \quad \therefore \sin \theta = \frac{x}{A}$$

   and $\sin^2 \theta + \cos^2 \theta = 1$

   $$\therefore \cos \theta = \sqrt{1 - \sin^2 \theta}$$

   $$= \sqrt{1 - \frac{x^2}{A^2}}$$

   $$= \frac{1}{A}\sqrt{A^2 - x^2}$$

From equation 10.3, $v = A\omega \cos\theta$

$$= A\omega \frac{1}{A}\sqrt{A^2 - x^2}$$

$$\therefore \quad \boxed{v = \omega\sqrt{A^2 - x^2}} \qquad \textbf{(10.4) velocity in SHM}$$

Acceleration (a)

Since $a = \dfrac{dv}{dt}$

$$a = \frac{d}{dt} A\omega \cos(\omega t)$$

$$\therefore \quad a = -A\omega^2 \sin(\omega t)$$

Since $x = A\sin(\omega t)$ (10.2)

$$a = -\omega^2 x$$

Notes

1. Since ω is constant, the acceleration is directly proportional to the displacement but with opposite sign. This leads to an alternative definition of simple harmonic motion, namely that simple harmonic motion is a periodic motion in which the acceleration is directly proportional to the displacement and is always directed toward the equilibrium position. This is illustrated in Figure 10.4.

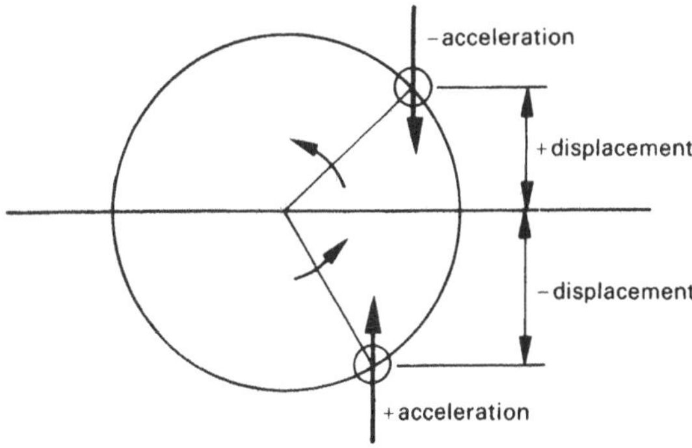

Fig. 10.4 *Sign of acceleration and displacement in SHM*

2. It is usual to ignore the minus sign for the acceleration, since it is always the case that the acceleration is directed toward the equilibrium position. Hence these equations become:

$$\boxed{a = A\omega^2 \sin(\omega t)} \qquad \textbf{(10.5)}$$

$$\boxed{a = \omega^2 x} \qquad \textbf{(10.6)}$$

acceleration in SHM

3. The acceleration is a maximum when $\sin(\omega t) = 1$, i.e.

$$a_{(max)} = A\omega^2.$$

Frequency (f)

The frequency is the number of completed cycles per second, i.e. the number of revolutions made per second by the crank (in the scotch yoke mechanism).

Since $\omega = 2\pi N_S$ (where N_S is the rotational speed in revolutions per second)

and $f = N_S$ (by definition)

$$f = \frac{\omega}{2\pi}$$

(10.7) frequency in SHM

Also since $a = \omega^2 x$ where a = magnitude of the acceleration

$$\therefore \quad \omega = \sqrt{\frac{a}{x}}$$

$$\therefore \quad f = \frac{1}{2\pi}\sqrt{\frac{a}{x}}$$

(10.8) frequency in SHM

Period (T)

Equation 10.1 may be applied to determine the period, that is:

$$T = \frac{1}{f}$$

where f is given by equation 10.8.

Example 10.1

A body moving with simple harmonic motion has amplitude 600 mm and period 1.5 seconds. Determine:

(a) frequency
(b) maximum velocity
(c) maximum acceleration
(d) velocity and acceleration when the displacement is one-half the amplitude.

Solution

(a) From equation 10.1:

$$f = \frac{1}{T} = \frac{1}{1.5} = \mathbf{0.667\ Hz}$$

(b) From equation 10.7:

$$\omega = 2\pi f = 2\pi \times 0.667 = 4.19 \text{ rad/s}$$

Now $v_{(max)} = A\omega$

$$\therefore v_{(max)} = 0.6 \times 4.19 = \mathbf{2.51\ m/s}$$

(c) Now $a_{(max)} = A\omega^2$

$$\therefore a_{(max)} = 0.6 \times 4.19^2 = \mathbf{10.53\ m/s^2}$$

(d) When $x = \frac{A}{2}$, $x = 0.3$ m.

From equation 10.4:

$v = \omega\sqrt{A^2 - x^2} = 4.19\sqrt{0.6^2 - 0.3^2} = \mathbf{2.18\ m/s}$

From equation 10.6:

$a = \omega^2 x = 4.19^2 \times 0.3 = \mathbf{5.27\ m/s}$ (note half the maximum acceleration).

Example 10.2

A mass of 6 kg moving with simple harmonic motion has velocity 4.5 m/s and acceleration 30 m/s^2 when displaced 300 mm from the equilibrium position.

Determine:

(a) frequency
(b) amplitude
(c) maximum velocity
(d) maximum inertia force.

Solution

(a) From equation 10.6:

$$a = \omega^2 x$$
$$\therefore 30 = \omega^2 \times 0.3$$
$$\therefore \omega^2 = 100$$
$$\therefore \omega = 10 \text{ rad/s}$$

From equation 10.7:

$$f = \frac{\omega}{2\pi} = \frac{10}{2\pi} = \mathbf{1.59\ Hz}$$

Alternatively from equation 10.8:

$$f = \frac{1}{2\pi}\sqrt{\frac{a}{x}} = \frac{1}{2\pi}\sqrt{\frac{30}{0.3}}$$
$$= \mathbf{1.59\ Hz} \quad \text{(as before)}$$

(b) From equation 10.4:

$$v = \omega\sqrt{A^2 - x^2}$$
$$\therefore A^2 - x^2 = \left(\frac{v}{\omega}\right)^2$$
$$\therefore A^2 = \left(\frac{4.5}{10}\right)^2 + 0.3^2$$
$$\therefore A = \mathbf{0.541\ m}$$

(c)
$$v_{(max)} = A\omega$$
$$= 0.541 \times 10$$
$$= \mathbf{5.41\ m/s}$$

(d)
$$a_{(max)} = A\omega^2 \qquad \text{or} \qquad a_{(max)} = \frac{0.541}{0.3} \times 30$$
$$= 0.541 \times 10^2 \qquad\qquad = 54.1 \text{ m/s}^2$$
$$= 54.1 \text{ m/s}^2$$

$$F_i = ma = 6 \times 54.1 = \mathbf{325\ N}$$

10.3 Vibrating spring-mass system

Consider a spring of stiffness (spring constant) k Nm supporting a mass m kg in equilibrium position ①, Figure 10.5. The mass is then pulled down a further distance x by an externally applied force F, position ②. The external force is then removed suddenly so the mass vibrates in free vibration with natural frequency f and initial amplitude A equal to the initial displacement x.

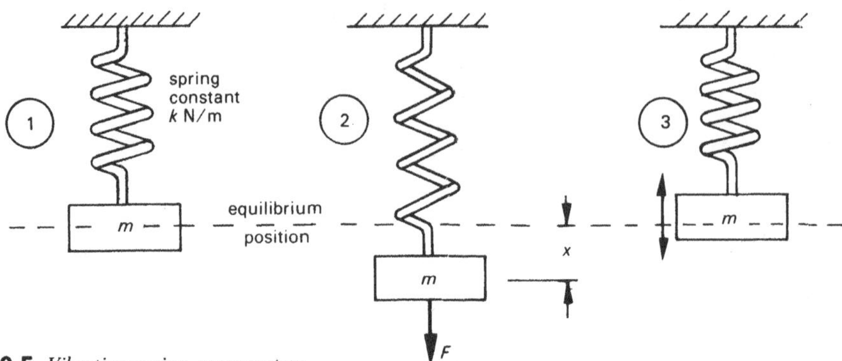

Fig. 10.5 *Vibrating spring-mass system*

In the ideal case (neglecting the effect of air resistance), the vibration is simple harmonic motion because it satisfies the two requirements of simple harmonic motion:

1. The acceleration is always directed toward the equilibrium position. This follows because the weight of the block and the spring force are balanced in this position. When the block is below equilibrium position, the spring force is greater than the weight force, the net applied force is upward and therefore the acceleration is directed upward. When the mass is above equilibrium position, the weight force is greater than the spring force, so that the net applied force is downward and therefore the acceleration is directed downward.
2. The net force on the mass varies directly with the displacement. This follows because the weight force remains constant so that the only variable force is that due to the spring. For a spring, the force varies directly with the displacement (Hooke's law) and the force–extension diagram is a straight line as shown in Figure 10.6. Therefore the net force on the mass (which is the difference between the spring force and the weight force) varies directly with extension (or displacement).

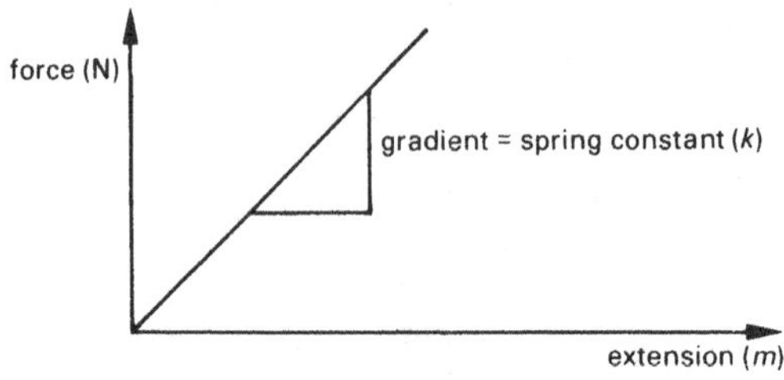

Fig. 10.6 *Force-extension diagram for a spring*

Hence the mass vibrates in simple harmonic motion. This vibration is a free one and the initial amplitude is the initial displacement from the equilibrium position (x). In the ideal case (undamped), the amplitude remains constant but in the actual case the amplitude of the vibration will diminish with time due to the presence of damping forces. However the frequency remains constant in both cases and is of course the natural frequency of the system.

The frequency, velocity, acceleration and displacement may be calculated using equations 10.1–10.8 derived for simple harmonic motion, using the equivalent angular velocity derived below.

Equivalent angular velocity (ω)

The weight force and spring force are balanced at the equilibrium position and therefore for any displacement x from the equilibrium position the force on the mass is due only to a change in the spring force (since the weight force does not change with displacement).

Spring force $F = kx$.

But $F = ma$

$$\therefore a = \frac{kx}{m}$$

But from equation 10.6:

$$a = \omega^2 x$$

$$\therefore \omega^2 x = \frac{kx}{m}$$

$$\therefore \quad \boxed{\omega = \sqrt{\frac{k}{m}}} \qquad \textbf{(10.9) equivalent angular velocity}$$

This is the equivalent angular velocity of a vibrating spring-mass system. Using the equivalent angular velocity, all the equations for simple harmonic motion previously derived may be used to calculate the frequency, period, velocity, acceleration or displacement.

Springs in series and parallel

Springs in series and parallel are shown in Figure 10.7.

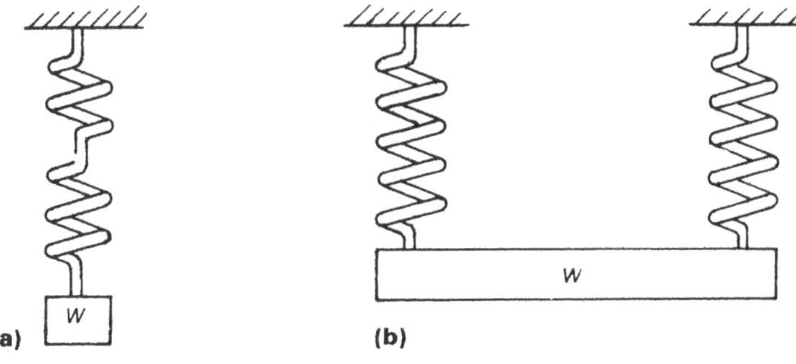

Fig. 10.7 *Springs in (a) series and (b) parallel*

Series springs

Consider two springs with spring constants k_1 and k_2 in series. The load F is common to both springs and the extensions are:

$$x_1 = \frac{F}{k_1}, \quad x_2 = \frac{F}{k_2}$$

$$\therefore x_1 + x_2 = F\left(\frac{1}{k_1} + \frac{1}{k_2}\right)$$

$$\therefore \frac{x_1 + x_2}{F} = \frac{1}{k_1} + \frac{1}{k_2}$$

But $\dfrac{F}{x_1 + x_2} = k_T$ (total spring constant)

$$\therefore \quad \frac{1}{k_T} = \frac{1}{k_1} + \frac{1}{k_2}$$

If there are N equal springs in series:

$$\frac{1}{k_T} = \frac{N}{k}$$

or $k_T = \dfrac{k}{N}$ equal springs in series

Parallel springs

If there are N equal springs in parallel and the load is spread evenly over each spring, the deflection of each spring is reduced in proportion to the number of springs. With parallel springs, the total deflection is the same as the deflection of each spring, hence:

$k_T = Nk$ equal springs in parallel

Example 10.3

A vertical spring with free length 160 mm is extended to 180 mm by a mass of 2.5 kg hung on to the end of it. The mass is then pulled down a further 25 mm and released, thus causing the mass to vibrate freely.

Determine the frequency and period of the vibration. Also determine the velocity and acceleration when the displacement is 15 mm above the equilibrium position.

Solution

First determine the spring constant k.

$$k = \frac{\text{force}}{\text{extension}} = \frac{2.5 \times 9.81}{0.02} = 1226 \text{ N/m}$$

Now from equation 10.9, the equivalent angular velocity is:

$$\omega = \sqrt{\frac{k}{m}} = \sqrt{\frac{1226}{2.5}} = 22.15 \text{ rad/s}$$

And from equation 10.7:

$$f = \frac{\omega}{2\pi} = \frac{22.15}{2\pi} = \mathbf{3.52\ Hz}$$

And from equation 10.1:

$$T = \frac{1}{f} = \frac{1}{3.52} = \mathbf{0.284\ s}$$

And from equation 10.4:

$$v = \omega\sqrt{A^2 - x^2}$$

$A = 25$ mm (initial displacement) and $x = 15$ mm.

$$\therefore v = 22.15\sqrt{0.025^2 - 0.015^2} = \mathbf{0.443\ m/s}$$

And from equation 10.6:

$$a = \omega^2 x = 22.15^2 \times 0.015 = \mathbf{7.36\ m/s^2}$$

10.4 Simple pendulum

A simple pendulum is one in which the mass may be considered as concentrated, that is of small size compared to the length of the pendulum, and the angular displacement is small (< 10°). The simple pendulum then closely approximates simple harmonic motion.

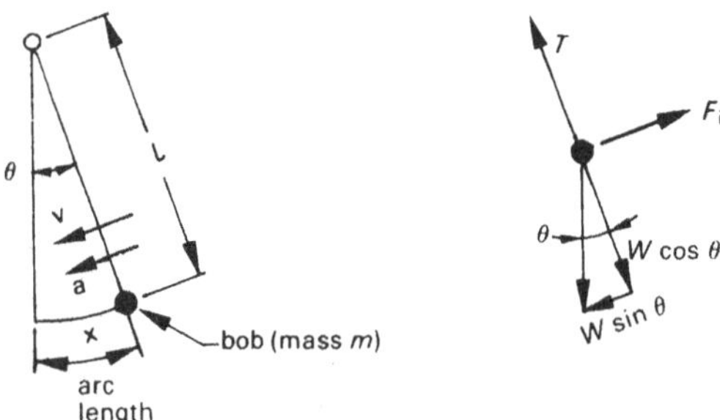

Fig. 10.8 *Simple pendulum*

Now $\frac{x}{l} = \theta$ (rad)

But if θ is small, θ (rad) $\simeq \sin\theta$

$$\therefore x = l\sin\theta \qquad \textbf{(1)}$$

From the equilibrium of the forces:

$$W\sin\theta = F_i = ma$$

$$\therefore mg\sin\theta = ma$$

$$\therefore a = g\sin\theta$$

$$\therefore \sin\theta = \frac{a}{g}$$

Substituting in (1):

$$x = l\frac{a}{g}$$

or

$$\boxed{a = \frac{g}{l}x}$$

(10.10) acceleration of a simple pendulum

Now g and l are both constant, therefore the acceleration is directly proportional to the displacement. Also the velocity is a maximum at the central (vertical) position and zero at the maximum displacement position, therefore the acceleration must always be directed toward the central position. Therefore the motion is simple harmonic motion.

Now for simple harmonic motion, equation 10.8 applies, that is:

$$f = \frac{1}{2\pi}\sqrt{\frac{a}{x}}$$

Substituting equation 10.10 above for a:

$$\boxed{f = \frac{1}{2\pi}\sqrt{\frac{g}{l}}}$$

(10.11) frequency of a simple pendulum

The velocity may be determined from equation 10.4:

$$v = \omega\sqrt{A^2 - x^2}$$

But $\omega = 2\pi f$ and $f = \dfrac{1}{2\pi}\sqrt{\dfrac{g}{l}}$

$$\therefore \omega = \sqrt{\frac{g}{l}}$$

$$\boxed{v = \sqrt{\frac{g}{l}(A^2 - x^2)}}$$

(10.12) velocity of a simple pendulum

Example 10.4

A simple pendulum length 1200 mm is displaced an angle of 6° from the vertical and released. Determine:

(a) frequency
(b) period
(c) maximum velocity
(d) maximum acceleration
(e) velocity and acceleration when the displacement is 3°.

Solution

(a) From equation 10.11,

$$f = \frac{1}{2\pi}\sqrt{\frac{g}{l}}$$

$$= \frac{1}{2\pi}\sqrt{\frac{9.81}{1.2}}$$

$$= \mathbf{0.455\ Hz}$$

(b) $T = \dfrac{1}{f} = \mathbf{2.2\ s}$

(c) Amplitude A = arc length $= r\theta$, $\theta(\text{rad}) = 0.105$

$$\therefore A = 1.2 \times 0.105 = 0.126 \text{ m}$$

From equation 10.12,

$$v_{(\max)} = \sqrt{\frac{g}{l}A^2} \quad \text{(when } x = 0\text{)}$$

$$= A\sqrt{\frac{g}{l}}$$

$$= 0.126\sqrt{\frac{9.81}{1.2}} = \mathbf{0.36\ m/s}$$

(d) From equation 10.10,

$$a = \frac{g}{l}x \quad \therefore a_{(\max)} = \frac{g}{l}A \quad \text{(when } x = A\text{)}$$

$$\therefore a_{(\max)} = \frac{9.81}{1.2} \times 0.126 = \mathbf{1.03\ m/s^2}$$

(e) When the displacement is 3°, the arc length x is

$$x = 1.2 \times 0.0524 = 0.0628 \text{ m}$$

From equation 10.12,

$$v = \sqrt{\frac{g}{l}(A^2 - x^2)} = \sqrt{\frac{9.81}{1.2}(0.126^2 - 0.0628^2)}$$

$$= \mathbf{0.312\ m/s}$$

From equation 10.10,

$$a = \frac{g}{l}x = \frac{9.81}{1.2} \times 0.0628$$

$$= \mathbf{0.513\ m}$$

10.5 Forced vibration

Consider the case of a forced vibration such as that caused by an out-of-balance rotating mass in a machine or mechanism mounted on elastic supports and constrained to move in one direction (one degree of freedom). The periodic disturbing force is the centrifugal force and the component of this force in the direction of the restraint varies sinusoidally with time (see Fig. 10.9).

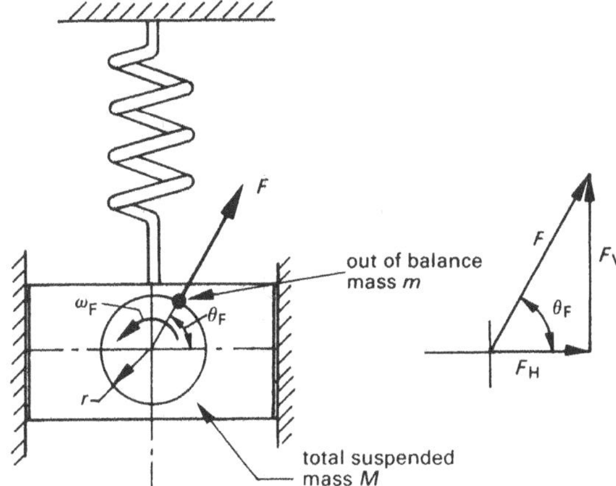

Fig. 10.9 *Forced vibration caused by an out-of-balance rotating mass*

Now the centrifugal force is given by:

$$F = mr\omega_F^2$$

where m = out-of-balance mass (kg)
r = radius at which mass m is located (m)
ω_F = angular velocity of the out-of-balance mass (rad/s) or the equivalent angular velocity of the forced vibration

The vertical component of the out-of-balance force is:

$$F_V = F \sin \theta_F$$

but $\theta_F = \omega_F t$.

Therefore the vertical out-of-balance force is:

$$F_V = F \sin(\omega_F t)$$

which varies sinusoidally with time.

Derivation of the equations for the motion of the system involves differential equations which are beyond the scope of this book. However it can be shown mathematically that the resultant motion consists of a sinusoidal component upon which is impressed a secondary motion. The secondary motion quickly damps out and the steady state motion is sinusoidal (simple harmonic) with the same frequency and period as the forced vibration but out of phase with it. That is to say, when θ_F has some value, the angle of the equivalent velocity vector of the resultant vibration will have some different value. The amplitude of the resultant simple harmonic motion is given by:

$$A = \frac{F}{M(\omega_F^2 - \omega^2)}$$

(10.13) amplitude of forced vibration

where A = amplitude of the resultant steady state vibration (m)
F = out-of-balance disturbing force (N)
M = total system mass (kg)
ω_F = angular velocity of the out-of-balance force or the equivalent angular velocity of the forced vibration (rad/s)
ω = equivalent angular velocity of the free vibration (rad/s)

Note: Because the resultant motion is out of phase with the forced vibration velocity vector, A can have a negative value. This occurs when $\omega_F < \omega$ and means that, when F is directed upward, the system is in fact below the equilibrium position. In general practice, the negative sign is ignored because the magnitude of the resultant motion is the only factor of importance.

Example 10.5

A 200 kg mass is supported by a spring which has a spring rate of 30 N/mm. The out-of-balance force has a maximum value of 90 N and frequency 1.5 Hz.

Determine the amplitude of the steady state vibration.

Solution

The equivalent angular velocity of the spring mass system is given by equation 10.9:

$$\omega = \sqrt{\frac{k}{M}} = \sqrt{\frac{30 \times 10^3}{200}} = 12.25 \text{ rad/s}$$

$$\text{Now } f_F = 1.5 \text{ Hz}$$
$$\therefore \omega_F = 1.5 \times 2\pi = 9.42 \text{ rad/s}$$

The mass M is 200 kg and the force F is 90 N.

Substituting in equation 10.13:

$$A = \frac{90}{200(9.42^2 - 12.25^2)}$$
$$= \mathbf{7.36\ mm} \qquad \text{(actually negative but the sign is ignored)}$$

Example 10.6

A machine of mass 80 kg is supported evenly on four compression springs which deflect 12 mm due to the weight. The machine rotor has an out-of-balance mass of 200 g located at 60 mm from the centre.

Determine the amplitude of the steady state vibration if the rotor turns at (a) 720 rpm, (b) 1440 rpm.

Solution

The total spring constant is $k = \dfrac{80 \times 9.81}{0.012} = 65.4 \text{ kN/m}$

The equivalent angular velocity of the free vibration is:

$$\omega = \sqrt{\frac{k}{M}} = \sqrt{\frac{65.4 \times 10^3}{80}} = 28.6 \text{ rad/s}$$

(a) $\omega_F = 720 \text{ rpm} = 75.4 \text{ rad/s}$

$$F = mr\omega_F^2 = 0.2 \times 0.06 \times 75.4^2 = 68.2 \text{ N}$$
$$\therefore A = \frac{68.2}{80(75.4^2 - 28.6^2)} \text{ (m)} = \mathbf{0.175\ mm}$$

(b) $\omega_F = 1440 \text{ rpm} = 150.8 \text{ rad/s}$

$$F = mr\omega_F^2 = 0.2 \times 0.06 \times 150.8^2 = 272.8 \text{ N}$$
$$\therefore A = \frac{272.8}{80(150.8^2 - 28.6^2)} \text{ (m)} = \mathbf{0.156\ mm}$$

It is evident therefore that in (b), despite the fact that the rotor is turning faster than it was in (a) and hence the out-of-balance force is greater, the amplitude of the resultant vibration is reduced.

10.6 Resonance

It will be seen from equation 10.13 that the amplitude of the resultant vibration will increase, the closer the value of ω_F is to ω and in theory will become infinite if $\omega_F = \omega$, since $\omega_F^2 - \omega^2 = 0$.

This condition is known as resonance and occurs whenever the frequency of the forced vibration is equal to the natural frequency of the free vibration. This frequency is also known as the resonant frequency. At the resonant frequency, there is no phase change between the natural and forced vibration so that they augment one another, and if there were no damping present, the amplitude would continue to increase toward infinity with each cycle. However, because of the presence of damping forces, the amplitude at resonance reaches a finite value which becomes less the greater the amount of damping. This is illustrated in Figure 10.10 in which the magnitude of the amplitude is plotted against the frequency of the forced vibration.

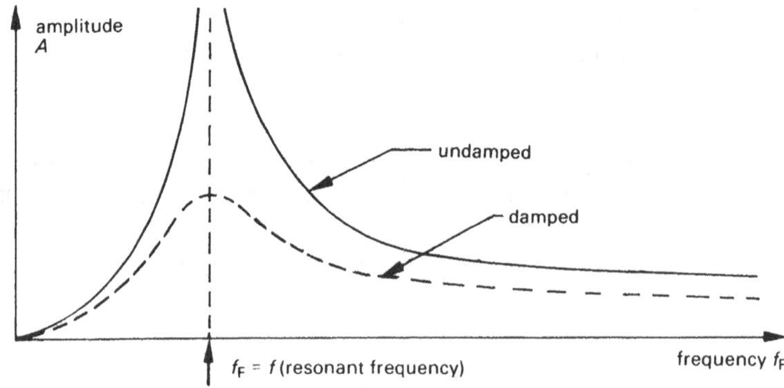

Fig. 10.10 *Magnitude of amplitude versus frequency for forced vibration*

Some examples of resonance which occur in practice are the tuning of musical instruments and exhaust systems for reciprocating engines, or a vehicle travelling over evenly spaced bumps in a smooth road at resonant speed.

In machine design, resonance is usually harmful as it causes excessive movement of the components which can easily cause fracture. Resonance may be avoided by:

1. changing the natural frequency of the system so that it is not close to the frequency of the forced vibration, for example by altering the spring rate or the mass;
2. provision of suitable damping devices;
3. minimising the disturbing force by suitable balancing.

Balancing is usually possible with rotating machines such as rotary compressors or pumps but not with reciprocating machines, such as reciprocating compressors, pumps or engines, because the piston and crank mechanism do not cause exact simple harmonic motion with a single frequency but in fact a number of higher order frequencies known as harmonics are present (see introduction to Chap. 11).

Transmissibility (T_R)

In the system shown in Figure 10.9, it is evident that the support spring transmits a reaction force R to the base on to which the end of the spring is mounted. If the system were fastened to the base without the spring, then clearly all the disturbing force would be transmitted to the base and the maximum value of the reaction R would be F. Because of the fact that a spring is used, the reaction force must be equal to the spring force and its maximum value may be calculated from the spring rate and the maximum spring deflection.

The maximum transmitted force (reaction) is therefore given by:

$$R = kA$$

From equation 10.13:

$$A = \frac{F}{M(\omega_F^2 - \omega^2)}$$

$$\therefore R = \frac{kF}{M(\omega_F^2 - \omega^2)}$$

$$= \frac{kF}{M\omega^2\left[\left(\frac{\omega_F}{\omega}\right)^2 - 1\right]}$$

But from equation 10.9:

$$\omega = \sqrt{\frac{k}{M}} \quad \text{or } k = M\omega^2$$

$$\therefore R = \frac{kF}{k\left[\left(\frac{\omega_F}{\omega}\right)^2 - 1\right]}$$

$$= \frac{F}{\left(\frac{\omega_F}{\omega}\right)^2 - 1}$$

The transmissibility (or magnification factor) is defined as the ratio of the maximum transmitted force to the maximum disturbing force, i.e.:

$$T_R = \frac{R}{F}$$

$$\therefore \quad \boxed{T_R = \frac{1}{\left(\frac{\omega_F}{\omega}\right)^2 - 1}}$$

(10.14) transmissibility

Example 10.7

A motor of mass 200 kg rotating at 1440 rpm is to be mounted on four springs so that only 5 per cent of any vibration force is transmitted to the foundation.

Determine:

(a) necessary natural frequency of the system
(b) necessary spring rate.

Solution

(a) $T_R = 5\% = 0.05$

$$\omega_F = 2\pi \times \frac{1440}{60} = 150.8 \text{ rad/s}$$

From equation 10.14:

$$\left(\frac{\omega_F}{\omega}\right)^2 - 1 = \frac{1}{T_R}$$

$$\therefore \omega = \frac{\omega_F}{\sqrt{\frac{1}{T_R} + 1}}$$

$$\therefore \omega = \frac{150.8}{\sqrt{\frac{1}{0.05} + 1}}$$

$$\therefore \omega = 32.9 \text{ rad/s}$$

$$\text{But } f = \frac{\omega}{2\pi} = \frac{32.9}{2\pi} = \mathbf{5.24\ Hz}$$

(b) From equation 10.9:

$$\omega = \sqrt{\frac{k}{M}}$$

$$\therefore k = \omega^2 M = 32.9^2 \times 200 = 216.5 \text{ N/mm}$$

Since there are four springs, the rate per spring = **54.1 N/mm** (springs in parallel).

Example 10.8

Draw a graph of transmissibility versus frequency ratio (f_F/f) for frequency ratios from 0 to 7. Draw conclusions from the graph.

Solution

The frequency ratio (f_F/f) is the ratio of the frequency of the forced vibration to the frequency of the natural vibration. Since $\omega = 2\pi f$, the frequency ratio will also be equal to the angular velocity ratio. Therefore equation 10.14 may also be written:

$$T_R = \frac{1}{\left(\frac{f_F}{f}\right)^2 - 1}$$

Substituting the frequency ratio in this equation the following values are obtained:

Frequency ratio	0	0.25	0.5	0.75	1	1.25	1.5	1.75	2	3	4	5	6	7
Transmissibility	−1	−1.07	−1.33	−2.28	∞	1.78	0.8	0.48	0.33	0.125	0.067	0.042	0.029	0.021

The values have been plotted in Figure 10.11 (ignoring the negative sign as previously mentioned).

Conclusions

1. When $f_F = f$, $T_R \to \infty$ (i.e. resonance).
2. Frequency ratios of less than 1.414 ($\sqrt{2}$) should be avoided in practice since they cause amplification rather than reduction in the transmissibility.
3. As the frequency ratio increases above $\sqrt{2}$, the transmissibility reduces, giving better cushioning of the vibration. In practice, ratios as high as 7 are often used as this gives really good cushioning with the transmissibility only about 2 per cent.

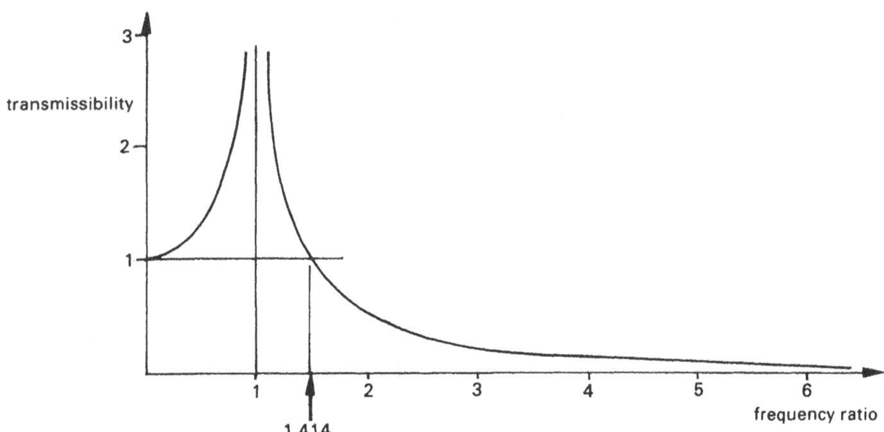

Fig. 10.11 *Transmissibility versus frequency ratio*

Problems

10.1 A body moves with simple harmonic motion of amplitude 300 mm and period 0.75 s.
Determine the maximum velocity and acceleration.
2.51 m/s, 21.1 m/s^2

10.2 A body moving with simple harmonic motion of amplitude 120 mm has an acceleration of 12 m/s^2 when 80 mm from the equilibrium position.
Determine:
(a) frequency
(b) period
(c) velocity at the 80 mm position.
(a) 1.95 Hz (b) 0.513 s (c) 1.1 m/s

10.3 For the body given in problem 10.2, calculate the velocity and acceleration for displacements from zero to the maximum in steps of 10 mm. Use the calculated values to draw a velocity–displacement graph and an acceleration–displacement graph.

10.4 A body moving with simple harmonic motion has a maximum velocity of 5 m/s and a maximum acceleration of 250 m/s^2.
Determine:
(a) amplitude
(b) frequency
(c) velocity and acceleration when the displacement is one-half of its maximum value
(d) displacement and acceleration when the velocity is one-half its maximum value.
(a) 100 mm (b) 7.96 Hz (c) 4.33 m/s, 125 m/s^2 (d) 86.7 mm, 217 m/s^2

10.5 A scotch yoke mechanism has a stroke of 125 mm and the crank rotates at 780 rpm. The follower has a mass of 2.4 kg.
Determine for the follower:
(a) amplitude
(b) frequency

(c) maximum velocity
(d) maximum acceleration
(e) inertia force when the crank angle is 45° from the mid-position.

(a) 62.5 mm (b) 13 Hz (c) 5.1 m/s (d) 417 m/s² (e) 708 N

10.6 A horizontal poppet valve of mass 250 g is spring-loaded against a cam which moves it in simple harmonic motion, that is the motion of the valve in opening and closing is one half of a sine curve. The valve lift is 10 mm and the time between opening and closing is 6 ms.

Determine:
(a) maximum velocity of the valve
(b) maximum acceleration of the valve
(c) spring force necessary in order to maintain a force of 200 N between the valve and the cam at the point of maximum acceleration.

(a) 5.24 m/s (b) 2740 m/s² (c) 885 N

10.7 A mass of 10 kg is placed on a tension spring of free length 120 mm. The extended length at equilibrium is 130 mm.

If the mass is now set into motion, determine the natural frequency and period of the vibration.

5 Hz, 0.2 s

10.8 A compressor is mounted on spring-loaded supports which deflect 25 mm due to the weight of the compressor.

Show that the natural frequency of the system may be determined without taking into account the mass of the compressor and determine this natural frequency.

3.15 Hz

10.9 A mass of 2 kg rests on a vertical compression spring whose spring constant is 50 N/mm. The mass is then pushed down a further 20 mm by a force which is then removed so that the system vibrates freely.

Determine:
(a) frequency
(b) period
(c) maximum velocity
(d) maximum acceleration
(e) maximum kinetic energy
(f) maximum potential energy (relative to the equilibrium position).

(a) 25.2 Hz (b) 0.04 s (c) 3.16 m/s (d) 500 m/s² (e) 10 J (f) 0.392 J

10.10 For the spring-mass system given in problem 10.9, determine the velocity and acceleration when the displacement is 12 mm from the equilibrium position.

2.53 m/s, 300 m/s²

10.11 A simple pendulum of length 1.5 m has a maximum displacement of 8° from the vertical position.

Determine:
(a) frequency
(b) period
(c) amplitude
(d) maximum velocity
(e) maximum acceleration.

(a) 0.407 Hz (b) 2.46 s (c) 209.4 mm (d) 0.536 m/s (e) 1.37 m/s²

10.12 For the pendulum given in problem 9.10, determine the velocity and acceleration when the bob is displaced 5° from the vertical position.

0.418 m/s, 0.856 m/s^2

10.13 A simple pendulum is to have a period of 2 s.

(a) Determine the length of cord required.

(b) Determine the maximum acceleration and hence inertia force if the bob has a mass of 1.25 kg and a maximum displacement of 10° from the vertical.

(c) Draw a vector force diagram to scale for the maximum displacement position and hence determine the cord tension.

(a) 994 mm (b) 1.71 m/s^2, 2.14 N (c) 12.1 N

10.14 A mass of 3 kg was attached to the end of a helical tension spring causing a static deflection of 15 mm. It was then pulled down a further 10 mm and released.

Determine:

(a) period of vibration

(b) frequency of vibration

(c) maximum velocity of the mass

(d) maximum acceleration of the mass.

(a) 0.246 s (b) 4.07 Hz (c) 0.256 m/s (d) 6.55 m/s^2

10.15 A mass of 4 kg is placed on a spring which deflects 20 mm under this load. A machine of mass 80 kg is then placed on four springs of the same type such that the weight is evenly distributed. The machine has an out-of-balance force of maximum value 50 N and frequency 2.5 Hz.

Determine the amplitude of the steady state vibration.

4.2 mm

10.16 A machine of mass 50 kg is supported evenly on three compression springs which deflect 10 mm due to the weight. The machine rotor has an out-of-balance mass of 120 g at 50 mm from the centre.

Determine the amplitude of the steady state vibration when the rotor turns at 360 rpm. Also determine the rotor speed at which resonance would occur.

0.387 mm, 299 rpm

10.17 For the system given in problem 10.16, determine:

(a) transmissibility

(b) maximum out of balance force transmitted to the foundation

(c) rotor speed to give a transmissibility of 0.05.

(a) 2.23 (b) 19 N (c) 1370 rpm

10.18 An electric motor of mass 120 kg rotates at 720 rpm. It is mounted on four isolating springs which transmit only 10 per cent of the motor out-of-balance force to the foundation.

Determine:

(a) natural frequency of the system

(b) spring constant of each isolator

(c) static deflection.

(a) 3.62 Hz (b) 15.5 N/mm (c) 19 mm

10.19 A rotating machine of mass 250 kg is supported by four springs each of which has a spring rate of 125 N/mm.

Determine:

(a) resonant frequency of the system
(b) shaft speed at which resonance occurs
(c) maximum out-of-balance mass at radius 40 mm so that the maximum out-of-balance force transmitted to the foundation at a rotor speed of 500 rpm does not exceed 50 N.

(a) 7.12 Hz (b) 427 rpm (c) 0.169 kg

10.20 A machine, of the same mass and mounted in the same way as the machine given in problem 10.19, has a shaft speed of 720 rpm. There is an out-of-balance mass of 0.5 kg located at 120 mm from the centre line of the shaft.

Determine:
(a) maximum out-of-balance force
(b) transmissibility
(c) maximum out-of-balance force transmitted to the foundation
(d) amplitude of the vibration.

(a) 341 N (b) 0.543 (c) 185 N (d) 0.37 mm

10.21 A motor of mass 100 kg rotating at 1760 rpm is mounted on four equal springs such that only 10 per cent of any out-of-balance force is transmitted to the foundation.

Determine:
(a) natural frequency of the system
(b) static deflection
(c) spring rate of each spring.

(a) 8.84 Hz (b) 3.18 mm (c) 77.2 N/mm

10.22 The out-of-balance effect on the rotor of a machine is equivalent to a mass of 600 g rotating at a radius of 50 mm. The total mass of the machine is 200 kg and the speed of the rotor is 1800 rpm.

Determine:
(a) the magnitude of the out-of-balance force
(b) the stiffness of an isolator to be fitted to the machine in order that only 10 per cent of the out-of-balance force is transmitted to the foundations of the machine.

(a) 1066 N (b) 646 N/mm

11

Balance and reaction of rotating masses

In Chapter 10 it was seen that vibration is caused by unbalanced reciprocating or rotary forces, that is forces which arise from the motion of a system of connected masses which have unbalanced linear or rotating components. The magnitude of the unbalance may be reduced by **balancing**, that is by providing equal and opposite forces to the out-of-balance forces.

For a system of connected masses in rotation only, good balance may be achieved with little difficulty but, where there is reciprocating motion as well as rotation, balance is more difficult. Indeed, complete balance of reciprocating masses is often impossible to achieve in practice unless they are arranged so as to oppose one another, for example in a horizontally opposed reciprocating engine. If the reciprocating masses move in simple harmonic motion, then balance may also be achieved by means of opposed rotating masses. However, in the usual piston and crank configuration, the piston does not move in simple harmonic motion and complete balance of the reciprocating component by means of rotating masses is impossible because of the higher order harmonics which are present.

In engines and compressors it is important to distinguish between out-of-balance forces caused by the motion of unbalanced masses and torsional power fluctuations caused by variation in gas pressure at different positions of the stroke. The torsional power fluctuations may be smoothed by constructing the engine or compressor so that there are a number of cylinders out of phase with one another, and by provision of a suitable flywheel.

Consideration of torsional power smoothing and the balance of reciprocating masses is beyond the scope of this chapter which deals only with systems of rotating connected masses in single or multi-plane configuration. Also this chapter considers the bearing reactions resulting from out-of-balance rotating motion. In all cases, bearing friction is considered negligible.

11.1 Static and dynamic balance

Static balance

A system of connected masses rotating about a common axis of rotation is in a condition of static balance when the centrifugal forces are in equilibrium, that is the vector sum of these forces is zero.

The reason for the term **static balance** is that, even though the system is a rotating one, it can readily be shown that if the centrifugal forces are in equilibrium, then the centre of mass (centre of gravity) of the system lies on the axis of rotation. Hence the system will be in balance at any angular position when at rest. Indeed static balance may be verified in practice by supporting the system in low friction bearings and checking that balance occurs at any angular position, i.e. there is no tendency for self-rotation. This also means that when static balance is achieved, the vector sum of the moments of the weight of each mass about the centre of rotation is zero.

Dynamic balance

A system of connected masses rotating about a common axis of rotation is in a condition of dynamic balance when both the centrifugal forces and the moments of these forces are in equilibrium, that is the sum of the forces and the moments of forces about any point is zero.

It is evident, therefore, that static balance is a necessary condition for dynamic balance but not the only condition. That is, a system in static balance is not necessarily in dynamic balance, whereas a system in dynamic balance must also be in static balance. Dynamic balance is therefore also known as **complete balance**.

The distinction between the two types of balance will be clear from Figure 11.1. Here two equal masses are fixed opposite one another at the same distance from the axis of rotation. In system (a) the masses lie in one plane (coplanar), whereas in system (b) the masses do not lie in the same plane (non-coplanar).

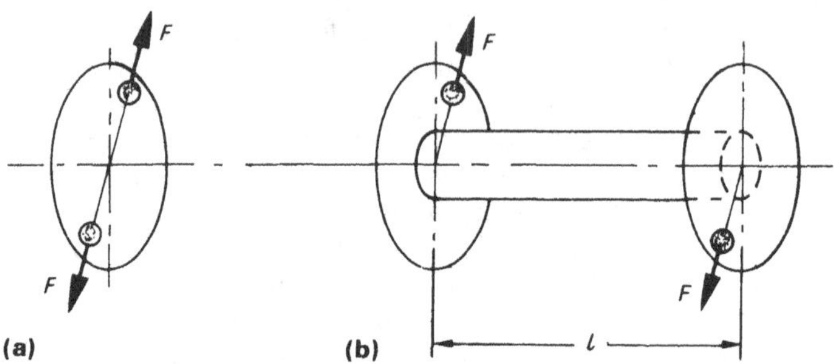

11.1 *(a) Coplanar and (b) non-coplanar rotating masses*

In both systems (a) and (b) the forces are balanced at any rotational speed and therefore both systems are in static balance. However in system (b) an unbalanced moment (or couple) of magnitude Fl occurs as the shaft rotates and this will cause a cyclic rocking motion just as undesirable as are unbalanced forces. System (b) is therefore not in dynamic balance even though it is in static balance.

In practice, exactly the same effect occurs when the wheels of a motor vehicle are balanced. If the out-of-balance masses were in the same plane, then a static balance would also provide dynamic balance. However the out-of-balance masses are not usually coplanar and, under these conditions, a static balance will not ensure complete balance at speed. This can only be achieved by use of a dynamic balancing machine in which the wheels are rotated while the balance is checked.

Summary

Dynamic balance occurs when the centrifugal forces and moments are in equilibrium and this can be verified experimentally only by rotating the system. If dynamic balance is achieved, then static balance also occurs, but the reverse does not apply except to a system of coplanar masses.

Static balance occurs when the centrifugal forces but not the moments are in equilibrium, which can be verified experimentally when the system is at rest. When static balance occurs, the centroid of the combined system of masses lies on the axis of rotation and the sum of the moments of the weight force of these masses about the centre of rotation is zero (at any angular position).

11.2 Coplanar balance: Single mass

As an introduction to the theory of balancing, consider the case of a single out-of-balance mass m_1, located at radius r_1, from the centre of rotation O and rotating at ω rad/s. It is clear that this mass can be balanced by a single balance mass m_2 located in the same plane and opposite to the out-of-balance mass m_1 and at radius r_2 as shown in Figure 11.2.

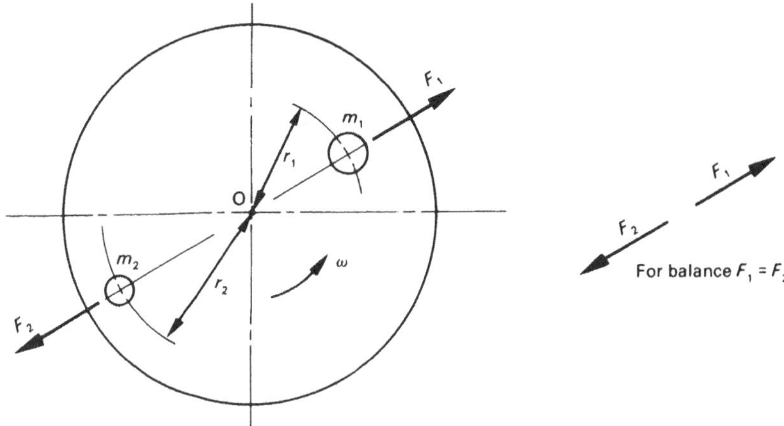

Fig. 11.2 *Coplanar balance of a single rotating mass*

Since the masses are coplanar, complete balance (static and dynamic) will occur when the algebraic sum of the centrifugal forces of both masses is zero.

The centrifugal force due to mass m_1: $F_1 = m_1r_1\omega^2$

The centrifugal force due to mass m_2: $F_2 = m_2r_2\omega^2$

For balance: $F_1 - F_2 = 0$ or $F_1 = F_2$

Hence $m_1r_1\omega^2 = m_2r_2\omega^2$

ω is common to both sides and cancels, hence the condition for balance (both dynamic and static) is:

$m_1r_1 = m_2r_2$

The term mr is called "reduced force" and given the symbol $\hat{F}$. It is not a true force because it does not have units of force but rather units, kg m. In fact any units may be used for mass and radius provided that consistency is maintained in each case.

Hence the condition for balance of coplanar masses is that the algebraic sum of the reduced forces in any direction is zero. That is:

$\Sigma\hat{F} = 0$

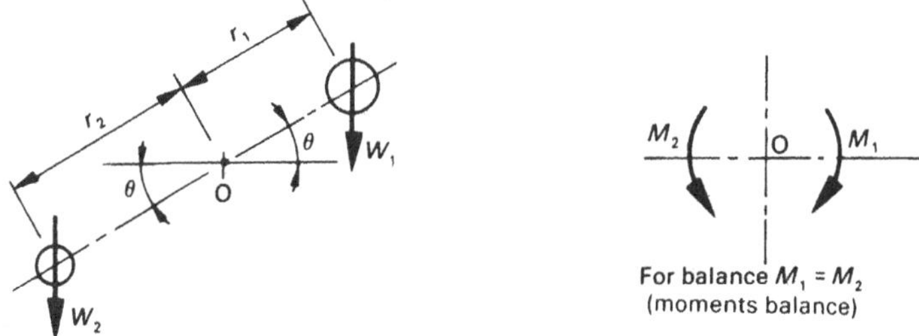

Fig. 11.3 *Masses balanced at rest*

It can readily be shown that, when the reduced forces are in equilibrium, the shaft will balance at rest (static balance). Consider Figure 11.3.

Calling the angle of inclination θ and taking moments about the centre O:

For balance at rest (static): $W_1r_1 \cos\theta = W_2r_2 \cos\theta$

Since θ is common, $\cos\theta$ cancels on both sides of this equation.

$\therefore W_1r_1 = W_2r_2$ or $m_1gr_1 = m_2gr_2$

Now g is common and cancels so that:

$m_1r_1 = m_2r_2$ (as before)

Note: Both the magnitude and radius of the balance mass cannot be unknown and one or the other must be specified. This is clarified in example 11.1.

Example 11.1

An unbalanced mass of 200 g is located at radius 320 mm from the centre of rotation.
Determine:

(a) necessary location of a coplanar mass of 400 g in order to provide balance
(b) necessary coplanar mass located at radius 800 mm in order to provide balance.

Solution

(a)
$$m_1r_1 = m_2r_2$$
$$\therefore 200 \times 320 = 400 \times r_2$$
$$\therefore r_2 = \mathbf{160\ mm}$$

(b)
$$m_1r_1 = m_2r_2$$
$$\therefore 200 \times 320 = m_2 \times 800$$
$$\therefore m_2 = \mathbf{80\ g}$$

Note: It is not necessary to convert mass units to kg or radius units to m, provided that consistency is maintained on both sides of the equation.

11.3 Coplanar balance: Multiple masses

Where a number of coplanar masses are rotating, it is evident that balance is achieved if the vector diagram for the reduced forces closes. Consider Figure 11.4.

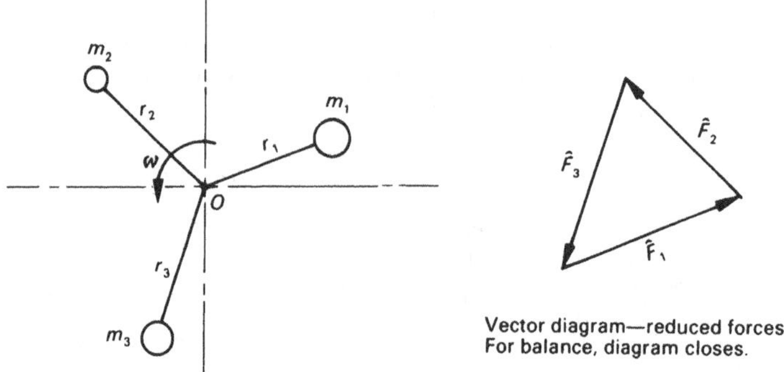

Fig. 11.4 *Balanced coplanar rotating masses*

Balance (static and dynamic) occurs when the vector diagram for the reduced forces closes. Hence the algebraic sum of the components of the reduced forces in any direction is also zero.

Notes

1. If the reduced force vector diagram closes, coplanar masses are in dynamic balance and hence must also be in static balance.
2. When a number of coplanar masses are to be balanced by a single mass, the location of the mass is opposite to the resultant of the reduced forces for the out-of-balance masses.
3. Both the radius and the magnitude of the balance mass cannot be unknown and one or the other must be specified. However, the angle made by the balance mass cannot be specified and is determined from the vector diagram.
4. Problems may be solved by any one of the three methods available for the solution of coplanar force systems: (a) graphical solution, (b) analytical solution and (c) computer solution.

Graphical solution

A vector diagram for the reduced forces is drawn to scale. The equilibrant reduced force to close the vector diagram is the balancing reduced force.

Analytical solution

The x and y components of the reduced forces are calculated and the resultant x and y components determined by summation. Hence the magnitude and direction of the resultant reduced force may be determined and the balancing reduced force is equal and opposite to this resultant (equilibrant).

Computer solution

A computer solution may readily be obtained by modification of the previously developed program FORCE 1 for coplanar forces. The new program named COPBAL is listed in Appendix 3.12. The program inputs are:

1. number of masses to be balanced
2. magnitude, radius and angle (degrees anticlockwise) of each mass
3. either the magnitude or radius of the balance mass.

The program then outputs the magnitude, radius and angle of the balance mass.

Note: Any units may be used for mass and radius provided consistency is maintained.

Example 11.2

Determine the radius r and the angle θ in order that a mass of 2 kg will balance the two out-of-balance coplanar masses of 1 kg and 1.5 kg located as shown in Figure 11.5, using (a) graphical, (b) analytical and (c) computer solutions.

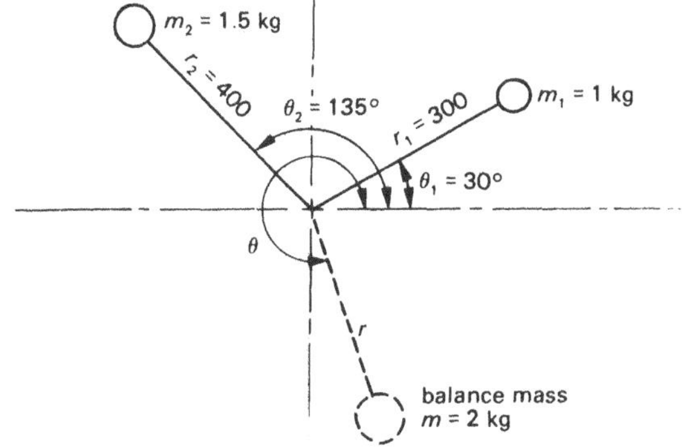

Fig. 11.5

Solution

$\hat{F}_1 = m_1 r_1 = 1 \times 300 = 300$ kg mm

$\hat{F}_2 = m_2 r_2 = 1.5 \times 400 = 600$ kg mm

(a) Graphical solution

The vector diagram for the reduced forces is as shown in Figure 11.6.

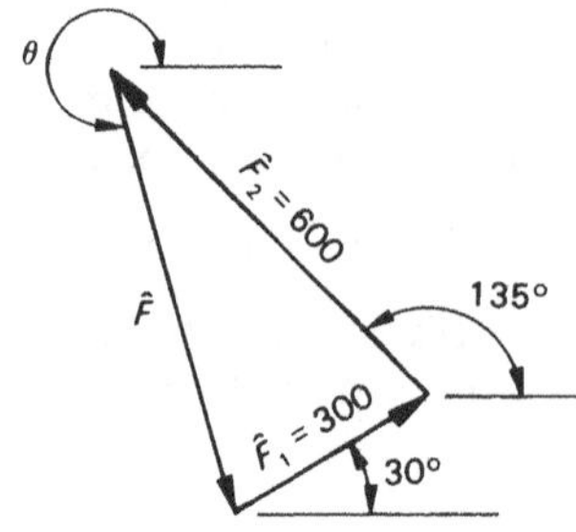

Fig. 11.6

Note that the balancing reduced force is the equilibrant reduced force, that is it is equal and opposite to the resultant of the out-of-balance reduced forces.

By measurement $\hat{F} = 597$ kg mm

and θ = **286°**

Since $\hat{F} = mr$ and $m = 2$ kg,

$$\therefore \quad r = \frac{597}{2} = \mathbf{298.5\ mm}$$

(b) Analytical solution

Reduced force (kg mm)	*x component*	*y component*
$\hat{F}_1 = 300$ $\hat{F}_2 = 600$	300 cos 30° = 259.8 600 cos 135° = −424.3	300 sin 30° = 150.0 600 sin 135° = 424.3
Total	−164.5	574.3

Combining the x and y components gives Figure 11.7.

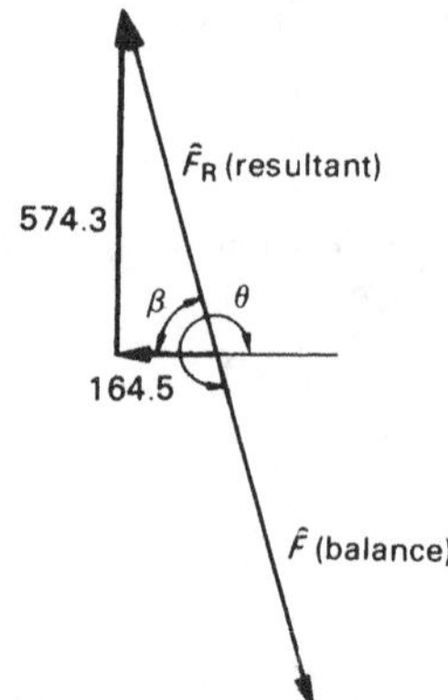

Fig. 11.7

By trigonometry $\hat{F}_R = 597.4$ kg mm and $\beta = 74°$.

Therefore the balancing reduced force $\hat{F} = 597.4$ kg mm at angle $\theta = 360 - 74 =$ **286°**.

Since $\hat{F} = mr$ and $m = 2$ kg,

$\therefore \quad r =$ **298.7 mm**

(c) Computer solution

The computer solution is given below the program listing in Appendix 3.12. The magnitude, radius and angle of the two unbalanced masses are input as:

1, 300, 30
1.5, 400, 135

The output from the program run gives the balance mass as follows:

mass	radius	angle
2	298.67	285.98

11.4 Bearing reactions for coplanar masses

When a number of rotating masses are coplanar, there will be no reaction at the bearing due to centrifugal force if the masses are in balance. In this case, the bearing reaction has a constant value equal to the weight of the masses and acts vertically upward for any rotational position of the masses. However, if the masses are unbalanced, there is a resultant centrifugal force which causes an equal and opposite reaction force at the bearing which changes direction as the masses rotate. The total reaction at the bearing will therefore vary from a minimum value when the centrifugal force acts upward (opposite to the weight) to a maximum value when the centrifugal force acts downward (additive to the weight).

Consider Figure 11.8 where two out-of-balance coplanar masses are rotating about a centre.

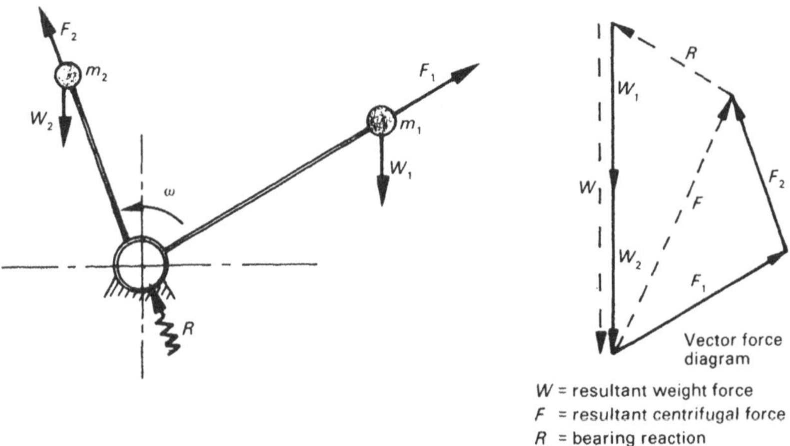

Fig. 11.8 *Bearing reaction for coplanar rotating masses*

The bearing reaction R may be determined by drawing a vector force diagram. For the particular position of the shaft shown, the bearing reaction acts upward and to the left.

As the shaft rotates, the resultant centrifugal force F remains constant in magnitude but varies in direction. The resultant weight W remains constant in magnitude and direction. The maximum bearing reaction occurs when F and W have the same direction and the minimum bearing reaction occurs when F and W have opposite directions as shown in Figure 11.9.

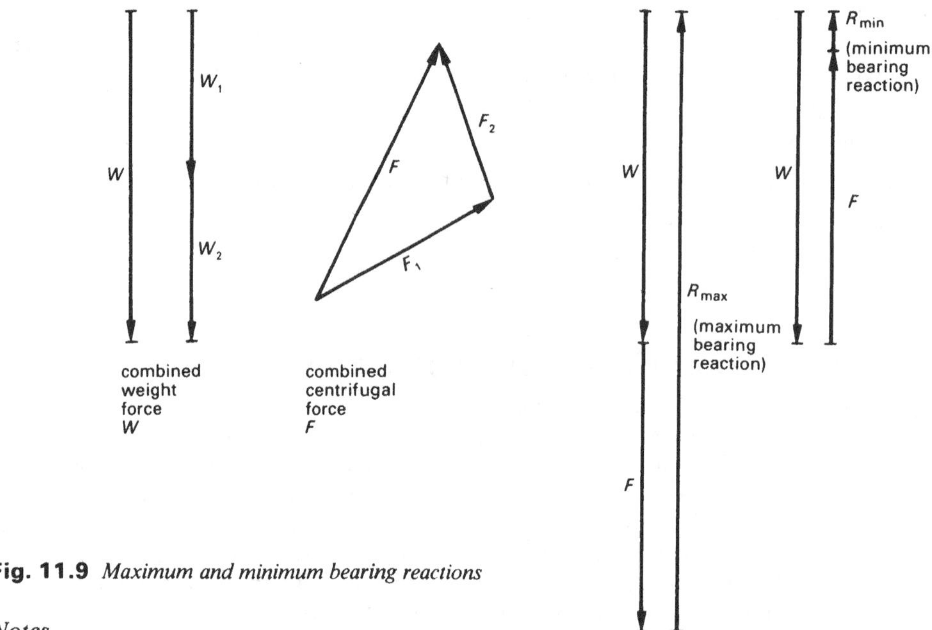

Fig. 11.9 *Maximum and minimum bearing reactions*

Notes

1. Shaft positions when maximum and minimum bearing reactions occur are 180° apart from one another.
2. Centrifugal force increases as the square of the speed so that even at moderate shaft speeds the centrifugal force due to out-of-balance masses may be far greater than the weight of these masses and hence the weight may often be neglected.

Example 11.3

For the shaft shown in Figure 11.10 rotating at 90 rpm, determine:

(a) bearing reaction in position shown when the angle θ made by mass m_1 is 45°

(b) maximum bearing reaction and the angle θ made by mass m_1 when the maximum reaction occurs

(c) minimum bearing reaction and the angle θ made by mass m_1 when the minimum reaction occurs.

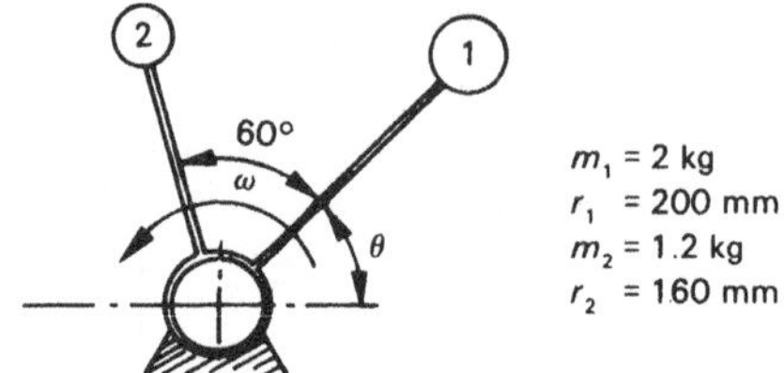

Fig. 11.10

Solution

At 90 rpm $\omega = \pi \times \dfrac{90}{30} = 9.42$ rad/s

$$F_1 = m_1 r_1 \omega^2 = 2 \times 0.2 \times 9.42^2 = 35.53 \text{ N}$$
$$F_2 = m_2 r_2 \omega^2 = 1.2 \times 0.16 \times 9.42^2 = 17.05 \text{ N}$$

Also
$$W_1 = m_1 g = 2 \times 9.81 = 19.62 \text{ N}$$
$$W_2 = m_2 g = 1.2 \times 9.81 = 11.77 \text{ N}$$

(a) The bearing reaction may be determined either graphically, analytically or by computer. The graphical solution is shown in Figure 11.11.

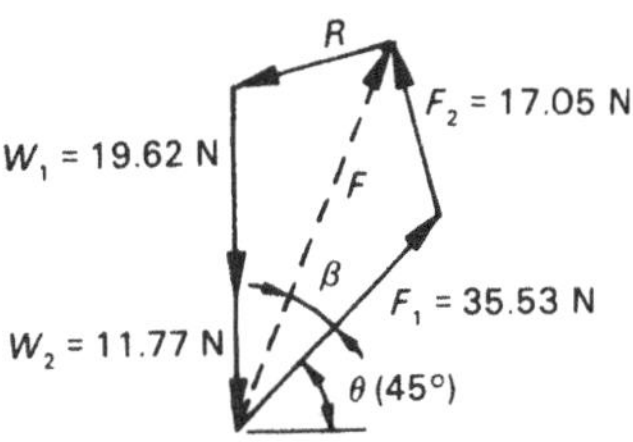

Fig. 11.11

The bearing reaction is the equilibriant force R. Scaling from the diagram:

$$\boldsymbol{R = 23.1 \text{ N @ } 206°}$$

(b) The maximum bearing reaction occurs when the resultant centrifugal force F acts vertically downward and is therefore additive to the weight. From the diagram:

$$F = 46.5 \text{ N @ } \beta = 18.5°$$

Hence the maximum bearing reaction is:

$$R_{max} = 46.5 + 19.62 + 11.77 = \mathbf{77.9 \text{ N}}$$

This occurs when $\theta + \beta = 270°$ (vertically down)

i.e. when $\theta = 270 - 18.5 = \mathbf{251.5°}$

(c) The minimum bearing reaction occurs when the resultant centrifugal force F acts vertically upward.

Hence the minimum bearing reaction is:

$$R_{min} = 46.5 - 19.62 - 11.77 = \mathbf{15.1 \text{ N}}$$

This occurs when $\theta + \beta = 90°$ (vertically up)

i.e. when $\theta = 90 - 18.5 = \mathbf{71.5°}$ (which is 180° from the maximum position as expected)

Computer solution

A computer program to determine the bearing reactions for coplanar rotating masses was obtained by modification of the balancing program COPBAL. The program named COPRE is listed in Appendix 3.13 and has the following inputs:

1. number of rotating masses
2. shaft speed in rpm
3. magnitude (kg), radius (m) and angle (degrees anticlockwise) of each mass in turn.

The program outputs are:

1. magnitude and angle of the bearing reaction for the input position of the masses
2. maximum bearing reaction and angular position of mass 1 when this occurs (mass 1 being the first mass input)
3. minimum bearing reaction and angular position of mass 1 when this occurs.

The solution to example 11.3 is given below the program listing in Appendix 3.13 and should be self-explanatory.

11.5 Balance of non-coplanar masses

Consider four masses m_1, m_2, m_3 and m_4 located on a rotating shaft as shown in Figure 11.12.

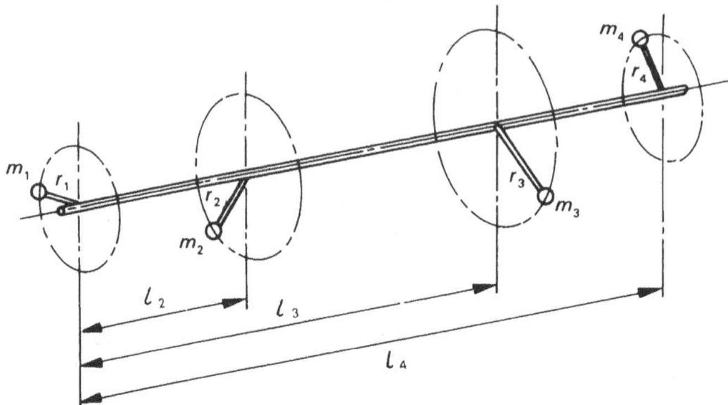

Fig. 11.12 *Non-coplanar rotating masses*

It is evident that for static balance the vector diagram drawn for the reduced forces must close. For dynamic balance, not only must this condition be met, but also the sum of the moments of the reduced forces about any reference line perpendicular to the shaft must also be zero. The analytical method of taking moments in three dimensions is rather complicated, so a graphical method known as **Dalby's method** is often used. In this method, the moments of the reduced forces are treated as vector quantities and a vector diagram drawn for the reduced moments will close if the masses are in dynamic balance. The method is applied as follows:

1. Choose a reference plane in the plane of one of the unknown masses.
2. Calculate the reduced force for each known mass as before, i.e.

   $\hat{F}_1 = m_1 r_1, \quad \hat{F}_2 = m_2 r_2$, and so on.
3. Calculate the moment of each reduced force with respect to the reference plane using the convention that distances to the right of the reference plane are positive and those to the left are negative. For example, in the system shown in Figure 11.12, if the reference plane were the plane of m_1, then:

   $\hat{M}_1 = 0, \quad \hat{M}_2 = \hat{F}_2 l_2, \quad \hat{M}_3 = \hat{F}_3 l_3$ and $\hat{M}_4 = \hat{F}_4 l_4$.

   If the reference plane were the plane of m_2, then:

   $\hat{M}_1 = -\hat{F}_1 l_2, \quad \hat{M}_2 = 0, \quad \hat{M}_3 = \hat{F}_3(l_3 - l_2)$ and $\hat{M}_4 = \hat{F}_4(l_4 - l_2)$.
4. Draw a vector diagram for the moment of each reduced force. The direction of the reduced moment vector is the same as that of the reduced force vector if the reduced moment is positive. If the reduced moment is negative, the moment vector has the

opposite direction to the force vector. For dynamic balance, the diagram must close. Hence an unknown reduced moment may be found and, from it, the corresponding unknown reduced force may be calculated (see Fig. 11.13).

5. Draw a vector diagram for the reduced forces. This diagram must also close and hence the other unknown reduced force may be determined (see Fig. 11.13).
6. From the value of the reduced forces found in steps 4 and 5 the required balance masses may be calculated.

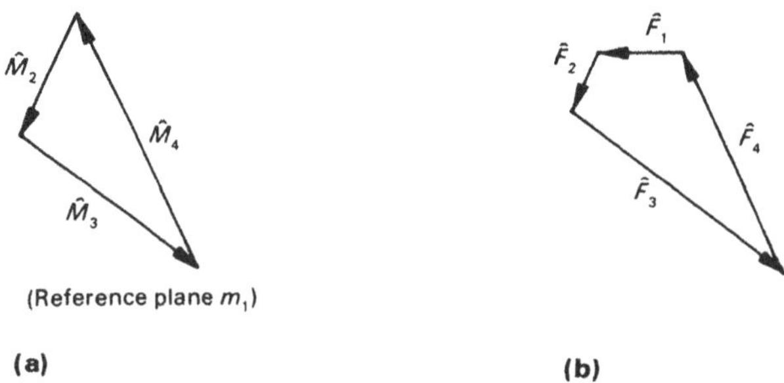

Fig. 11.13 *(a) Reduced moment and (b) reduced force vector diagrams*

Notes

1. There are always less vectors in the reduced moment vector diagram than in the reduced force vector diagram. The number less is equal to the number of masses in the reference plane.
2. Corresponding reduced force and moment vectors in each diagram are parallel, i.e. $\hat{F}_1$ is parallel to $\hat{M}_1$, $\hat{F}_2$ parallel to $\hat{M}_2$, and so on.
3. Two axially displaced rotating masses may be in static balance if they lie in a common plane which contains the rotational axis, but cannot be in dynamic balance.

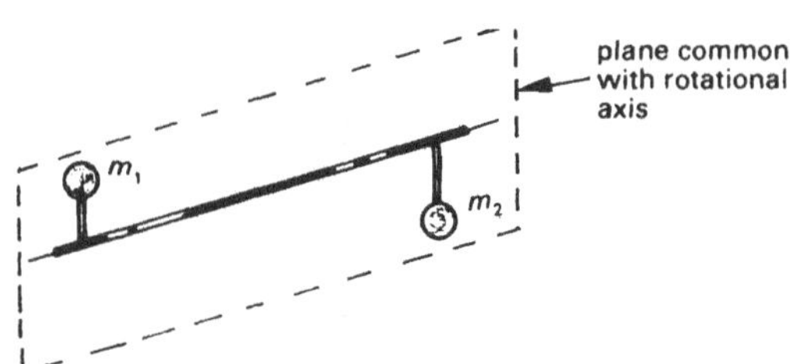

Fig. 11.14 *Two axially displaced masses. Only static balance is possible*

4. Three axially displaced rotating masses can only be in dynamic balance if they all lie in a common plane which contains the rotational axis. This follows because the reduced moment vector diagram has only two vectors relative to a reference plane containing one of the masses and these two vectors cannot be in equilibrium unless they are equal and opposite.
5. Four or more axially displaced rotating masses do not have to be coplanar with the axis of rotation in order to be in dynamic balance.

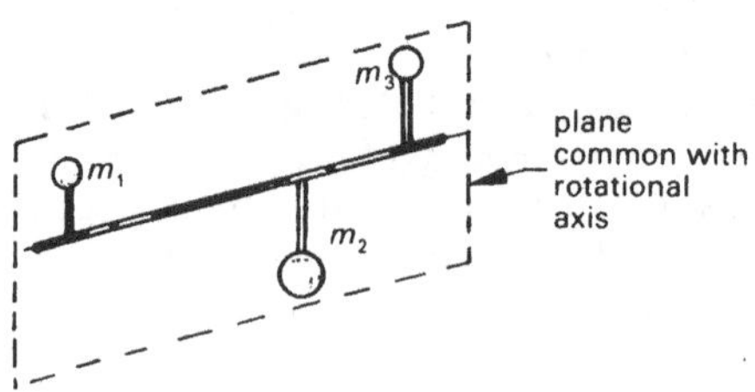

Fig. 11.15 *Three axially displaced masses can be in dynamic balance only if they are coplanar with the axis of rotation*

6. Two axially displaced balance masses are sufficient to balance any number of axially displaced rotating masses. This is the general problem in dynamic balancing where the two balance masses are unknown and the other masses are given. As in Dalby's method, one mass may be determined from the reduced moment equilibrium and the other from the reduced force equilibrium. Example 11.4 illustrates the general problem and method of solution.

Example 11.4

Determine the unknown masses m_2 and m_4 located at radius 127 mm in order to dynamically balance the three masses m_1, m_3 and m_5 located at radius 102 mm as shown in Figure 11.16.

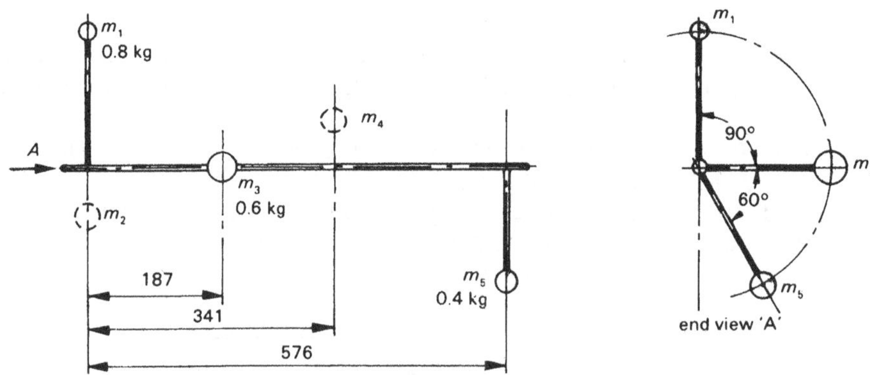

Fig. 11.16

Solution

Calculate the reduced forces (in kg mm) as follows:

$$\hat{F}_1 = 0.8 \times 102 = 81.6$$
$$\hat{F}_2 = m_2 \times 127 = 127\, m_2$$
$$\hat{F}_3 = 0.6 \times 102 = 61.2$$
$$\hat{F}_4 = m_4 \times 127 = 127\, m_4$$
$$\hat{F}_5 = 0.4 \times 102 = 40.8$$

Choose the plane containing the unknown mass m_1 as the reference plane.

Relative to this plane the moments of the reduced forces (in kg mm^2) are as follows:

$\hat{M}_1 = 0$ (reference plane)
$\hat{M}_2 = 0$ (reference plane)
$\hat{M}_3 = 61.2 \times 187 = 11.4 \times 10^3$
$\hat{M}_4 = 127\, m_4 \times 341 = 43.3 \times 10^3 \times m_4$
$\hat{M}_5 = 40.8 \times 576 = 23.5 \times 10^3$

For clarity it is useful to tabulate the data as shown below:

Number	m (kg)	r (mm)	l (mm)	θ (°)	$\hat{F}$ (kg mm)	$\hat{M}$ (kg mm^2 × 10^{-3})
1	0.8	102	0	90	81.6	0
2	m_2	127	0	θ_2	127 m_2	0
3	0.6	102	187	0	61.2	11.4
4	m_4	127	341	θ_4	127 m_4	43.3 m_4
5	0.4	102	576	300	40.8	23.5

The reduced moment vector diagram may now be drawn as shown in Figure 11.17(a), from which,

$$\hat{M}_4 = 30.8 \times 10^3 \quad \text{and} \quad \boldsymbol{\theta_4 = 139°}$$

$$\text{but } \hat{M}_4 = 43.3 \times 10^3 \times m_4$$

$$\therefore \quad m_4 = \frac{30.8 \times 10^3}{43.3 \times 10^3} = \mathbf{0.71\ kg}$$

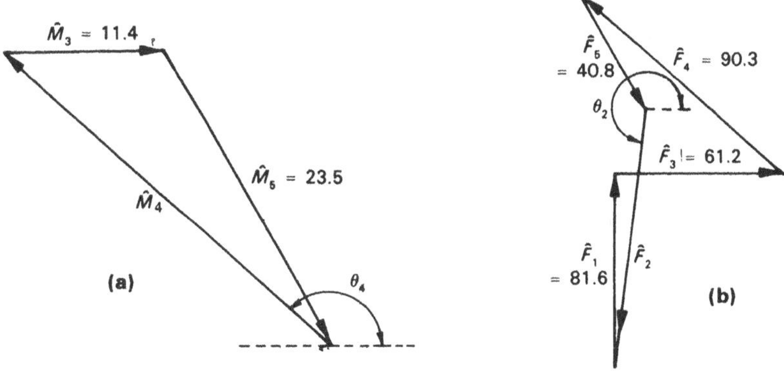

Fig. 11.17 *(a) Reduced moment and (b) reduced force vector diagrams*

Now $\hat{F}_4 = 127\, m_4 \quad \therefore \hat{F}_4 = 90.3$ (kg mm)

The reduced force vector diagram may now be drawn as shown in Figure 11.17(b), from which,

$$\hat{F}_2 = 107 \quad \text{and} \quad \boldsymbol{\theta_2 = 262°}$$

$$\text{but } \hat{F}_2 = 127\, m_2$$

$$\therefore \quad \boldsymbol{m_2 = 0.84\ kg}$$

Computer solution

The computer program NCBAL listed in Appendix 3.14 was written to solve dynamic balancing problems involving up to eight axially displaced rotating masses and two balancing masses. When using the program, the user must decide which of the two balance mass planes will be used as the reference plane, and apply correct signs accordingly to axial lengths depending upon whether they are measured in front of (+) or behind (−) the reference plane. The program inputs are:

1. the number of known masses to be balanced
2. mass, radius, angle and axial distance of each known mass in turn (any sequence may be used)
3. radius and axial distance of the balance mass not in the reference plane
4. radius of the balance mass in the reference plane.

The program outputs are:

1. mass, radius, angle and axial distance of the balance mass not in the reference plane
2. mass, radius and angle of the balance mass in the reference plane.

Notes

1. If the mass rather than the radius of either of the balancing masses is specified, any arbitrary radius (say 100 mm) may be used and the balance mass at this radius determined from the program. It is then simply a matter of ratios to determine the corresponding radius for the given mass.
2. If the mass rather than the axial distance is specified for the balancing mass not in the reference plane, any arbitrary axial distance (say 100 mm) may be used and the balance mass at this distance obtained from a first run of the program. It is then simply a matter of ratios to determine the corresponding axial distance for the given mass. The program must then be rerun using the calculated axial distance, as the first run will not give the correct value for the balance mass in the reference plane.

The computer solution to example 11.4 using:

(a) reference plane in the plane of m_1, and
(b) reference plane in the plane of m_4

is given below the program listing in Appendix 3.14 and should be self-explanatory.

11.6 Bearing reactions for non-coplanar masses

When unbalanced, axially displaced masses rotate about a common shaft which is supported by two bearings, the bearing reactions will vary as the shaft rotates. There is a variable component due to centrifugal forces and also a fixed component due to the weight of the masses. The resultant reaction at each bearing varies in both magnitude and direction as the shaft rotates and will have minimum and maximum values during each complete revolution.

The reaction at each bearing for any given position of the shaft may be determined by analytical, graphical or computer methods. The calculation of maximum and minimum values and the position of the shaft for which these occur is generally difficult and best done by computer.

Note that the component of the reaction due to weight does not change as the shaft rotates and is always vertical. If it is required that the weight of the shaft itself be included as well as the weight of the unbalanced masses, this may be done readily by treating the weight of the shaft as a vertical load acting at the mid-point of the shaft (centre of mass position).

Example 11.5

Determine the reactions at A and B for the unbalanced rotating masses shown in Figure 11.18 when they are in the position shown and rotating at 90 rpm. Use analytical, graphical and computer methods. Also determine the maximum and minimum reaction at each bearing and the position of m_1 when they occur using the computer method.

Data: $m_1 = 1.5$ kg $\quad m_2 = 1.2$ kg
$r_1 = 300$ mm $\quad r_2 = 250$ mm
$\theta_1 = 90°$ $\quad \theta_2 = 0°$
$l_1 = 350$ mm $\quad l_2 = 600$ mm $\quad l = 950$ mm

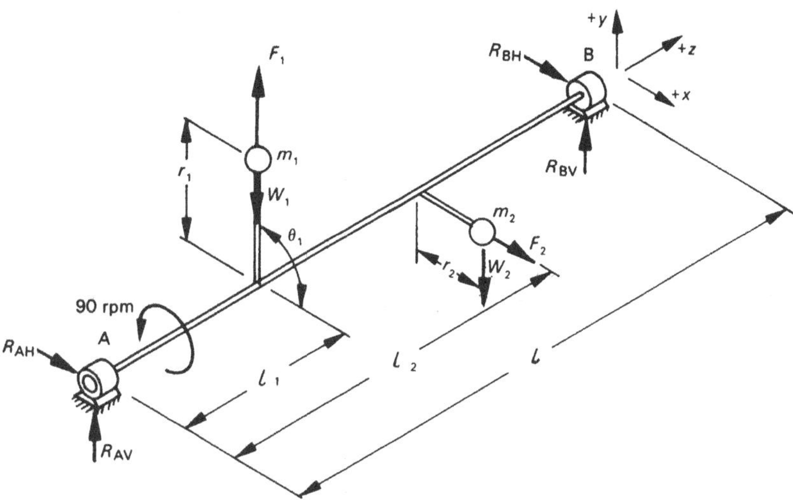

Fig. 11.18

Solution

Analytical method

$$\omega = \frac{\pi \times 90}{30} = 9.425 \text{ rad/s}$$

$$F_1 = m_1 r_1 \omega^2 = 1.5 \times 0.3 \times 9.425^2 = 40 \text{ N}$$

$$F_2 = m_2 r_2 \omega^2 = 1.2 \times 0.25 \times 9.425^2 = 26.65 \text{ N}$$

$$W_1 = m_1 g = 14.72 \text{ N} \qquad W_2 = m_2 g = 11.77 \text{ N}$$

Taking moments about a horizontal line through A and denoting clockwise moments as positive:

$$(-F_1 + W_1)l_1 + W_2 l_2 - R_{BV} l = 0$$

$$\therefore (-40 + 14.72) \times 0.35 + 11.77 \times 0.6 - R_{BV} \times 0.95 = 0$$

$$\therefore R_{BV} = -1.88 \text{ N} \quad (\downarrow)$$

Similarly, taking moments about a vertical line through A:

$$F_2 l_2 + R_{BH} l = 0$$

$$\therefore 26.65 \times 0.6 + R_{BH} \times 0.95 = 0$$

$$\therefore R_{BH} = -16.83 \text{ N} \quad (\leftarrow)$$

Since $\Sigma F_V = 0$ and using the conventional positive directions for the x, y and z axes (as shown in Fig. 11.18):

$$F_1 - W_1 - W_2 + R_{BV} + R_{AV} = 0$$

$$\therefore 40 - 14.72 - 11.77 - 1.88 + R_{AV} = 0$$

$$\therefore R_{AV} = -11.63 \text{ N} \quad (\downarrow)$$

Similarly $\Sigma F_H = 0$

$$F_2 + R_{BH} + R_{AH} = 0$$

$$\therefore 26.65 - 16.83 + R_{AH} = 0$$

$$\therefore R_{AH} = -9.82 \text{ N} \quad (\leftarrow)$$

The horizontal and vertical components of each reaction may now be combined to obtain the total reaction and the angle with the horizontal axis (x axis) as shown in Figure 11.19:

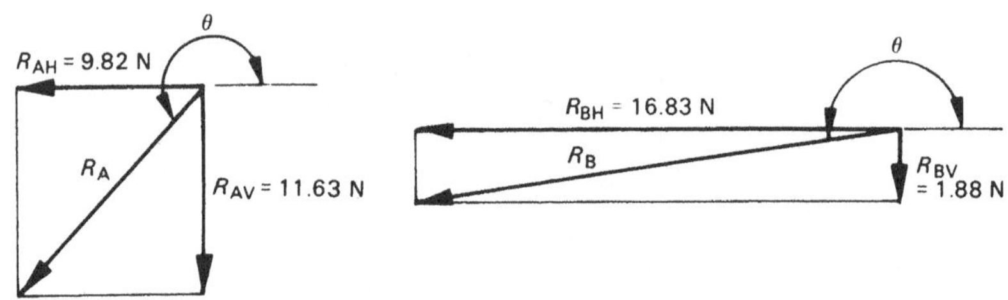

Fig. 11.19

From which, the reaction at A is **15.2 N @ 229.8°**

and the reaction at B is **16.9 N @ 186.4°**

Graphical method

The graphical method is most conveniently done by initially ignoring the weight of the masses. Dalby's method is then used, treating the bearing reactions as unknown reduced forces required for balance. Having determined the reduced force at each bearing, the actual force may readily be calculated and this is, in fact, the reaction at the bearing due to rotation of the unbalanced masses (centrifugal force). The reaction at each bearing due to the weight of the masses may readily be determined either graphically (string polygon) or by simple calculation. The total reaction at each bearing may then be determined by a vector force diagram drawn for each bearing.

Using the plane of bearing A as the reference plane:

Plane	m (kg)	r (mm)	$\hat{F}$ (kg mm)	l (mm)	$\hat{M}$ (kg mm^2)
A	m_A	r_A	$m_A r_A$	0	0
1	1.5	300	450	350	157.5×10^3
2	1.2	250	300	600	180×10^3
B	m_B	r_B	$m_B r_B$	950	$\hat{F}_B \times 950$

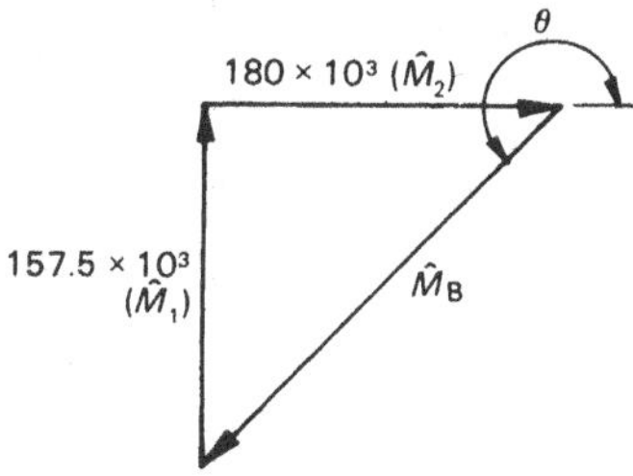

Fig. 11.20

The reduced moment polygon is as shown in Figure 11.20, from which:

$$\hat{M}_B = 239 \times 10^3 \text{ kg mm}$$

Now $\hat{F}_B \times 950 = 239 \times 10^3$

$$\therefore \hat{F}_B = 251.6 \text{ kg mm @ } 221°$$

The actual force at B may readily be computed from:

$$F = m r \omega^2 = \hat{F} \omega^2$$

$$\therefore F_B = 251.6 \times 10^{-3} \times 9.425^2$$

$$= 22.3 \text{ N @ } 221°$$

The reduced force polygon may now be drawn (Fig. 11.21).

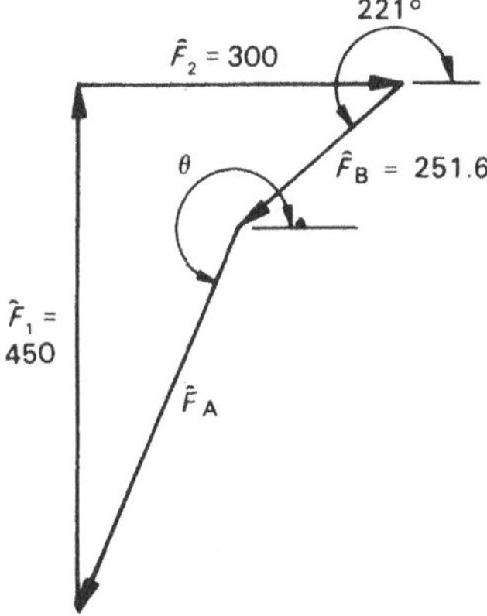

Fig. 11.21

From which:

$$\hat{F}_A = 305 \text{ kg mm} = m_A r_A$$

$$\therefore F_A = 305 \times 10^{-3} \times 9.425^2$$

$$= 27.1 \text{ N @ } 249°$$

Now taking into account the weight of each mass gives Figure 11.22.

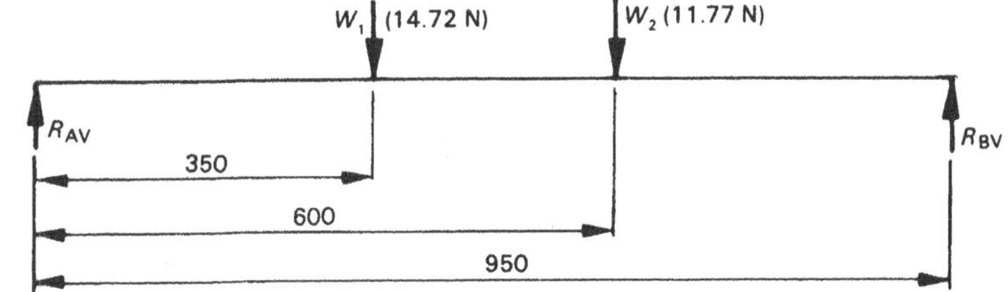

Fig. 11.22

By any of the methods of statics it will be found that:

$R_{BV} = 12.86$ N and $R_{AV} = 13.63$ N

Adding these components to the reaction due to unbalance gives Figure 11.23, from which:

$\boldsymbol{R_A = 15.2}$ **N @ 230°** and $\boldsymbol{R_B = 16.9}$ **N @ 186°** (as before)

Fig. 11.23

Computer method

The program NCREACT as listed in Appendix 3.15 was written to determine reactions when rotating masses are attached to a shaft supported by two bearings. The program calculates the reaction in magnitude and direction at each bearing (rounded off to two decimal places) for any given position of the shaft. If required, the program will then determine the maximum and minimum bearings reactions and the position of mass 1 (to the nearest degree) when they occur.

Notes

1. Mass 1 is the first mass input and may in fact be any of the masses.
2. Maximum and minimum values of the reactions are determined by positioning mass 1 at zero degrees (horizontal) and rotating the shaft incrementally by one degree at a time. On a microcomputer this part of the program may take several minutes to run.
3. If it is required that the weight of the shaft itself is considered, this may readily be done by treating the shaft as a point mass concentrated at the centre of mass position, i.e. on the rotational axis.
4. The program will cope with up to eight masses. Any of these may be overhung (outboard of the bearings).

The solution to example 11.5, including calculation of the maximum and minimum reactions, is given below the program listing in Appendix 3.15 and should be self-explanatory.

Computer calculated values are:

reaction at B **16.93 N @ 186.34°**

reaction at A **15.21 N @ 229.79°**

Maximum and minimum reactions and angle made by mass 1 when they occur:

	Max.	*Angle*	*Min.*	*Angle*
Bearing B	**35.22 N**	**319°**	**9.51 N**	**139°**
Bearing A	**40.72 N**	**291°**	**13.46 N**	**111°**

Problems

In these problems the angular displacements are measured anticlockwise from the horizontal (x axis) and the shafts are viewed from the left-hand end.

11.1 An out-of-balance mass of 1.5 kg is located at a distance of 450 mm from a centre of rotation.
Determine:
(a) the maximum bearing reaction at a rotational speed of 80 rpm
(b) where a mass of 800 g would have to be located for balance
(c) what mass would be necessary at a radius of 600 mm for balance.
(a) 62.1 N (b) 844 mm opposite (c) 1.125 kg

11.2 Two coplanar rotating masses of mass 400 g and 600 g are at radius 300 mm and 250 mm and angular displacement 45° and 210° respectively.
Determine the radius and angular displacement of a balance mass of 300 g.
154 mm, 348°

11.3 A shaft has two coplanar masses of magnitude 900 g and 700 g at radius 500 mm and 400 mm and angular displacement 0° and 120° respectively.
Determine the bearing reaction in this position when:
(a) the shaft is at rest
(b) the shaft rotates at 45 rpm.
(a) 15.7 N @ 90° (b) 12.4 N @ 124°

11.4 A flat circular rotating plate has two holes of diameter 30 mm and 50 mm drilled in it at radius 220 mm and 150 mm and angular displacement 30° and 135°.
Determine for balance:
(a) radius and angular displacement of a hole, diameter 60 mm
(b) diameter and angular displacement of a hole located at radius 200 mm.
(a) 104.4 mm, 284.4° (b) 43.36 mm, 284.4°

11.5 For the unbalanced plate given in problem 11.4, determine the magnitude and location out-of-balance force on the bearing at a rotational speed of 420 rpm if the plate is made of material of relative density 4.6 and has a thickness of 20 mm.
52.54 N, 284.4°

11.6 Three coplanar rotating masses are arranged as follows:

Mass (kg)	*Radius* (mm)	*Angular displacement*
1.2	250	20°
0.8	320	140°
1.5	280	230°

Determine the angular displacement and radius of a balance mass of 1 kg.
16.5°, 192 mm

11.7 A circular plate of mass 5.8 kg has an unbalanced mass of 400 g attached to it at radius 200 mm. The assembly is then rotated at 240 rpm.
Determine:
(a) bearing reaction and its angle when the angle made by the unbalanced mass is 45°
(b) maximum bearing reaction.
(a) 43.7 N, 145° (b) 111.35 N

11.8 Solve problem 11.7 again if the plate has added an additional out-of-balance mass of 200 g located at radius 300 mm and at 120° (anticlockwise) to the 400 g mass.

(a) 17.3 N @ 87.1° (b) 108.3 N

11.9 Determine the unknown quantities (a) to (d) in the table below for complete balance of the rotating shaft.

Number	*Mass* (g)	*Axial distance along shaft* (mm)	*radius* (mm)	*Angular displacement*
1	(a)	0 (reference plane)	400	(b)
2	(c)	200	300	(d)
3	200	400	300	90°
4	120	400	200	300°

(a) 102.5 g (b) 73° (c) 273.4 g (d) 253°

11.10 Determine the unknown quantities (a) to (d) in the table below for complete balance of the rotating shaft.

Number	*Mass* (kg)	*Axial distance along shaft* (mm)	*Radius* (mm)	*Angular displacement*
1	(a)	0	100	(c)
2	2	500	200	180°
3	3	1000	200	270°
4	(b)	1500	100	(d)

(a) 3.33 kg (b) 4.22 kg (c) 36.9° (d) 71.6°

11.11 Four non-coplanar masses are arranged along a shaft as follows:

Number	*Mass* (kg)	*Axial distance along shaft* (mm)	*Radius* (mm)	*Angular displacement*
1	3.75	0	60	90°
2	(a)	400	72	(c)
3	2.5	900	78	0°
4	2.0	(b)	66	(d)

Determine the unknown quantities (a) to (d) for dynamic balance.

(*Hint*: Make the m_2 plane the reference plane.)

(a) 4.58 kg (b) 1405 mm (c) 253° (d) 137°

11.12 If the balance masses 1 and 2 were not placed on the shaft given in problem 11.9 but instead bearings were located at these positions, determine the bearing reactions at a rotational speed of 95 rpm when the masses are in the given position.

$R_1 = 1.4$ N @ 32°, $R_2 = 2.8$ N @ 212°

11.13 Determine the maximum and minimum bearing reactions for the rotating masses given in problem 11.12 and the angular position of mass 1 when they occur.

	R_{max}	*Angle*	R_{min}	*Angle*
Bearing 1	7.2 N	287°	0.92 N	107°
Bearing 2	14.4 N	287°	1.84 N	107°

11.14 If the balance masses 1 and 4 were not placed on the shaft given in problem 11.10 but instead bearings were located at these positions, determine the bearing reactions when the masses are in the position shown and the shaft is rotating at 120 rpm. Neglect the mass of the shaft itself.

R_1 = 68.85 N @ 52.3°, R_4 = 91.8 N @ 76.7°

11.15 Repeat problem 11.14 assuming the shaft itself has a mass of 12 kg with centre of mass located midway between the two bearings.

R_1 = 121 N @ 69.6°, R_4 = 150 N @ 81.9°

11.16 Determine the maximum and minimum bearing reactions and the position of mass 2 when they occur, for the rotating masses given in problem 11.14.

	R_{max}	*Angle*	R_{min}	*Angle*
Bearing 1	75.5 N	233°	29.75 N	53°
Bearing 4	92.7 N	198°	40.4 N	18°

PART III

STRENGTH OF MATERIALS

12

Stress and strain

The subject of strength of materials is devoted to the determination of the stress and strain in members subject to various loading conditions. Loads are the external forces and moments acting on the member which cause internal stresses within the material as well as deformation or strain.

In everyday language the words "stress" and "strain" are often used in an imprecise way, sometimes synonymously, for example when a person is said to be under mental stress and strain. Similarly in a tug-of-war, the words "take the strain" would be more correctly phrased as "take the load". In strength of materials there is a precise definition of the meaning of stress and strain as used throughout engineering.

In mechanical design, stress analysis of loaded members is usually more important than strain analysis because excessive stress is the usual mode of failure. However, in some cases the loads may produce stresses which are allowable but produce excessive strains and, in these cases, strain analysis becomes just as important as stress analysis. Also, since there is a precise relationship between stress and strain, measurement of strain may be used as a non-destructive means of determining stress—as is done with strain gauges.

In this chapter, only simple stress and strain analysis is considered. In many practical engineering situations, different types of stress may occur simultaneously in a loaded material. In such cases, the combined stresses may produce a resultant stress which is greater than the separate stresses (in the same way that forces may combine to produce a larger resultant). The analysis of combined stresses is the subject matter of Chapter 16.

12.1 Stress *(f)*

Stress is the *internal* force per unit area in a member and hence may be thought of as internal pressure. Consider the force F applied to the solid shown in Figure 12.1.

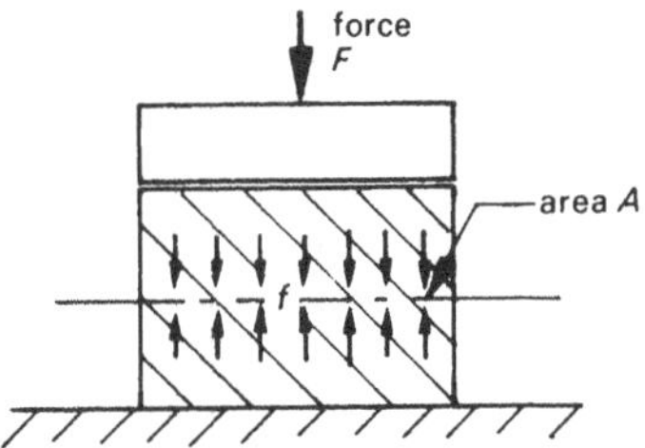

Fig. 12.1 *Axial stress*

The internal force per unit area or stress f is given by:

$$f = \frac{F}{A}$$ **(12.1) axial stress**

The units of stress are the same as for pressure, that is, pascals. However in strength of materials, the Pa is too small a unit so MPa is consistently used. A stress of 1 MPa occurs when a force of 1 N is applied to an area of 1 mm^2 so that in stress calculations it is convenient to use mm^2 for area and N for force and thereby derive stress directly in MPa.

Types of stress

Although there are many types of stress, all may be classified into one of two main groups: axial stress or shear stress.

Axial stress occurs when the force (or component of the force) is perpendicular to the plane of stress as is the case in Figure 12.1. **Shear stress** occurs when the force (or component of the force) is parallel to the plane of stress as shown in Figure 12.2.

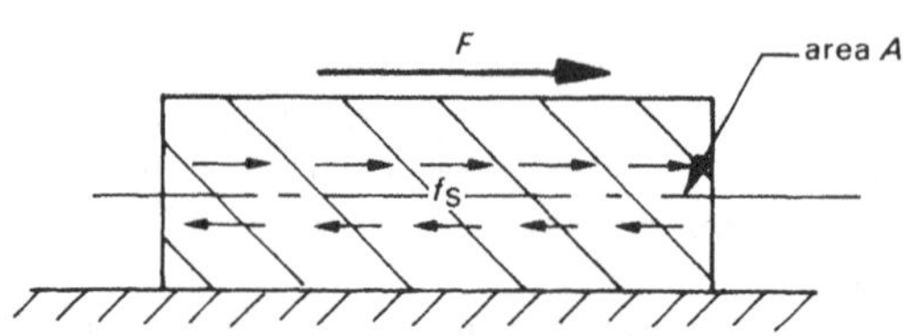

Fig. 12.2 *Shear stress*

The subscript *s* will be used to denote shear stress and the same basic relationship applies for shear stress:

$$f_s = \frac{F}{A}$$

(12.2) shear stress

Within the broad categories of axial and shear stress, there are further subdivisions according to the nature of the stress. The most common of these in engineering are as follows:

1. Axial stress
   (a) tension (direct)
   (b) compression (direct)
   (c) bearing or crushing
   (d) bending.
2. Shear stress
   (a) direct
   (b) torsional
   (c) bending.

These various types of stress are illustrated in Figure 12.3 which shows the applied force *F* and a small element of the material under stress.

The distinction between tension and compression stress is purely one of sign or direction. Tension or tensile stress occurs when the member is stretched or extended and is usually given a positive sign. Compression stress occurs when the member is pushed or compressed and is usually given a negative sign.

Crushing or bearing stress is a localised compression stress and strictly speaking also involves some localised shear but is usually considered as an axial stress. Crushing becomes important when dealing with soft materials (such as footings bearing on soil) or when the load is applied over a small area (such as the base of a sailing yacht mast). Crushing failure is evidenced by localised indentation of the material at the point of application of the load.

Axial bending stress occurs when a member is subject to bending moment (as for example in a simply supported beam). The result is that one side of the member is placed in compression and the other side in tension.

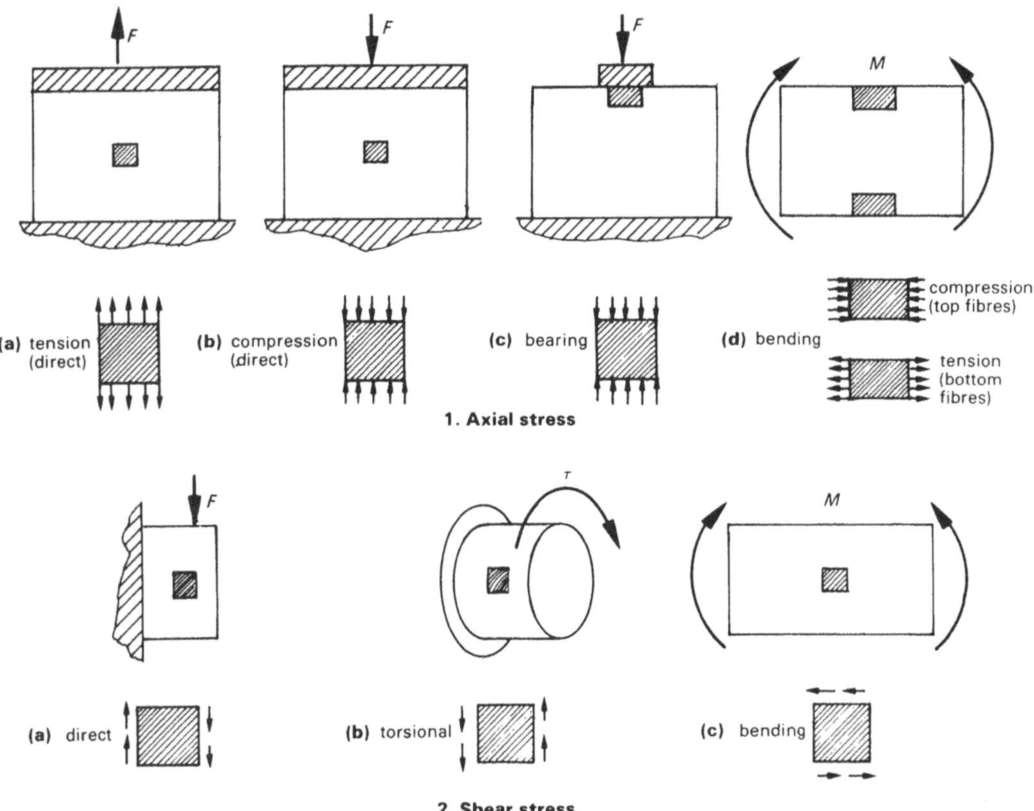

Fig. 12.3 *Types of stress*

Direct shear stress occurs when a force acts to the shear material (such as the punching out of a sheet metal blank or the cold cropping of a bar). Torsional shear occurs when a member is subject to a torque (turning moment) which tends to twist the member (as occurs in torsion bar suspension systems). Shear stress also occurs with bending of a member because there is a relative sliding of the fibres of the material along the longitudinal axis which results in shear strain and hence shear stress.

Stress area of bolts

When determining the stress in the unthreaded shank of a bolt, screw or stud, the cross-sectional area is used. However, in the threaded section, the cross-sectional area under stress is different to the shank area. Because the thread forms a helix, the precise area under stress will be somewhat greater than the root area. However for simplicity, and because there is only a small difference in any case, in this book the root area will always be used (see Fig. 12.4).

It should be noted that in the design of threaded fasteners the effect of stress concentration caused by the thread and tightening loads should be considered in conjunction with the applied load acting on the root area of the fastener.

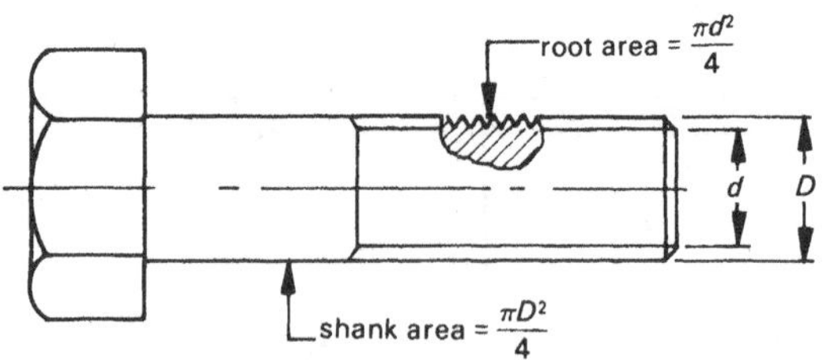

Fig. 12.4 *Shank and root areas of a threaded fastener*

Stress area of welds

When determining the stress in a butt weld, the area under stress is normally considered to be equal to the area of the unwelded plate (although in practice the weld is usually somewhat thicker than the plate). For a fillet weld, the stress area is usually taken as the theoretical area across the throat of the weld, that is, 0.707 × weld size × weld length (see Fig. 12.5).

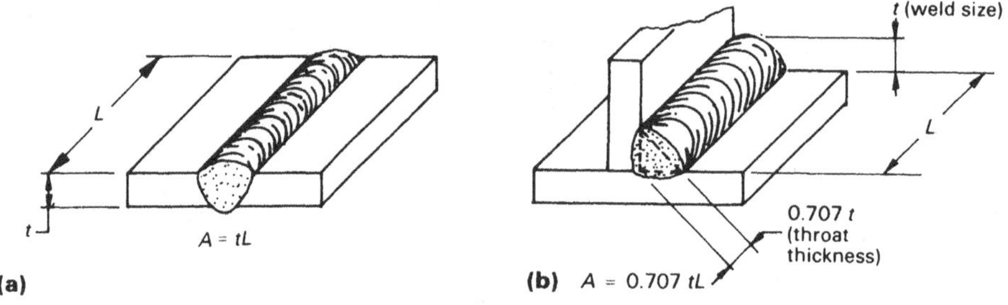

Fig. 12.5 *Stress area for (a) butt and (b) fillet welds*

Example 12.1

A circular steel pipe, OD 50 mm, carries a load of 4 kN in the centre and is supported by timber supports as shown in Figure 12.6.

Determine the required width b of each support if the average bearing stress in the timber is not to exceed 1 MPa.

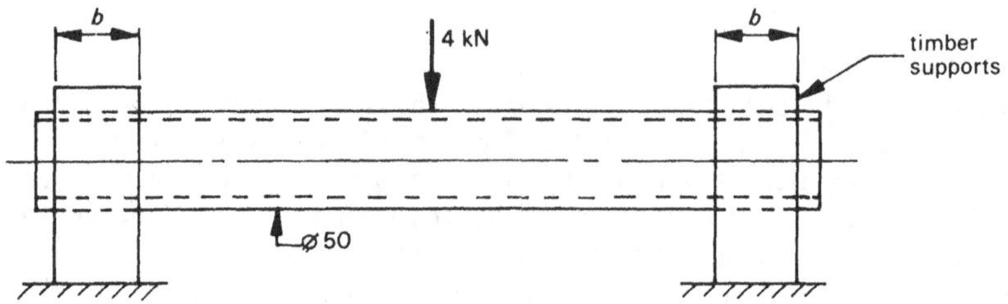

Fig. 12.6

Solution

The force at each support is 2 kN (reaction).

For a curved section such as a circular pipe resting on its side, the average bearing stress is determined by using the projected area at the support (not the total area around the periphery of the section).

In this case, the projected area A at each support is:

$$A = b \times 50 \text{ mm}^2$$

The bearing (or crushing) stress is:

$$f_c = \frac{F}{A} = \frac{2 \times 10^3}{b \times 50} \text{ MPa}$$

Since the allowable bearing stress is 1 MPa,

$$1 = \frac{2 \times 10^3}{b \times 50}$$

$$\therefore b = \frac{2 \times 10^3}{50}$$

$$= \mathbf{40 \text{ mm}}$$

12.2 Strain (ϵ)

The only way a material can withstand load is to deflect or extend. It is impossible for a material to be stressed without changing size. Sometimes the deflection is visible—for example when a guitar string is plucked, a flexible ruler is bent, or when a helical spring is deformed. In many other cases, however, the deflection may be too small to be visible—for example when a person walks on a concrete path. In this case the path deflects but the deflection is too small to be detected (except with sensitive instruments).

The extension per unit length is called **strain**. For example, consider a force F applied to a bar as shown in Figure 12.7. The bar extends an amount x and the strain is:

$$\epsilon = \frac{x}{L}$$

(12.3) axial strain

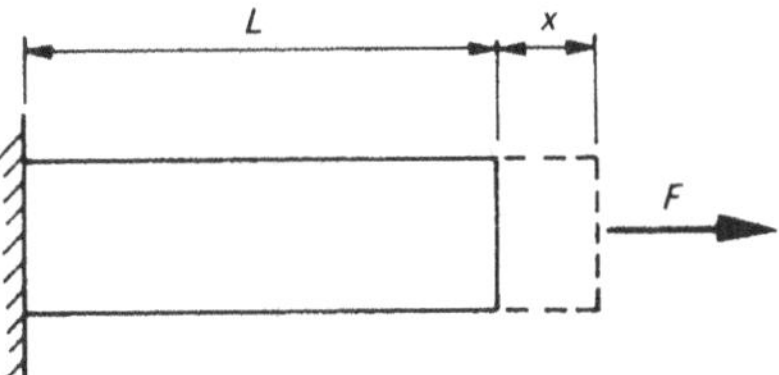

Fig. 12.7 *Axial strain*

Note: Strain has no units since it is a ratio of lengths and the length L is the original length (that is the length before the extension occurs).

Shear strain (ϵ_s)

Shear strain is defined in a similar way to axial strain. Consider a force F applied to a block of material as shown in Figure 12.8.

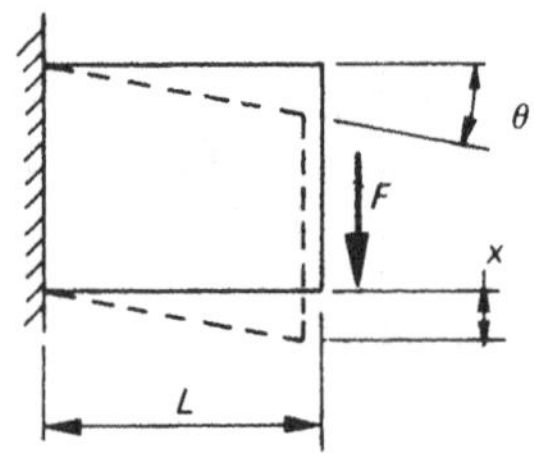

Fig. 12.8 *Shear strain*

The extension is x and the shear strain is:

$$\epsilon_S = \frac{x}{L}$$

Now $\frac{x}{L} \doteqdot \tan\theta \doteqdot \theta$ since θ is the very small angle of deformation. Hence the shear strain may also be written:

$$\boxed{\epsilon_S = \frac{x}{L} = \theta} \qquad \textbf{(12.4) shear strain}$$

12.3 Relationship between stress and strain

If an axial or shear stress test is performed on a material specimen it will be ıound that most common engineering materials used to support loads produce a graph of general shape as shown in Figure 12.9.

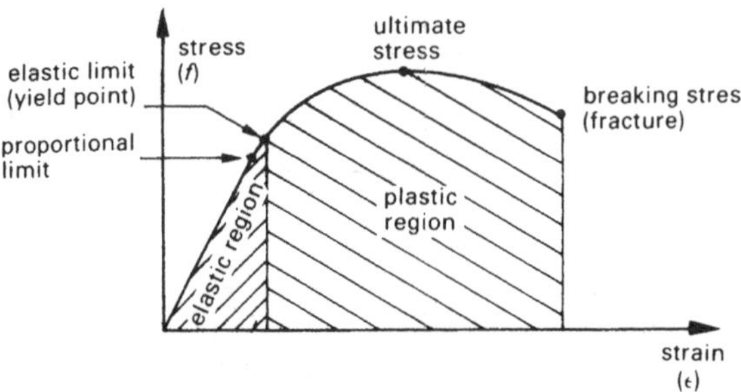

Fig. 12.9 *Stress-strain diagram (typical)*

There are two distinct and important regions—an elastic region which always occurs first and a plastic region which follows. The point of change between the two regions (sometimes difficult to identify in practice) is known as the **elastic limit**.

In the linear part of the elastic region (up to the proportional limit) the stress is directly proportional to the strain; load is directly proportional to the extension for a given material section. This is known as Hooke's law.

Hence $\frac{f}{\epsilon}$ = constant

Hooke's law applies for both axial and shear stress and the constant of proportionality is given a special name and symbol. In the case of axial loading it is called the **modulus of elasticity**, E, and in the case of shear loading it is called the **modulus of rigidity**, G. Hence:

$$E = \frac{f}{\epsilon} = \frac{FL}{xA}$$

(12.5) modulus of elasticity

$$G = \frac{f_S}{\epsilon_S} = \frac{FL}{xA}$$

(12.6) modulus of rigidity

Notes

1. Materials which do not obey Hooke's law (do not have a linear region) are mainly non-metallic and synthetic materials, such as most plastics and putties, as well as non-homogeneous materials such as soil.
2. The modulus of elasticity was formerly known as Young's modulus and is still sometimes referred to by this name; also sometimes as Young's modulus of elasticity.
3. Since strain is dimensionless, E and G have the same units as stress, that is MPa or GPa.
4. Surprisingly, the value of E and G for engineering materials does not vary significantly with usual treatment of the material. For example, there is only variation of a few per cent in the value of E for steel with change in carbon content, alloy composition, method of heat treatment or degree of cold working. Typical values of E are given in Table 12.1.

Table 12.1 Moduli of elasticity and rigidity for common engineering materials (average values)

	E GPa	G
Aluminium	70	Take as 0.4 E in all cases
Brass	95	
Bronze	100	
Copper	110	
Concrete	15	
Steel (and wrought iron)	200	
Timber	12	

5. The modulus of rigidity for materials which have uniform properties in all directions is very close to 40 per cent of the modulus of elasticity for the same material and may be taken so for all practical purposes. That is:

   $G = 0.4\,E$
6. On a stress-strain diagram, the slope of the line up to the proportional limit is the modulus of elasticity (axial load) or modulus of rigidity (shear load). The steeper the slope, the higher the value of the modulus.
7. There is no relationship between the modulus of elasticity and the strength of the material as measured by the maximum stress it can withstand without fracture (ultimate stress).

For example, cast iron has a high modulus of elasticity in tension but a very low strength in tension.

8. In practice, there is often little difference between the proportional limit, the elastic limit and the yield point. If a material is deformed elastically, the deformation is not permanent and the material returns to its initial (undeformed) condition if the load is removed.
9. The typical stress-strain diagram shown in Figure 12.9 is based upon stress calculated using the original area of the material and strain calculated on the initial length. If stress is calculated on the actual minimum area of the material as deformation occurs, it will be found to increase continually toward the point of fracture.

Example 12.2

A punch of diameter 30 mm is used to punch a hole in an aluminium sheet of thickness 2.5 mm. The clearance between the punch and the die is 0.25 mm per side (see Fig. 12.10).

Determine:

(a) depth of penetration of the punch in the sheet before permanent deformation occurs
(b) punch force necessary to complete the shearing operation.

The ultimate shear strength of the aluminium (USS) is 260 MPa and the elastic limit in shear is 180 MPa.

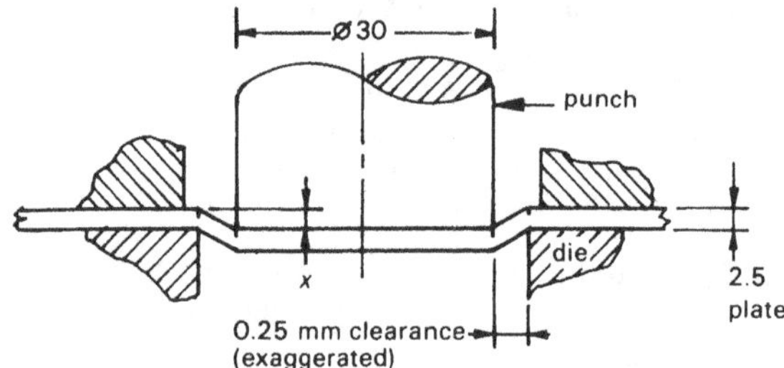

Fig. 12.10

Solution

(a) From Table 12.1, E for aluminium is 70 GPa.
Now $G = 0.4E = 0.4 \times 70 = 28$ GPa
At the elastic limit point, $f_S = 180$ MPa

$$G = \frac{f_S}{\epsilon_S} \quad \therefore \epsilon_S = \frac{f_S}{G}$$

$$\therefore \epsilon_S = \frac{180}{28 \times 10^3} = 0.00643$$

$$\text{But } \epsilon_S = \frac{x}{L}, \quad \therefore x = \epsilon_S L$$

$$= 0.00643 \times 0.25$$

$$= \mathbf{1.61 \times 10^{-3}\ mm}$$

(b) At the point of rupture, $f_S = 260$ MPa (USS)

$$\text{and } f_S = \frac{F}{A} \quad \therefore F = f_S A = 260 \times \pi \times 30 \times 2.5 \quad \text{(N)}$$

$$= \mathbf{61.3\ kN}$$

Note that the area in shear is the cylindrical area (πdt) and not the surface area $\left(\pi \frac{d^2}{4}\right)$.

12.4 Non-uniform stress distributions

In many cases (particularly with direct stress), the stress may be considered to be uniformly distributed over a section of the member. For example, consider the gradually tapered member shown in Figure 12.11 subject to an axial compression load.

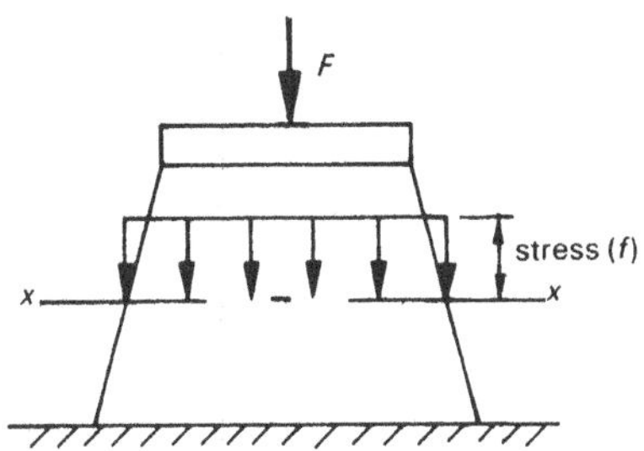

Fig. 12.11 *Uniformly distributed stress*

Experiment confirms that the stress will be uniformly distributed (closely enough) at section *X-X*. This means that the maximum stress at this section is also equal to the average stress.

When the stress is not direct but occurs from bending or torsion, then the stress distribution will not be uniformly distributed over the section. Two examples of this are shown in Figure 12.12.

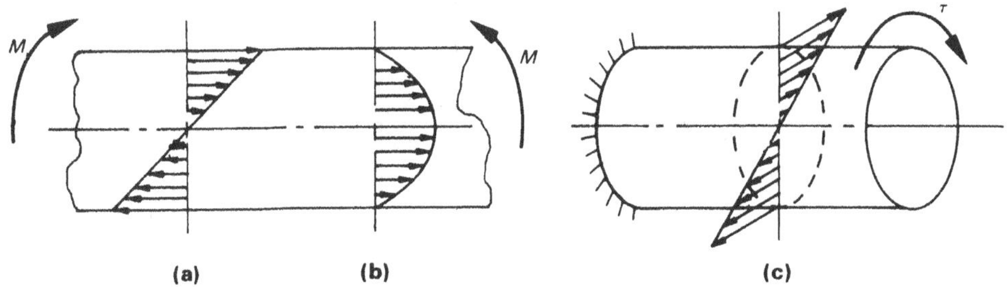

Fig. 12.12 *Non-uniform stress distribution: (a) axial stress and (b) shear stress caused by bending, (c) shear stress caused by torsion*

In such cases, it is necessary to determine the maximum stress rather than the average stress because failure may occur if the maximum stress exceeds the ultimate stress at some position in the section even though the average stress at the section may be well below that which would cause failure.

Another example of non-uniform stress distribution occurs when there is a sudden change in the section of a member such as a step with a sharp corner, a circlip groove in a shaft or a notch or cavity in a bar. These are known as "stress raisers" or "stress concentrators" because they caused a highly localised increase in stress. An example of this is shown in Figure 12.13.

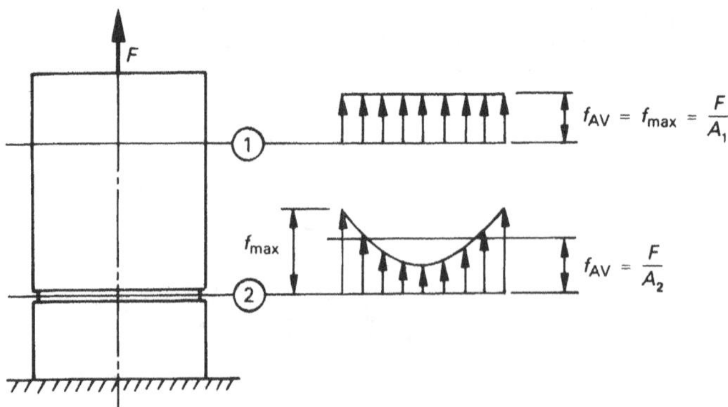

Fig. 12.13 *Stress distributions in a grooved bar*

Stress concentrators are of particular importance where the member is subjected to cyclic loading since fatigue failure is initiated at a stress concentration point. The analysis of stress concentration and fatigue failure is normally part of engineering design and in this book stress concentration effects are not considered.

Factor of safety

When determining the size and shape of sections to withstand given load conditions, it is important to ensure that the maximum calculated stress (design stress) is below the ultimate stress. The degree of safety is measured by the factor of safety (FOS) defined as:

$$\text{FOS} = \frac{\text{ultimate material strength}}{\text{maximum load stress}}$$

As thus defined, the FOS is based on ultimate strength. In some cases it may be more appropriate to use the yield stress rather than the ultimate stress, for example when plastic deformation is not permissible. When so calculated, the FOS will be lower than when it is based on ultimate strength since the yield point is lower than the ultimate strength.

The more exact the design calculation, the lower the FOS which may be used. For example, in aircraft design the material weight is of utmost importance and hence design calculations must be very precise and low factors of safety are used (typically about 1.5). Such a precise approach is usually not merited in general engineering design and consequently higher factors of safety are used (typically about 5–10 based on UTS).

The determination of an appropriate factor of safety is not considered part of strength of materials theory and is treated in texts dealing with mechanical design. Hence in this book, for the solution of problems, the FOS will always be given if it is required.

Example 12.3

For the screw jack illustrated in Figure 12.14, determine:

(a) maximum compressive load F
(b) gear wheel thickness y at maximum load for safe shear in the thread.

Use a factor of safety of 5 with an ultimate compressive strength of 450 MPa and an ultimate shear stress of 270 MPa for the screw and gear wheel.

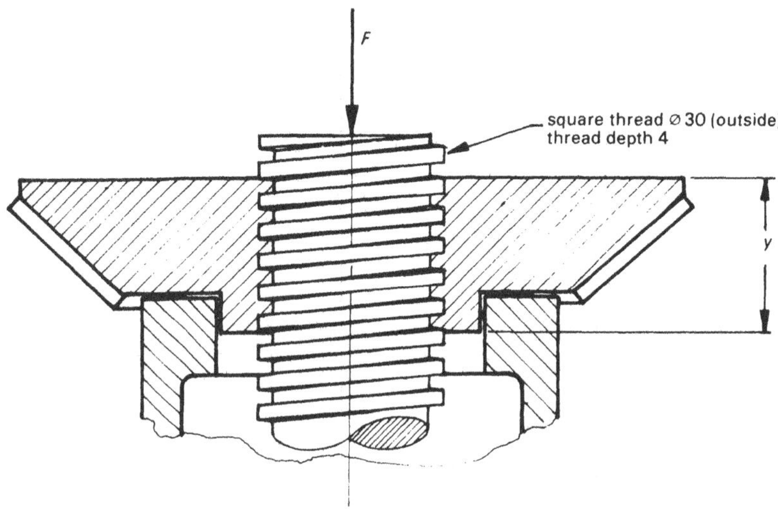

Fig. 12.14

Solution

The root diameter of the thread is $30 - 4 \times 2 = 22$ mm

(a) Now $f = \dfrac{F}{A}$

$$\therefore \frac{450}{5} = \frac{F}{\pi \times \dfrac{22^2}{4}}$$

$$F = \mathbf{34.2\ kN}$$

(b) Using t for the required metal thickness in shear:

$$\frac{270}{5} = \frac{34.2 \times 10^3}{\pi \times 22 \times t}$$

$\therefore t = 9.16$ mm

Since the thread is of square profile:

$$y = 2t = \mathbf{18.3\ mm}$$

Note: The pitch of the thread has no effect on the answer.

12.5 Pressure stress

Pressure stress results when a fluid under pressure is contained in a vessel. If the fluid is a gas there is the very real danger that high pressures can result in failure and explosion. In fact bombs are pressure vessels designed to fail when detonated.

Nowadays, the design and maintenance of pressure vessels are subject to stringent codes and regulations and design of such vessels should never be undertaken unless these are followed. However an understanding of the basic principles involved is very useful and can be simply done if the vessel is considered as having a thin wall, that is, the diameter to thickness ratio is greater than 10: 1. Under these conditions, the stress may be considered uniformly distributed over the sections of the vessel.

Axial stress

Consider a spherical pressure vessel as shown in Figure 12.15.

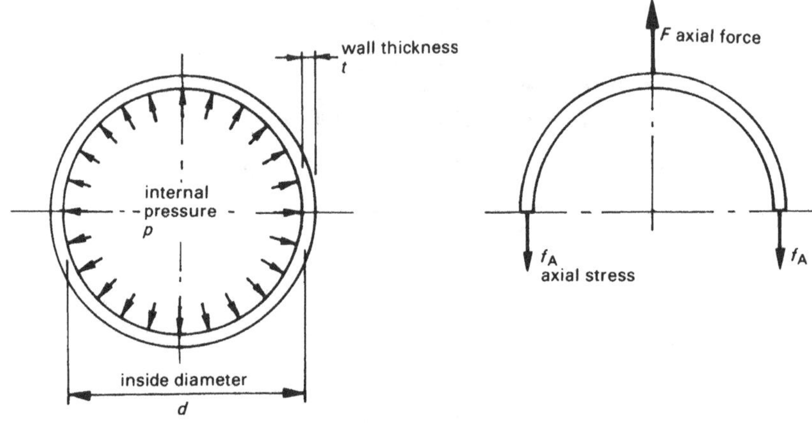

Fig. 12.15 *Axial stress in a spherical pressure vessel*

The force due to pressure on the walls acts normally to the wall and hence changes direction around the periphery. However, the total force in any direction may be obtained by considering the pressure as acting on the projected area of the vessel. Hence the total force tending to split the vessel is:

$$F = p \times \frac{\pi d^2}{4}$$

and the area over which the axial stress acts is:

$$A = \frac{\pi}{4}(d + 2t)^2 - \frac{\pi d^2}{4} = \frac{\pi}{4}(d^2 + 4dt + 4t^2 - d^2)$$

$$= \pi(dt + t^2)$$

Since $d \geqslant t$, the t^2 term is small and customarily neglected.

Hence $A = \pi dt$

which is also the circumference multiplied by the thickness since circumference $= \pi d$.

Hence the axial stress is:

$$f_A = \frac{F}{A} = \frac{p \times \pi \frac{d^2}{4}}{\pi dt} = \frac{pd}{4t}$$

$$\boxed{f_A = \frac{pd}{4t}}$$ **(12.7) axial pressure stress**

Note that in this formula, the pressure p is gauge pressure and f_A has the same units as p if d and t are both in the same units (mm).

Hoop stress

If the pressure vessel is a cylinder rather than a sphere, there is also an axial stress on the longitudinal seam. This stress is known as hoop stress. Refer to Figure 12.16.

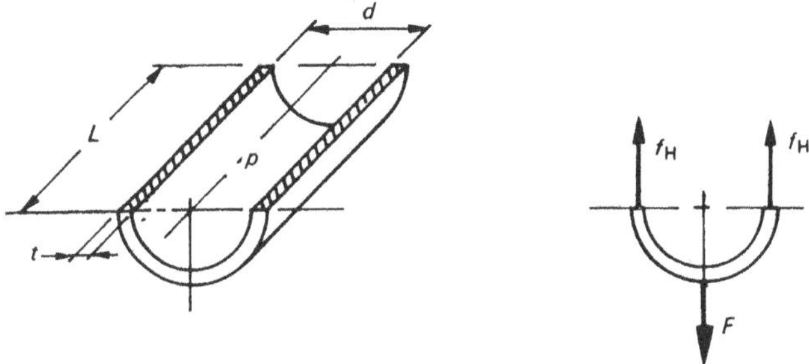

Fig. 12.16 *Hoop stress in a cylindrical pressure vessel*

In this case, the force $F = p \times d \times L$

and the area under stress $A = 2Lt$

$$\therefore f_H = \frac{F}{A} = \frac{pdL}{2Lt}$$

$$\boxed{f_H = \frac{pd}{2t}}$$ **(12.8) hoop stress due to pressure**

Again p is gauge pressure and it will be seen that the stress on the longitudinal seam of a cylindrical pressure vessel is twice as great as the stress on the axial seam. That is a cylindrical pressure vessel of uniform diameter and wall thickness will withstand only one-half the pressure of a spherical vessel of the same diameter and wall thickness.

Example 12.4

Determine the stress in the welded joints of the high pressure air receiver shown in Figure 12.17 when it is filled with air at a gauge pressure of 2.4 MPa.

The receiver is constructed of 10 mm steel plate, $E = 200$ GPa.

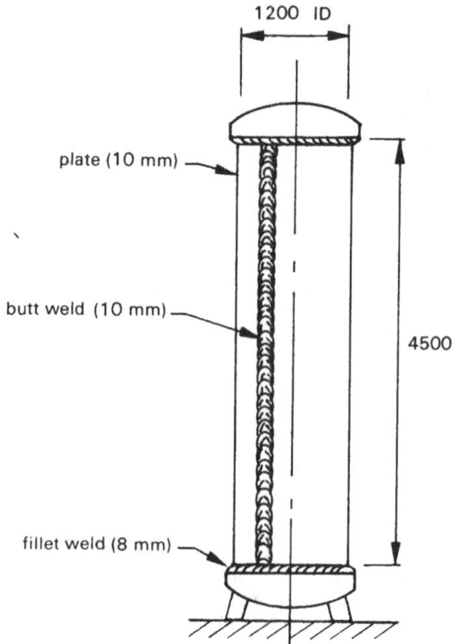

Fig. 12.17

Solution

For the axial stress in the fillet weld use equation 12.7 with t = throat thickness of the weld, i.e., $t = 0.707 \times 8$ mm.

For the hoop stress in the butt weld, use equation 12.8 with t = thickness of the butt weld = thickness of the plate = 10 mm.

$$f_A = \frac{pd}{4t} = \frac{2.4 \times 1200}{4 \times 0.707 \times 8} = \mathbf{127\ MPa}$$

$$f_H = \frac{pd}{2t} = \frac{2.4 \times 1200}{2 \times 10} = \mathbf{144\ MPa}$$

12.6 Thermal stress

When a solid material is subjected to a change in temperature, there is a corresponding change in size. If the material is not restrained (that is, if it is free to expand or contract), the change in length of any dimension may be determined by the following formula:

(12.9) thermal expansion

$$\boxed{x = \alpha L \Delta T}$$

where x = change in length (mm)
L = original length (mm)
ΔT = change in temperature, i.e. $(T_2 - T_1)$ and may be expressed in °C or K
α = coefficient of linear expansion (/°C) or (/K)

Notes

1. The length increases with increase in temperature and decreases with decrease in temperature.
2. α has units mm/mm°C which is also written /°C. Since the change in temperature is the same in °C or K, this may also be written /K.
3. This formula is valid only over a restricted temperature range because α varies with temperature. If the temperature range is less than about 200°C, the difference in α is small and a single value may be used. If the temperature range is greater, then a mean value of α should be used.

If the material is fully or partially restrained in any dimension, an axial stress will occur. This stress may readily be calculated if the degree of restraint is known. For example if the restraint is complete so that no expansion or contraction can occur, the stress is given by:

$$E = \frac{f}{\epsilon} \quad \therefore f = \epsilon E$$

Since $\epsilon = \dfrac{x}{L}$

and $x = \alpha L \delta T$ **(12.9)**

then $f = \alpha E \Delta T$

It is interesting to note that for full restraint, the thermal stress is independent of both the cross-sectional area and the length.

Example 12.5

A stainless steel strap of inside diameter 1158 mm and thickness 10 mm is to be fitted over a wheel of diameter 1160 mm.

Determine:

(a) minimum increase in the temperature of the strap so that it just fits over the wheel
(b) tensile stress in the strap when it has cooled back to room temperature
(c) bearing stress on the inside of the strap when it has cooled back to room temperature.

Assume the wheel to be rigid and for stainless steel $\alpha = 12 \times 10^{-6}$/°C, E = 200 GPa.

Solution

(a) The diametral change is given by:

$$x_d = \alpha d \Delta T$$

where d = mean diameter of the strap namely 1168 mm.

$$\therefore 2 = 12 \times 10^{-6} \times 1168 \times \Delta T$$

$$\therefore \boldsymbol{\Delta T = 143°C}$$

(b) The tensile stress at room temperature is:

$f_H = \frac{x_d}{d}E$ or $f_H = \alpha E \Delta T$

$= \frac{2}{1168} \times 200 \times 10^3$ $= 12 \times 10^{-6} \times 200 \times 10^3 \times 143$

$=$ **343 MPa** $=$ **343 MPa**

(c) The bearing stress between the strap and the wheel is equivalent to the pressure which would need to be exerted on the inside of the strap to give the same tensile stress as was calculated in (b) above.

Now $f_H = \frac{pd}{2t} \quad \therefore p = \frac{2f_H t}{d}$

$$\therefore p = \frac{2 \times 343 \times 10}{1160}$$

$=$ **5.91 MPa**

12.7 Rotational stress

It was seen in Chapter 7 that a rotating body experiences an inwardly directed centripetal acceleration and in Chapter 9 that this acceleration gives rise to an outwardly directed inertia force called the centrifugal force. This force is given by equations 8.4 and 8.5:

$$F_C = mr\omega^2 \quad \text{or} \quad F_C = m\frac{v^2}{r}$$

As a result of the centrifugal force, there is a tensile stress since the rotating mass must be fixed in some way to the centre of rotation. This stress may be calculated at any section by determining the centrifugal force from either of the equations above, bearing in mind that r is the distance from the centre of rotation to the centre of mass (CG) of that part of the rotating mass which is located outward of the section under consideration.

The method is illustrated in example 12.6.

Example 12.6

A mass of 2 kg is attached to a grooved rod as shown in Figure 12.18 and rotated about a horizontal axis.

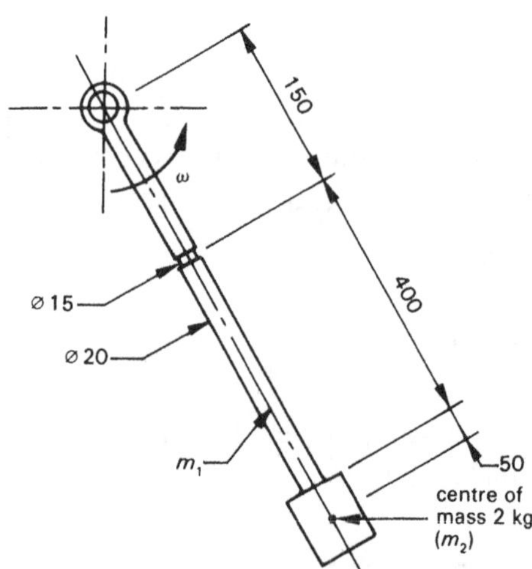

Fig. 12.18

Determine the limiting rotational speed in rpm so that the maximum stress in the groove does not exceed 50 MPa. Assume the density of steel is 7800 kg/m^3 and that the groove causes a stress concentration $f(\text{max}) = 2f(\text{mean})$.

Solution

The mass of the rod below the groove is given by:

$$m_1 = \rho V_1$$

$$= 7800 \times \pi \times \frac{0.02^2}{4} \times 0.4$$

$$= 0.98 \text{ kg}$$

The centrifugal force at the groove is:

$$F_C = m_1 r_1 \omega^2 + m_2 r_2 \omega^2$$
$$= (0.98 \times 0.35 + 2 \times 0.6)\omega^2$$
$$= 1.543\ \omega^2$$

The maximum stress $f(\text{max}) = 2 \times f(\text{mean})$

$$\therefore 50 = 2 \times \frac{F_C + W}{\pi \times \frac{15^2}{4}} \qquad \left(\text{where } W = 2.98 \times 9.81 = 29 \text{ N}\right)$$

$$\therefore F_C = 4389 \text{ N}$$
$$\therefore 4389 = 1.543\ \omega^2$$
$$\therefore \omega = 53.3 \text{ rad/s}$$
$$= \mathbf{509\ rpm}$$

Rotating rings

If a ring rotates, the centrifugal force causes a stress in the ring which is indeed a hoop stress equivalent to that caused by internal pressure in a pressure vessel. If the ring is a thin wall one ($r/t > 5$), this stress may be assumed to be uniformly distributed over the thickness of the ring. The magnitude of the stress may be determined as follows (see Fig. 12.19).

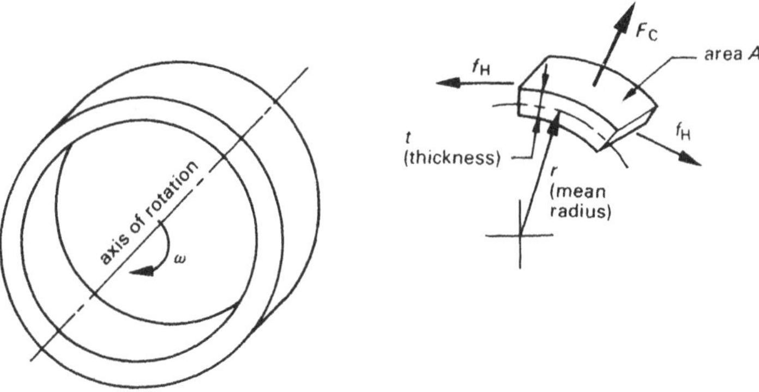

Fig. 12.19 *Rotating thin ring*

The equivalent pressure is $p = \dfrac{F_C}{A}$.

Now $F_C = \dfrac{mv^2}{r}$

and for the small section of ring illustrated:

$$m = \rho V = \rho A t$$

$$\therefore F_C = \rho A t \frac{v^2}{r}$$

$$p = \frac{F_C}{A} = \frac{\rho A t \frac{v^2}{r}}{A} = \rho t \frac{v^2}{r}$$

But from equation 12.8:

$$f_H = \frac{pd}{2t} = \frac{pr}{t}$$

Substituting:

$$f_H = \rho t \frac{v^2}{r} \times \frac{r}{t}$$

$$\therefore \quad \boxed{f_H = \rho v^2} \qquad \textbf{(12.10) hoop stress in a rotating ring}$$

where f_H = hoop stress caused by the rotation (Pa)
ρ = density of the ring material (kg/m^3)
v = linear rotational speed of the ring (m/s)

Note: Equation 12.10 is also valid for a belt running over a pulley.

Example 12.7

A cast iron ring, OD 2 m, ID 1.7 m, rotates at 720 rpm.

Determine the mean stress and the maximum stress in the ring if the density of cast iron is 7200 kg/m^3.

Solution

The mean radius $r = 0.925$ m and $t = 0.15$ m.

$\dfrac{r}{t} = 6.17$ which is > 5, therefore the ring may be regarded as a thin wall one.

$$\text{Now } v = r\omega$$

$$= r \times \frac{\pi N}{30}$$

$$v(\text{mean}) = 0.925 \times \frac{\pi \times 720}{30}$$

$$= 69.74 \text{ m/s}$$

Substituting in equation 12.10:

$$f_H(\text{mean}) = \rho v^2$$

$$= 7200 \times 69.74^2 \quad \text{(Pa)}$$
$$= \mathbf{35\ MPa}$$
$$v(\text{max}) = 1.0 \times \frac{\pi \times 720}{30}$$
$$= 75.4 \text{ m/s}$$
$$f_H(\text{max}) = 7200 \times 75.4^2 \quad \text{(Pa)}$$
$$= \mathbf{40.9\ MPa}$$

12.8 Series bars: Axial stress

Frequently in practice, bars of the same or different material may be joined and subjected to an axial load as shown in Figure 12.20. This is an example of series bars where the load is common but the stress and extension in each bar is different.

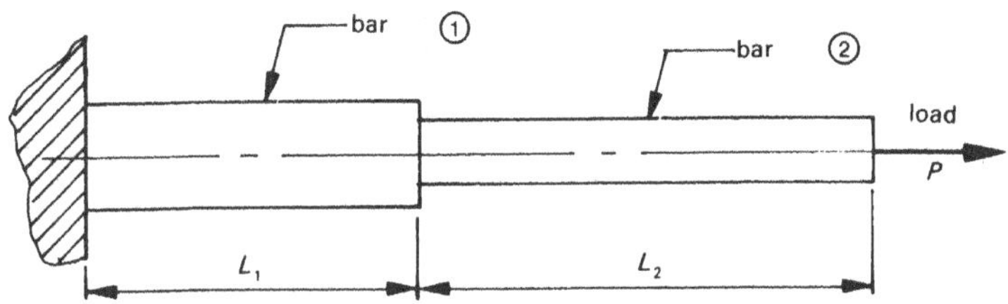

Fig. 12.20 *Series bars*

In order to analyse the stress and extension which occurs, it will be assumed that the bars are made of different materials with different moduli of elasticity and that the deformation occurs elastically (within the elastic limit). This gives the most general analysis which will also be applicable when the bars are of the same material (e.g. a stepped shaft).

The load P is the same in each bar and the total extension due to the applied load is the sum of the individual extensions of each bar. That is:

$$P_1 = P_2 = P \quad \text{(applied load)} \qquad \textbf{(1)}$$
$$x_1 + x_2 = x \quad \text{(total extension)} \qquad \textbf{(2)}$$

These two basic relationships allow the stress and extension in each bar to be determined.

From (1), $f_1 A_1 = f_2 A_2$

i.e. $$\frac{f_1}{f_2} = \frac{A_2}{A_1}$$

Thus it will be seen that the stress in each bar is inversely proportional to the area of the bar and is independent of the material from which the bar is made (its modulus of elasticity).

The extension of each bar is:

$$x_1 = \frac{PL_1}{A_1 E_1} \qquad x_2 = \frac{PL_2}{A_2 E_2}$$

From (2) the total extension $x = x_1 + x_2$

Note: In the above analysis no allowance has been made for the actual joining of the bars. It may well be that the maximum stress will occur in each material as a result of the joint itself as illustrated in Example 12.8.

Example 12.8

As shown in Figure 12.21, a steel rod diameter 25 mm and length 2 m is screwed into a brass rod, diameter 30 mm and length 3 m, and subjected to an axial tensile load of 20 kN.

Given that E (steel) = 200 GPa and E (brass) = 95 GPa, determine:

(a) stress in each bar at a full section
(b) total extension
(c) average stress in each bar at the threaded joint if the thread has an outside diameter of 25 mm and a root diameter of 20 mm.

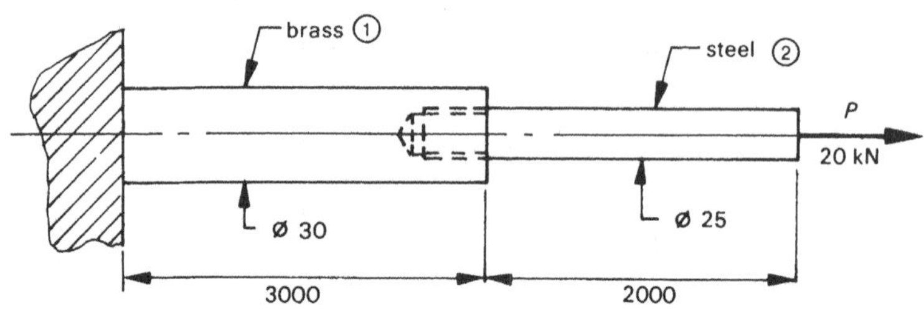

Fig. 12.21

Solution

(a) $f_2 = \dfrac{P}{A_2} = \dfrac{20 \times 10^3}{\pi \times \dfrac{25^2}{4}} = \mathbf{40.7\ MPa}$

$f_1 = \dfrac{P}{A_1} = \dfrac{20 \times 10^3}{\pi \times \dfrac{30^2}{4}} = \mathbf{28.3\ MPa}$

(b) $x_2 = \dfrac{f_2 L_2}{E_2} = \dfrac{40.7 \times 2000}{200 \times 10^3} = 0.407 \text{ mm}$

$x_1 = \dfrac{f_1 L_1}{E_1} = \dfrac{28.3 \times 3000}{95 \times 10^3} = 0.894 \text{ mm}$

$x \text{ (total)} = x_1 + x_2 = \mathbf{1.3\ mm}$

(c) At the screwed section in bar ②, the effective diameter is 20 mm. Hence:

$f_2 = \dfrac{20 \times 10^3}{\pi \times \dfrac{20^2}{4}} = \mathbf{63.7\ MPa}$

For bar ①, the effective area at the screwed section is:

$A_1 = \dfrac{\pi}{4}(30^2 - 25^2) = 215.98 \text{ mm}^2$

$f_1 = \dfrac{20 \times 10^3}{215.98} = \mathbf{92.6\ MPa}$

Example 12.9

As shown in Figure 12.22, a steel bolt 20 mm in diameter just fits inside a bronze tube 30 mm outside diameter and length 600 mm. The bolt has a pitch of 1.25 mm and after just making contact at the end of the tube is tightened a further one-quarter turn.

Determine the stress at a full section of the bolt and tube and the extension in length of the bolt and the contraction in length of the tube.

Assume E (steel) = 200 GPa, E (bronze) = 100 GPa.

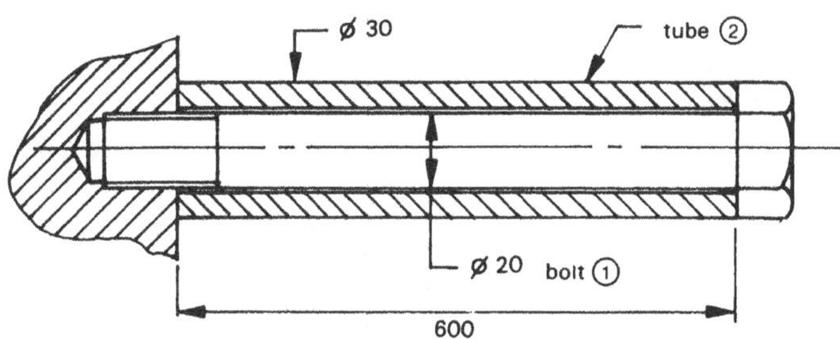

Fig. 12.22

Solution

At first sight this may not appear to be a problem involving series materials. However, it is clear that the load is common to both the bolt and the tube and the total movement of the nut is taken in the extension of the bolt and the compression of the tube. Therefore this problem may be dealt with by the method outlined previously for series materials, except that the load is not known but rather the total extension is given.

Since the extension of the bolt and the compression of the tube equal the total movement of the nut:

$$x_1 + x_2 = 0.3125 \text{ mm (one-quarter turn)}$$

Now $A_1 = \pi \times \dfrac{20^2}{4} = 314 \text{ mm}^2, \quad A_2 = \dfrac{\pi}{4}(30^2 - 20^2) = 393 \text{ mm}^2, \quad x_1 + x_2 = 0.3125$

$$\therefore \frac{PL}{A_1 E_1} + \frac{PL}{A_2 E_2} = 0.3125$$

$$\therefore \frac{P \times 600}{314 \times 200 \times 10^3} + \frac{P \times 600}{393 \times 100 \times 10^3} = 0.3125$$

$$\therefore 24.8 \times 10^{-6} P = 0.3125$$

$$\therefore P = 12.6 \text{ kN}$$

$$f_1 = \frac{P}{A_1} = \frac{12\,600}{314} = \mathbf{40.1\ MPa}$$

$$f_2 = \frac{P}{A_2} = \frac{12\,600}{393} = \mathbf{32.1\ MPa}$$

$$x_1 = \frac{f_1 L}{E_1} = \frac{40.1 \times 600}{200 \times 10^3} = \mathbf{0.12\ mm}$$

$$x_2 = \frac{f_2 L}{E_2} = \frac{32.1 \times 600}{100 \times 10^3} = \mathbf{0.193\ mm}$$

Check $x_1 + x_2 = 0.313$ mm

which agrees with the total extension of $\dfrac{1.25}{4} = 0.3125$ mm.

12.9 Parallel bars: Axial stress

When an axial load is applied to several bars such that the load is shared between them but the extension is the same for each, this is an example of bars in parallel. In some cases the materials may be effectively joined to form a composite—for example with laminated materials or reinforcing rods in a concrete beam or column. The general case of parallel bars is illustrated in Figure 12.23.

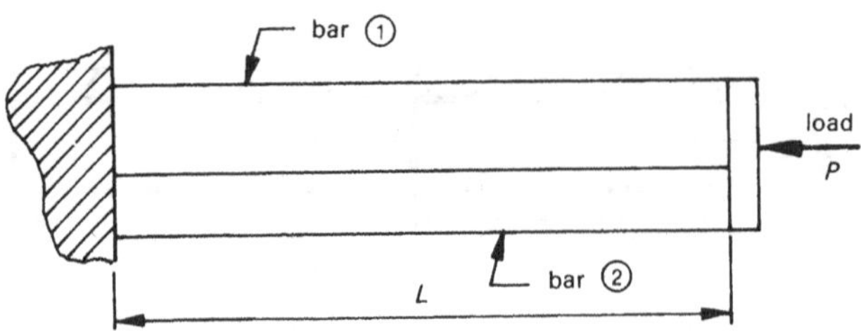

Fig. 12.23 *Parallel bars*

It is evident that the length is common and the extension is also common but the load is *not* common and the total load is the sum of the loads in each bar. Hence the basic relationships are:

$L_1 = L_2 = L$ (length)

$P_1 + P_2 = P$ (applied load) **(1)**

$x_1 = x_2 = x$ (extension) **(2)**

But $x_1 = \dfrac{P_1 L}{E_1 A_1}$ and $x_2 = \dfrac{P_2 L}{E_2 A_2}$

From (2):

$$\frac{P_1 L}{E_1 A_1} = \frac{P_2 L}{E_2 A_2}$$

$$\therefore \frac{P_1}{A_1 E_1} = \frac{P_2}{A_2 E_2}$$

$$\therefore \frac{f_1}{E_1} = \frac{f_2}{E_2}$$

$$\text{or } \frac{f_1}{f_2} = \frac{E_1}{E_2}$$

Thus the stress in each bar is directly proportional to its modulus of elasticity and is independent of its area.

Now $f_1 = \dfrac{E_1}{E_2} f_2$

It is useful to define a dimensionless number $n = \dfrac{E_1}{E_2}$ (the ratio of the moduli of elasticity).

$$\therefore f_1 = n f_2$$

From (1):

$$P_1 + P_2 = P$$

$$\therefore f_1A_1 + f_2A_2 = P$$

$$\therefore f_1A_1 + \frac{f_1}{n}A_2 = P$$

or

$$\boxed{f_1 = \frac{P}{A_1 + \frac{A_2}{n}}}$$

(12.11) stress in parallel bars

Note: The quantity $\frac{A_2}{n}$ may be considered as an "equivalent area" to be added to the area of bar ① in order to determine the stress in bar ①.

Example 12.10

As shown in Figure 12.24, a steel rod diameter 25 mm and length 500 mm is placed inside a brass sleeve inside diameter 26 mm, outside diameter 34 mm and of the same length. The assembly is subjected to an axial compression load of 60 kN.

Determine the stress in the rod and sleeve and the total compression of the assembly (assuming no buckling).

Assume E (brass) = 95 GPa
and E (steel) = 200 GPa.

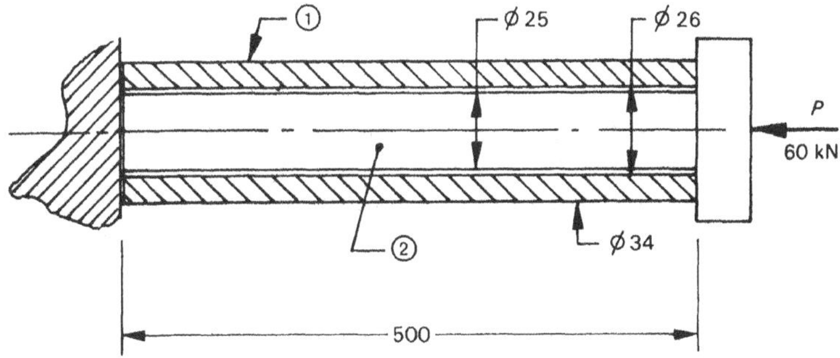

Fig. 12.24

Solution

$$n = \frac{E_1}{E_2} = \frac{95}{200} = 0.475$$

$$A_1 = \frac{\pi}{4}(34^2 - 26^2) = 377 \text{ mm}^2$$

$$A_2 = \frac{\pi}{4} \times 25^2 = 491 \text{ mm}^2$$

From equation 12.11:

$$f_1 = \frac{P}{A_1 + \frac{A_2}{n}}$$

$$= \frac{60 \times 10^3}{377 + \dfrac{491}{0.475}}$$

$$= \mathbf{42.5\ MPa}$$

$$f_2 = \frac{f_1}{n} = \frac{42.5}{0.475} = \mathbf{89.5\ MPa}$$

The total compression of the assembly may now be determined by calculating either x_1 or x_2. Both will be calculated here as a check of correctness.

$$x_1 = \frac{f_1 L}{E_1} = \frac{42.5 \times 500}{95 \times 10^3} = \mathbf{0.224\ mm}$$

$$x_2 = \frac{f_2 L}{E_2} = \frac{89.5 \times 500}{200 \times 10^3} = \mathbf{0.224\ mm} \quad \text{(checks)}$$

The total compression of the assembly is therefore 0.224 mm.

Problems

For these problems use values of E from Table 12.1 and $G = 0.4E$ in all cases.

12.1 A steel tube, 300 mm long, outside diameter 25 mm and wall thickness 2 mm, rests on an aluminium plate and is subject to a compressive load of 10 kN.

Determine:

(a) compressive stress in the tube
(b) compression in length of the tube
(c) bearing stress on the plate
(d) required wall thickness of the tube so that the bearing stress on the plate does not exceed 50 MPa.

(a) 69.2 MPa (b) 0.104 mm (c) 69.2 MPa (d) 2.88 mm

12.2 For the rod end assembly illustrated in Figure P12.2, determine:

(a) tensile stress in rod
(b) shear stress in pin
(c) bearing stress between rod and pin
(d) bearing stress between pin and yoke.

(a) 153 MPa (b) 76.4 MPa (c) 60 MPa (d) 150 MPa

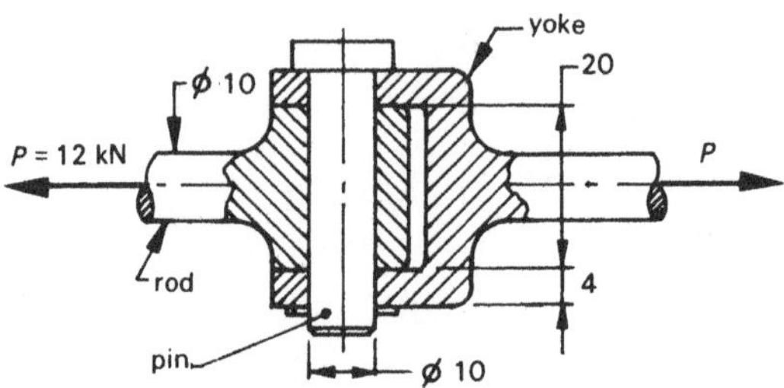

Fig. P12.2

12.3 A steel shaft, diameter 300 mm and length 6 m, is equally supported on two plain bearings.

Determine the required width of the bearings so that the static bearing stress due to the weight of the shaft does not exceed 500 kPa. Assume the relative density of steel is 7.8.

108 mm

12.4 A bracket support shown in Figure P12.4 is welded to the column with 8 mm fillet welds and subject to a load $P = 80$ kN.

Determine the required length of bracket L so that the average shear stress in the weld does not exceed 30 MPa.

236 mm

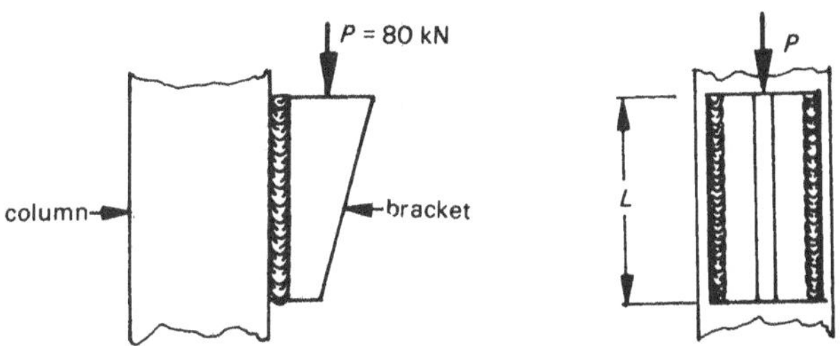

Fig. P12.4

12.5 The structural bracket shown in Figure P12.5 is riveted to the column with 24 mm diameter rivets.

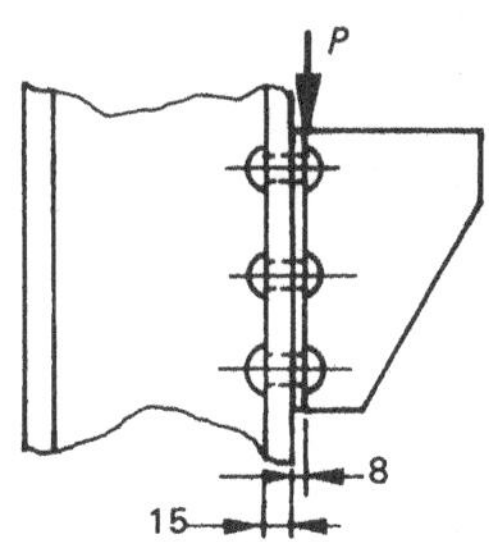

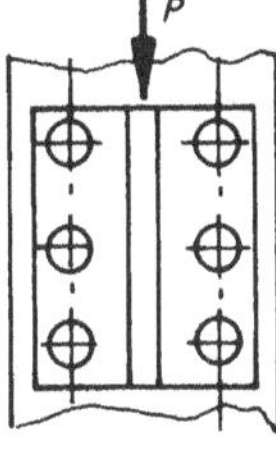

Fig. P12.5

Determine the maximum safe load P which may be supported by the bracket if the ultimate strength of the rivets in shear is 320 MPa and in bearing is 1100 MPa. Use a factor of safety of 4 and assume that the load acts close to the column so there is negligible tension in the rivets.

217 kN (shear, bearing = 317 kN)

12.6 A load $P = 25$ kN is applied to the brass rod shown in Figure P12.6.

Determine:

(a) average shear stress in the rod

(b) shear strain.

(a) 35.4 MPa (b) 0.931×10^{-3}

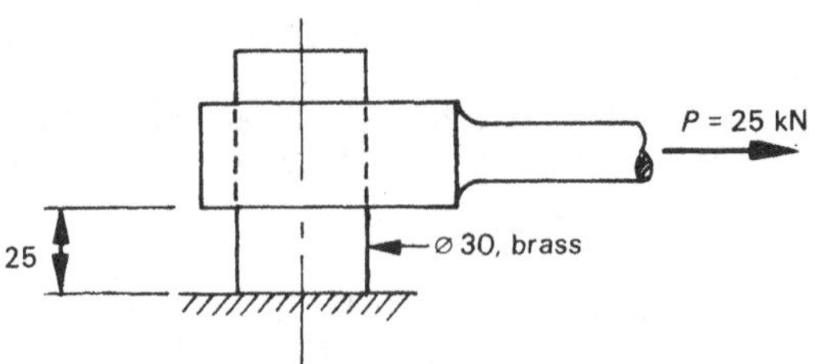

Fig. P12.6

12.7 Determine the force necessary to punch a slot of dimensions shown in Figure P12.7 in a steel plate of thickness 3 mm given the ultimate shear stress of steel = 240 MPa.

70 kN

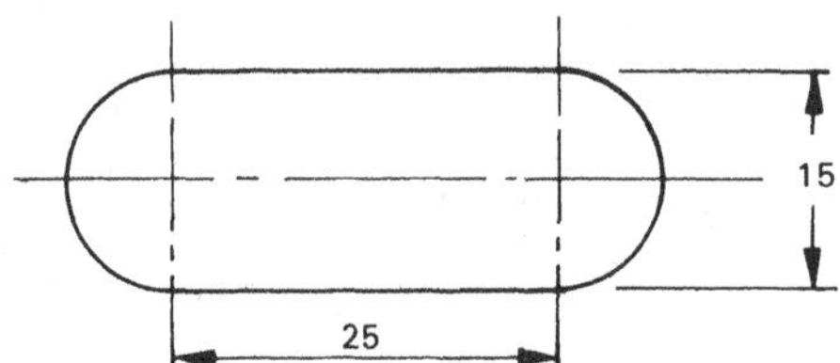

Fig. P12.7

12.8 A copper rod diameter 12 mm and length 80 mm is rigidly fixed at the ends so that it is unable to expand.

Determine the axial stress in the rod when the temperature is increased by 80°C. The coefficient of linear expansion of copper is 14×10^{-6}/°C.

123 MPa

12.9 The steel bolt illustrated in Figure P12.9 passes through fixed supports. The nut is screwed to just contact the right-hand support after which the rod is heated to 100°C above room temperature.

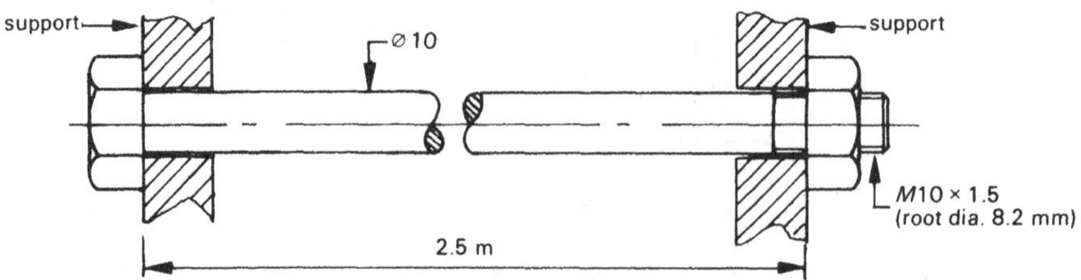

Fig. P12.9

Determine:

(a) number of turns of the nut required to just contact the support again

(b) tensile stress in the rod after cooling down to room temperature if the deflection of the supports is negligible

(c) average tensile stress across the root of the thread.

The coefficient of linear expansion of steel is 11×10^{-6}/°C.

(a) 1.83 turns (b) 220 MPa (c) 327 MPa

12.10 A steel tube, outside diameter 650 mm, is heated and just fits over a disc of diameter 600 mm when the difference in temperature between the tube and the disc is 100°C.

Determine:

(a) inside diameter of the tube at room temperature

(b) stress in the tube when tube and disc are at the same temperature

(c) circumferential pressure (bearing stress) between tube and disc when they are both at the same temperature.

The coefficient of linear expansion of steel is $11 \times 10^{-6}/°C$.

(a) 599.3125 mm (b) 220 MPa (c) 18.3 MPa

12.11 A spherical diving vessel 2 m in diameter is constructed of 10 mm steel plate, butt welded.

Determine the maximum depth below sea water to which the bell may be lowered so the stress in the welds does not exceed 80 MPa. The relative density of sea water may be taken as 1.05.

155 m

12.12 For the steel pressure vessel shown in Figure P12.12, determine:

(a) maximum safe gas pressure so that the stress in the butt weld does not exceed 90 MPa

(b) minimum size of fillet weld required

(a) 1.54 MPa (b) 8.5 mm

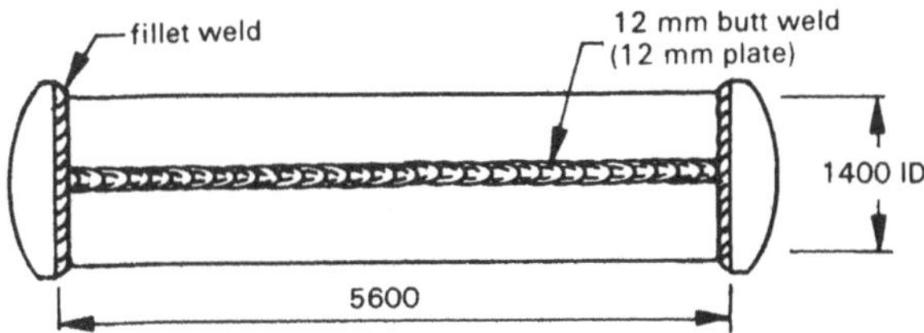

Fig. P12.12

12.13 Determine the maximum rotational speed of a cast iron ring, 2 m OD and 40 mm wall thickness, so that the stress does not exceed 30 MPa. The relative density of cast iron may be taken as 7.2.

616 rpm

12.14 A steel pressure vessel, inside diameter 750 mm and wall thickness 4 mm, contains gas at a pressure of 600 kPa. The vessel rotates about the centre line axis at a speed of 1800 rpm.

Determine the mean hoop stress in the vessel given the relative density of steel is 7.8.

95.6 MPa

12.15 Determine the maximum safe rotational speed for the steel rod shown in Figure P12.15 given the relative density of steel is 7.8 and the UTS is 400 MPa. Use a factor of safety of 5.

1094 rpm

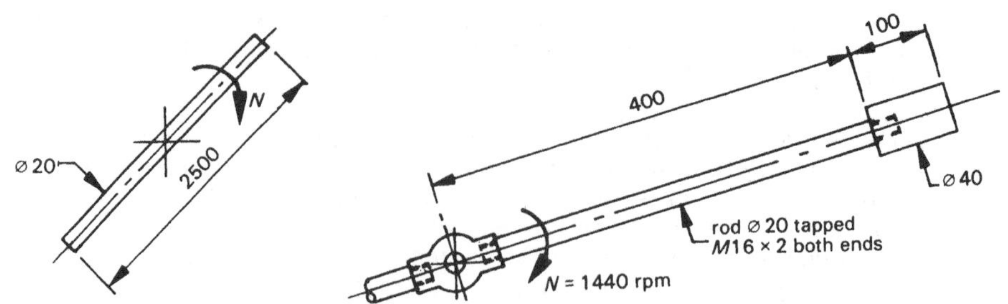

Fig. P12.15 Fig. P12.16

12.16 Determine the average stress at the root area of the outer and inner threads for the rotating steel shaft assembly illustrated in Figure P12.16. The relative density of steel is 7.8 and the M16 × 2 thread has a root diameter of 13.5 mm.

70.1 MPa, 101 MPa

12.17 The stepped steel rod shown in Figure P12.17 is subject to a load of 50 kN. Determine the maximum stress in the rod if the step causes concentration $f(\text{max}) = 1.5f(\text{mean})$. Also determine the extension of the rod caused by the load.

106 MPa, 0.316 mm

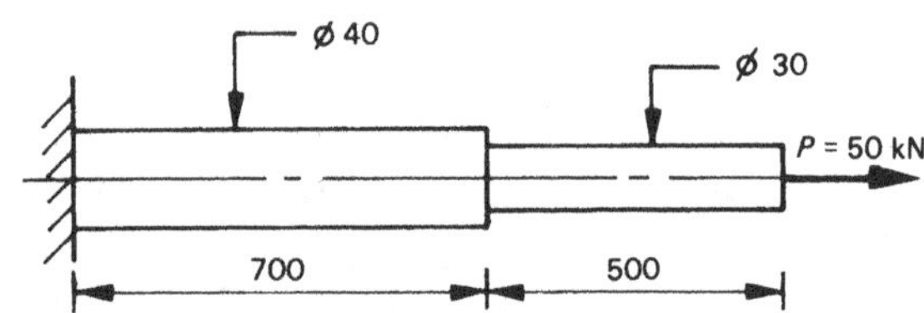

Fig. P12.17

12.18 Determine the axial stress in the copper and brass rods shown in Figure P12.18 when a load of 40 kN is applied. Also determine the compression in length caused by the load.

31.8 MPa, 20.4 MPa, 0.112 mm

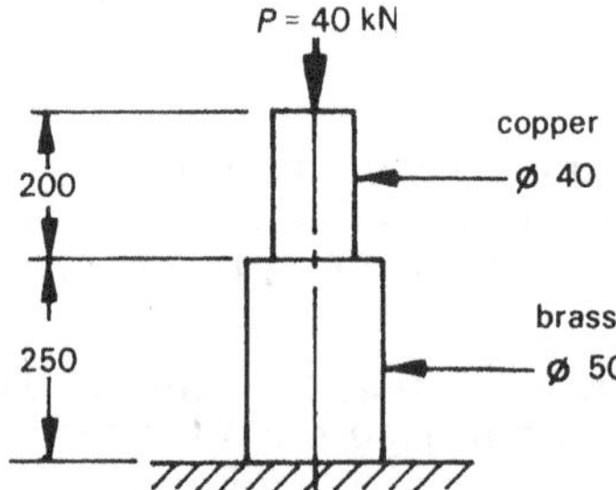

Fig. P12.18

12.19 A steel tube, length 300 mm, ID 17 mm and wall thickness 4 mm, is fitted over a steel stud, 16 mm in diameter threaded M 16 (pitch 2 mm, root diameter 13.5 mm). A washer is placed on the stud and the nut tightened until the assembly is firm after which the nut is tightened a further 45°.

Determine:

(a) stress in stud at full section

(b) stress in stud at the root section of the thread

(c) stress in tube
(d) extension of stud
(e) compression of tube.
(a) 94.5 MPa (b) 133 MPa (c) 72 MPa (d) 0.142 mm (e) 0.108 mm

12.20 Repeat problem 12.19 if the tube is made of brass rather than steel.
(a) 64.0 MPa (b) 89.9 MPa (c) 48.8 MPa (d) 0.096 mm (e) 0.154 mm

12.21 Two steel channels 76 × 38 mm (A = 853 mm² each) and a steel strap 75 × 20 mm are welded together to form an assembly as shown in Figure P12.21 and subject to an axial load P of 100 kN.

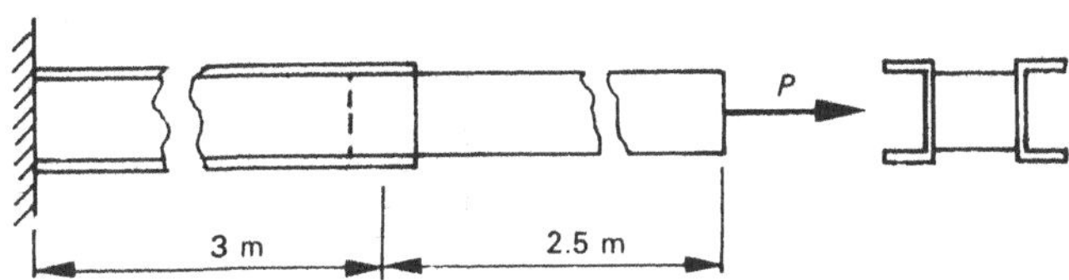

Fig. P12.21

Determine:
(a) stress in channel
(b) stress in strap
(c) extension of channel
(d) extension of strap
(e) total extension
(f) total length of 10 mm fillet weld required to join strap and channel if the stress in the weld is not to exceed 60 MPa.
(a) 58.6 MPa (b) 66.7 MPa (c) 0.879 mm (d) 0.834 mm
(e) 1.713 mm (f) 236 mm

12.22 An aluminium alloy tube (OD 20 mm, ID 12 mm) is adhesive bonded to a brass rod 12 mm diameter as shown in Figure P12.22 and subject to a pull P = 8 kN.

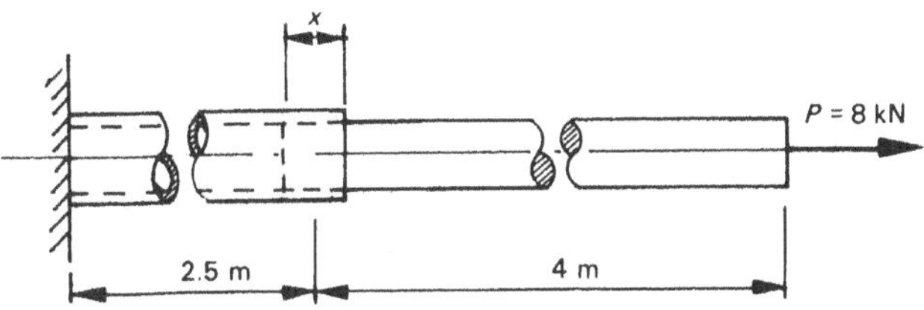

Fig. P12.22

Determine:
(a) stress in tube
(b) stress in rod
(c) extension of tube
(d) extension of rod
(e) total extension
(f) overlap length x so that the shear stress in the adhesive does not exceed 7 MPa.
(a) 39.8 MPa (b) 70.7 MPa (c) 1.42 mm (d) 2.98 mm (e) 4.4 mm
(f) 30.3 mm

12.23 A bronze sleeve, wall thickness 5 mm and length 500 mm, is shrunk over a steel rod, diameter 15 mm and of the same length as the tube and subject to an axial tensile load of 50 kN.

Determine:

(a) axial stress in the rod
(b) axial stress in the sleeve
(c) extension.

(a) 150 MPa (b) 75 MPa (c) 0.375 mm

12.24 A 2 m long reinforced concrete column of square section 250 $\times$ 250 mm is reinforced with four steel rods diameter 15 mm. The column is subjected to a compressive load of 400 kN.

Determine:

(a) compressive stress in the steel rods
(b) compressive stress in the concrete
(c) compression in length.

(a) 74.9 MPa (b) 5.62 MPa (c) 0.75 mm

12.25 A steel bar, diameter 20 mm, length 400 mm, is placed inside a bronze tube, inside diameter 22 mm, wall thickness 4 mm and length 400.15 mm. Rigid plates are placed over the ends of the assembly which is then subject to an axial compressive load of 50 kN.

Assuming compression without buckling, determine:

(a) stress in bar
(b) stress in tube
(c) compression in length of the assembly.

(a) 79 MPa (b) 77 MPa (c) 0.308 mm

13

Strain energy and dynamic loads

In previous chapters, stresses and strains were considered for members in static equilibrium, that is members loaded at some previous time and where the load was now constant. A load which does not vary is a **static load**, also called a dead load. Such loads are common in engineering, for example loads due to weight of a member or fixed weights resting on a member. However, it is also very often the case that the loads may vary and such loads are known as **live loads**. If the change in loading is such that inertia effects are negligible, the load can be considered as gradually applied and treated as if it were a static load with negligible error. Examples of this are when a bolt is tightened slowly with a spanner or when a slow-moving hydraulic ram is used to apply force.

However if the live load varies rapidly, so that the effects of inertia are significant, the load is considered as a **dynamic load**. This can occur if the load is suddenly applied, for example releasing a weight on a member or the engagement of a dog clutch between two shafts, one of which is at rest and the other of which is moving. Dynamic loads may also occur by impact, for example when a falling weight is used to drive in a pile or when a nail is struck with a hammer.

Dynamic loads are characterised by initial instability and vibration which cause higher maximum stresses and strains than would occur if the load were static. Although maximum values may occur only momentarily, if the material is stretched beyond its elastic limit, permanent deformation will result. Indeed if the momentary stress exceeds the ultimate, fracture will occur.

In mechanical design, dynamic loads are often taken into account by application of a suitable dynamic loading factor (which may also be included as part of the safety factor). In this chapter, a theoretical treatment of dynamic axial loads and axial stresses is given. The principles developed are also applicable to dynamic torsional or bending loads and will assist in the understanding of design procedures. Analysis is by the energy method and therefore it is necessary first to define and analyse strain energy. It will be assumed that the material is stressed below the elastic limit (more precisely, the proportional limit) and the strain energy associated with plastic deformation is not considered, so that the term "strain energy" unqualified will always mean elastic strain energy.

13.1 Strain energy (*U*)

Strain energy is the energy of deformation of a material and is equal to the work done by the force which deforms the material. If the material is elastic, the energy is recoverable, that is when the deforming force is removed the material resumes its original shape and dimensions and, in so doing, releases the strain energy stored in it.

A practical example of the use of elastic strain energy is in archery where the archer provides the force and the energy to strain the bow and this energy is released to the arrow when the deforming force is removed. In engineering, elastic strain energy is most often used with springs which store energy when the spring is deformed and release it when the spring recovers its original position.

Consider a typical force-extension diagram for an elastic material as shown in Figure 13.1.

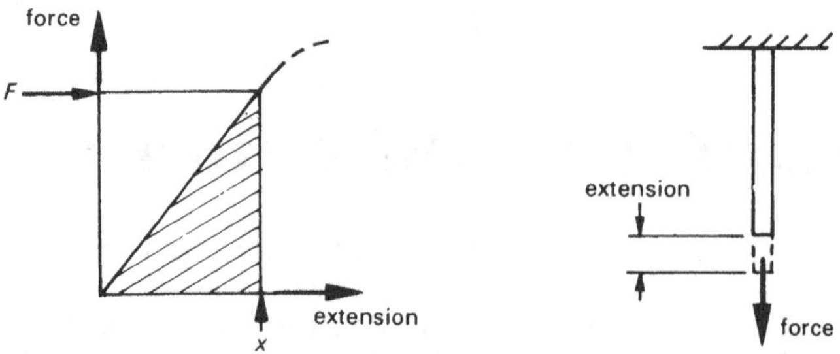

Fig. 13.1 *Typical force-extension diagram for an elastic material*

As the force increases there is a corresponding extension of the material. At any particular value of the force F, when the corresponding extension is x, the work done is the area under the diagram up to this point (shown shaded in Fig. 13.1). This work is equal to the strain energy U. That is:

$$U = \tfrac{1}{2}Fx$$ **(13.1) axial strain energy**

where F = maximum value of the force (N)
x = extension (mm)
U = strain energy (Nmm)

Notes

1. If the force were constant, twice as much energy would be involved but because the force varies from zero to F, the average force is $\tfrac{1}{2}F$, and consequently the energy is one-half that which would result from a constant force.
2. The units of strain energy are Nmm if the force is in N and the extension in mm. In order to convert to J, divide by 1000.

Equation 13.1 may be written in two alternative forms.

Since $x = \dfrac{FL}{EA}$

Substituting, $U = \dfrac{F^2 L}{2EA}$

Multiplying top and bottom of the RHS by A (area):

$$U = \frac{F^2 LA}{A^2 2E}$$

But $\dfrac{F}{A} = f$ (stress)

$$\therefore \quad U = \frac{f^2 AL}{2E}$$ **(13.2) axial strain energy**

Also since $f = \dfrac{Ex}{L}$

Substituting in 13.2:

$$U = \frac{E^2x^2}{L^2} \times \frac{AL}{2E}$$

$$\boxed{U = \frac{x^2EA}{2L}}$$ **(13.3) axial strain energy**

In these formulas:

f = axial stress (MPa)
x = axial extension (mm)
A = cross-sectional area of the bar or rod (mm^2)
L = length of bar or rod (mm)
E = modulus of elasticity (MPa)
U = strain energy (Nmm)

Example 13.1

A 50 kN axial load is gradually applied to a brass rod, diameter 20 mm and length 5 m. Determine:

(a) stress
(b) extension
(c) strain energy, using all three formulas.

E for brass is 95 GPa (Table 12.1).

Solution

$$A = \frac{\pi d^2}{4} = \frac{\pi \times 20^2}{4} = 314.2\ \text{mm}^2$$

(a) $f = \dfrac{P}{A} = \dfrac{50 \times 10^3}{314.2} = \mathbf{159\ MPa}$

(b) $x = \dfrac{FL}{E} = \dfrac{159 \times 5000}{95 \times 10^3} = \mathbf{8.38\ mm}$

(c) Using equation 13.2:

$$U = \frac{f^2AL}{2E}$$
$$= \frac{159^2 \times 314.2 \times 5000}{2 \times 95 \times 10^3}$$
$$= 209 \times 10^3\ \text{Nmm}$$
$$= \mathbf{209\ J}$$

Using equation 13.3:

$$U = \frac{x^2EA}{2L}$$
$$= \frac{8.38^2 \times 95 \times 10^3 \times 314.2}{2 \times 5000}$$
$$= 209 \times 10^3\ \text{Nmm}$$
$$= \mathbf{209\ J} \quad \text{(as before)}$$

Using equation 13.1:

$$U = \tfrac{1}{2}Fx$$
$$= \tfrac{1}{2} \times 50 \times 10^3 \times 8.38 \quad \text{(Nmm)}$$
$$= \mathbf{209\ J} \quad \text{(as before)}$$

13.2 Series and parallel bars

The equations so far derived for strain energy were derived for a single material of uniform cross-section. When this is not the case, the total strain energy is the sum of the individual strain energies in each section of material. This may be calculated by determining the stress or extension in each section of material as previously treated in Chapter 12 (Sects 12.8 and 12.9). The method is illustrated in example 13.2.

Example 13.2

Determine the strain energy resulting from a load of 25 kN applied to the grooved steel bar shown in Figure 13.2. Compare this to the strain energy which would be stored in a bar of the same material with uniform diameter equal to the groove diameter (20 mm) and with the same length (2 m). Assume $E = 200$ GPa.

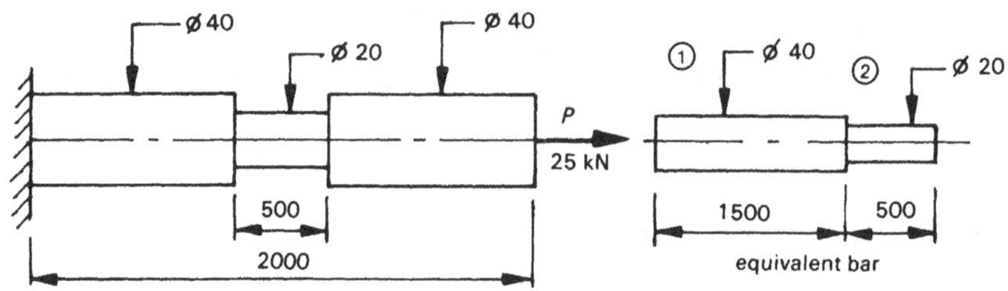

Fig. 13.2

Solution

This bar may be considered as an equivalent series bar as shown to the right in Figure 13.2.

$$A_1 = \frac{\pi \times 40^2}{4} = 1257 \text{ mm}^2$$

$$A_2 = \frac{\pi \times 20^2}{4} = 314 \text{ mm}^2$$

$$f_1 = \frac{P}{A_1} = \frac{25\,000}{1257} = 19.9 \text{ MPa}$$

$$f_2 = \frac{P}{A_2} = \frac{25\,000}{314} = 79.6 \text{ MPa}$$

Using equation 13.2 for each bar:

$$U = \frac{f_1^2 A_1 L_1}{2E} + \frac{f_2^2 A_2 L_2}{2E}$$

$$= \frac{19.9^2 \times 1257 \times 1500 + 79.6^2 \times 314 \times 500}{2 \times 200 \times 10^3} \quad \text{(Nmm)}$$

$$= \mathbf{4.35\ J}$$

Alternatively:

$$x_1 = \frac{f_1 L_1}{E} = \frac{19.9 \times 1500}{200 \times 10^3} = 0.1492 \text{ mm} \qquad x_2 = \frac{f_2 L_2}{E} = \frac{79.6 \times 500}{200 \times 10^3} = 0.199 \text{ mm}$$

$$x(\text{total}) = 0.3482 \text{ mm}$$

$$U = \tfrac{1}{2}Fx = \tfrac{1}{2} \times 25\,000 \times 0.3482 \quad \text{(Nmm)}$$

$$= \mathbf{4.35\ J} \quad \text{(as before)}$$

If the bar were of uniform 20 mm diameter over its entire length, the stress would be 79.6 MPa and the strain energy would be:

$$U = \frac{79.6^2 \times 314 \times 2000}{2 \times 200 \times 10^3} \quad \text{(Nmm)}$$

$$= \mathbf{9.95\ J}$$

It is evident, therefore, that the grooved bar stores less than one-half the elastic strain energy of the uniform section bar, even though the maximum stress is the same. This has important implications in the design of members to absorb energy and shock loads.

13.3 Resilience

Various properties of a material are related to the amount of strain energy which can be stored in the material. For example, in the materials used for motor vehicle bodies an important property is that they be able to absorb large amounts of energy (shock loading) without fracture. The property which measures this is called the **toughness** of the material and may be defined as the total strain energy per unit volume of the material up to the point of fracture.

Of course it would be ideal if the material could absorb the required energy by elastic deformation only, so that after a collision the panels resumed their former shape. The measure of the ability of a material to absorb energy by elastic deformation is called **resilience** and is the most important property of materials to be used for springs or shock buffers.

Since the area under the force–extension diagram is the strain energy, it is clear from Figure 13.3 that the difference between toughness and resilience may be measured by the two areas shown.

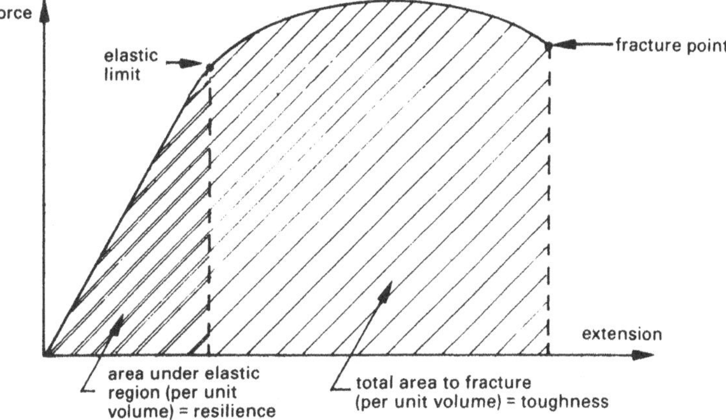

Fig. 13.3 *Toughness and resilience*

Notice that resilience and toughness are not necessarily related to other properties such as tensile strength or elasticity. This is made clear in Figure 13.4 which is the force–extension diagram for two different materials. Material A has a high strength and modulus of elasticity but low resilience. Material B has a lower strength and modulus of elasticity but higher resilience.

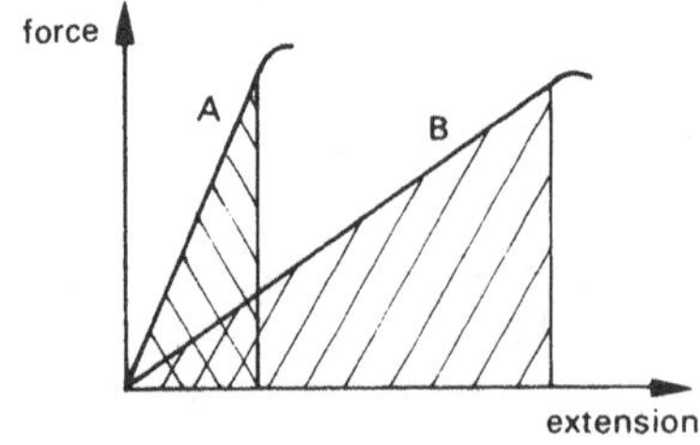

Fig. 13.4 *Material B has a lower strength than material A but a higher resilience*

Modulus of resilience (E_R)

The resilience of materials may be compared by calculation of the modulus of resilience which is defined as the maximum amount of elastic strain energy per unit volume of the material. This is the amount of strain energy per unit volume up to the elastic limit (yield point).

Mathematically $E_R = \dfrac{U}{V}$

For axial loads, equation 13.2 may be used:

$$U = \frac{f^2 A L}{2E}$$

$$\therefore \quad \boxed{E_R = \frac{f_E^2}{2E}} \qquad \textbf{(13.4) modulus of resilience}$$

where f_E = stress at elastic limit (MPa)
E = modulus of elasticity (MPa)
E_R = modulus of resilience (MPa)

Example 13.3

Determine the modulus of resilience in tension for the materials whose properties are listed below.

	Modulus of elasticity (GPa)	*Tensile elastic limit (MPa)*
(a) Mild steel (1020)	190	250
(b) Alloy steel (4140)	200	1100
(c) Brass	95	200
(d) Glass	65	9

Solution

Using equation 13.4:

(a) Mild steel $E_R = \frac{250^2}{2 \times 190 \times 10^3}$ (MPa)

$= \mathbf{165\ kPa}$

(b) Alloy steel $E_R = \frac{1100^2}{2 \times 200 \times 10^3}$ (MPa)

$= \mathbf{3025\ kPa}$

(c) Brass $E_R = \frac{200^2}{2 \times 95 \times 10^3}$ (MPa)

$= \mathbf{210\ kPa}$

(d) Glass $E_R = \frac{9^2}{2 \times 65 \times 10^3}$ (MPa)

$= \mathbf{0.62\ kPa}$

Note the extremely low value of the modulus for glass and the high value of the modulus for alloy steel.

13.4 The energy method

In Chapter 9 (in the Mechanics section) it was seen that the energy method provides an alternative method of solution of problems in dynamics. Also in strength of materials the energy method provides an alternative method which is particularly useful in problems involving dynamic loads occurring in elastic systems. For such systems it is convenient to locate the system boundary so that no work transfer occurs across the boundary, either into the system (by external forces) or out of the system (by friction). Under these conditions (with no losses), the conservation of energy principle for mechanical energy yields the following equation:

$$\boxed{U_1 + PE_1 + KE_1 = U_2 + PE_2 + KE_2}$$

(13.5) energy equation (elastic systems)

where U = strain energy (J)
PE = potential energy (J)
KE = kinetic energy (J)
and the subscripts $_1$ and $_2$ refer to the initial and final conditions respectively.

Notes

1. This equation is valid provided the only forms of energy are mechanical energy and strain energy.
2. Since the strain energy is calculated assuming only elastic deformation then this equation is valid only below the yield point of the material.
3. This equation neglects energy losses (such as friction and impact losses). Typically these losses are small and by neglecting losses the stress is overestimated rather than underestimated which is safer.

Example 13.4

A load of 5 kg is lowered on to a vertical compression spring and then released.

If the spring constant (spring rate) is 9 N/mm, determine the maximum deflection of the spring caused by the load using the energy method and compare this to the static deflection.

Solution

Figure 13.5 shows the initial and final conditions and the load–extension diagram for the spring.

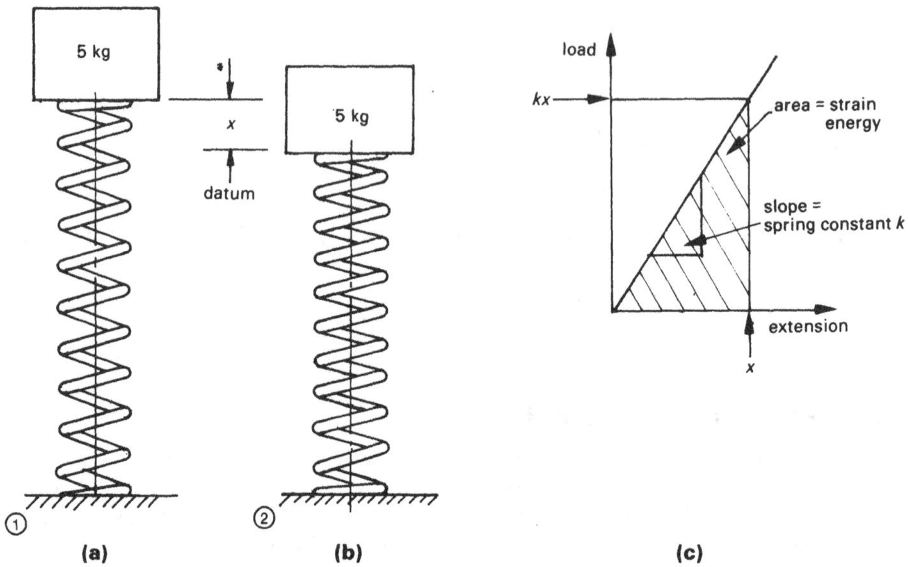

Fig. 13.5 *(a) Initial condition, (b) final condition, (c) load-extension diagram for a spring*

The mass is at rest both at the initial and final conditions so $KE_1 = 0$ and $KE_2 = 0$. Also if position ② is taken as datum for potential energy, $PE_2 = 0$. Since in position ① the spring is not extended, $U_1 = 0$. Hence the mechanical energy equation is simply:

$$U_2 = PE_1$$

Now for a spring, the maximum force is kx where k is the spring constant or spring rate and x the deflection. Since strain energy is the area under the force–extension diagram, the strain energy stored in a spring at any deflection x is:

$$U = \tfrac{1}{2}kx \times x = \tfrac{1}{2}kx^2$$

Hence the energy equation is:

$$\tfrac{1}{2}kx^2 = mgx$$

$$\text{or } x = \frac{2mg}{k}$$

$$= \frac{2 \times 5 \times 9.81}{9}$$

$$= \mathbf{10.9\ mm}$$

The static deflection of the spring is:

$$x = \frac{F}{k} \quad \left(k = \frac{F}{x}\right)$$

Hence $x = \dfrac{5 \times 9.81}{9} = \mathbf{5.45\ mm}$

Note that the static deflection (that is the deflection at rest) is only one-half the maximum deflection.

13.5 Suddenly applied loads

The situation where a mass just rests on a spring and is then released is an example of a suddenly applied load, because the full load is applied even when the deflection is zero. Hence when the mass is released, the spring first deflects past the static equilibrium position because the mass still has kinetic energy when it reaches this position. Hence a vibration of the spring is set up which damps out with time due to the presence of natural damping forces. Figure 13.6 illustrates various positions of the system.

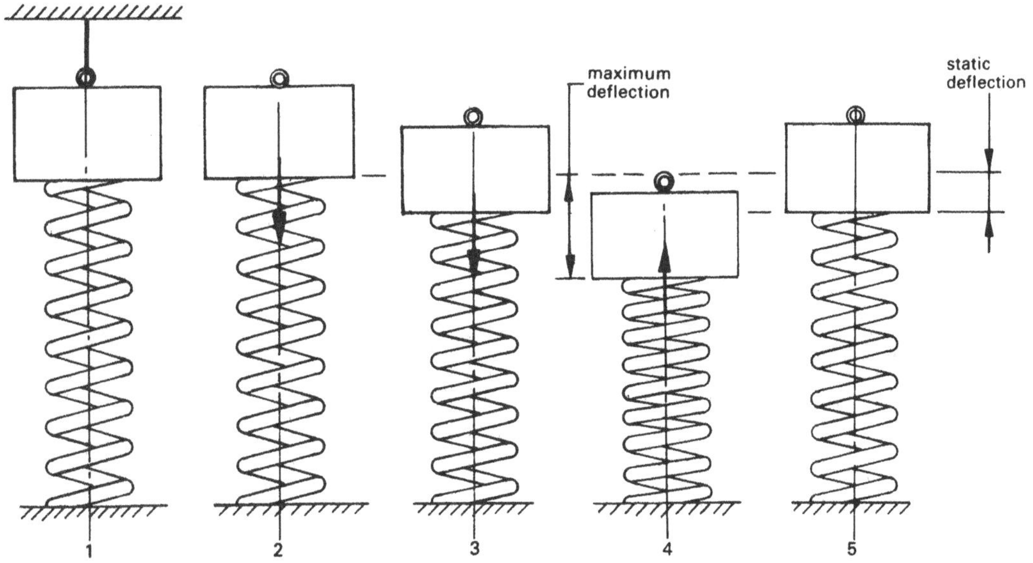

Fig. 13.6 *Deflection of a spring-mass system*

The positions shown are:

1. Load held just in contact with the spring.
2. Load released, initial velocity zero, initial acceleration g.
3. Static equilibrium position is reached but the mass is still moving (possesses kinetic energy) and hence moves past this position.
4. Load comes to rest but accelerates upward because the spring force is greater than the weight force.
5. Final position after a number of oscillations, the static equilibrium position.

Note that the static equilibrium position finally reached by the mass is the same position as would be reached if the load were gradually applied (mass increased gradually from zero to the final value).

The situation outlined above is typical of any suddenly applied load on an elastic material, which is, therefore, characterised by the following:

(a) Full load is applied before the material has deflected.
(b) Vibration occurs which damps out in time due to natural damping and the system achieves static equilibrium.
(c) The final extension of the material is the same as would occur if the load had been applied gradually from the start.
(d) The maximum stress and deflection are twice the static stress and deflection. This was demonstrated in example 13.4 but is true for any suddenly applied axial load. This is because of the triangular shape of the load–extension diagram with a gradually applied load which absorbs only half the energy as with a suddenly applied load where the load is constant.

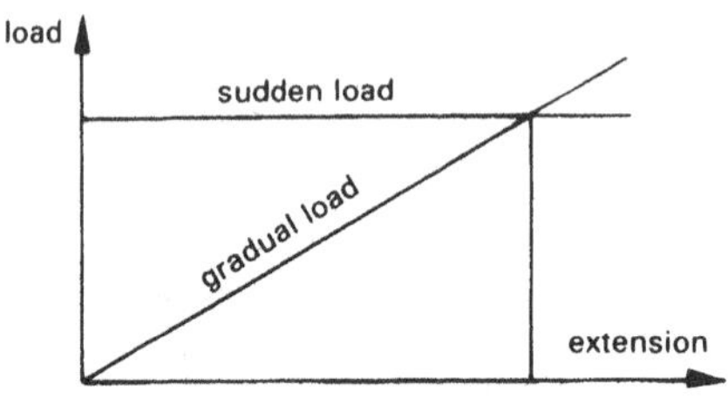

Fig. 13.7 *Load-extension diagram for a sudden and a gradual load*

Example 13.5

A mass of 5 kg is attached to a copper wire, length 1.5 m, diameter 0.8 mm, and rests on a support as shown in Figure 13.8.

If the support suddenly collapses, determine the maximum stress and extension in the wire (E copper = 110 GPa).

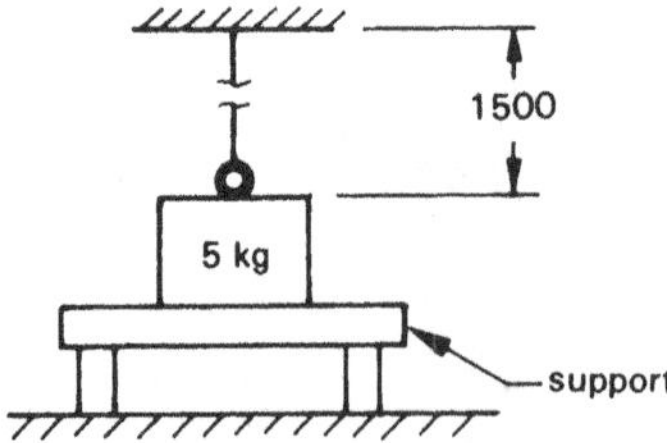

Fig. 13.8

Solution

Static stress
$$f = \frac{F}{A} = \frac{5 \times 9.81}{\pi \times \frac{0.8^2}{4}} = 97.6 \text{ MPa}$$

Static deflection
$$x = \frac{fL}{E} = \frac{97.6 \times 1500}{110 \times 10^3} = 1.33 \text{ mm}$$

Hence maximum stress $f(\text{max}) = 97.6 \times 2 = \mathbf{195.2\ MPa}$
Maximum extension $x(\text{max}) = 1.33 \times 2 = \mathbf{2.66\ mm}$

13.6 Impact or dynamic loads

In addition to the static and suddenly applied load, there is a third type of load, the impact or dynamic load. This type of load is characterised by the fact that the load is in motion at the instant when it first contacts the resisting material. Just as with the suddenly applied load, the maximum stress and deflection are greater than that resulting from the same load gradually applied.

Calculations involving impact or dynamic loads are best solved by the energy method. As previously stated, it is customary to neglect any energy loss (e.g. due to friction or heating). The method is illustrated in example 13.6.

Example 13.6
Solve example 13.4 if the mass is released from a height of 15 mm above the spring.

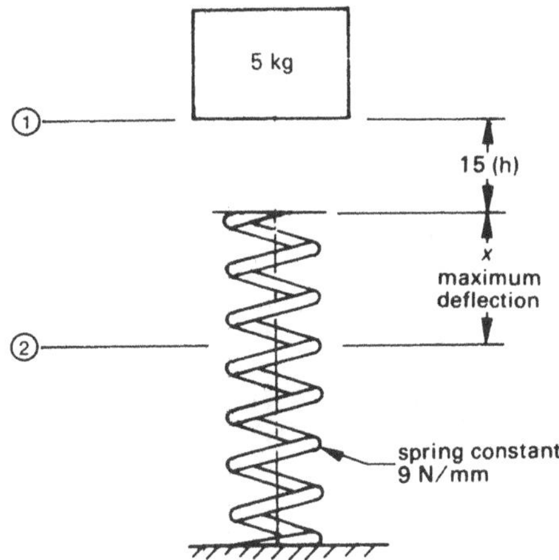

Fig. 13.9

Solution
It was shown in example 13.4 that the energy equation reduces to

$$U_2 = PE_1$$

and the strain energy of a spring is

$$U = \tfrac{1}{2}kx^2$$

Hence

$$\tfrac{1}{2}kx^2 = mg(h + x)$$

$$\therefore \tfrac{1}{2} \times 9 \times x^2 = 5 \times 9.81 \times (15 + x)$$

$$\therefore 4.5x^2 = 49.05x + 735.75$$

$$\therefore x^2 - 10.9x - 163.5 = 0$$

This is a quadratic equation which may be solved using the quadratic formula:

$$x = \frac{-b \pm \sqrt{b^2 - 4ac}}{2a}$$

$$= \frac{10.9 \pm \sqrt{(-10.9)^2 - 4(1 \times -163.5)}}{2}$$

$$= \frac{10.9 \pm 27.8}{2}$$

$$= 19.35 \quad \text{or} \quad -8.45$$

The negative answer is obviously impossible, hence the maximum deflection of the spring is **19.35 mm**.

Note that the kinetic energy of the mass does not appear in the energy equation because the mass is at rest in both positions ① and ②.

13.7 Axial stress due to dynamic loads

Consider Figure 13.10 where a mass m is initially a distance h above a shoulder before being released. After release, the mass falls and strikes the shoulder and the rod extends a distance x before coming to rest momentarily. This is the maximum deflection position and hence the position of maximum stress.

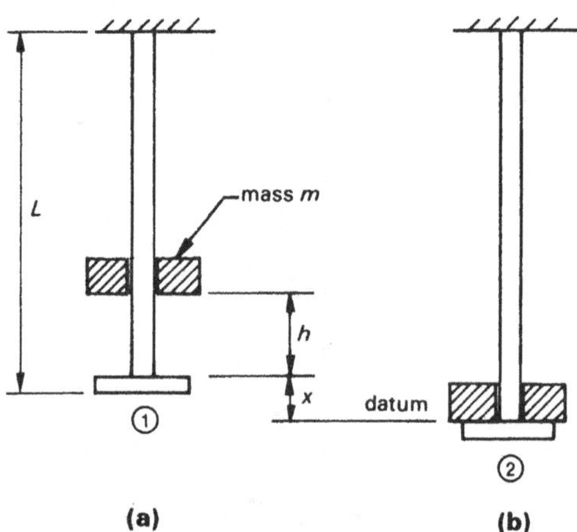

Fig. 13.10 *Dynamic axial load: (a) initial condition, (b) maximum deflection (and stress) condition*

Neglecting air friction and impact losses, the energy equation is

$$U_1 + PE_1 + KE_1 = U_2 + PE_2 + KE_2$$

Now $KE_1 = 0$ (rest), $KE_2 = 0$ (rest) and $PE_2 = 0$ (datum).
Also $U_1 = 0$ (rod is not loaded if the self-weight load is negligible).
Hence the energy equation becomes simply:

$$PE_1 = U_2$$

Using equation 13.2 for strain energy:

$$mg(h + x) = \frac{x^2 EA}{2L} \qquad \textbf{(1)}$$

This formula may be simplified by substituting in terms of the static deflection x_{st}.

$$x_{st} = \frac{mgL}{EA} \quad \therefore mg = \frac{x_{st}EA}{L}$$

Substituting in (1) and cancelling the common term $\left(\frac{EA}{L}\right)$:

$$x_{st}(h + x) = \frac{x^2}{2}$$

$$\therefore x^2 - 2x_{st}x = 2hx_{st}$$

Dividing both sides by x_{st}^2 and adding 1:

$$\left(\frac{x}{x_{st}}\right)^2 - \frac{2x}{x_{st}} + 1 = \frac{2h}{x_{st}} + 1$$

$$\therefore \left(\frac{x}{x_{st}} - 1\right)^2 = 1 + \frac{2h}{x_{st}}$$

$$\therefore \frac{x}{x_{st}} = 1 + \sqrt{1 + \frac{2h}{x_{st}}}$$

Since stress and strain (extension) are directly proportional:

$$\boxed{\frac{f}{f_{st}} = \frac{x}{x_{st}} = 1 + \sqrt{1 + \frac{2h}{x_{st}}}}$$

(13.6) dynamic stress and extension (axial impact loads)

Notes

1. The ratio $\frac{f}{f_{st}} = \frac{x}{x_{st}}$ is the ratio of the maximum dynamic stress and extension to the static stress and extension respectively. Since these are ratios and hence dimensionless, the units of height h from which the load drops must be the same as the units of extension x (that is mm if the extension is in mm).
2. Equation 13.6 may also be used in the case of a suddenly applied load. For a suddenly applied load $h = 0$ and the right-hand side of the equation is:

   $$1 + \sqrt{1 + 0} = 2$$

   This shows that the stress and extension caused by a suddenly applied load is twice the static stress and extension (as derived before).
3. Because equation 13.6 was derived using the strain energy formula for elastic deformation, it is valid only if the maximum stress developed is less than the elastic limit stress (yield point). That is, the material deformation must lie within the elastic region.

Example 13.7

A mass of 100 kg is released from a height of 50 mm above a shoulder on an alloy steel rod (E 200 GPa) as shown in Figure 13.11.

Assuming the extension is within the elastic region, determine the maximum stress and extension in the rod.

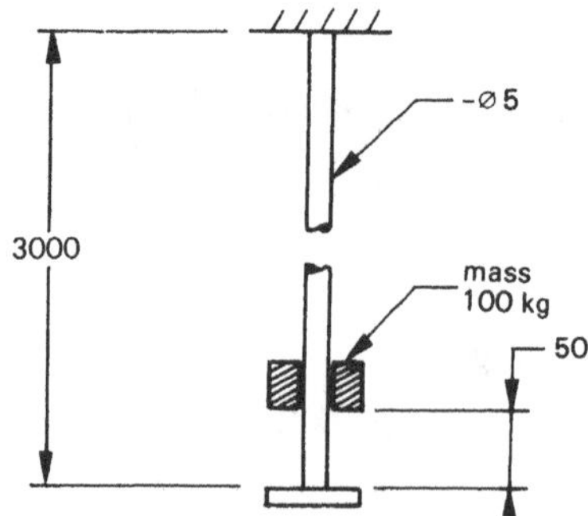

Fig. 13.11

Solution

The static stress is:

$$f_{st} = \frac{W}{A} = \frac{100 \times 9.81}{\pi \times \frac{5^2}{4}} = 49.96 \text{ MPa}$$

The static deflection is:

$$x_{st} = \frac{f_{st} L}{E} = \frac{49.96 \times 3000}{200 \times 10^3} = 0.7494 \text{ mm}$$

From equation 13.6, the ratio of the dynamic stress and extension to the static stress and extension is:

$$1 + \sqrt{1 + \frac{2h}{x_{st}}} = 1 + \sqrt{1 + \frac{2 \times 50}{0.7494}} = 12.595$$

Hence the dynamic stress is:

$$f = f_{st} \times 12.595 = 49.96 \times 12.595 = \mathbf{629\ MPa}$$

And the dynamic extension is:

$$x = x_{st} \times 12.595 = 0.7494 \times 12.595 = \mathbf{9.44\ mm}$$

Computer solution

The computer program DYNLOAD listed in Appendix 3.16 was written to solve problems involving dynamic axial loads on circular sections.

The program inputs are:

1. length and diameter of the rod
2. load mass
3. modulus of elasticity of the material
4. height of the load on release (distance of fall).

The program outputs are:

1. maximum stress and extension
2. static stress and extension.

Note: The program may be used for suddenly applied loads if the height of the load is input as zero.

The solution to examples 13.5 and 13.7 are given below the program listing in Appendix 3.16 and should be self-explanatory.

13.8 Axial stress due to suddenly stopped loads

A slightly different type of dynamic load to those already considered occurs when a mass attached to a cable or rod is being lowered by the cable or rod at some speed and then a brake applied so that the mass is suddenly stopped. A typical instance of this occurs in elevating systems such as hoists or winches when a brake is applied to the winch drum as the load is moving down.

Consider Figure 13.12 in which a mass m attached to a cable is moving down with velocity v when the brake is applied to the drum mass at position ①. Even if the brake action were instantaneous, the mass would continue to move down because of its inertia and in so doing stretch the cable until momentarily stopping at position ② which is the point of maximum stress and extension.

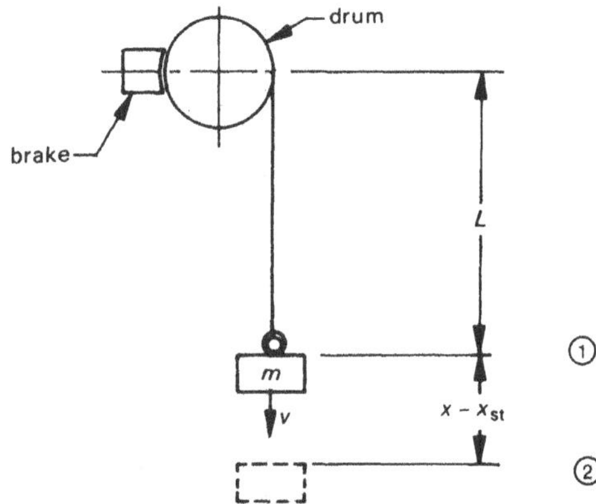

Fig. 13.12 *Suddenly stopped load*

If x is the total maximum extension of the cable at ②, then at ① the cable is extended by an amount x_{st} (the static extension). Hence the additional stretch in the cable between ① and ② is $(x - x_{st})$.

In order to determine the maximum possible stress and extension, brake action will be assumed instantaneous so that all the energy of the load must be transformed into strain energy in the cable. Also the kinetic energy of the cable itself will be neglected. With these assumptions the energy equation (13.5) applies:

$$U_1 + KE_1 + PE_1 = U_2 + KE_2 + PE_2$$

But $KE_2 = 0$ (rest) and $PE_2 = 0$ (datum).
Hence:

$$U_1 + KE_1 + PE_1 = U_2$$

Substituting:

$$\frac{x_{st}^2 EA}{2L} + \tfrac{1}{2}mv^2 + mg(x - x_{st}) = \frac{x^2 EA}{2L}$$

But $x_{st} = \dfrac{mgL}{EA}$. If this substitution is made and the algebra arranged in a similar manner to the derivation of equation 13.6, the result is:

$$\boxed{\frac{f}{f_{st}} = \frac{x}{x_{st}} = 1 + \frac{v}{\sqrt{g x_{st}}}}$$

(13.7) dynamic stress and extension (suddenly stopped axial load)

Note: In order to maintain consistent units on the right-hand side of this equation, since g has units m/s^2, the units of v must be m/s and the units of x must be m.

It is interesting to compare equations 13.7 and 13.6. In equation 13.7, the mass is moving with velocity v and could therefore be considered to have fallen from an equivalent height h_E related to v by the equation:

$$v = \sqrt{2gh_E}$$

Substituting in equation 13.7:

$$\frac{f}{f_{st}} = \frac{x}{x_{st}} = 1 + \sqrt{\frac{2h_E}{x_{st}}}$$

which is almost the same as equation 13.6:

$$\frac{f}{f_{st}} = \frac{x}{x_{st}} = 1 + \sqrt{1 + \frac{2h}{x_{st}}}$$

In fact in most cases h_E will be large compared to x_{st}, so the use of equation 13.6 with equivalent height h_E for a suddenly stopped load will result in small error (on the safe side in any case).

Example 13.8

A load of 5 t attached to the end of a steel cable is lowered at 0.9 m/s by a rotating winch drum. When the free length of the cable is 18 m, the brake is applied to the drum (suddenly).

Determine the maximum stress and extension in the cable if its modulus of elasticity is 110 GPa and the cross-sectional area of the wires in the cable is 1590 mm^2.

Note: Due to wire twist, E for a cable or wire rope made up of twisted strands is considerably less than that of the unformed wire.

Solution

$$f_{st} = \frac{W}{A} = \frac{5000 \times 9.81}{1590} = 30.85\ \text{MPa}$$

$$x_{st} = \frac{f_{st} L}{E} = \frac{30.85 \times 18\,000}{110 \times 10^3} = 5.048\ \text{mm}$$

From equation 13.7, the ratio of the dynamic stress and extension to the static stress and extension is:

$$1 + \frac{v}{\sqrt{g x_{st}}} = 1 + \frac{0.9}{\sqrt{9.81 \times 5.048 \times 10^{-3}}} = 5.044$$

Hence the dynamic stress is:

$$f = 30.85 \times 5.044 = \mathbf{156\ MPa}$$

and the dynamic extension is:

$$x = 5.048 \times 5.044 = \mathbf{25.5\ mm}$$

Computer solution

The computer program DYNLOAD is also applicable to suddenly stopped loads. The only difference in the application of the program is that the user inputs the velocity of the load mass rather than the height from which the mass falls.

The solution to example 13.8 using the program is given below the program listing in Appendix 3.16 and should be self-explanatory.

Problems

For these problems assume E(steel) = 200 GPa (unless stated otherwise). Neglect energy losses in all cases.

13.1 An axial tensile load of 15 kN is applied to a steel rod diameter 12 mm and length 240 mm.

Determine:

(a) stress
(b) extension
(c) strain energy using both formulas 13.1 and 13.2
(d) work done by the load.

(a) 133 MPa (b) 0.159 mm (c) 1.19 J (d) 1.19 J

13.2 For the steel rod shown in Figure P13.2, determine:

(a) compressive stress in each section
(b) total amount by which the rod shortens
(c) strain energy stored in the rod.

(a) 31.8 MPa, 127 MPa (b) 0.051 mm (c) 0.255 J

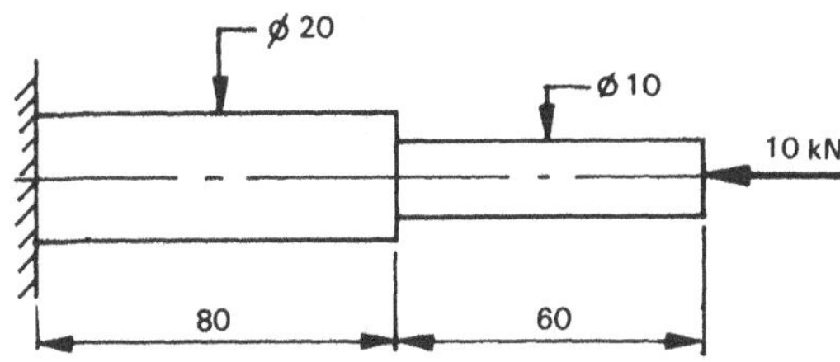

Fig. P13.2

13.3 Determine the percentage strain energy stored in the rod given in problem 13.2 to that of a uniform bar of the same total length but diameter 10 mm throughout.

57.2%

13.4 If the stepped steel bar shown in Figure P13.4 had the same length but uniform diameter, determine what this diameter would be in order to store the same amount of axial strain energy for any applied axial load.

96.6 mm

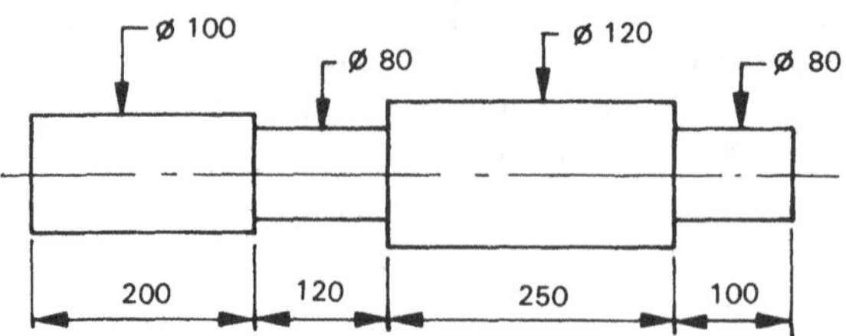

Fig. P13.4

13.5 Describe what is meant by toughness and resilience. Determine the modulus of resilience in tension for a steel which has a modulus of elasticity 207 GPa and a yield point 420 MPa.

426 kPa

13.6 A vertical tension spring has a spring rate 15 N/mm and a maximum extension of 20 mm.

(a) Determine the maximum energy which may be stored in the spring.
(b) If a mass of 8 kg attached to the spring is suddenly released, determine the maximum deflection which will result.
(c) Determine the extension of the spring in (b) when equilibrium is restored.

(a) 3 J (b) 10.5 mm (c) 5.23 mm

13.7 A mass of 12 kg is released from a height of 24 mm above a vertical compression spring which has a spring constant 18 N/mm.

Determine the maximum deflection which will occur and also the deflection at equilibrium.

25.4 mm, 6.54 mm

13.8 A mass of 20 kg rests just above a collar at the lower end of a vertical steel shaft, diameter 12 mm and length 800 mm.

Determine the maximum stress and extension if the mass is suddenly released.

3.47 MPa, 0.014 mm

13.9 If the mass given in problem 13.8 were released from a height of 50 mm above the collar, determine:

(a) maximum stress and extension
(b) maximum height from which the mass could be released in order that the maximum stress does not exceed 150 MPa and the corresponding extension.

(a) 210 MPa, 0.84 mm (b) 25.3 mm, 0.6 mm

13.10 A mass of 80 kg rests just above a collar at the lower end of a vertical steel rod, diameter 5 mm and length 5 m.

Determine:

(a) stress and extension if the load is suddenly released
(b) maximum height from which the load can be dropped so that the stress does not exceed 180 MPa and the corresponding maximum extension

(c) stress and extension at equilibrium.

(a) 80 MPa, 2 mm (b) 5.63 mm, 4.5 mm (c) 40 MPa, 1 mm

13.11 A load of 3 t is lowered by a steel cable at a uniform speed of 1.2 m/s.

If the brake is suddenly applied when the free length of cable is 15 m, determine the maximum stress and extension. The cable has a cross-sectional area of 1250 mm^2 and modulus of elasticity 110 GPa.

183 MPa, 24.9 mm

13.12 Determine the maximum velocity of the load given in problem 13.11 so that the maximum stress in the cable does not exceed 200 MPa. All other conditions are the same.

1.33 m/s

13.13 Determine the effective cable diameter necessary for a hoist designed for the following conditions:

Maximum load (total)	2600 kg
Minimum free length of cable	4 m
Maximum speed of load	4.2 km/h
Ultimate tensile strength of cable	1400 MPa
Modulus of elasticity of cable	110 GPa
Factor of safety	5

Assume instantaneous brake action for design purposes.

42.5 mm

14

Centroid and second moment of area

It was seen in Chapter 8 that in mechanics it is often necessary to use the centre of mass (also known as the centre of gravity) and the second moment of mass (also known as the mass moment of inertia), particularly in problems involving rotating masses.

In strength of materials it is often necessary to use similar concepts based on area rather than mass. These are known as the centroid (or centre of area) and the second moment of area (or area moment of inertia). These concepts are developed in this chapter and their use to solve problems involving beams and columns are the subject of subsequent chapters.

14.1 Centroid (C)

The centroid is the geometrical centre of area. Its position on any plane area may be determined experimentally by cutting the area from a uniform sheet of material and finding the point of balance (by trial and error). This experiment is illustrated in Figure 14.1.

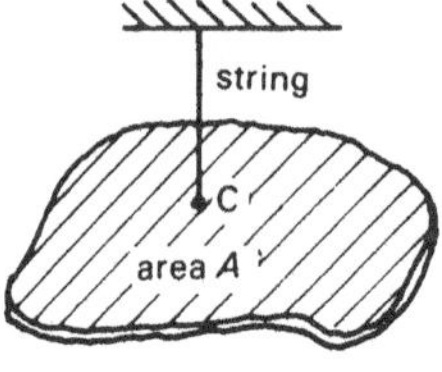

Fig. 14.1 *Experimental determination of the balance point (centroid)*

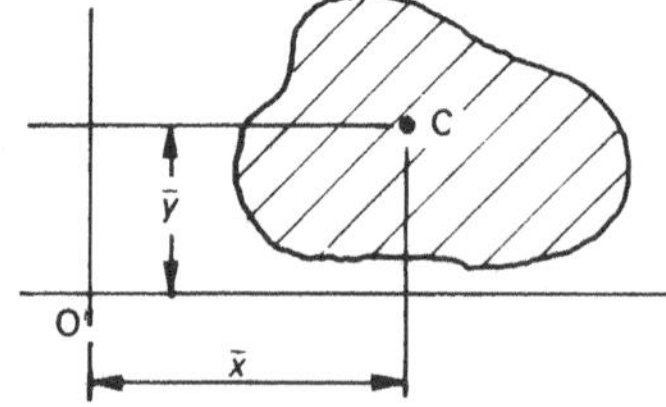

Fig. 14.2 *Use of the bar notation to locate the centroid*

In fact this experiment determines the location of the centre of mass on the horizontal plane but, because the sheet of material is uniform (in thickness and density), this location is the same as that of the centroid.

The experiment outlined above is a useful practical one when the area is of complex shape and also helps visualise the meaning of centroid. However it is important that the experiment does not lead to confusion regarding the concepts of centroid (which is based purely upon the two-dimensional shape of the area) and centre of mass (which depends also upon the density and thickness of the material).

The position of the centroid of any area must be located relative to a frame of reference. Using the conventional x-y frame of reference, the bar notation is often used to denote the horizontal and vertical distance to the centroid as shown in Figure 14.2.

14.2 Location of the centroid: Analytical method

Reflection on the experiment illustrated in Figure 14.1 leads to the conclusion that, if the area has an axis of symmetry, then the centroid must lie somewhere on that axis. If the area does not have an axis of symmetry, then the centroid lies on a line about which the moment of the areas on both sides of the line are numerically equal. That is, the *algebraic* sum of the moment of the areas about the line is zero when the correct sign of the moment is included. Consider Figure 14.3.

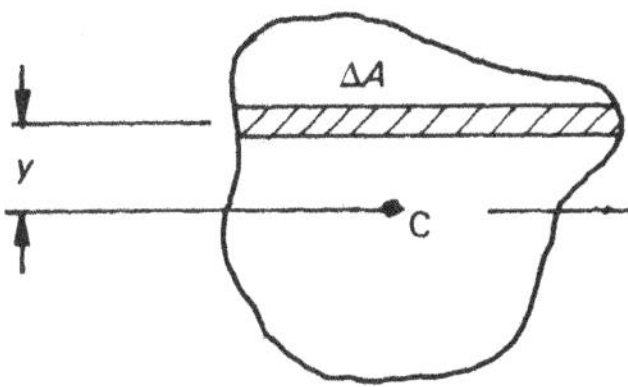

Fig. 14.3 *First moment of area of a small strip*

The location of the horizontal centroidal axis occurs when:

$$\Sigma y \Delta A = 0$$

or in the limit

$$\int y \, dA = 0$$

The expression $\Sigma y \Delta A$ or $\int y \, dA$ is called the **first moment of area** and given the symbol Q.
That is:

$$Q = \Sigma y \Delta A = \int y \, dA \qquad \textbf{(14.1) first moment of area}$$

If the position of the centroid of an area is not known, a trial and error method (as in the experiment) would be necessary by choosing a line and checking if $Q = 0$ about that line. If it is not, the line must be repositioned and the new position checked.

A direct procedure for determining the centroid location is to use a reference axis as shown in Figure 14.4 and locate the centroid with reference to this axis.

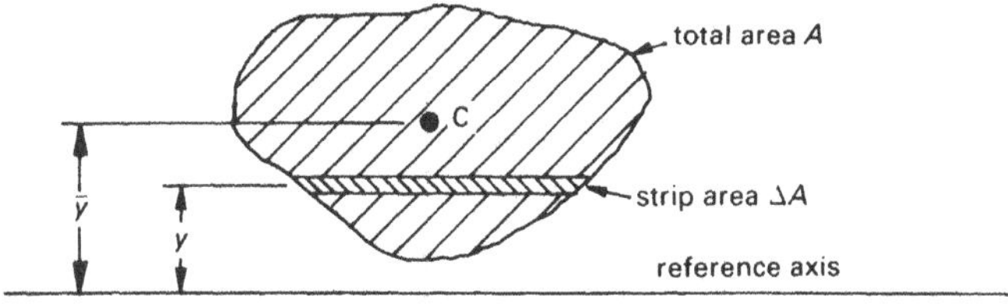

Fig. 14.4 *Centroid location relative to a reference axis*

Now $Q = \Sigma y \Delta A = \bar{y} A$

Hence

$$\bar{y} = \frac{Q}{A} \qquad \textbf{(14.2) centroid location}$$

Hence if the first moment of area (Q) can be determined relative to the reference axis and the total area is known, the centroid location may be determined immediately (without trial and error). This method is illustrated in example 14.1.

Example 14.1

Determine the location of the centroid relative to the x-y frame of reference for the unequal angle section shown in Figure 14.5.

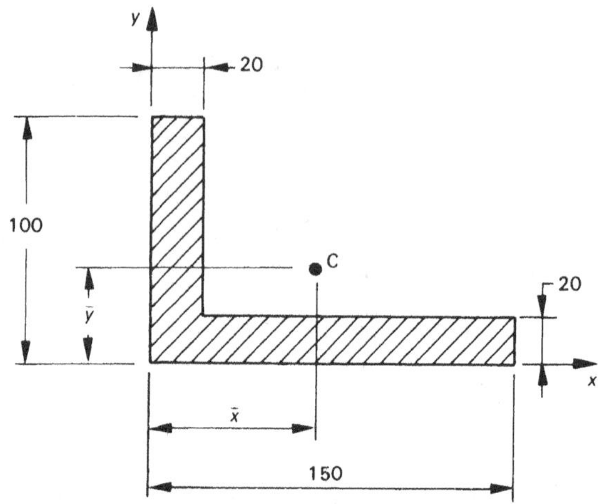

Fig. 14.5

Solution

The angle may be broken into two separate areas A_1 and A_2. These areas are rectangular and are symmetrical in both horizontal and vertical directions, hence their centroid locations may be determined by inspection and are as shown in Figure 14.6.

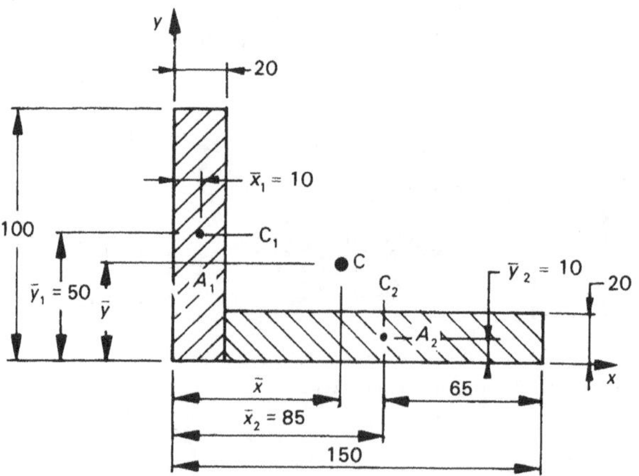

Fig. 14.6

The areas are as follows:

$$A_1 = 100 \times 20 = 2000\ \text{mm}^2 \qquad A_2 = 130 \times 20 = 2600\ \text{mm}^2$$

$$A = A_1 + A_2 = 4600\ \text{mm}^2$$

Taking moments about the x axis:

$$A\bar{y} = A_1\bar{y}_1 + A_2\bar{y}_2$$

$$\therefore 4600\bar{y} = 2000 \times 50 + 2600 \times 10$$

$$\therefore \bar{y} = \mathbf{27.4\ mm}$$

Similarly taking moments about the y axis:

$$A\bar{x} = A_1\bar{x}_1 \times A_2\bar{x}_2$$

$$\therefore 4600\bar{x} = 2000 \times 10 + 2600 \times 85$$

$$\therefore \bar{x} = \mathbf{52.4\ mm}$$

14.3 Centroid location for common shapes

It has already been stated that if an area has an axis of symmetry then the centroid must lie along this axis. Therefore it is clear that for a circle, the centroid is located at the centre, and for a rectangle, square or flange beam section, the centroid is located at the midpoint (in both horizontal and vertical directions).

For non-symmetrical sections, the location of the centroid may be determined by the methods of calculus. For commonly used non-symmetrical sections such as a triangle, segment of a circle or angle section, the position of the centroid may be determined from handbooks or manufacturers' tables. The centroid location for some non-symmetrical shapes is shown in Figure 14.7.

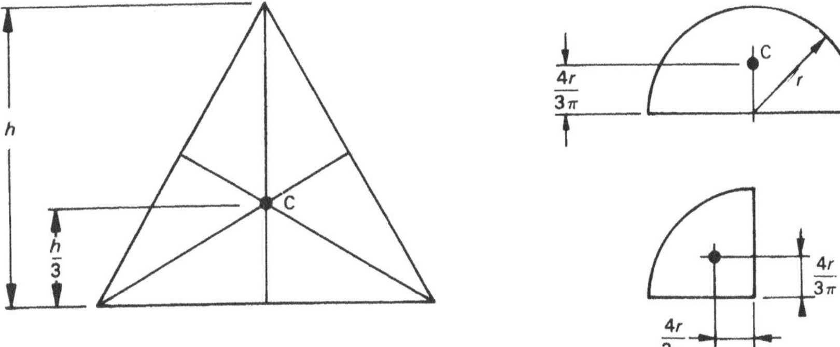

Fig. 14.7 *Centroid location for some non-symmetrical sections*

If a section contains a void such as a hole, the centroid may be determined in the same general way as outlined in example 14.1 except that the first moment of area of the void is negative. This method is illustrated in example 14.2.

Example 14.2

Determine the location of the centroid relative to the x and y axes for the area shown in Figure 14.8.

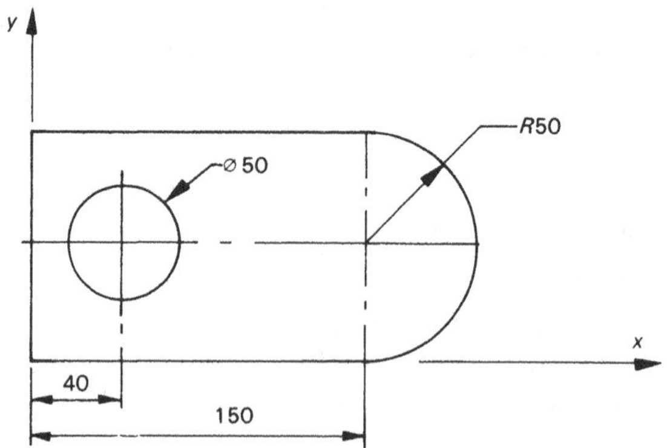

Fig. 14.8

Solution

Since the section has a horizontal axis of symmetry, the vertical position to the centroid may be determined by inspection. That is:

$\bar{y}$ = **50 mm**

In order to determine the horizontal distance to the centroid, the section may be broken into three separate areas, A_1, A_2 and A_3 as shown in Figure 14.9.

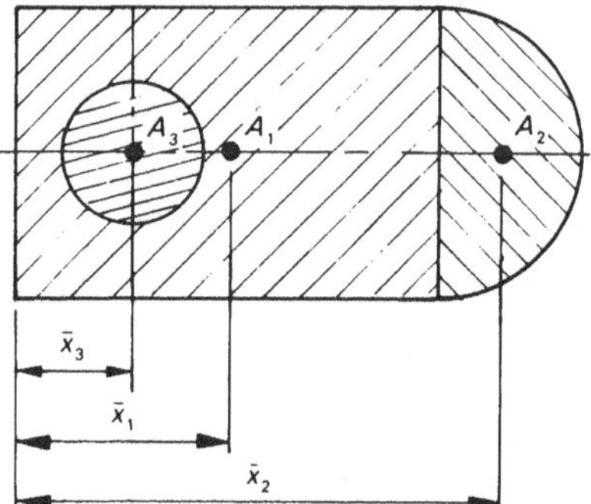

Fig. 14.9

$$A_1 = 150 \times 100 = 15\,000 \text{ mm}^2, \bar{x}_1 = 75 \text{ mm}$$

$$A_2 = \frac{\pi \times 50^2}{2} = 3927 \text{ mm}^2, \bar{x}_2 = 150 + \frac{4}{3} \times \frac{50}{\pi} = 171.22 \text{ mm}$$

$$A_3 = \frac{\pi \times 50^2}{4} = 1963.5 \text{ mm}^2, \bar{x}_3 = 40 \text{ mm}$$

$$A = A_1 + A_2 - A_3 = 16\,963.5 \text{ mm}^2$$

$A\bar{x} = A_1\bar{x}_1 + A_2\bar{x}_2 - A_3\bar{x}_3$ (note negative sign for A_3 because it is a hole)

$$16\,963.5 \times \bar{x} = 15\,000 \times 75 + 3927 \times 171.22 - 1963.5 \times 40$$

$$\bar{x} = \mathbf{101.3 \text{ mm}}$$

14.4 Second moment of area (*I*)

The second moment of area is also known as the area moment of inertia and is analogous in area terms to the second moment of mass or mass moment of inertia. The second moment of area is defined as:

$$I = \Sigma y^2 \Delta A = \int y^2 dA \qquad \textbf{(14.3) second moment of area}$$

Notes

1. The units of second moment of area are mm^4.
2. Since the second moment of area is a concept based on area rather than mass, it depends only upon the member's sectional shape and size and is independent of the material from which the member is made.
3. The second moment of area (like any other moment) is specified relative to an axis about which the moment is taken. This is the reference axis which should always be stated. The most common axes used are the x and y axes passing through the centroid of the section.
4. If the section has area distributed on both sides of the reference axis, each area contributes in the same way to the second moment of area of the whole section. That is to say, even though an area may be on the negative side of the reference axis, it still has a positive second moment of area because a negative value of y becomes positive when squared. This contrasts of course with first moment of area which may be positive or negative depending upon which side of the axis the area is located.
5. If the section contains voids or hollows, these are equivalent to a negative second moment of area when calculating the second moment of area of the entire section.

Calculation of *I*

The second moment of area for regular sections such as rectangular or circular may be calculated by the methods of calculus. As an example, the method will be illustrated for second moment of area about a horizontal axis passing through the centroid for a rectangle of width b and height h.

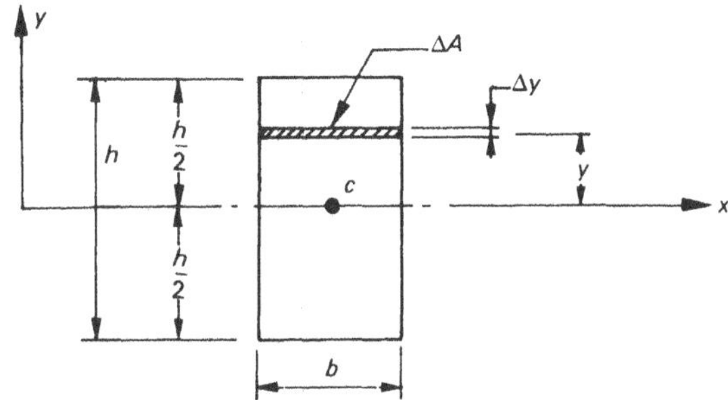

Fig. 14.10 *Second moment of area for a rectangle*

The small area ΔA is: $b\ \Delta y$
Using equation 14.3:

$$I_{CX} = \int y^2 dA = \int_{-\frac{h}{2}}^{\frac{h}{2}} by^2\, dy = \left[\frac{by^3}{3}\right]_{-\frac{h}{2}}^{\frac{h}{2}}$$

Substituting:

$$I_{CX} = \frac{b}{3}\left[\left(\frac{h}{2}\right)^3 - \left(-\frac{h}{2}\right)^3\right]$$

$$= \frac{b}{3}\left[\frac{h^3}{8} + \frac{h^3}{8}\right]$$

$$= \frac{b}{3} \times \frac{h^3}{4}$$

$$\boxed{I_{CX} = \frac{bh^3}{12}}$$

(14.4) second moment of area for a rectangle about centroid

Using the same method, the calculation for a circular section yields:

$$\boxed{I_{CX} = \frac{\pi d^4}{64}}$$

(14.5) second moment of area for a circle about the centre

When the section is non-symmetrical about a centroidal axis, I_{CX} about each side of the axis will be different but the total I_{CX} may still be calculated. Some common sections and their second moment of area are shown in Figure 14.11.

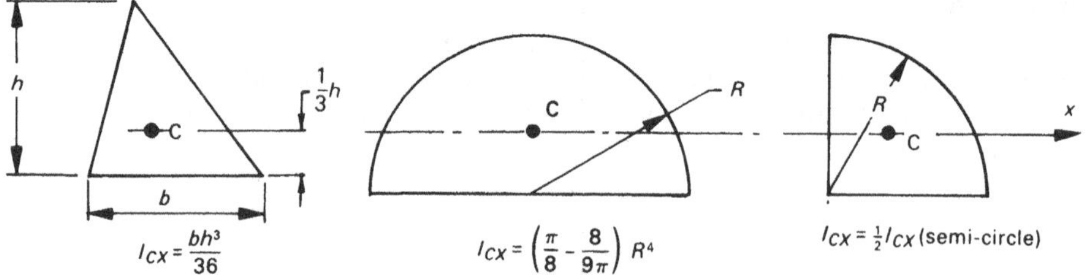

Fig. 14.11 *Second moment of area for some non-symmetrical sections*

Notes

1. For a rectangular section, I_{CX} varies directly with the breadth but as the cube of the height. Hence if b is doubled, I_{CX} doubles; but if h is doubled, I_{CX} increases eightfold.
2. For a circular section, I_{CX} varies as the fourth power of the diameter. Hence if d is doubled, I_{CX} increases 16-fold.

Example 14.3

A rectangular beam, 15 × 250 mm, has a 40 mm hole drilled through from one side as shown in Figure 14.12.

Determine the second moment of area about the x and y axes passing through the centre of the hole.

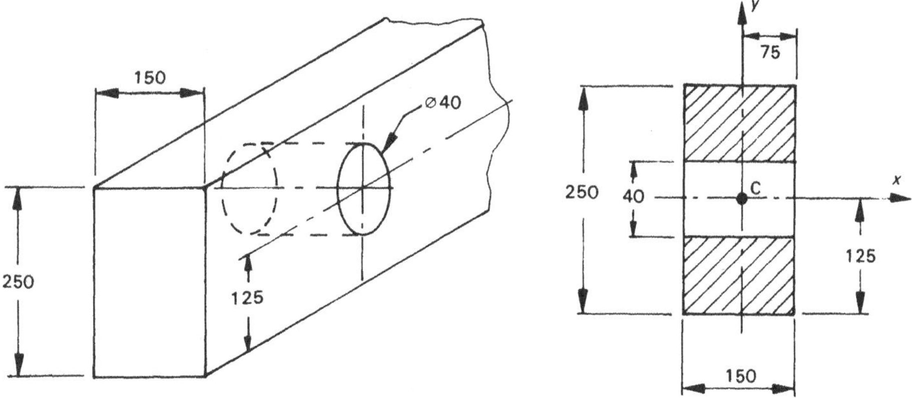

Fig. 14.12

Solution

About the x axis:

	Solid section	*Hole*
b (mm)	150	150
h (mm)	250	40

and about the y axis:

b (mm)	250	40
h (mm)	150	150

Using equation 14.4, $I_{CX} = \dfrac{bh^3}{12}$ and remembering that I is negative for the hole:

$$I_{CX} = \frac{150 \times 250^3}{12} - \frac{150 \times 40^3}{12} = \mathbf{194.5 \times 10^6\ mm^4}$$

$$I_{CY} = \frac{250 \times 150^3}{12} - \frac{40 \times 150^3}{12} = \mathbf{59.1 \times 10^6\ mm^4}$$

14.5 Parallel axis theorem

The parallel axis theorem for second moment of mass (see equation 8.13) has an equivalent for calculation of second moment of area. The equation is:

$$I_{XX} = I_{CX} + Ax^2 \qquad \textbf{(14.6) parallel axis theorem (area)}$$

where I_{XX} = second moment of area about any axis x-x (mm^4)
I_{CX} = second moment of area about an axis parallel to x-x and passing through the centroid of the area (mm^4)
A = area (mm^2)
x = perpendicular distance between the centroidal axis and the x-x axis (mm)

These symbols are clarified in Figure 14.13.

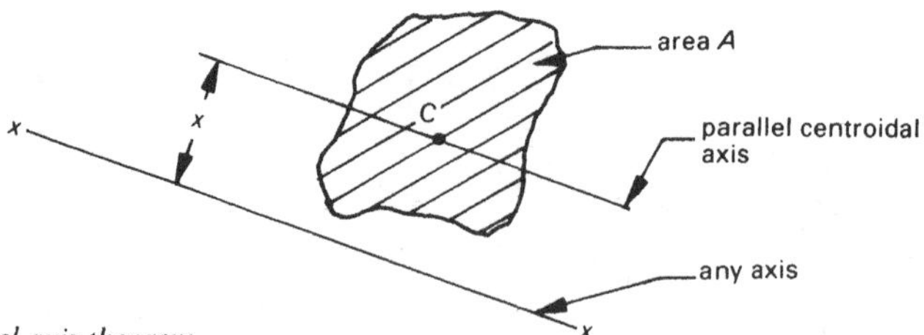

Fig. 14.13 *Parallel axis theorem*

Example 14.4

Handbook data for a rolled flange beam (410 UB 610) are shown in Figure 14.14.

Check the values quoted for I_{CX} and I_{CY} by calculation (ignoring the corner fillets).

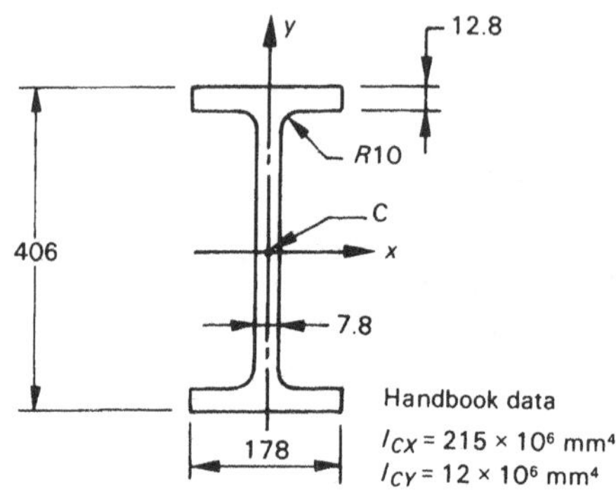

Fig. 14.14

Solution

In order to calculate I_{CX}, one method is to break the sectional area into component parts as shown in Figure 14.15.

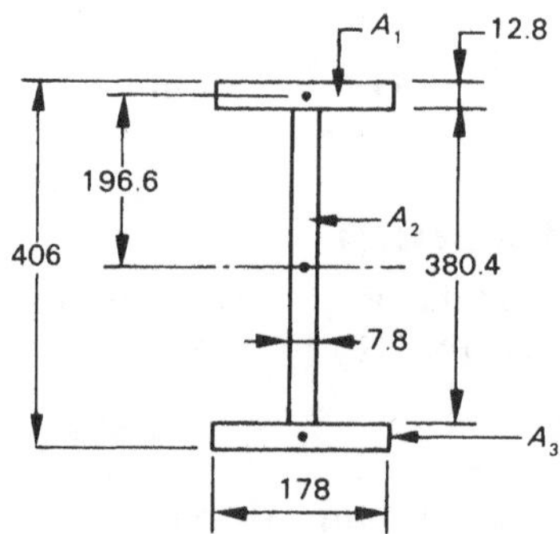

Fig. 14.15

The areas A_1 and A_3 are equal and are both located at the same distance from the centroid. Hence they contribute the same amount to the second moment of area for the section.

$A_1 = A_3 = 178 \times 12.8 = 2278.4\ \text{mm}^2,$

$$I_{CX} = 2\left[\frac{bh^3}{12} + Ax^2\right] + \frac{bh^3}{12}$$

(areas 1 and 3) (area 2)

$$= 2\left[\frac{178 \times 12.8^3}{12} + 2278.4 \times 196.6^2\right] + \frac{7.8 \times 380.4^3}{12}$$

$= \mathbf{212 \times 10^6\ mm^4}$ (which is slightly less than the handbook value because the corner fillets were neglected)

An alternative method is to break the section into positive and negative areas as shown in Figure 14.16.

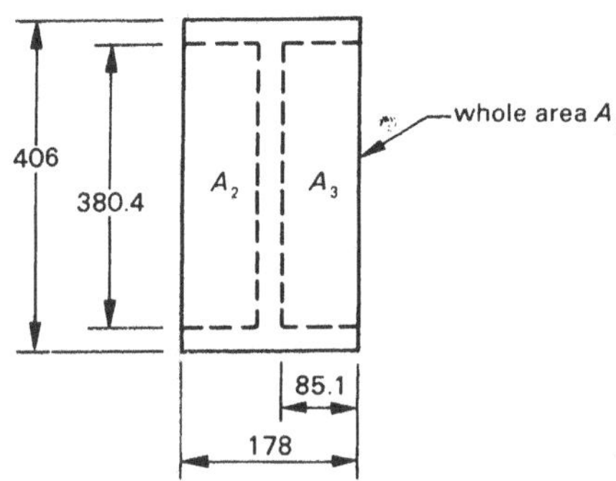

Fig. 14.16

$$I_{CX} = \frac{bh^3}{12} - 2\left[\frac{bh^3}{12}\right]$$

(area 1) (areas 2 and 3)

$$= \frac{178 \times 406^3}{12} - 2\left[\frac{85.1 \times 380.4^3}{12}\right]$$

$= \mathbf{212 \times 10^6\ mm^4}$ (as before)

In order to calculate I_{CY}, it is possible to use the positive and negative area method but easier to use positive areas only. Referring to Figure 14.15:

$$I_{CY} = 2\left[\frac{bh^3}{12}\right] + \frac{bh^3}{12}$$

(areas 1 and 3) (area 2)

$$= 2\left[\frac{12.8 \times 178^3}{12}\right] + \frac{380.4 \times 7.8^3}{12}$$

$= 12 \times 10^6\ \text{mm}^4$ (which agrees with the handbook value)

Note: It is interesting to consider why neglecting the corner fillets affects the calculated value of I_{CX} more than that of I_{CY}.

14.6 Radius of gyration (k)

If the area of a section could be considered as concentrated at a point, then the distance of that point from the centroidal axis in order to give the same second moment of area is known as the **radius of gyration**. The larger the radius of gyration the "stiffer" the section for a given area, that is the higher the second moment of area.

For example, consider the simple rectangular sections shown in Figure 14.17.

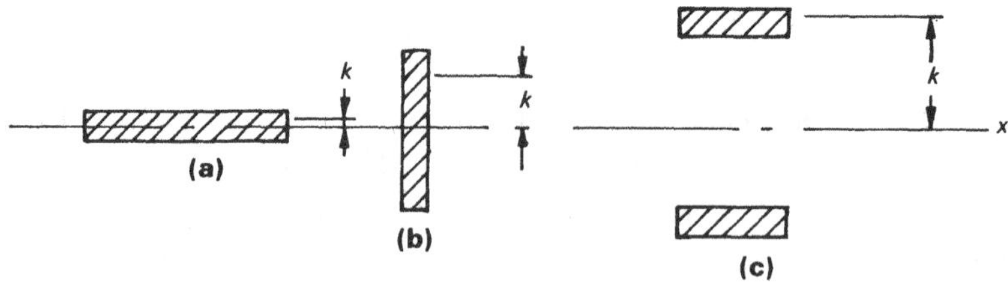

Fig. 14.17 *Radius of gyration for various sections with the same area*

All the sections shown have the same area but the radius of gyration is very different for each. Area (c) has the highest value of I and k (about the x-x axis) and hence is the stiffest section (most resistant to bending).

Note that the radius of gyration is not located at the centroid of the area above the axis but is further out from this position. The exact position is given by:

$$k = \sqrt{\frac{I}{A}}$$ **(14.7) radius of gyration (area)**

Example 14.5

The handbook values of radius of gyration for the section given in example 14.4 are as follows:

$k_{CX} = 168$ mm, $k_{CY} = 39.7$ mm

and the area is given as 7600 mm^2.

Check these values of radius of gyration by calculation.

Solution

$$k_{CX} = \sqrt{\frac{I_{CX}}{A}} = \sqrt{\frac{215 \times 10^6}{7600}} = \mathbf{168\ mm}$$

$$k_{CY} = \sqrt{\frac{I_{CY}}{A}} = \sqrt{\frac{12 \times 10^6}{7600}} = \mathbf{39.7\ mm}$$

Therefore the calculated values agree with the handbook values.

14.7 Comparison of mass and area concepts

Table 14.1 gives a comparison of mass and area concepts which are related to the second moment of mass or area.

Table 14.1 Comparison of mass and area concepts related to I

	Mass	*Area*
Units I	kg m^2	mm^4
I (concentrated mass or area)	$I = mr^2$	$I = Ay^2$
I (general)	$I = \Sigma r^2 \Delta m$ $= \int r^2\, dm$	$I = \Sigma y^2 \Delta A$ $= \int y^2\, dA$
Parallel axis theorem	$I_{XX} = I_{CG} + mx^2$	$I_{XX} = I_{CX} + Ax^2$
Radius of gyration	$k = \sqrt{\frac{I}{m}}$	$k = \sqrt{\frac{I}{A}}$

14.8 Standard properties of sections

The calculation of centroid location and second moment of area is not necessary if standard rolled sections are being used. Since the engineer cannot vary the dimensions of standard sections and must choose from those which are commercially available, unnecessary calculation is eliminated by using a table of standard properties from which the necessary data may be extracted as required.

In Australia, structural rolled sections readily available are: flat bar, plate, round bar, flange beams, flange columns, equal angles, unequal angles and channels. For these sections, handbooks and manufacturers' tables give the following data: dimensions, mass per m length, area, position of the centroid, second moment of area about the centroid, and radius of gyration.

Although no calculation of the properties of standard sections is necessary if the tables are available, if the standard sections are machined, welded, riveted or otherwise joined to form a new sectional shape, the properties of this section will need to be calculated from the available data modified in accordance with the changes made. The procedure is illustrated in example 14.6.

Example 14.6

A tee section is produced from angle and flat bar welded together as shown in Figure 14.18. The properties of the angle are shown to the right of this figure.

Determine the centroid location vertically from the base of the section and the second moment of area about the horizontal centroidal axis.

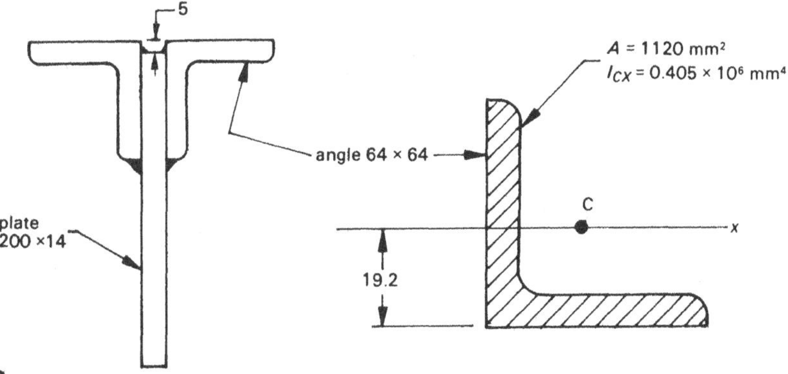

Fig. 14.18

Solution

The vertical location of the centroid of the angle from the base of the plate in the composite section is:

$$200 + 5 - 19.2 = 185.8 \text{ mm}$$

The area of the plate is:

$$200 \times 14 = 2800 \text{ mm}^2$$

and the total area of the section is:

$$2800 + 1120 \times 2 = 5040 \text{ mm}^2$$

Taking the first moment of area about the base of the composite section:

$$2 \times 1120 \times 185.8 + 2800 \times 100 = 5040 \times \bar{y}$$

$$\therefore \bar{y} = \mathbf{138.1\ mm}$$

The vertical distance between the centroid of the composite section and the centroid of the plate is:

$$y = 138.1 - 100 = 38.1 \text{ mm}$$

and for the angles is:

$$y = 185.8 - 138.1 = 47.7 \text{ mm}$$

Using the parallel axis theorem, the second moment of area of the composite section may now be calculated:

$$I_{CX} = \frac{14 \times 200^3}{12} + 2800 \times 38.1^2 + 2(0.405 \times 10^6 + 1120 \times 47.7^2)$$

$$= \mathbf{19.3 \times 10^6\ mm^4}$$

14.9 Computer solution

The computer program CENTINT listed in Appendix 3.17 was developed for determining the centroid and second moment of area for a single section or composite section built up of solid sectional areas or voids. The program can deal with up to ten areas of any of the twelve types of sectional shape shown in Figure 14.19, joined in any combination to form a composite section. Note that area type 12 is any section with known properties, e.g. a standard rolled section.

In Figure 14.19, the distance y must be input by the user and it locates the section relative to a reference axis.

The program inputs are:

1. whether or not the section is symmetrical
2. the area type (number), whether solid or void (S or V) and the number of such areas in the composite section
3. dimensions of the area or, if type 12 standard section is used, the known area and second moment of area
4. location of the area relative to the reference axis (dimension y in Fig. 14.19).

The above data must be re-input for each area in turn.

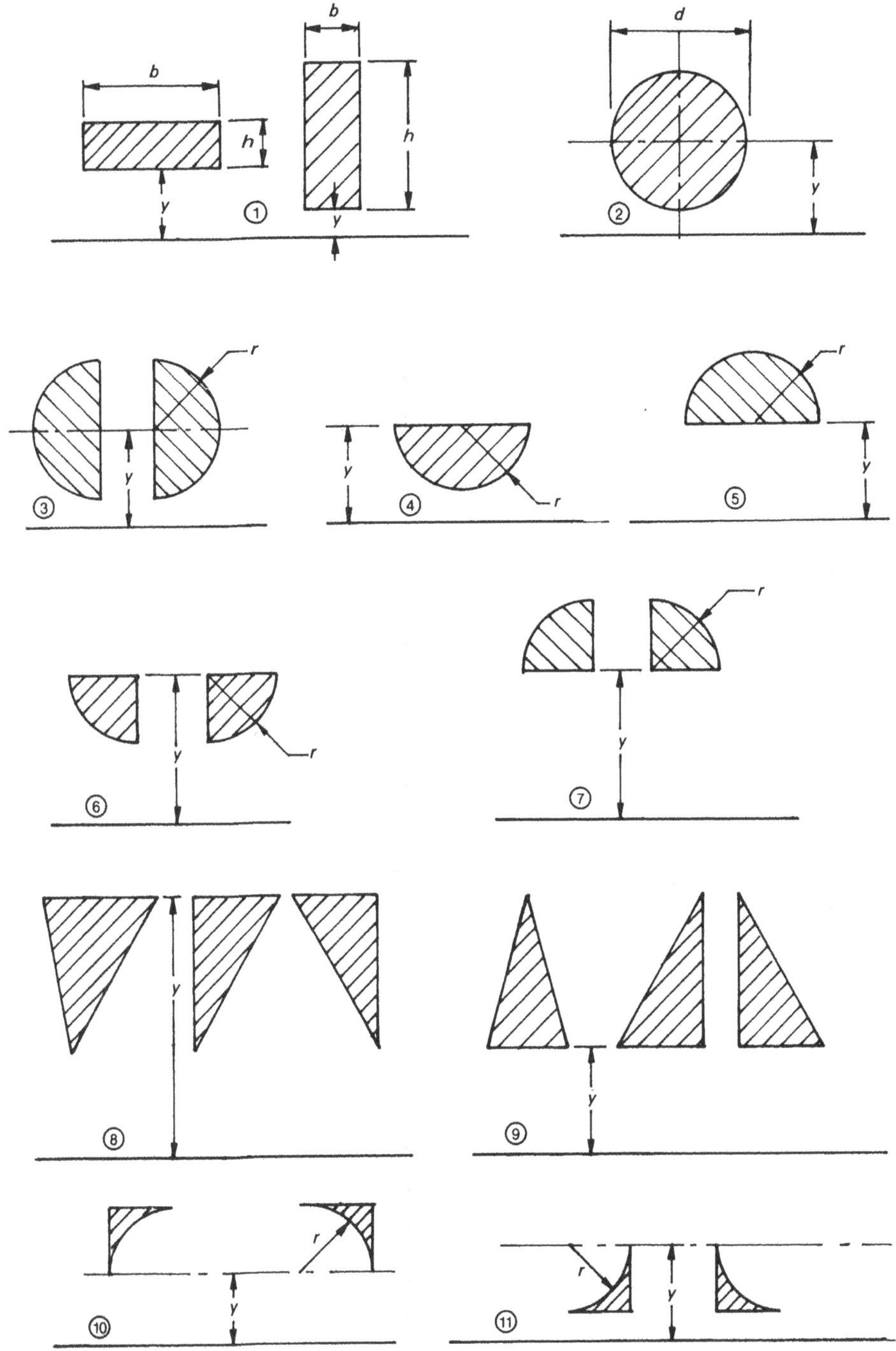

⑫ Any area with known properties

Fig. 14.19 *Area shapes identified by number*

The program outputs for the composite section are:

1. area
2. centroidal distance relative to the reference axis
3. second moment of area about the centroidal axis
4. second moment of area about the reference axis
5. radius of gyration about the centroidal axis
6. radius of gyration about the reference axis.

The solution to examples 14.1, 14.2, 14.3, 14.4 and 14.6 are given below the program listing in Appendix 3.17 and should be self-explanatory.

Notes

1. All dimensions are in mm.
2. Any reference axis may be used and the distance y may be zero or negative. The reference axis may also be the centroidal axis for the composite section.
3. Identical areas are those which have the same shape and are symmetrical with regard to the reference axis.
4. The data must be re-input if the frame of reference is the y axis rather than the x axis.

14.10 Polar moment of inertia (J)

The polar moment of inertia is a very similar concept to the area moment of inertia (second moment of area) except that the reference axis is the z axis rather than the x or y axis. Mathematically it is defined as:

$$J = \Sigma r^2 \Delta A = \int r^2 dA \qquad \textbf{(14.8) polar moment of inertia}$$

The distinction between the two concepts is clarified in Figure 14.20.

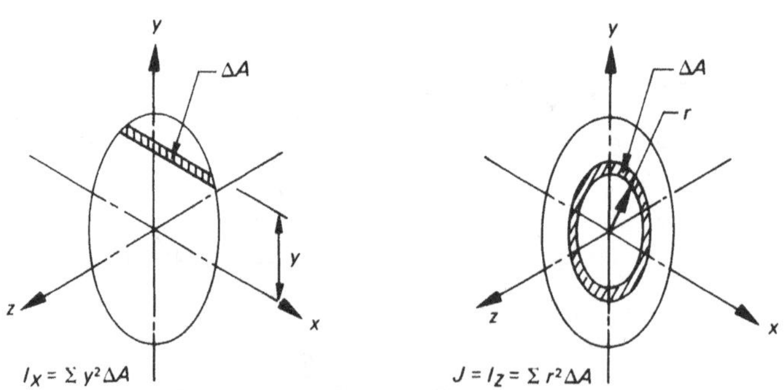

Fig. 14.20 *Distinction between area moment of inertia and polar moment of inertia*

Polar moment of inertia for a circular shaft

An equation for the polar moment of inertia of a circular section may be derived as follows:

Referring to Figure 14.21:

$$\Delta A = 2\pi r \Delta r$$

$$J = \Sigma r^2 \Delta A = \Sigma 2\pi r^3 \Delta r$$

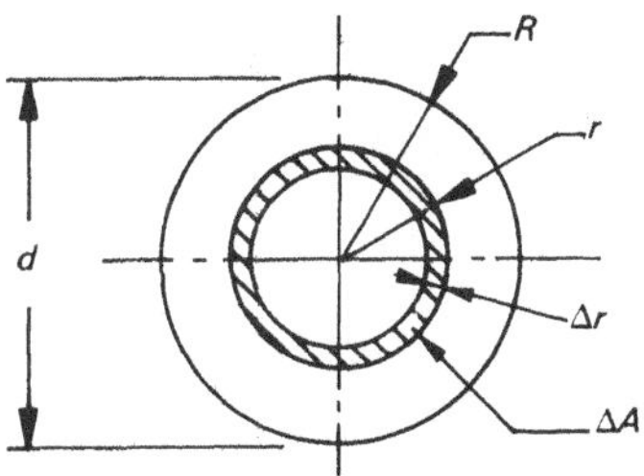

Fig. 14.21

In the limit:

$$J = \int_0^R 2\pi r^3 dr$$

$$= 2\pi \left[\frac{r^4}{4}\right]_0^R$$

$$= \frac{\pi R^4}{2}$$

But $d = 2R$

$$J = \frac{\pi \times d^4}{2 \times 16}$$

$$\therefore \quad \boxed{J = \frac{\pi d^4}{32}}$$

(14.9) polar moment of inertia for a circular section

Notes

1. The polar moment of inertia is an area-based concept just as is the second moment of area and has the same units (mm^4).
2. The polar moment of inertia of a circular section is exactly twice the second moment of area of the same section.

Example 14.7

Determine the polar moment of inertia for a tube section, outside diameter 30 mm and wall thickness 2 mm.

Solution

$$J = \frac{\pi D^4}{32} - \frac{\pi d^4}{32}$$

$$= \frac{\pi}{32}[D^4 - d^4]$$

$$= \frac{\pi}{32}[30^4 - 26^4]$$

$$= \mathbf{34.7 \times 10^3\ mm^4}$$

Problems

14.1-14.6 For each of the symmetrical sections shown in Figures P14.1-P14.6, determine:

(a) area
(b) second moment of area about the horizontal centroidal axis
(c) radius of gyration about the horizontal centroidal axis
(d) second moment of area about the vertical centroidal axis
(e) radius of gyration about the vertical centroidal axis.

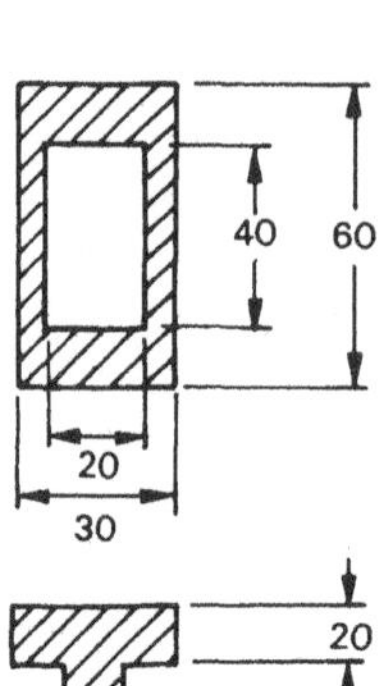

Fig. P14.1

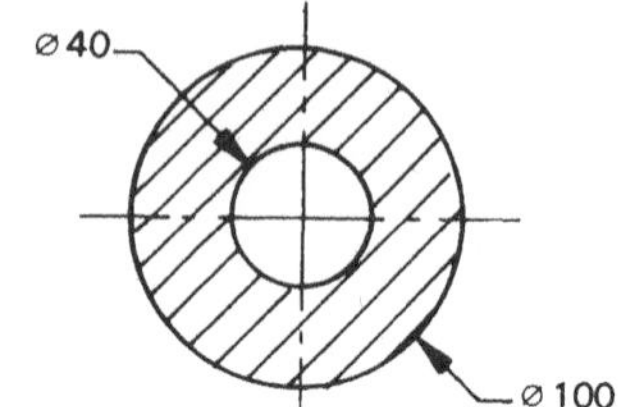

Fig. P14.2

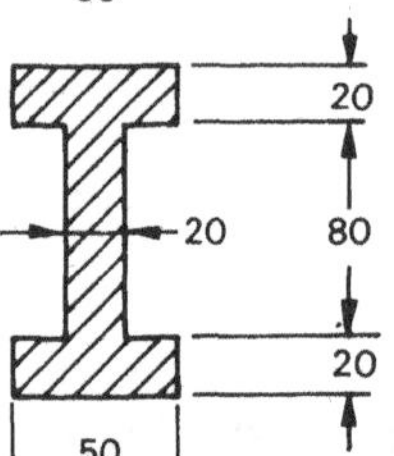

Fig. P14.3

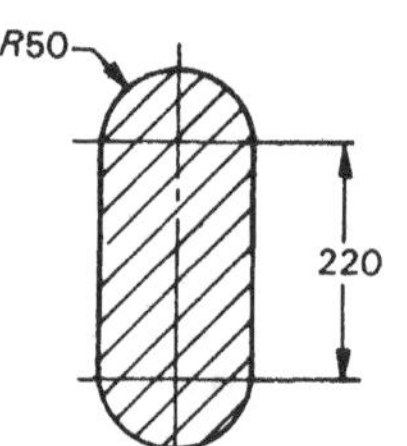

Fig. P14.4

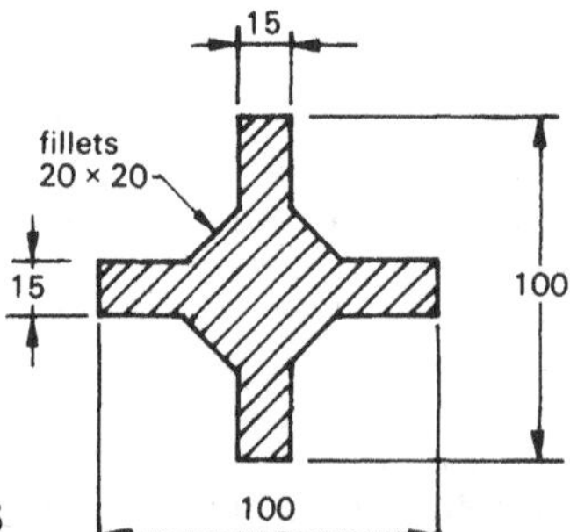

Fig. P14.5

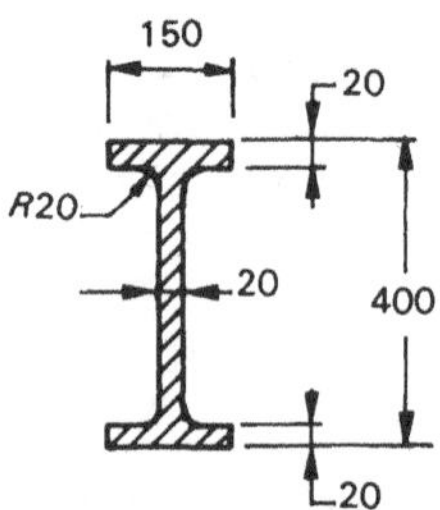

Fig. P14.6

Solutions

Problem	(a) $A \times 10^3$ mm²	(b) $I_{CX} \times 10^6$ mm⁴	(c) k mm	(d) $I_{CY} \times 10^6$ mm⁴	(e) k mm
14.1	1.0	0.433	20.8	0.108	10.4
14.2	6.6	4.78	26.9	4.78	26.9
14.3	3.6	5.92	40.6	0.47	11.4
14.4	29.9	225	86.9	23.2	27.9
14.5	3.575	1.45	20.2	1.45	20.2
14.6	13.54	305	150	11.6	29.2

14.7–14.9 For each of the non-symmetrical sections shown in Figures P14.7–P14.9, determine:
(a) area
(b) vertical centroidal distance from the reference axis x–x
(c) second moment of area about horizontal centroidal axis
(d) radius of gyration about the horizontal centroidal axis
(e) second moment of area about the reference axis x–x
(f) radius of gyration about the reference axis x–x.

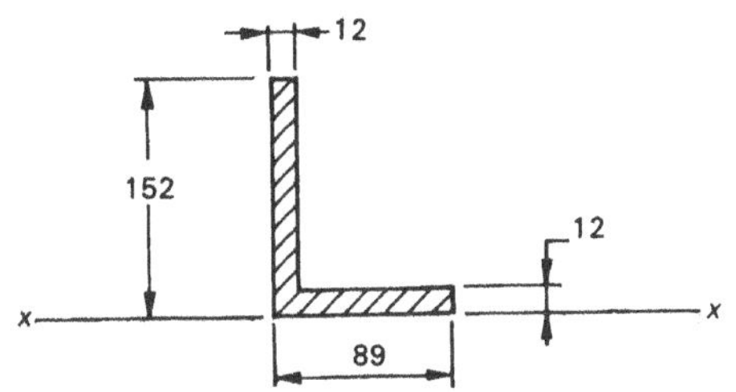

Fig. P14.7

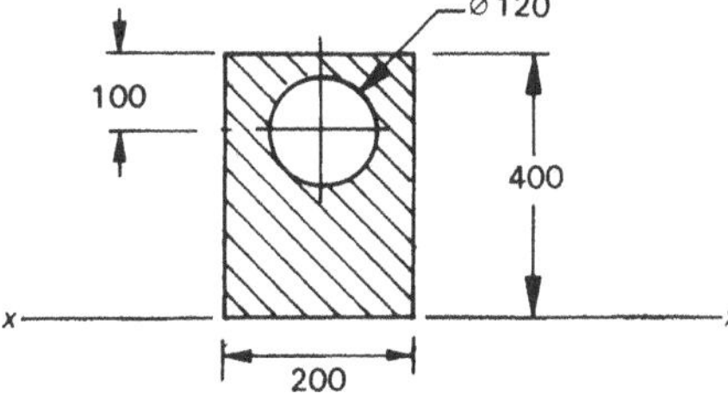

Fig. P14.8

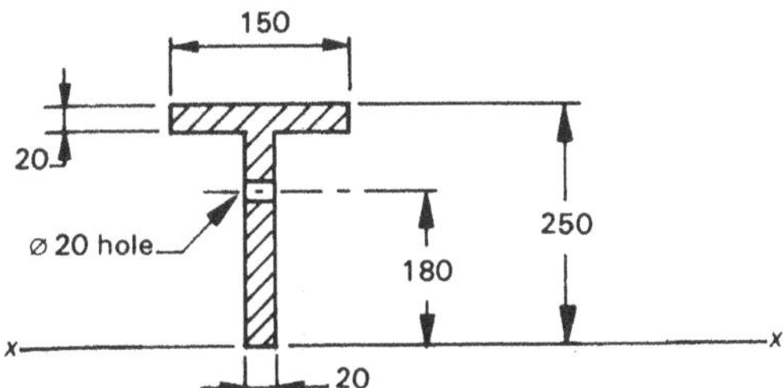

Fig. P14.9

Solutions

Problem	(a) $A \times 10^3$ mm²	(b) $\bar{y}$ mm	(c) $I_{CX} \times 10^6$ mm⁴	(d) k mm	(e) $I_{XX} \times 10^6$ mm⁴	(f) k mm
14.7	2.75	52.5	6.53	48.7	14.1	71.6
14.8	68.7	183.5	925	116	3239	217
14.9	7.2	163.5	48.6	82.2	241	183

14.10–14.12 For each of the composite sections shown in Figures P14.10–P14.12, determine:
(a) area
(b) vertical centroidal distance from the base of the section
(c) second moment of area about the horizontal centroidal axis
(d) radius of gyration about the horizontal centroidal axis.

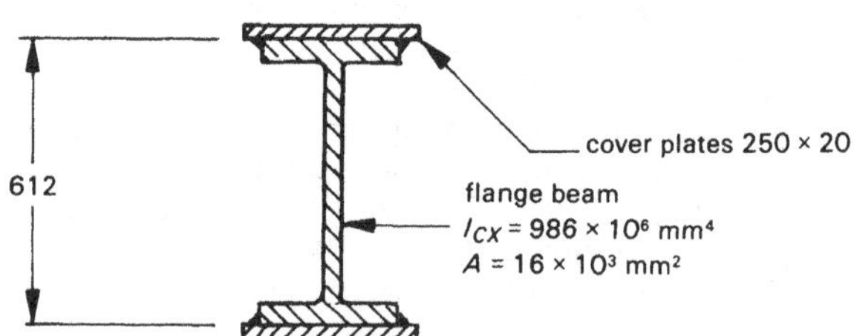

Fig. P14.10

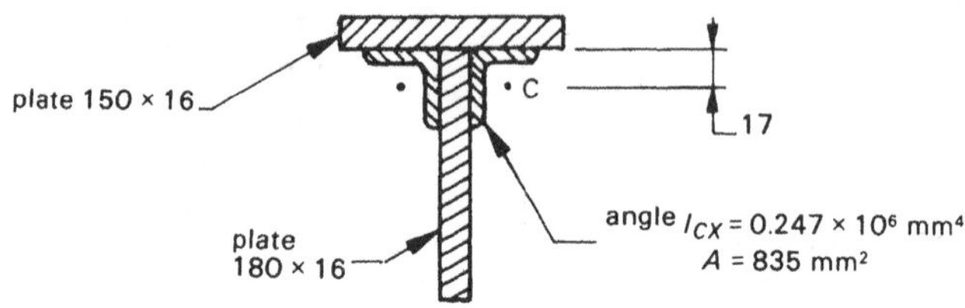

Fig. P14.11

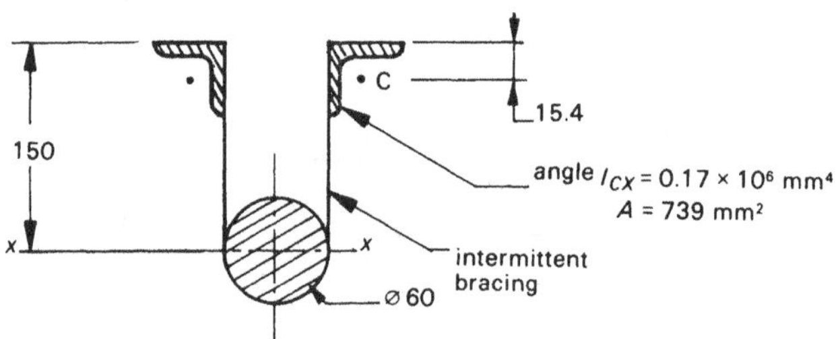

Fig. P14.12

Solutions

Problem	(a) $A \times 10^3$ mm^2	(b) $\bar{y}$ mm	(c) $I_{CX} \times 10^6$ mm^4	(d) k mm
14.10	26.0	326	1985	276
14.11	6.95	141.4	21.9	56.2
14.12	4.305	46.2	18.6	65.7

15

Bending and shear in beams and shafts

A **beam** is a member with transverse loads, that is loads perpendicular to the axis of the member. The loads may be one or more of the following:

1. concentrated force
2. distributed force
   (a) uniform
   (b) non-uniform
3. moment or couple.

The beam support system may be one of the following:

1. simple supports
2. cantilever
3. other (e.g. statically indeterminate).

This book deals only with concentrated, uniformly distributed or moment loads on simply supported beams and cantilevers, other types of loading and supports are not considered. Previous chapters dealt with determination of reactions and the drawing of shear force and bending moment diagrams for these loads and support systems. These are necessary first steps before the bending stress and shear stress in the beam may be determined.

A **shaft** is a member subject primarily to torsional loads, that is a torque applied in a plane perpendicular to the axis. Shafts are usually circular and rotate and transmit power, but for the purpose of stress analysis the shaft may be considered as a static member since a given torque will produce the same stress whether the shaft is rotating or stationary. Previous chapters dealt with torque distribution diagrams which show the variation in torque along the axis of the shaft and are a necessary first step before the torsional stress at any location may be determined.

In practice, a shaft or beam is often subject to combined loading; for example a shaft may be in simultaneous bending, shear and torsion. In this chapter, determination of the individual stresses is considered; the combined effect of stresses is the subject of Chapter 16.

15.1 Bending of beams

Consider a beam of uniform section subject to bending as a result of a concentrated load, distributed load, bending moment or couple, or any combination of these, acting on it. In the unloaded condition assume the beam is horizontal and straight as shown in Figure 15.1(a).

If the bending moment is positive, the beam sags and assumes a curved shape as shown in Figure 15.1(b).

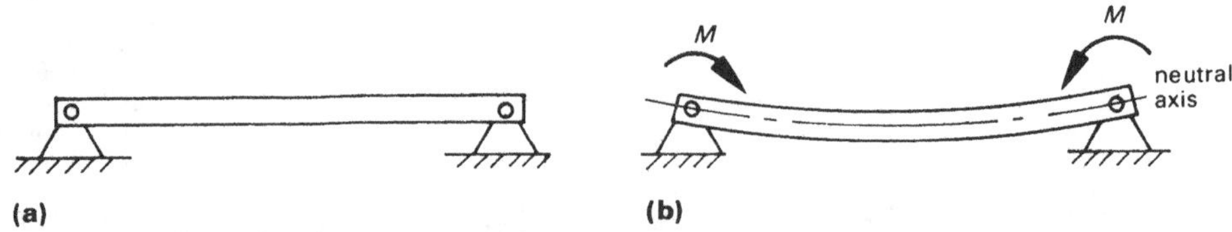

Fig. 15.1 *Bending of a beam: (a) unloaded, (b) loaded*

Experimental evidence shows:

1. The top fibres are shortened and the lower fibres extended, that is the top section of the beam is in axial compression and the lower section in axial tension.
2. There is a centre section called the neutral axis which does not change in length and therefore carries no stress.
3. If the bending moment is constant over any part of the beam, the deformed shape in this part is a curve with constant radius of curvature (i.e. a circle).

15.2 Axial bending stress in beams

Consider a segment of the loaded beam as shown in Figure 15.2.

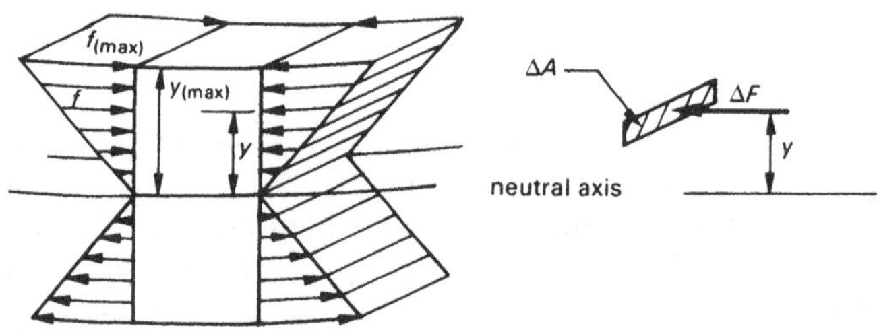

Fig. 15.2 *Bending stress distribution in a segment of a beam*

Above the neutral axis there is compression and below the neutral axis there is tension. Since the neutral axis is a curve with a certain radius of curvature, it follows that the stress distribution is linear since the deformation (strain) varies directly with the distance from the centre of curvature.

The relationship between the stress and the bending moment may be derived as follows:

The external bending moment M is resisted by the sum of the internal bending moments due to the small force ΔF acting on each small area of the beam ΔA. That is:

$$M = \Sigma \, \Delta F y$$

$$= \Sigma f \Delta A y \qquad \textbf{(1)}$$

Since the stress distribution is linear:

$$\frac{f}{y} = \frac{f_{(\max)}}{y_{(\max)}}$$

$$\text{or } f = \frac{f_{(\max)}}{y_{(\max)}} \times y$$

Substituting in (1):

$$M = \Sigma \frac{f_{(max)}}{y_{(max)}} y \,\Delta A\, y$$

$$= \frac{f_{(max)}}{y_{(max)}} \Sigma y^2 \,\Delta A$$

$$= \frac{f}{y} \Sigma y^2 \,\Delta A$$

But from equation 14.3:

$$\Sigma y^2 \Delta A = I \quad \text{(second moment of area)}$$

$$\therefore M = \frac{f}{y} I$$

The subscript b is usually used with f to denote bending stress.

Hence:

$$f_b = \frac{My}{I}$$

(15.1) bending stress

In the design of beams it is often necessary to choose a standard section from those which are available in order to give an allowable bending stress. In this case, equation 15.1 cannot be used directly because it has two unknowns (y and I) and it is more convenient to rewrite the equation as:

$$f_b = \frac{M}{Z}$$

(15.2) bending stress

In equations 15.1 and 15.2:

f_b = axial stress in bending (MPa)

y = distance from the neutral axis to the position at which stress f_b occurs (mm)

I = total second moment of area of the beam section about the neutral axis at the plane being considered (mm^4)

Z = section modulus $= \dfrac{I}{y_{(max)}}$ (mm^3)

M = bending moment at the plane being considered (Nmm)

Notes

1. In most cases the maximum stress is required in which case y is the distance from the neutral axis to the outer fibre of the section. It is always the case when section modulus (equation 15.2) is used.
2. The neutral axis passes through the centroid (centre of area) of the section at the plane being considered.
3. The stress is tension on one side of the neutral axis and compression on the other side. A positive bending moment (sag) will put the top section in compression and the bottom section in tension, whereas the opposite occurs for a negative bending moment.
4. If the section is not symmetrical about the neutral axis, the maximum stress occurs at the outer fibre which is the greatest distance from the neutral axis (see Fig. 15.3).

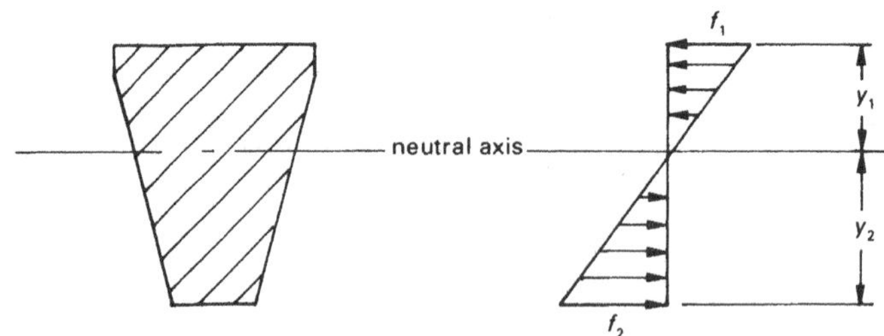

Fig. 15.3 *Bending stress distribution in a non-symmetrical section; $f_2 > f_1$ because $y_2 > y_1$*

5. If the material is weaker in tension than in compression (e.g. cast iron or concrete), the section design may take this into account (see Fig. 15.4).
6. When choosing a standard section shape for a particular application (e.g. flange beam), it is convenient to use handbooks which list properties such as section modulus and second moment of area for the section as well as dimensions and area.

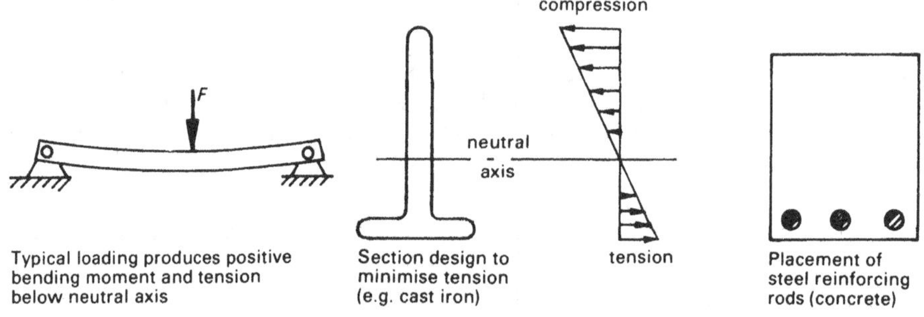

Fig. 15.4 *Design of section for material weaker in tension that in compression*

Example 15.1

A beam which spans 6 m is to be designed for a uniformly distributed load of 5 kN/m.

Determine the required section modulus of a flange beam for this application if the bending stress is not to exceed 90 MPa.

Solution

The loading, reactions and bending moment diagram are as shown in Figure 15.5.

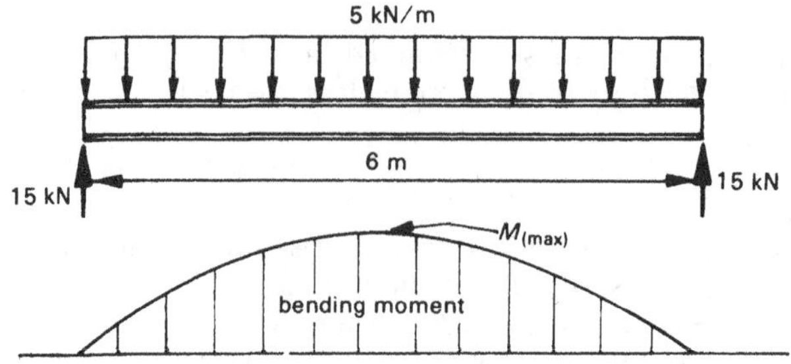

Fig. 15.5

The maximum bending moment occurs at the centre of the span and has a value:

$$M = 15 \times 3 - 5 \times \frac{3^2}{2} = 22.5\ \text{kNm} = 22.5 \times 10^6\ \text{Nmm}$$

Using equation 15.2:

$$f_b = \frac{M}{Z} \therefore Z = \frac{M}{f_b} = \frac{22.5 \times 10^6}{90} = \mathbf{250 \times 10^3\ mm^3}$$

From handbook values it will be found that a suitable beam section is:
200 UB 30 for which $Z = 280 \times 10^3\ \text{mm}^3$.
This is greater than the section modulus required but the next smaller size is:
200 UB 25 for which $Z = 232 \times 10^3\ \text{mm}^3$, which is too small.

Example 15.2

A timber beam as shown in Figure 15.6 is constructed of 200 × 25 boards glued into a tee section and carries a concentrated load of 2 kN.

Determine if the beam is safe in bending if the allowable stresses are 6.2 MPa in compression and 4.5 MPa in tension. Also show the distribution of bending stress over the most highly stressed section.

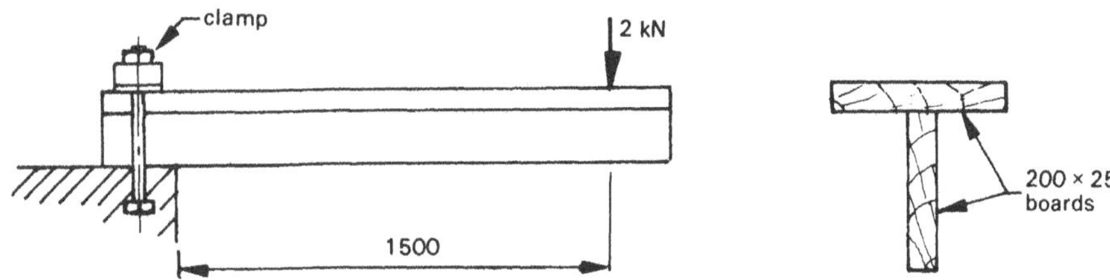

Fig. 15.6

Solution

It is first necessary to determine the properties of the section, namely the centroid location (neutral axis) and the second moment of area. Using the methods outlined in Chapter 14, it will be found that:

$$\bar{y} = 156.2\ \text{mm} \quad \text{(from the base of the section)}$$

$$I_{CX} = 48.6 \times 10^6\ \text{mm}^4$$

The maximum bending stress occurs where the bending moment is a maximum (since the beam is of uniform section throughout its length), namely at the support where:

$$M = 2 \times 10^3 \times 1500 = 3 \times 10^6\ \text{Nmm}$$

The distance from the neutral axis to the top of the beam is:

$$y = 225 - 156.2 = 68.8\ \text{mm}$$

and to the bottom of the beam is 156.2 mm.

The maximum tension stress at the top fibre is:

$$f_b = \frac{My}{I} = \frac{3 \times 10^6 \times 68.8}{48.6 \times 10^6} = \mathbf{4.25\ MPa}$$

The maximum compression stress at the bottom fibre is:

$$f_b = \frac{My}{I} = \frac{3 \times 10^6 \times 156.2}{48.6 \times 10^6} = \mathbf{9.64\ MPa}$$

Therefore the beam is safe in tension but unsafe in compression.

The distribution of bending stress over the most highly stressed section (at the support) is as shown in Figure 15.7.

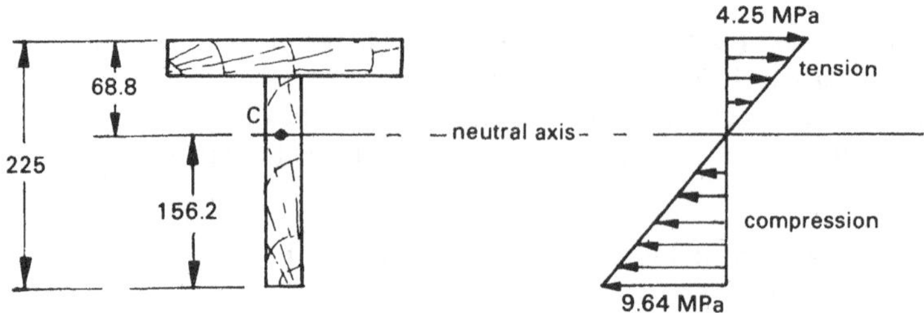

Fig. 15.7

15.3 Curvature and bending moment

It has previously been stated that in any section of a beam in which the bending moment is constant, the deformed curved shape will be the arc of a circle. The relationship between the radius of curvature of this circle and the bending moment will now be derived.

Consider a segment of the beam as shown in Figure 15.8.

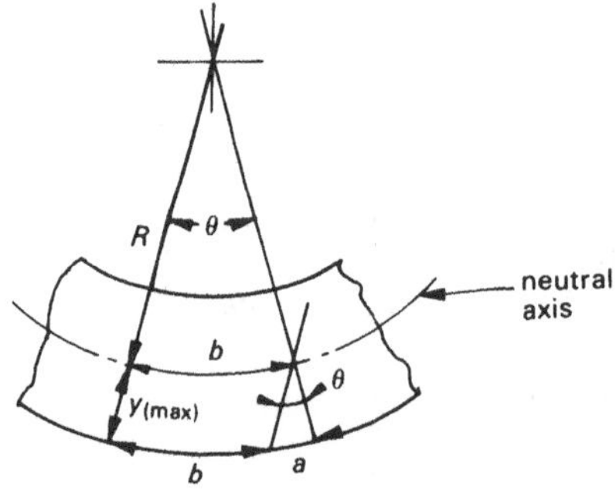

Fig. 15.8 *Curvature of a segment of a beam*

From similarity:

$$\frac{a}{y_{(max)}} = \frac{b}{R} = \theta \text{ (rad)}$$

$$\therefore \frac{a}{b} = \frac{y_{(max)}}{R}$$

But $\frac{a}{b} = \epsilon_{(max)}$ (the maximum strain) and $\epsilon_{(max)} = \frac{f_{(max)}}{E}$

$$\therefore \frac{f_{(max)}}{E} = \frac{y_{(max)}}{R}$$

$$\text{or } f_{(max)} = E\frac{y_{(max)}}{R} = \frac{My_{(max)}}{I}$$

$$\therefore \frac{E}{R} = \frac{M}{I}$$

or

$$\boxed{R = \frac{EI}{M}}$$ **(15.3) radius of curvature of a beam**

Radius of curvature may also be related to stress as follows:

From equation 15.1, $f_b = \dfrac{My}{I} \quad \therefore \dfrac{M}{I} = \dfrac{f_b}{y}$

and from equation 15.3, $\dfrac{M}{I} = \dfrac{E}{R} \quad \therefore \dfrac{f_b}{y} = \dfrac{E}{R}$

$$\therefore \quad \boxed{f_b = \frac{Ey}{R}}$$ **(15.4) bending stress**

Notes

1. R has units mm if E is in MPa, I in mm^4 and M in Nmm.
2. Equation 15.3 was derived for constant bending moment only. However the equation is valid if R is the instantaneous radius of curvature and M the moment at any section, and thus may be applied to beams in which the bending moment varies along the axis of the beam.
3. If the bending moment is zero (e.g. at the support of a simply supported beam or at the free end of a cantilever), the radius of curvature is infinite. That is the elastic curve of the beam is straight in this region.

Example 15.3

A hoist system with a winch drum, diameter 600 mm, uses steel cable made of strands of hardened steel wire.

Determine the maximum diameter of the strands of wire so that the stress due to bending of the wire over the drum does not exceed 100 MPa. The modulus of elasticity of the cable is 110 GPa.

Solution

From equation 15.4:

$$f_b = \frac{Ey}{R}$$

$$\therefore y = \frac{f_b R}{E}$$

$$= \frac{100 \times 300}{110 \times 10^3} = 0.273 \text{ mm}$$

Since y is the distance from the centroid to the outer fibre, this answer is the maximum radius of the wire. Therefore the maximum wire diameter is **0.546 mm**.

Example 15.4

Determine the radius of curvature at the support of the timber beam given in example 15.2 if E for timber is 12 GPa.

Solution

The bending moment at the support is:

$$M = 3 \times 10^6 \text{ Nmm}$$

and $I_{CX} = 48.6 \times 10^6 \text{ mm}^4$

Using equation 15.3:

$$R = \frac{EI}{M} = \frac{12 \times 10^3 \times 48.6 \times 10^6}{3 \times 10^6} = 194.4 \times 10^3 \text{ mm}$$

$\therefore$ $\mathbf{R = 194.4 \text{ m}}$

15.4 Shear stress in beams

If a shear load is applied very close to the support of a block of material, it is usually assumed that the shear stress is uniformly distributed over the section as shown in Figure 15.9(a). However, if the load is moved away from the support so that the block becomes a cantilever beam, bending as well as shear occurs and it is invalid to consider the shear stress as uniformly distributed. In fact, the distribution will be as shown in Figure 15.9(b).

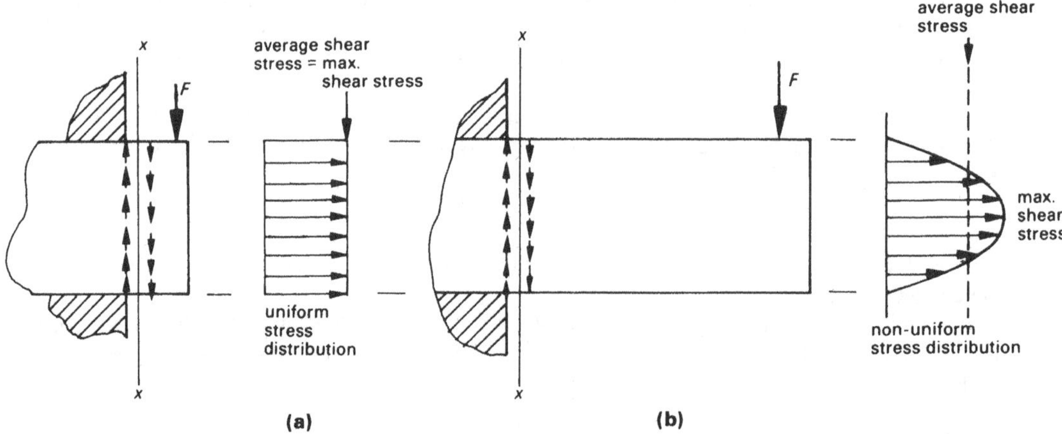

Fig. 15.9 *(a) Direct shear and (b) shear in bending*

In practice completely uniform distribution of shear stress does not occur because shear without bending is impossible since there will always be some clearance between the load and the support and hence some bending.

Furthermore, the vertical shear stress is also accompanied by an equal horizontal shear stress. This is clear from consideration of a small element of material cut at section x-x as shown in Figure 15.10.

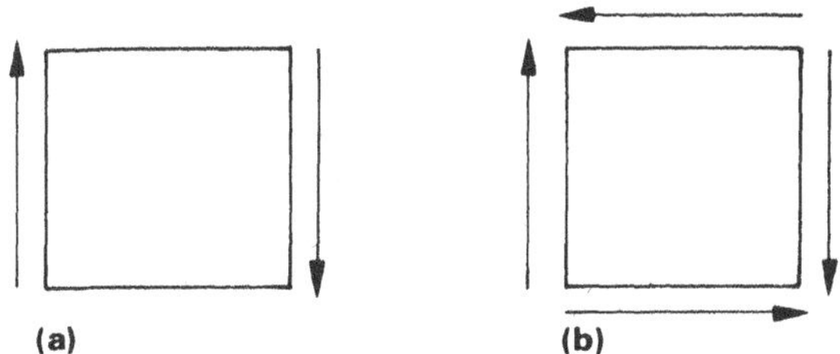

Fig. 15.10 *Small element at section x–x: (a) vertical shear stress only, (b) vertical and horizontal shear stress*

The element (a) cannot be in equilibrium because there is a clockwise couple which is unbalanced. The element can only be in equilibrium if an equal and opposite couple also acts on it as shown by element (b).

The existence of horizontal shear stress is easily demonstrated using a beam made up of a number of planks which are not fastened together so that relative sliding can occur as shown in Figure 15.11.

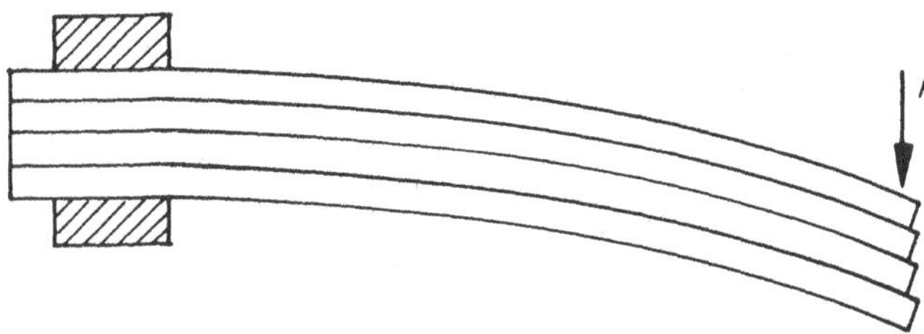

Fig. 15.11 *Demonstration of horizontal shear*

When the load is applied, the planks will slide relative to one another. If this sliding is prevented, for example by glueing the planks together, then the planks would act as a single beam and the glue joints would be in shear stress.

Sign of shear stress

To be consistent with the sign convention adopted previously in connection with shear force and bending moment diagrams for beams, shear stress is considered positive if the vertical shear stress produces a clockwise couple on the element under consideration, and negative if the couple is anticlockwise. Hence positive shear stress acts on the element shown in Figure 15.10.

Note that the horizontal (or resisting) shear stress is always of opposite sign to the vertical (or applied) shear stress.

15.5 Derivation of the shear stress formula

If the bending moment along any length of a beam is constant, then this length of the beam is in **pure bending** and there is no shear force present and no shear stress. Several examples of pure bending are illustrated in Figure 15.12.

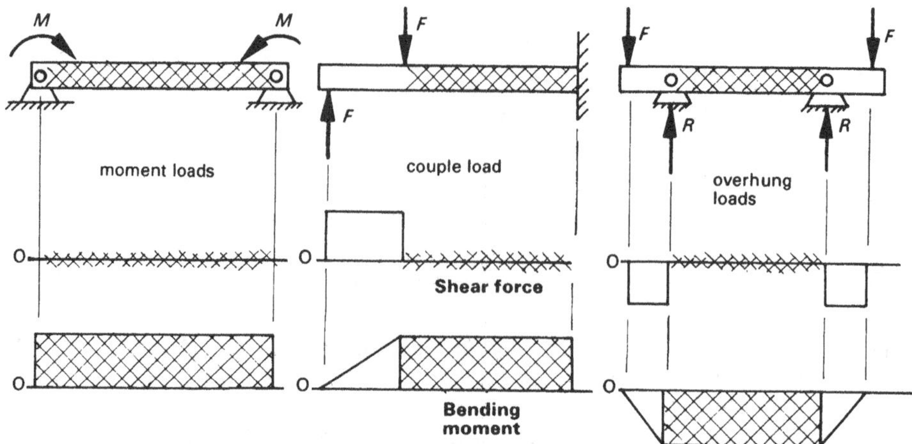

Fig. 15.12 *Pure bending (without shear) occurs in the shaded sections*

In the general case, however, the bending moment varies along the axis of the beam and shear force will occur. The axial stress due to bending on either side of a small length of the beam will be different and equilibrium can occur only if a horizontal shear force is also present as shown in Figure 15.13.

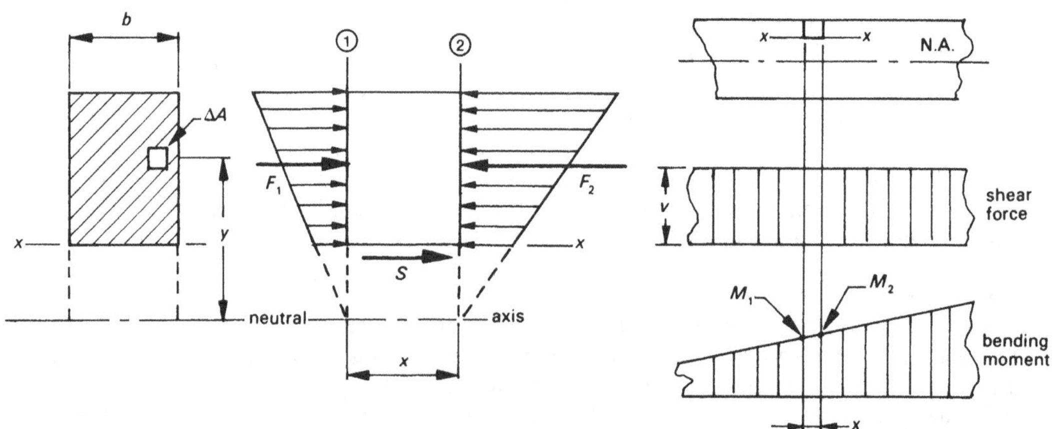

Fig. 15.13 *Horizontal shear force S occurs along x–x because $F_2 > F_1$ since $M_2 > M_1$*

Consider a small area ΔA in plane ① of the beam of width b where the bending moment varies from M_1 to M_2 along the length x.

Since $f_b = \dfrac{My}{I}$ and $F = f_b A$,

$\therefore \Delta F_1 = \dfrac{M_1 y}{I} \Delta A$ and $F_1 = \Sigma \dfrac{M_1 y}{I} \Delta A$ i.e. $F_1 = \dfrac{M_1}{I} \Sigma y \Delta A$

The quantity $\Sigma y \Delta A$ is in fact the first moment of area Q already encountered in Chapter 14 and defined by equation 14.1.

$$\therefore F_1 = \frac{M_1 Q}{I}$$

Similarly, $F_2 = \dfrac{M_2 Q}{I}$

From the equilibrium of the horizontal forces, the shear force S is:

$$S = F_2 - F_1 = \frac{Q}{I}(M_2 - M_1) \qquad \textbf{(1)}$$

Now the slope of the bending moment diagram at any section of a beam is equal to the shear force at that section (refer to the right-hand side of Fig. 15.13).

Calling this vertical shear force V,

$$V = \frac{M_2 - M_1}{x} \quad \text{or} \quad M_2 - M_1 = Vx$$

Substituting in (1):

$$S = \frac{Q}{I} Vx$$

or using the symbol S' for the horizontal shear force per unit length:

$$S' = \frac{S}{x} = \frac{QV}{I}$$

(15.5) horizontal shear force per unit length in a beam

Now divide both sides of this equation by b (the width of the beam):

$$\frac{S}{xb} = \frac{QV}{Ib}$$

But xb = area over which the horizontal shear force S acts and

$\dfrac{S}{xb}$ = horizontal shear stress (also the vertical shear stress) f_S acting at plane x-x.

That is:

$$f_S = \frac{QV}{Ib}$$

(15.6) shear stress in beams

In equations 15.5 and 15.6:

f_S = shear stress (horizontal and vertical) at plane x-x (MPa)
Q = first moment of the cross-sectional area above plane x-x about the neutral axis (mm^3)
S' = horizontal shear force per unit length at plane x-x (Nmm)
I = second moment of area of the entire cross-section of the beam about the neutral axis (mm^4)
b = width of the beam resisting shear at plane x-x (mm)
V = vertical shear force (N)

Notes

1. Although equation 15.6 was derived for horizontal shear stress in beams it also gives the vertical shear stress, because at any part of the beam the horizontal and vertical shear stresses are equal.
2. Equation 15.6 may be used to determine the shear stress at any plane and also the maximum shear stress. The average shear stress over any section is:

$$\bar{f}_S = \frac{V}{A}$$

3. The shear stress at the top and bottom fibres is zero since $Q = 0$ at the outer limits of the section. Maximum shear stress generally occurs at the neutral axis (except in special cases such as the H beam and cross beam). This is, of course, opposite to axial bending stress which is always a maximum at the outer fibres and zero at the neutral axis.
4. If the section is symmetrical about the neutral axis, the shear stress distribution is also symmetrical.

Example 15.5

Determine the shear stress at planes 1 to 6 for the rectangular section beam shown in Figure 15.14 at a position where the shear force is 400 kN. Hence draw the shear stress distribution to scale over the section. Also show the average shear stress on the diagram.

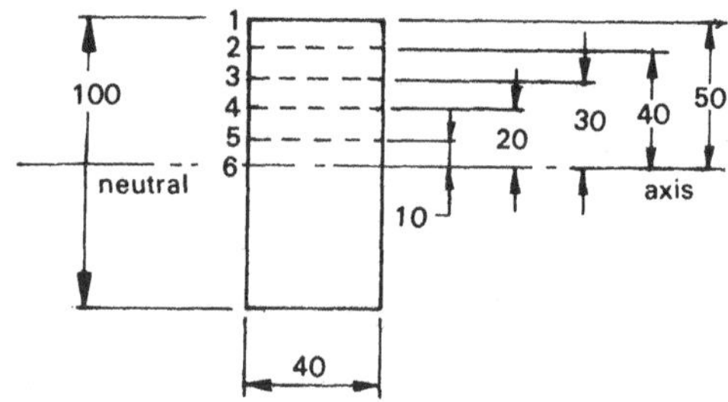

Fig. 15.14

Solution

The area of the section is $40 \times 100 = 4000 \text{ mm}^2$

The average shear stress $\bar{f}_S = \dfrac{V}{A} = \dfrac{400 \times 10^3}{4000} = 100 \text{ MPa}$

The second moment of area about the neutral axis is:

$$I = \frac{bh^3}{12} = \frac{40 \times 100^3}{12} = 3.33 \times 10^6 \text{ mm}^4$$

Using equation 15.6 (with $A\bar{y} = Q$):

$$f_S = \frac{VA\bar{y}}{Ib} = \frac{400 \times 10^3 \times A\bar{y}}{3.33 \times 10^6 \times 40} = 0.003A\bar{y} \text{ MPa}$$

Calculating A and $\bar{y}$ for each of the planes 1 to 6 and substituting in this equation, the table below may be derived.

	Plane					
	1	2	3	4	5	6
A (mm)	0	400	800	1200	1600	2000
$\bar{y}$ (mm)	50	45	40	35	30	25
f_S (MPa)	0	54	96	126	144	150

The shear stress distribution may now be drawn to scale as shown in Figure 15.15.

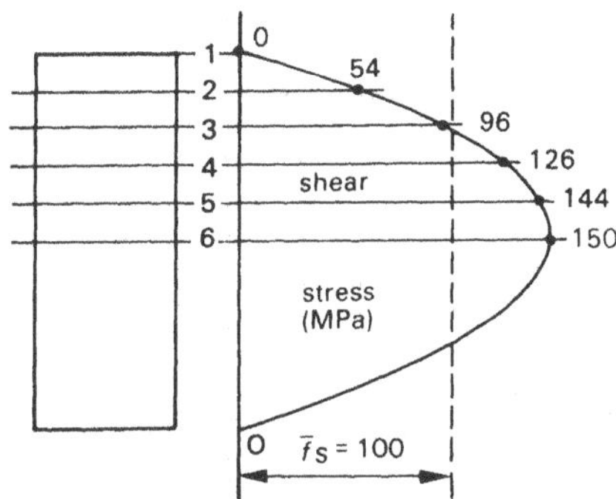

Fig. 15.15

It will be seen that the shear stress distribution is parabolic in shape and that the maximum shear stress occurs at the neutral axis where it is 1.5 times as great as the average shear stress. Hence a 50 per cent error would result in using average shear stress rather than the maximum.

15.6 Distribution of shear stress over a section

In example 15.5 the shear stress distribution for a rectangular section beam was derived and plotted. This is relatively easy to do with rectangular beam sections because the width b at any shear plane is constant and the first moment of area above the shear plane is readily calculated. However, with more complex section shapes, the calculation is rather more difficult but the principles are the same. The calculation of shear stress distribution and the ratio of maximum to average shear stress for some chosen sections yields distributions as shown in Figure 15.16.

Note the sudden change in shear stress which occurs in sections such as the universal beam section at the join between the flange and the web because the width b changes suddenly at this position.

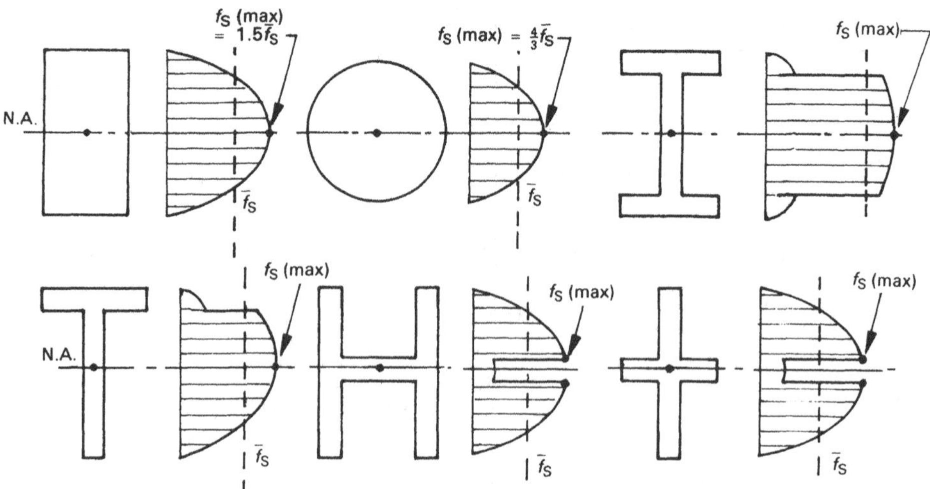

Fig. 15.16 *Shear stress distribution for some section shapes*

15.7 Shear stress in fabricated beams

It will normally be found that for beams made of homogeneous material, the axial bending stress is more critical than the shear stress and failure will usually occur by excessive axial stress rather than by excessive shear stress. If the material is weaker in shear than in axial stress then shear stress would be the critical factor but materials such as cast iron, concrete and ceramics which are weak in shear are also weak in tension and beams made of such materials will generally fail in tension rather than in shear.

However, when a beam is fabricated using bonded, riveted, welded or bolted connections, then the shear stress in the connection should be determined as this is likely to be a critical factor in the strength of the beam. An example of the calculation is given in example 15.6 below.

Example 15.6

A steel beam as shown in Figure 15.17 is fabricated from 100 × 20 steel section using intermittent 8 mm welds. The beam is designed to span 6 m and carry a uniformly distributed load of 8 kN/m.

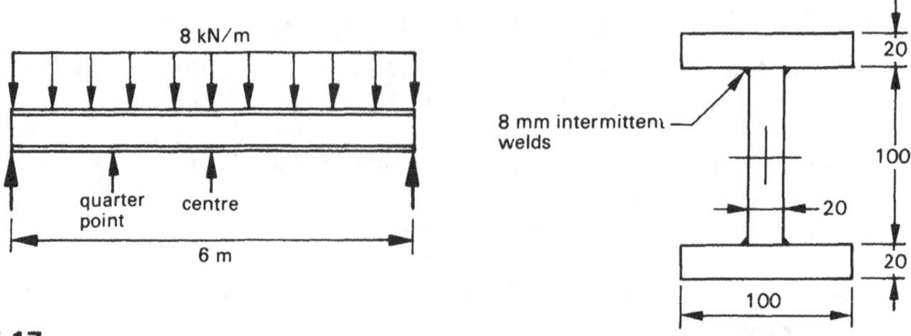

Fig. 15.17

Determine the required weld length of weld per m at (a) centre, (b) quarter points, (c) ends of the span if the shear stress in the welds is not to exceed 100 MPa.

Solution
The shear force diagram is shown in Figure 15.18.

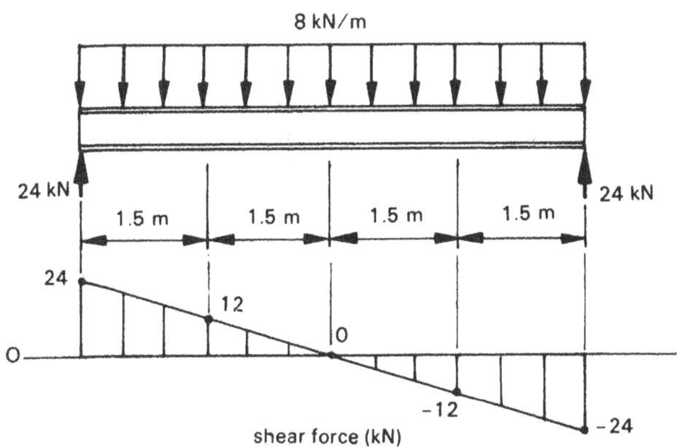

Fig. 15.18

The second moment of area of the beam about the neutral axis is:

$$I = \frac{100 \times 140^3}{12} - 2\left[\frac{40 \times 100^3}{12}\right] = 16.2 \times 10^6 \text{ mm}^4$$

At the plane of the weld, $Q = 20 \times 100 \times 60 = 120 \times 10^3 \text{ mm}^3$

$$b = 0.707 \times 8 \times 2 = 11.312 \text{ mm}$$

From equation 15.6, $f_S = \frac{QV}{Ib} = \frac{120 \times 10^3 \times V}{16.2 \times 10^6 \times 11.312}$

$$= 0.655 \times 10^{-3} V \quad \text{(MPa)}$$

(a) Shear force at the centre of the beam, $V = 0$.
Therefore $f_S = 0$ and no weld is required.

(b) Shear force at the quarter points, $V = 12$ kN.
Therefore $f_S = 0.655 \times 10^{-3} \times 12 \times 10^3$
$= 7.86$ MPa

Since the allowable weld stress is 100 MPa, the weld length per m is:

$$1000 \times \frac{7.86}{100} = \mathbf{79 \text{ mm}}$$

(c) Shear force at the ends of the beam, $V = 24$ kN.
Therefore $f_S = 15.72$ MPa and the required weld length is **158 mm**.

Note
An alternative method of solving this problem is to calculate the horizontal shear force per mm length of the beam using equation 15.5, that is,

$$S' = \frac{QV}{I} = \frac{120 \times 10^3 \times V}{16.2 \times 10^6} = 7.41 \times 10^{-3} V \quad \text{N/mm}$$

Hence at the quarter points, $V = 12$ kN and the shear load per m is:

$$S' = 7.41 \times 10^{-3} \times 12 \times 10^3 \times 1000 = 88.9 \text{ kN}$$

Since the shear stress in the weld is $f_S = \dfrac{S'}{A}$

where A is the area of weld in shear (per m length of beam),

$$\therefore 100 = \frac{88.9 \times 10^3}{0.707 \times 8 \times 2 \times L}$$

$$L = \mathbf{79\ mm} \quad \text{(as before)}$$

15.8 Shafts in torsion

Consider a circular shaft acted upon by a torsional load or torque as shown in Figure 15.19.

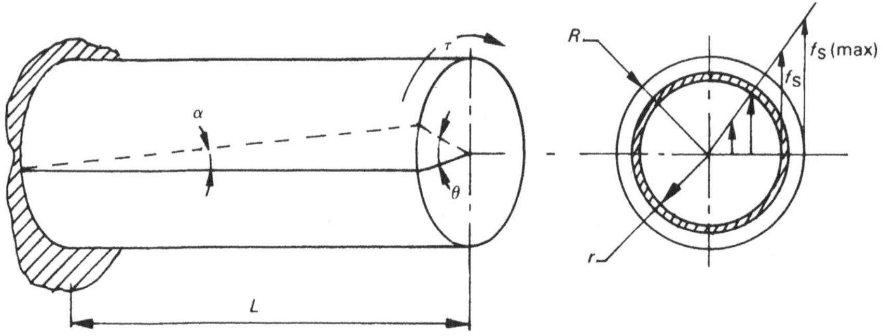

Fig. 15.19 *Torsion in a circular shaft*

It is evident that the shaft will twist in response to the load as shown by the dashed line. Assuming that the deformation is within the elastic limit, the following observations apply:

1. The angle of twist (θ) at any plane or cross-section increases in direct proportion to the distance from the support so that a straight line marked on the undeformed shaft will trace out a helix (angle α) after deformation.
2. If a section is taken through the shaft, it will be evident that the centre of the shaft is undeformed and the outer surface has the greatest deformation. The deformation increases uniformly (in a straight line) from the centre to the outside, hence the stress also increases uniformly from zero at the centre to a maximum, $f_{S(max)}$ at the outer surface.

Torsional stress formula

An equation relating the torsional shear stress and the torque may be derived as follows:

Consider a small annular area ΔA at radius r where the stress is f_S.

By similar triangles,

$$\frac{f_S}{f_{S(max)}} = \frac{r}{R} \quad \therefore f_S = \frac{r}{R} \times f_{S(max)}$$

The force due to this stress is: $\Delta F = f_{S(max)} \times \dfrac{r}{R} \times \Delta A$

Hence the corresponding torque is:

$$\Delta\tau = f_{S(max)} \times \frac{r}{R} \times \Delta A \times r$$

$$= \frac{f_{S(max)}}{R} r^2 \Delta A$$

Summing over the shaft from the centre to the outside:

$$\tau = \frac{f_{S(max)}}{R} \Sigma r^2 \Delta A$$

The quantity $\Sigma r^2 dA$ is called the polar moment of inertia, J (see Chap. 14, sect. 14.10).

$$\therefore \tau = \frac{f_{S(max)}}{R} J$$

or $$\frac{f_{S(max)}}{R} = \frac{\tau}{J}$$

Hence at any radius r, since $\frac{f_S}{r} = \frac{f_{S(max)}}{R}$

$$\frac{f_S}{r} = \frac{\tau}{J}$$

or $$\boxed{f_S = \frac{\tau r}{J}}$$ **(15.7) torsional shear stress**

Angle of twist formula

The angle of twist at any cross-section of the shaft may be derived as follows:

Consider a small length of the shaft at distance ΔL from the support as shown in Figure 15.20.

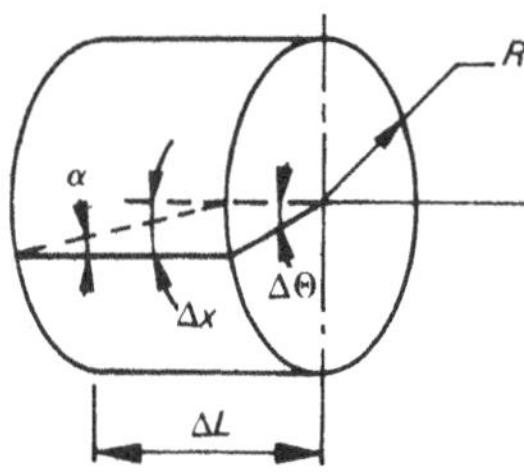

Fig. 15.20 *Twist in a small length of shaft*

The helix angle α is a measure of the shear strain and is related to the shear stress by the modulus of elasticity in shear (modulus of rigidity).

Now $$\frac{f_{(max)}}{\epsilon_S} = G$$

$$\therefore \frac{f_{S(max)}}{\alpha} = G$$

$$\therefore \quad \alpha = \frac{f_{S(max)}}{G} \qquad (1)$$

But $\dfrac{\Delta x}{\Delta L} = \alpha \quad \therefore \Delta x = \alpha \Delta L$

and $\dfrac{\Delta x}{R} = \Delta\theta \quad \therefore \Delta x = R\Delta\theta$

$$\therefore \alpha \Delta L = R\Delta\theta$$

or $\alpha = R\dfrac{\Delta\theta}{\Delta L}$

But $\dfrac{\Delta\theta}{\Delta L}$ is constant because the angle of twist increases in direct proportion to the distance from the support and is equal to $\dfrac{\theta}{L}$.

$$\therefore \alpha = \frac{R\theta}{L}$$

Substituting in (1):

$$\frac{f_{S(max)}}{G} = \frac{R\theta}{L}$$

or $\dfrac{f_{S(max)}}{R} = \dfrac{G\theta}{L}$

But $\dfrac{f_{S(max)}}{R} = \dfrac{\tau}{J}$ (from equation 15.7)

$$\therefore \frac{\tau}{J} = \frac{G\theta}{L}$$

or

$$\boxed{\theta = \frac{\tau L}{GJ}}$$

(15.8) angle of twist in torsion

In equations 15.7 and 15.8:

f_S = torsional shear stress at any radius (MPa)
r = radius at which the torsional shear stress occurs (mm)
θ = angle of twist over a given length of shaft (rad)
L = length of shaft (mm)
τ = applied torque (Nmm)
J = polar moment of inertia of the shaft (mm^4)
G = modulus of rigidity (MPa)

Notes

1. Equations 15.7 and 15.8 are also valid for a hollow shaft using r as the maximum radius and J the polar moment of inertia for a hollow shaft which is $\frac{\pi}{32}(D^4 - d^4)$.
2. Equations 15.7 and 15.8 may also be written as follows:

$$\frac{\tau}{J} = \frac{f_S}{r} = \frac{G\theta}{L}$$

Example 15.7

A torque of 15 kNm is applied to a bronze shaft 8 m in length.

Determine the maximum shear stress and angle of twist if the shaft is:

(a) solid, diameter 100 mm
(b) hollow, outside diameter 100 mm, inside diameter 50 mm.

Assume G for bronze is 40 GPa.

Solution

The torque $\tau = 15 \text{ kNm} = 15 \times 10^6 \text{ Nmm}$

(a) Solid shaft $J = \dfrac{\pi d^4}{32} = \dfrac{\pi \times 100^4}{32} = 9.8175 \times 10^6 \text{ mm}^4$

The radius at which maximum stress occurs is 50 mm.

$$\text{From (15.7) } f_S = \frac{\tau r}{J}$$

$$= \frac{15 \times 10^6 \times 50}{9.8175 \times 10^6}$$

$$= \mathbf{76.4\ MPa}$$

$$\text{From (15.8) } \theta = \frac{\tau L}{GJ}$$

$$= \frac{15 \times 10^6 \times 8000}{40 \times 10^3 \times 9.8175 \times 10^6}$$

$$= 0.3056 \text{ rad}$$

$$= \mathbf{17.5^\circ}$$

$$\text{(b) Hollow shaft } J = \frac{\pi}{32}(D^4 - d^4)$$

$$= \frac{\pi}{32}(100^4 - 50^4)$$

$$= 9.2 \times 10^6 \text{ mm}^4$$

The radius at which maximum stress occurs is the same as for the solid shaft, 50 mm.

Substituting in equation 15.7,

$$f_S = \frac{15 \times 10^6 \times 50}{9.2 \times 10^6}$$

$$= \mathbf{81.5\ MPa}$$

Substituting in equation 15.8,

$$\theta = \frac{15 \times 10^6 \times 8000}{40 \times 10^3 \times 9.2 \times 10^6}$$

$$= 0.3261 \text{ rad}$$

$$= \mathbf{18.7^\circ}$$

Note that the hollow shaft has 75 per cent of the mass of the solid shaft but 94 per cent of the strength. It is evident, therefore, that a hollow shaft is more efficient in torsion than a solid shaft, just as a flange beam is more efficient in bending than a solid rectangular section. This is because, in both cases, the outer fibres of the section carry the greatest stress and the centre section carries no stress. Area around the centre therefore contributes mass without proportional load-carrying capability.

Example 15.8

A motor develops 90 kW maximum power at 1800 rpm. The output shaft is keyed to a flywheel/clutch.

Determine the shaft diameter required if the allowable shear stress in the shaft (unkeyed) is 60 MPa.

Note: Reduce the allowable stress by 25 per cent to allow for the keyway.

Solution

Torque and power are related by equation 9.7,

$$P = \tau\omega$$

$$\therefore \tau = \frac{P}{\omega} = \frac{90 \times 10^3}{\pi \times \frac{1800}{30}} = 477.5 \text{ Nm} = 477.5 \times 10^3 \text{ Nmm}$$

It is not possible to substitute in equation 15.7, $f_S = \frac{\tau r}{J}$, directly because there are two unknowns, r and J. However the equation may readily be transposed as follows:

$$f_S = \frac{\tau r}{J} = \frac{\tau d/2}{\pi d^4/32} = \frac{16\tau}{\pi d^3}$$

$$\therefore d = \sqrt[3]{\frac{16\tau}{\pi f_S}}$$

The substitution may now be made:

$$\therefore d = \sqrt[3]{\frac{16 \times 477.5 \times 10^3}{\pi \times 60 \times 0.75}}$$

$$= \mathbf{37.8\ mm}$$

15.9 Torsion in shafts with varying torque

When a shaft has a number of power take-off or power input positions, for example when it is fitted with pulleys or gears, the torque varies along the length of the shaft and hence the shear stress and angle of twist also vary. The maximum stress and angle of twist are best determined by first drawing a torque distribution diagram (refer to sect. 5.10).

As previously adopted, the convention for torque is that torque is positive if in the clockwise direction when viewed from the left-hand end of the shaft.

Example 15.9

The geared steel shaft shown in Figure 15.21 has an input power of 45 kW at gear C and rotates at 200 rpm. The power taken off at each of the gears A, B and D is shown.

The gears are attached to the shaft by keys and the shaft is supported by bearings (not shown) which absorb negligible power. The bearings are located close to the gears so there is negligible bending of the shaft. Allowable shear stress in the shaft is 80 MPa and G may be taken as 80 GPa.

(a) Draw the torque distribution diagram for the shaft.
(b) Determine the shaft diameter required to the nearest whole mm size (decreasing allowable stress by 25 per cent to account for the keyways).
(c) Determine the angle of twist in each portion of the shaft and hence the total twist of gear C relative to gear A and also the twist of gear D relative to gear A.

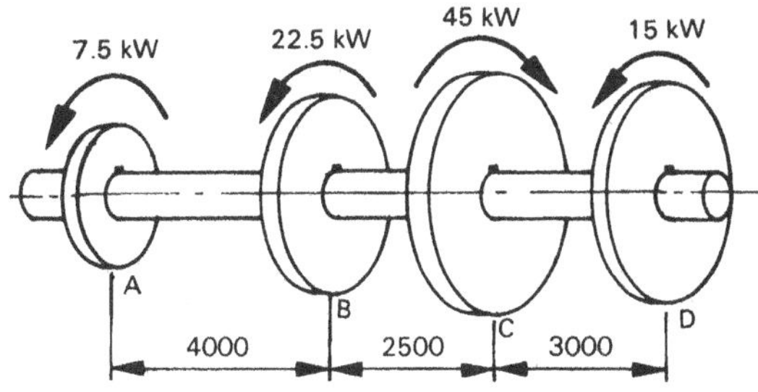

Fig. 15.21

Solution

(a) Viewing the shaft from the left-hand end (gear A), clockwise torque and angular twist are positive.

$$\text{Since } P = \tau\omega \quad \therefore \tau = \frac{P}{\omega} = \frac{P}{\pi \times \frac{N}{30}} = \frac{30P}{\pi N}$$

Now $N = 200$ rpm

$$\therefore \tau = \frac{30P}{\pi \times 200} = 0.0477 \text{ P}$$

At pulley A	$\tau = 0.0477 \times 7500$	=	358 Nm
B	$\tau = 0.0477 \times 22\,500$	=	1074 Nm
C	$\tau = 0.0477 \times -45\,000$	=	−2148 Nm
D	$\tau = 0.0477 \times 15\,000$	=	716 Nm
	Check total torque		0

Working from the left-hand end of the shaft, the torque in each section may now be calculated:

Position	*Torque* (Nm)
A_L	0
$A_R - B_L$	358
$B_R - C_L$	358 + 1074 = 1432
$C_R - D_L$	1432 − 2148 = −716
D_R	−716 + 716 = 0

The torque distribution diagram may now be drawn (Fig. 15.22).

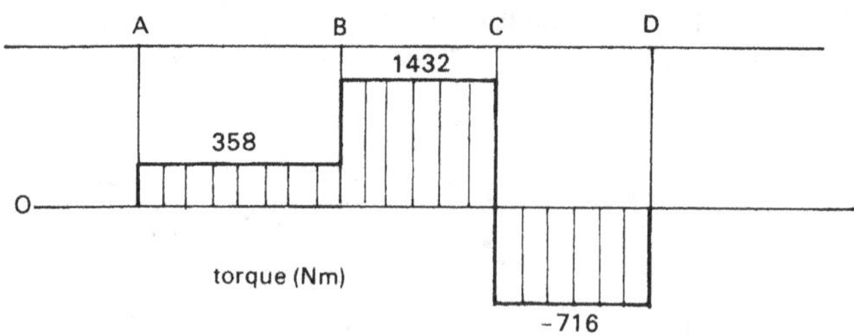

Fig. 15.22

(b) In order to determine the shaft diameter, transpose equation 15.7 as shown in example 15.8:

$$d = \sqrt[3]{\frac{16\tau}{\pi f_S}}$$

The maximum torque is $\tau = 1432$ Nm.
The allowable stress is multiplied by 0.75 because of the keyways.
Substituting,

$$d = \sqrt[3]{\frac{16 \times 1432 \times 10^3}{\pi \times 80 \times 0.75}}$$

$$= 49.5 \text{ mm } \textbf{say 50 mm}$$

(c) The angle of twist in each section of the shaft may now be determined using equation 15.8:

$$\theta = \frac{\tau L}{GJ}$$

In this equation τ is in Nmm and L in mm, so that if τ is in Nm and L in m the equation is:

$$\theta = \frac{\tau L \times 10^6}{GJ}$$

For a 50 mm circular section:

$$J = \frac{\pi d^4}{32} = \frac{\pi \times 50^4}{32} = 614 \times 10^3 \text{ mm}^4$$

and $G = 80 \text{ GPa} = 80 \times 10^3 \text{ MPa}$

$$\therefore \theta = \frac{\tau L \times 10^6}{80 \times 10^3 \times 614 \times 10^3}$$

$$\theta = \frac{\tau L}{49\,087}$$

Substituting in this equation, the various angles may now be calculated as tabulated below:

Position	*Torque* (Nm)	*Length* (m)	θ (rad)	$\theta°$
A_L	0	—	0	0
$A_R - B_L$	358	4	0.0292	1.67
$B_R - C_L$	1432	2.5	0.0729	4.18
$C_R - D_L$	−716	3.0	−0.0438	−2.51
D_R	0	—	0	0

Twist of C relative to A is 1.67 + 4.18 = **5.85°**

Twist of D relative to A is 5.85 − 2.51 = **3.34°**

Problems

15.1 A cantilever beam has a tee section shape and is loaded as shown in Figure P15.1.

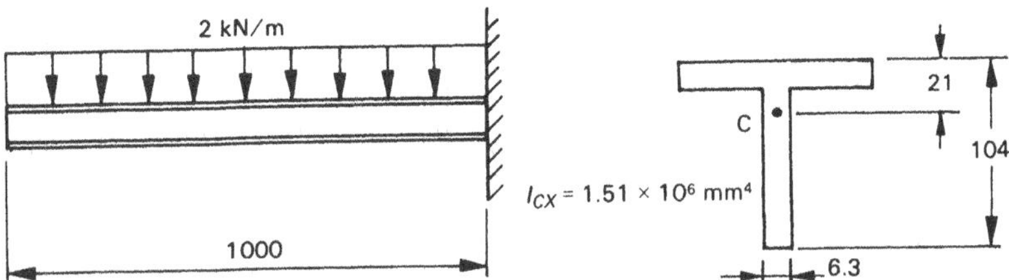

Fig. P15.1

Determine the maximum tension and compression stress due to bending.

13.9 MPa tension, 55 MPa compression

15.2 If the tee section shown in Figure P15.1 was used as a simply supported beam carrying the same load over the whole span, determine the maximum span in order that the axial stress (tension or compression) does not exceed 90 MPa.

2.56 m

15.3 Determine the maximum shear stress (at the neutral axis) for the beam given in problem 15.1.

4.56 MPa

15.4 Determine the required section modulus of a beam to span 4 m and carry a uniformly distributed load of 6 kN/m if the bending stress is not to exceed 110 MPa.

109×10^3 mm³

15.5 If the beam given in problem 15.1 is made of steel E = 200 GPa, determine the radius of curvature at the support.

302 m

15.6 A steel cable is made of wires 0.6 mm in diameter and modulus of elasticity 110 GPa.

Determine the minimum diameter of a drum over which the cable is wound in order that the stress due to bending in the wire does not exceed 120 MPa.

550 mm

15.7 A band saw uses a 0.8 mm thick hardened steel band. The pulleys at each end are 900 mm in diameter and adjusted so that the tension in the straight portion of the band is 50 MPa.

Determine the maximum tensile stress in the bent portion of the band. Assume $E = 200$ GPa.

228 MPa

15.8 The cantilever beam as shown in Figure P15.8 is of rectangular section and is subject to a concentrated load of 5 kN.

Determine:

(a) maximum tensile stress and where it occurs

(b) maximum shear stress and where it occurs.

(a) 360 MPa top surface at support (b) 1.5 MPa neutral axis along span

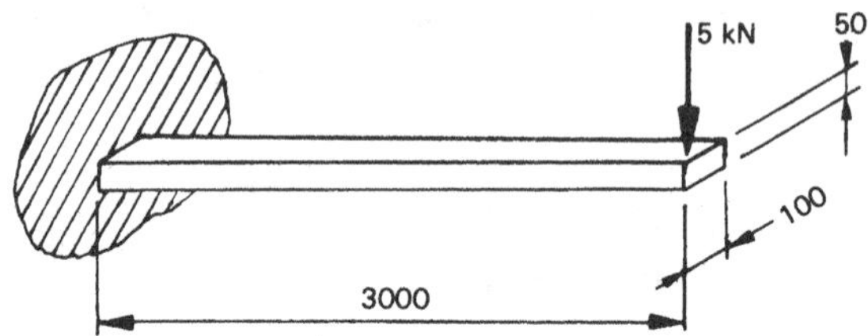

Fig. P15.8

15.9 Repeat problem 15.8 if the beam were simply supported and carried the load at the centre.

(a) 90 MPa bottom surface at centre (b) 0.75 MPa neutral axis along span

15.10 A timber beam as shown in Figure P15.10 is built up using three 100×20 mm planks glued together. It spans 4 m and carries a concentrated load of 15 kN at the midpoint.

Determine:

(a) second moment of area about the neutral axis

(b) shear stress in the glued joint.

(a) 16.2×10^6 mm^4 (b) 2.78 MPa

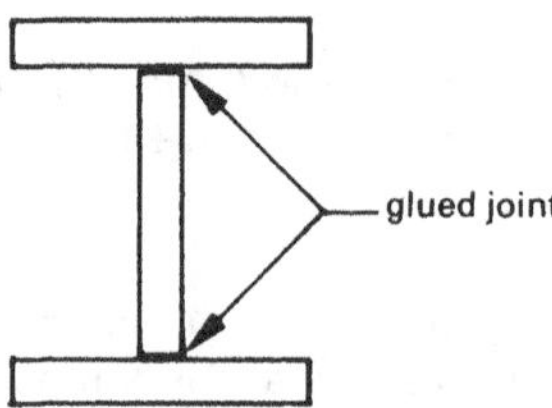

Fig. P15.10

15.11 If the beam in problem 15.10 were joined with 3 mm nails rather than glue, determine the required spacing of the nails so that the shear stress in the nails does not exceed 100 MPa.

12.7 mm

15.12 A timber beam, section 300 × 150 mm (height × width), spans 6 m and carries four concentrated loads of 12 kN equally spaced as shown in Figure P15.12.

(a) Draw the shear force and bending moment diagrams.
(b) Determine the maximum shear stress and state where it occurs.
(c) Determine the maximum bending stress and state where it occurs.

(b) 0.4 MPa neutral axis along 2 m from each end
(c) 10.7 MPa outer top and bottom surface along central 2 m section.

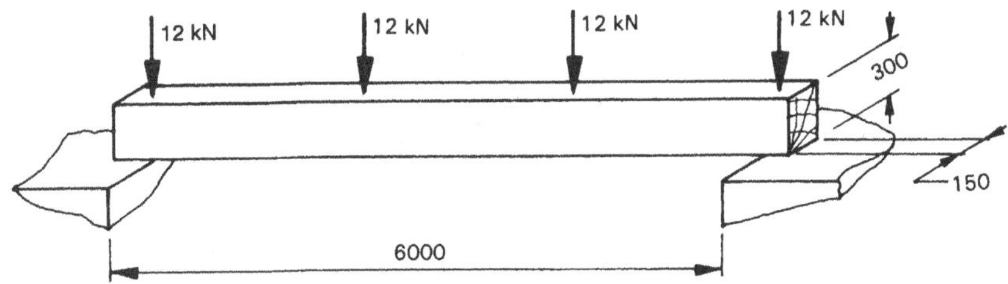

Fig. P15.12

15.13 Repeat problem 15.12 if the beam had a circular rather than rectangular section, diameter 250 mm.

(b) 0.326 MPa neutral axis along 2 m from each end
(c) 15.6 MPa outer top and bottom surface along central 2 m section.

15.14 A flange beam section 200 UB 25 has cover plates welded to the flanges as shown in Figure P15.14. The welds are 6 mm intermittent of length 24 mm and the permissible shear stress in the welds is 120 MPa.

If the beam is loaded as shown, determine the required weld spacing at (a) ends and (b) quarter points of the span.

(a) 118 mm (b) 236 mm

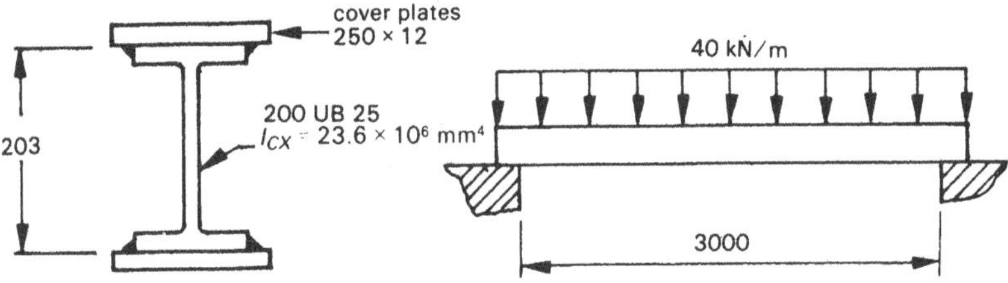

Fig. P15.14

15.15 Repeat problem 15.14 if rivets, diameter 10 mm, are used to attach the cover plates instead of welds. The shear stress in the rivets is not to exceed 80 MPa.

Determine the rivet spacing for each pair of rivets on the top flange or bottom flange at (a) ends and (b) quarter points.

(a) 60.4 mm (b) 120.8 mm

15.16 A tee section steel beam is fabricated from 100 × 20 mm plates welded continuously as shown in Figure P15.16.

Determine:

(a) second moment of area about the neutral axis
(b) maximum shear force which may be applied to the beam so that the shear stress in the welds does not exceed 110 MPa.

(a) 5.33×10^6 mm^4 (b) 82.9 kN

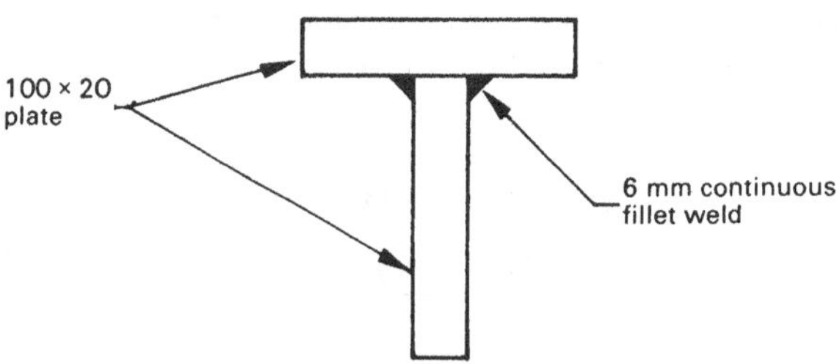

Fig. P15.16

15.17 For the composite bonded beam carrying loads as shown in Figure P15.17, draw the shear force and bending moment diagrams.

Hence determine:

(a) maximum shear stress in the bonded joints
(b) maximum axial stress due to bending.

(a) 0.128 MPa (b) 3.26 MPa

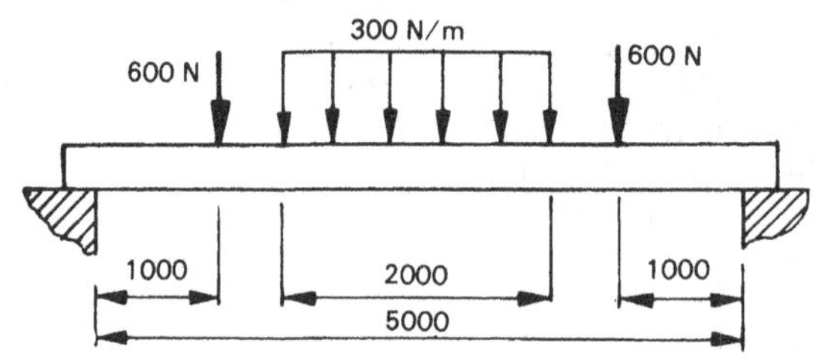

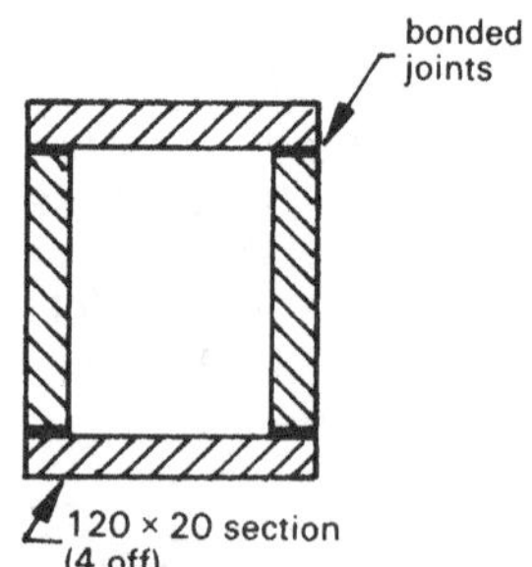

Fig. P15.17

15.18 A shaft, diameter 45 mm, transmits 200 kW at 1600 rpm.

Determine:

(a) maximum shear stress
(b) angle of twist in degrees per 20 shaft diameters length if $G = 80$ GPa.

(a) 66.7 MPa (b) 1.91°

15.19 A machine shaft is 30 mm in diameter and length 2500 mm.

Determine:

(a) maximum torque which may be applied to the shaft so that the shear stress does not exceed 50 MPa
(b) angle of twist at maximum torque if the modulus of rigidity is 80 GPa
(c) power which may be transmitted by the shaft at 1440 rpm.

(a) 265 Nm (b) 5.97° (c) 40 kW

15.20 Repeat problem 15.19 if the shaft is hollow with outside diameter 45 mm and inside diameter 35 mm, all other factors being the same.

(a) 567 Nm (b) 3.98° (c) 85.5 kW

15.21 A steel shaft is to transmit 1 MW at 240 rpm.

Determine the required shaft diameter if the following conditions are to be satisfied:

(a) The shear stress is not to exceed 80 MPa.

(b) The angle of twist is not to exceed one degree per 15 diameters shaft length (assuming $G = 80$ GPa).

163 mm ((a) 136 mm (b) 163 mm)

15.22 The tailshaft of a motor vehicle is made of steel tube, outside diameter 80 mm and wall thickness 4 mm.

Determine:

(a) maximum torque which may be transmitted so that the shear stress does not exceed 70 MPa

(b) angle of twist per m length when the shaft is transmitting maximum torque (assuming $G = 80$ GPa).

(a) 2.42 kNm (b) 1.25°

15.23 The shaft given in problem 5.13 is 15 mm diameter and made of steel ($G = 80$ GPa).

Determine:

(a) maximum shear stress

(b) angle of twist of C relative to A and D relative to A.

(a) 211 MPa (b) −2.16°, −14.3°

15.24 The shaft given in problem 5.17 is 18 mm diameter and made of steel ($G = 80$ GPa).

Determine:

(a) angle of twist of pulley C relative to pulley D

(b) angle of twist of pulley E relative to pulley D.

(a) 0.652° (b) −0.141°

16

Combined stress

In the analysis of stress so far, each stress has been calculated individually. In many cases, however, stresses occur simultaneously in a member, e.g. most sections of a beam are usually subjected to combined bending and shear stress.

When simultaneous stresses occur, it is possible that the combined effect produces a stress greater than any of the stresses occurring separately, in the same way that a number of forces may produce a resultant force greater than any of the individual forces. In such cases, serious underestimation of the stress may occur if the combined stress is not calculated.

The analysis of combined stress is simplified using the principle of superposition, which states that the combined stress may be determined by summation of each stress calculated separately, that is as if each stress occurred alone. Note that summation of the stresses must be in vector terms having due regard to the nature and direction of the stress and is not simply arithmetic summation of the stresses.

In the most complex cases, a member may be subjected to axial, normal and shear stress in three dimensions. A three-dimensional analysis is beyond the scope of this book which deals only with analysis of coplanar combined stresses, that is, a two-dimensional analysis. Although a two-dimensional approach may appear restrictive, in practice most situations may be adequately analysed by this method because there is usually a plane of maximum stress (which may be determined by inspection or by trial).

16.1 Combined bending and axial stress

One of the simplest cases of combined stress occurs when a member is subjected to simultaneous bending and axial stress. Bending produces tension and compression stresses which are axial and therefore may be added algebraically to the externally applied axial stress.

Consider a cantilever beam as shown in Figure 16.1 loaded simultaneously by a bending moment M and an axial load F.

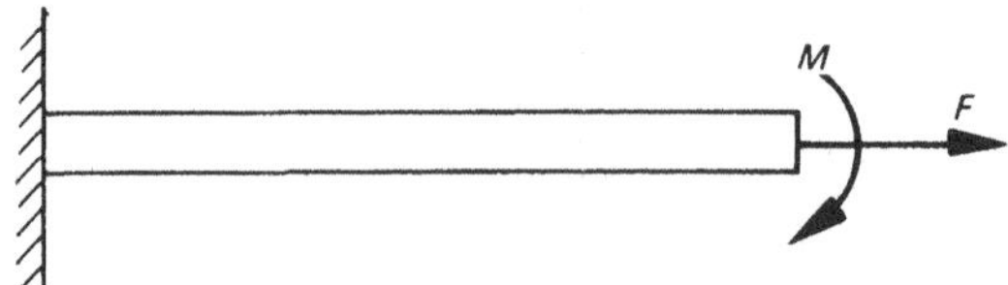

Fig. 16.1 *Simultaneous bending and axial loads on a cantilever beam*

The axial load produces axial tension stress which is constant along the length of the beam. The bending moment produces bending stress, that is tension above the neutral axis and compression below and these stresses have a maximum value at the outer fibres of the section. Using the principle of superposition, the bending stress and axial tension stress may be considered separately and then combined. This is shown in Figure 16.2.

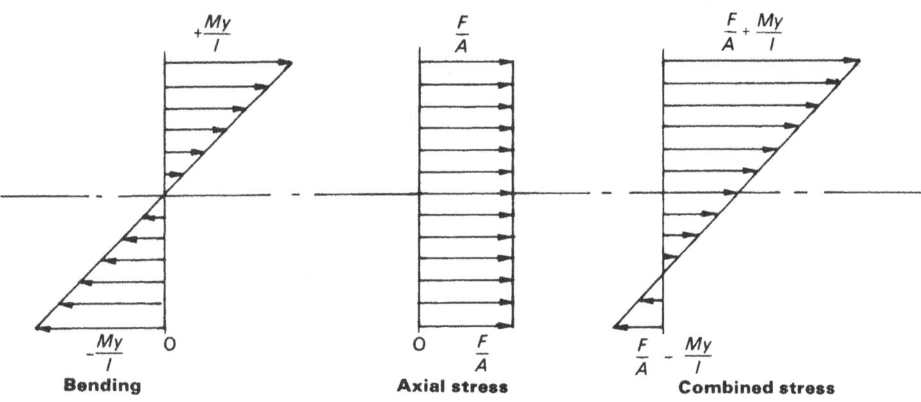

Fig. 16.2 *Combined bending and axial stress*

Note that the neutral axis (that is the axis passing through the centroid of the section) is no longer the plane of zero stress as it would be if bending moment alone acted on the beam.

Example 16.1

For the cantilever beam shown in Figure 16.3, determine the maximum combined tension and compression stress due to the applied load and the load due to the weight of the beam itself. Also determine the distance from the support beyond which there is no combined compression stress in the beam.

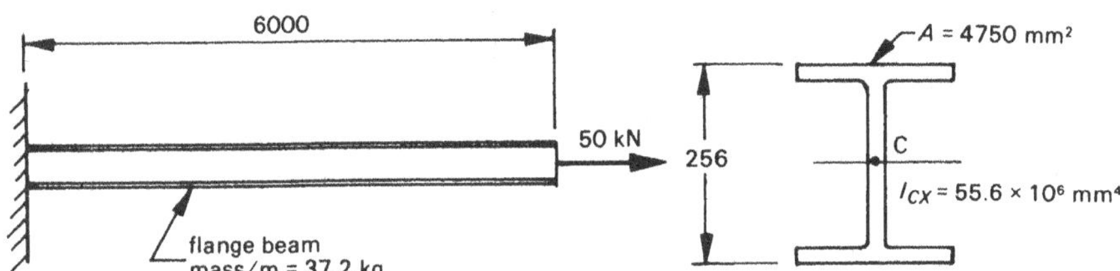

Fig. 16.3

Solution

The distributed load due to the weight of the beam itself is:

$$37.2 \times 9.81 = 365 \text{ N/m}$$

The bending moment is a maximum at the support and is:

$$M = \frac{WL^2}{2} = \frac{365 \times 6^2}{2} = 6570 \text{ Nm}$$

The maximum bending stress (at the support) is given by:

$$f_b = \frac{My}{I} = \frac{6570 \times 10^3 \times 128}{55.6 \times 10^6} = 15.1 \text{ MPa}$$

The direct tension stress is uniform over the span and is given by:

$$f = \frac{F}{A} = \frac{50 \times 10^3}{4750} = 10.5 \text{ MPa}$$

The stress distribution at the most highly stressed section (the support) may now be drawn (Fig. 16.4.).

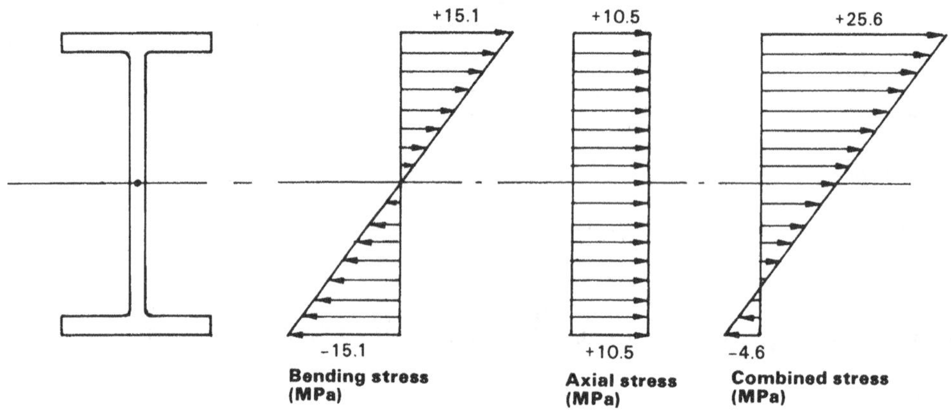

Fig. 16.4

Hence the maximum combined tension stress is **25.6 MPa** and the maximum combined compression stress is **4.6 MPa** and these occur at the top and bottom fibres respectively at the support.

The position at which no combined compression stress occurs is to the right of the support where the bending stress is 10.5 MPa or less, that is, where the bending moment is less than:

$$M = \frac{10.5}{15.1} \times 6570 = 4570 \text{ Nm}$$

The position where this occurs is given by:

$$\frac{WL^2}{2} = 4570$$

$$\therefore \frac{365 \times L^2}{2} = 4570$$

$$\therefore L = 5 \text{ m} \quad \text{(from the free end of the beam).}$$

Hence there is no combined compression stress to the right of the position **1 m** from the support.

Example 16.2

The tube shown in Figure 16.5 is pinned at A and free to slide at support B (which may be considered frictionless).

Determine the load F in order that the maximum combined axial stress in plane C is 80 MPa.

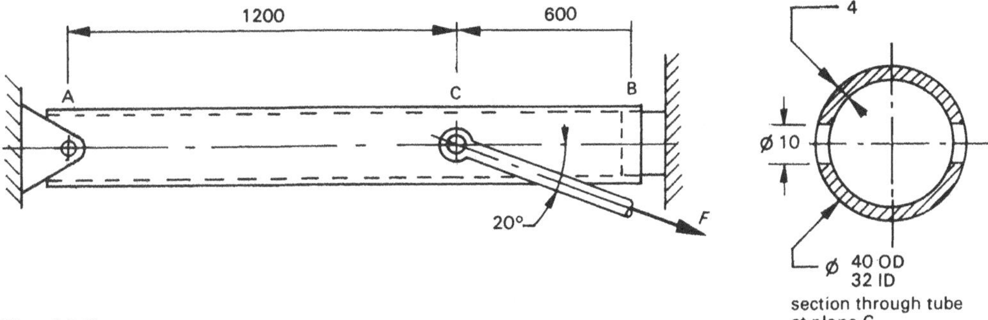

Fig. 16.5

Solution

In the plane C the cross-sectional area (closely enough) is:

$$A = \frac{\pi}{4}(40^2 - 32^2) - 2 \times 4 \times 10 = 372.4 \text{ mm}^2$$

and the second moment of area (closely enough) is:

$$I = \frac{\pi}{64}(40^4 - 32^4) - 2 \times \frac{4 \times 10^3}{12} = 73.5 \times 10^3 \text{ mm}^4$$

Load F has a horizontal component $F \cos 20°$ and a vertical component $F \sin 20°$. Taking moments about A, the reaction at B (vertical) may be determined:

$$R_{BV} = F \sin 20° \times \frac{1200}{1800} = \frac{2F}{3} \sin 20° \quad \text{(N)}$$

The bending moment at C is therefore:

$$M_C = R_{BV} \times 600 = \frac{2F}{3} \sin 20° \times 600 = 400F \sin 20° \quad \text{(Nmm)}$$

The maximum combined axial stress at plane C is therefore:

$$f = \frac{M_C y}{I} + \frac{F \cos 20°}{A}$$

$$= \frac{400F \sin 20° \times 20}{73.5 \times 10^3} + \frac{F \cos 20°}{372.4}$$

$$= 0.0372F + 0.0025F$$

$$= 0.0397F$$

Since $f(\text{max}) = 80$ MPa

$$80 = 0.0397F$$

$$\therefore F = 2014 \text{ N} = \mathbf{2.01\ kN}$$

16.2 Combined pressure and axial stress

If a vessel under pressure is also axially loaded by an applied force, then a combined stress occurs in the axial direction. Since the axial pressure stress and the load stress act in the same direction, the combined axial stress may be determined by algebraic summation.

Note that the hoop stress (stress on a longitudinal seam) is not affected since it is in a perpendicular direction.

Thin-wall vessel

Consider a section of a thin-wall vessel, internal pressure p and loaded by an axial force F as shown in Figure 16.6.

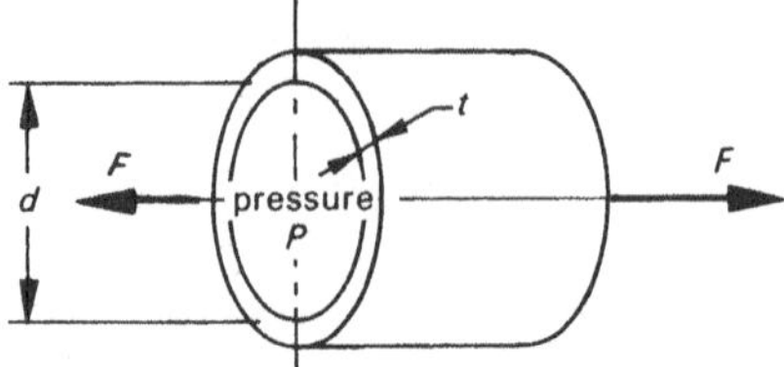

Fig. 16.6 *Axial and pressure stress in a thin-wall vessel*

The hoop stress is given by equation 12.8:

$$f_H = \frac{pd}{2t}$$

The axial stress due to pressure is given by equation 12.7:

$$f_A = \frac{pd}{4t}$$

The axial stress due to load F is given by (closely enough):

$$f = \frac{F}{\pi dt}$$

Hence the combined axial stress is:

$$f = \frac{pd}{4t} + \frac{F}{\pi dt}$$

Notes

1. If F is compressive rather than tensile the equation is:

   $$f = \frac{pd}{4t} - \frac{F}{\pi dt}$$

   and the axial stress is *reduced* by application of the external load.
2. Depending upon the value of F, the axial stress may be greater or less than the hoop stress.

Example 16.3

A pressure vessel diameter 1 m and wall thickness 10 mm contains fluid at a pressure of 700 kPa (gauge). An axial load F is applied to the vessel.

Determine:

(a) hoop stress
(b) combined axial stress when load F is 50 kN tensile
(c) magnitude of compressive load F in order that the combined axial stress is zero.

Solution

(a) $f_H = \dfrac{pd}{2t} = \dfrac{700 \times 1000}{2 \times 10} = \mathbf{35\ MPa}$

(b) Axial pressure stress is:

$$f_A = \frac{pd}{4t} = 17.5\ \text{MPa}$$

Axial stress due to load F is:

$$f = \frac{F}{\pi dt} = \frac{50 \times 10^3}{\pi \times 1000 \times 10} = 1.59\ \text{MPa}$$

Therefore the combined axial stress is $17.5 + 1.59 = \mathbf{19.1\ MPa}$

(c) Since the axial pressure stress f_A is 17.5 MPa tensile, the combined axial stress will be zero when the applied axial load stress is 17.5 MPa compressive.

$$\therefore 17.5 = \frac{F}{\pi dt}$$

$$\therefore F = 17.5 \times \pi \times 1000 \times 10 \quad \text{(N)}$$

$$= \mathbf{550\ kN} \quad \text{(compressive)}$$

16.3 Combined bending and torsion

A common problem in mechanical engineering occurs when a shaft transmitting torque is also subject to bending stress, for example, when a pulley on a shaft transmitting torque (and power) is located between bearings as shown in Figure 16.7.

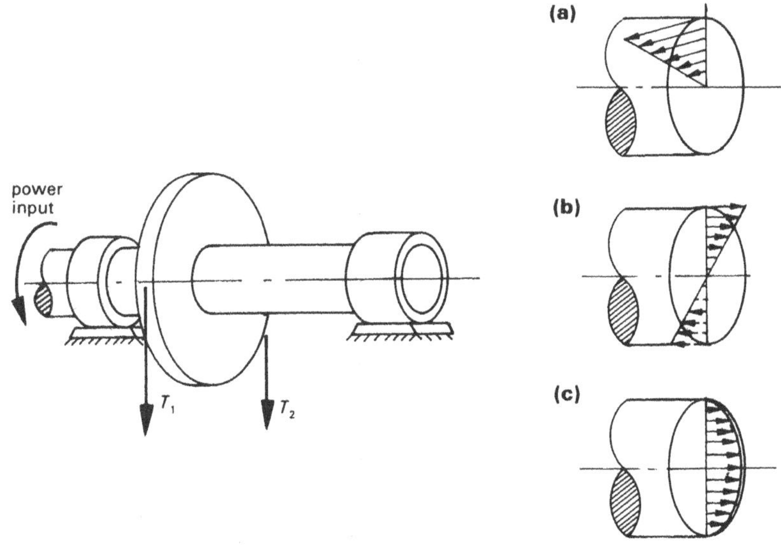

Fig. 16.7 *Bending and torsion in a circular shaft: (a) torsional shear stress, (b) bending stress, (c) direct shear stress (shear in bending)*

Here the torque in the shaft causes a torsional shear stress and the bending moment (resulting from the belt tensions and weight of the pulley and shaft) causes a bending stress. The torsional shear stress is a maximum on the outer surface of the shaft (Fig. 16.7(a)) and the bending stress is a maximum on the top and bottom fibres of the shaft (Fig. 16.7(b)). There is also a shear stress due to the direct shear load on the shaft and, since the shaft is supported between bearings and hence acts as a beam, this stress is a maximum at the centre of the shaft and zero at the outer surface (Fig. 16.7(c)). Because of this fact and the fact that, in engineering applications, the direct shear stress is always much smaller than the torsional shear stress, the direct shear stress does not need to be considered when calculating the maximum combined stress in the shaft.

A small element cut from the outer surface at the top and bottom of the shaft will therefore be stressed as shown in Figure 16.8.

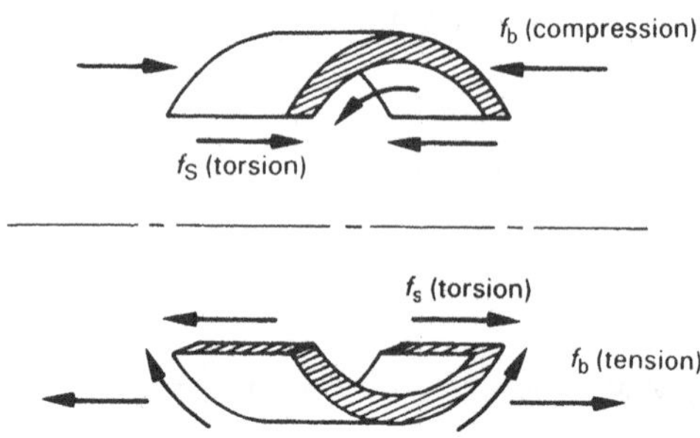

Fig. 16.8 *Stresses due to bending and torsion in a shaft*

The torsional shear stress is given by equation 15.7:

$$f_S = \frac{\tau r}{J}$$

The bending stress is given by equation 15.1:

$$f_b = \frac{My}{I}$$

The combined effect of torsional shear and bending occurring together may be determined from these equations by using an equivalent torque and an equivalent moment rather than the actual torque and moment. The equivalent torque and the equivalent moment may be calculated from the following equations (derived later in example 16.9).

$$\tau_E = \sqrt{\tau^2 + M^2}$$ **(16.1) equivalent torque**

$$M_E = \tfrac{1}{2}(M + \tau_E)$$ **(16.2) equivalent moment**

Of particular interest is the solid shaft which is used in the vast majority of applications. For a solid shaft:

$$I = \frac{\pi d^4}{64}, \quad J = \frac{\pi d^4}{32}, \quad r = \frac{d}{2}, \quad A = \frac{\pi d^2}{4}$$

Making these substitutions in equations 15.1 and 15.7 and using the equivalent torque and moment, the following equations are readily derived:

$$f_S = \frac{16\tau_E}{\pi d^3}$$ **(16.3) combined shear stress (circular shaft)**

$$f = \frac{32 M_E}{\pi d^3}$$ **(16.4) combined tension/compression stress (circular shaft)**

Notes

1. When applying these formulas to a shaft it is necessary to locate the maximum stress plane on the axis of the shaft. If the shaft is of uniform cross-section, then this plane is located where both the bending moment and torque are a maximum. This position can usually be determined by inspection of the bending moment and torque distribution diagrams.
2. If the position of maximum torque does not coincide with the position of maximum bending moment, several positions may require checking in order to determine the highest stress condition.
3. The plane of maximum stress often occurs at a pulley or gear. If these are keyed to the shaft, this also increases the likelihood of failure.
4. For a keyed shaft, it is common practice to reduce the allowable torque by 25 per cent compared to an unkeyed shaft of the same diameter in order to take into account shear stress concentration as a result of the keyway. That is, when calculating the required shaft diameter for a given loading condition, the allowable shear stress should be reduced by 25 per cent (multiplied by 0.75) if there is a keyway.
5. Although the mode of combined shear failure occurs more often in practice than combined tension or compression failure, safe design requires that both modes of failure should be checked.

Example 16.4

The pulley illustrated in Figure 16.9 is keyed to the shaft. The power input at the top coupling is 45 kW at 1800 rpm and the sum of the belt tensions at the pulley is 800 N. The weight of the system and bearing friction is negligible.

(a) Draw shear force, bending moment and torque distribution diagrams.
(b) Determine the shaft diameter required (allowing for the keyway) using a factor of safety of 8, an ultimate compressive and tensile stress of 480 MPa and ultimate shear strength equal to 60 per cent of the UTS.
(c) Check the maximum direct shear stress (shear in bending) and show that it is negligible.

Solution

Since $P = \tau\omega$, $\tau = \dfrac{P}{\omega} = \dfrac{45 \times 10^3}{\pi \times \dfrac{1800}{30}} = 238.7$ Nm

The bearing reactions (horizontal) are 400 N each which is therefore the shear force V over the length of the shaft (between the bearings).

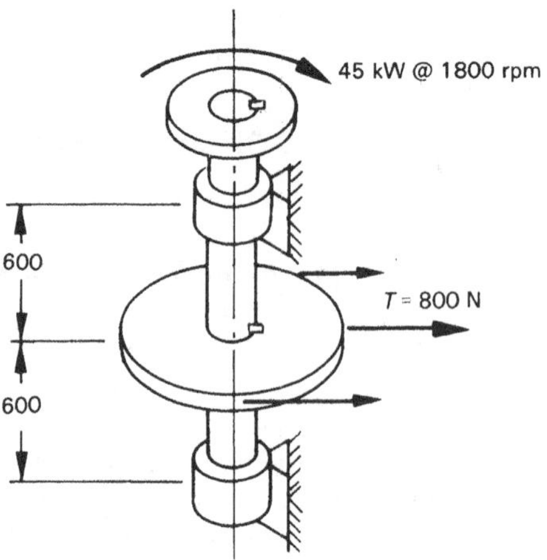

Fig. 16.9

The maximum bending moment M occurs at the centre and is:

$M = 400 \times 0.6 = 240$ Nm

(a) The diagrams may now be drawn as shown in Figure 16.10.

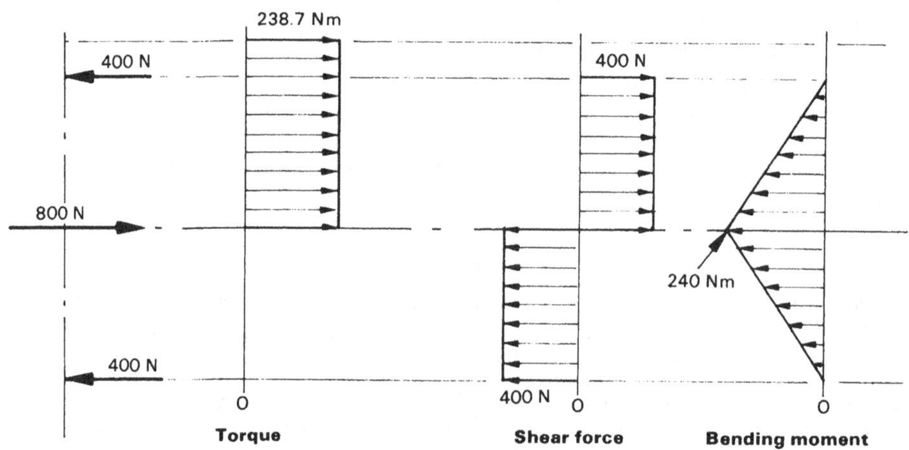

Fig. 16.10

(b) By inspection, maximum stress condition occurs at the pulley where the bending moment and torque are both maximum. Also this is the position where the keyway is located.

Now $M = 240$ Nm and $\tau = 238.7$ Nm

Using equation 16.1:

$$\tau_E = \sqrt{\tau^2 + M^2} = \sqrt{238.7^2 + 240^2} = 338.5 \text{ Nm}$$

Using equation 16.3:

$$f_S = \frac{16\tau_E}{\pi d^3} \quad \therefore d^3 = \frac{16\tau_E}{\pi f_S}$$

where f_S is the allowable shear stress $= \dfrac{0.6 \times 480 \times 0.75}{8} = 27$ MPa.

Substituting: $d^3 = \dfrac{16 \times 338.5 \times 10^3}{\pi \times 27}$

or $d = 40$ mm

Using equation 16.2:

$M_E = \frac{1}{2}(M + \tau_E) = \frac{1}{2}(240 + 338.5) = 289$ Nm

Using equation 16.4:

$f = \dfrac{32 M_E}{\pi d^3} \quad \therefore d^3 = \dfrac{32 M_E}{\pi f}$

where f is the allowable tension/compression stress $= \dfrac{480}{8} = 60$ MPa.

Substituting: $d^3 = \dfrac{32 \times 289 \times 10^3}{\pi \times 60}$

or $d = 36.6$ mm

It is evident that the combined tension/compression stress is less critical than the combined shear stress. The required shaft diameter is therefore **40 mm**.

(c) The average direct shear stress is:

$$\bar{f}_S = \frac{V}{A} = \frac{400}{\pi \times \dfrac{40^2}{4}} = 0.318 \text{ MPa}$$

For a circular shaft in bending:

$f_{S(max)} = \frac{4}{3}\bar{f}_S = $ **0.424 MPa**

As is usually the case, this stress is negligible compared to the other stresses.

Example 16.5

The shaft illustrated in Figure 16.11 has a power input of 3 kW at 1200 rpm applied by vee belts at pulley D. This power is taken off at pulley C with negligible power absorbed by the bearings. The ratio of the belt tensions is 3:1 for both pulleys which are attached to the shaft by means of taper lock bushes (without keys).

(a) Determine the belt tensions.
(b) Draw the shear force, bending moment and torque distribution diagrams taking into account the weight of the pulleys but neglecting the weight of the shaft.
(c) Determine the maximum combined shear stress.
(d) Determine the maximum combined tension/compression stress.

Solution

(a) The input torque at pulley D is:

$$\tau_D = \frac{P}{\omega} = \frac{3 \times 10^3}{\pi \times \dfrac{1200}{30}} = 23.87 \text{ Nm}$$

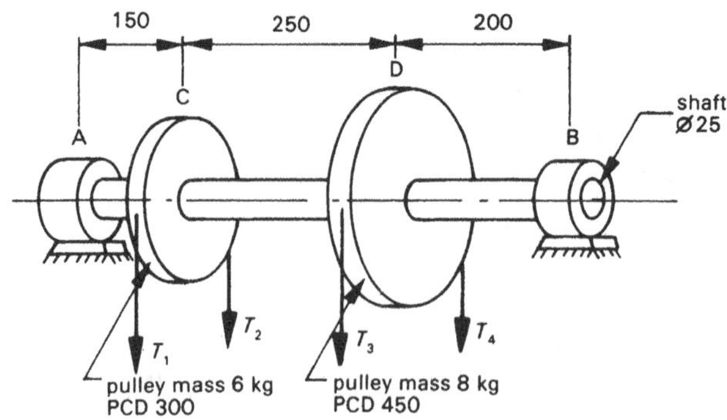

Fig. 16.11

Also $\tau_D = (T_4 - T_3)r = (T_4 - T_3) \times 0.225 = 23.87$ Nm

$\therefore T_4 - T_3 = 106$

But $\dfrac{T_4}{T_3} = 3 \quad \therefore T_4 = 3T_3$

$\therefore 2T_3 = 106, \quad \therefore T_3 =$ **53 N** and $T_4 =$ **159 N**

The torque at pulley C is the same as at pulley D since there is negligible bearing friction.

$\therefore (T_1 - T_2) \times 0.15 = 23.87$

$\therefore T_1 - T_2 = 159$

But $T_1 = 3T_2 \quad \therefore 2T_2 = 159 \quad \therefore T_2 =$ **79.6 N**

and $T_1 =$ **239 N**

Note: Since the same power is transmitted at both pulleys with the same belt tension ratio, T_2 and T_1 could also have been obtained by multiplying T_3 and T_4 by the ratio of the pulley diameters, 1.5.

(b) The load due to the weight of the pulleys is:

$W_C = 6 \times 9.81 = 58.9$ N and $W_D = 78.5$ N

The transverse load on the shaft at each pulley location is the sum of the belt tensions and the weight, i.e.

$F_C = 79.6 + 239 + 58.9 = 377.5$ N

$F_D = 53 + 159 + 78.5 = 290.5$ N

The shaft loading and reaction are as shown in Figure 16.12 together with the shear force, bending moment and torque distribution diagrams. The calculations are straightforward and are not given.

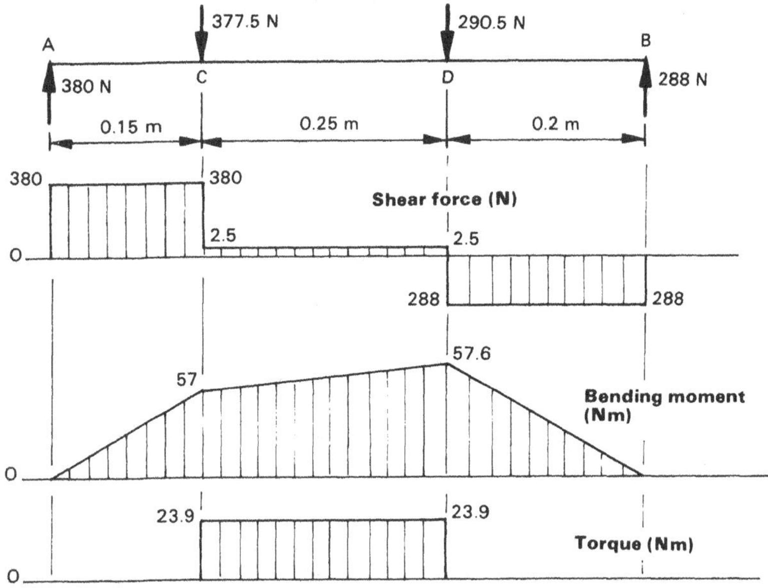

Fig. 16.12

(c) The most highly stressed position may now be located by inspection and is evidently at pulley D where the bending moment and torque are both maximum and have values:

$\tau = 23.9$ Nm, $M = 57.6$ Nm

Hence $\tau_E = \sqrt{\tau^2 + M^2} = \sqrt{23.9^2 + 57.6^2} = 62.36$ Nm

and the combined shear stress is:

$$f_S = \frac{16\,\tau_E}{\pi d^3} = \frac{16 \times 62.36 \times 10^3}{\pi \times 25^3} = \mathbf{20.3\ MPa}$$

(d) $M_E = \frac{1}{2}(M + \tau_E) = \frac{1}{2}(57.6 + 62.36) = 60$ Nm

and the combined tension/compression stress is:

$$f = \frac{32\,M_E}{\pi d^3} = \frac{32 \times 60 \times 10^3}{\pi \times 25^3} = \mathbf{39.1\ MPa}$$

16.4 Stress on an inclined plane

If an axial load F is applied to a member it produces an axial stress $f_x = \frac{F}{A}$ on a plane perpendicular to the axis of the member, that is on a plane whose perpendicular axis is parallel to the axis of the member. However, it is important to realise that there is also tension or compression and shear stress on a plane whose perpendicular axis is at some angle of inclination as shown in Figure 16.13.

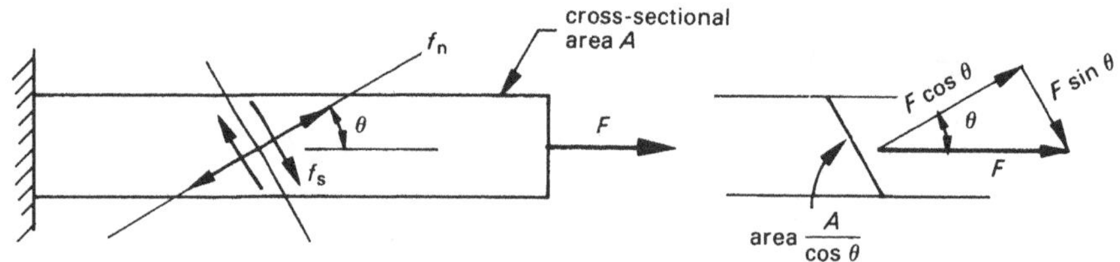

Fig. 16.13 *Stresses on an inclined plane*

Denoting the tension or compression stress on the inclined plane by f_n (normal stress) and the shear stress by f_s, it is evident that:

$$f_n = \frac{F\cos\theta}{A/\cos\theta} = \frac{F}{A}\cos^2\theta = f_x\cos^2\theta$$

and has a maximum value $f_n(\text{max}) = f_x$ when $\theta = 0°$.

$$f_s = \frac{F\sin\theta}{A/\cos\theta} = \frac{F}{A}\sin\theta\cos\theta = f_x\sin\theta\cos\theta$$

and has a maximum value $f_s(\text{max}) = 0.5f_x$ when $\theta = 45°$.

It is clear that f_n is always less than f_x, so tension or compression failure will not occur on an inclined plane unless the material has a flaw or weakness on such a plane. However, even though there is no shear force acting, there is shear stress on inclined planes and shear failure may occur. Indeed this is often the case in practice, for example when a tension or compression test is performed on a material it is often observed that failure occurs on an inclined plane (particularly if the material is weak in shear).

Example 16.6

A cast-iron rod, diameter 20 mm, is subject to a compressive load *F*.

Determine the maximum allowable value of *F* if the following stresses are not to be exceeded: tension 35 MPa, compression 130 MPa, shear 55 MPa.

Solution

$$A = \frac{\pi \times 20^2}{4} = 314 \text{ mm}^2$$

In compression:

$$f_x = \frac{F} {A} \quad \therefore F = f_x A = 130 \times 314 \text{ (N)} = 40.8 \text{ kN}$$

In shear, $f_s(\text{max})$ occurs on a plane inclined at 45° and is:

$$f_s(\text{max}) = 0.5f_x = 0.5 \times \frac{F}{314}$$

The allowable shear stress is 55 MPa.

$$\therefore 55 = 0.5 \times \frac{F}{314} \quad \therefore F = 34.5 \text{ kN}$$

It is therefore clear that shear failure will occur before compression failure and the maximum allowable load is:

F = **34.5 kN**

Example 16.7

An axial tensile load of 5 kN is applied to a bar of diameter 10 mm.

Draw a graph showing variation in normal and shear stress on a plane whose perpendicular axis is inclined at angles varying from 0° to 90° to the axis of the bar.

Solution

$$\text{Axial stress, } f_x = \frac{F}{A} = \frac{5000}{\frac{\pi \times 10^2}{4}} = 63.7 \text{ MPa}$$

Normal stress on an inclined plane is:

$$f_n = f_x \cos^2 \theta = 63.7 \cos^2 \theta$$

Shear stress on an inclined plane is:

$$f_s = f_x \sin \theta \cos \theta = 63.7 \sin \theta \cos \theta$$

By substituting in these equations with angles from 0° to 90°, in steps of 10°, yields results which when plotted appear as shown in Figure 16.14.

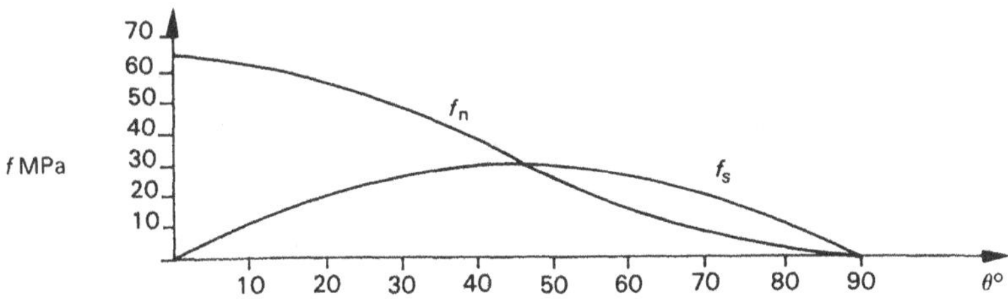

Fig. 16.14

From the graph it is clear that $f_n(\max) = f_x$ and $f_s(\max) = 0.5 f_x$ (when $\theta = 45°$).

16.5 Combined stress (general case)

In the general case, forces may act on a member so as to produce tension or compression stress in both the x and y directions and shear stress (both horizontal and vertical).

For example, consider the structural member illustrated in Figure 16.15 upon which forces F_1 and F_2 act and produce stresses as shown on a small element of the member.

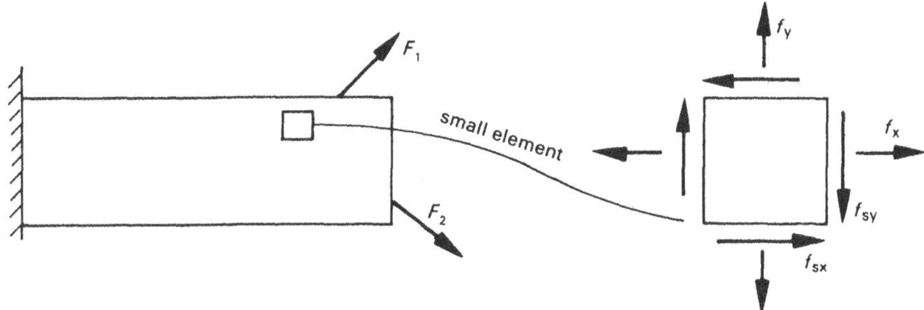

Fig. 16.15 *General case of plane stress*

The stresses have been assumed positive, that is f_x and f_y are both tensile and the vertical shear stress f_{sy} produces a clockwise couple.

In order to analyse the combined effect of the stresses, consider a plane whose perpendicular axis is inclined at angle θ to the x axis of the member as shown in Figure 16.16.

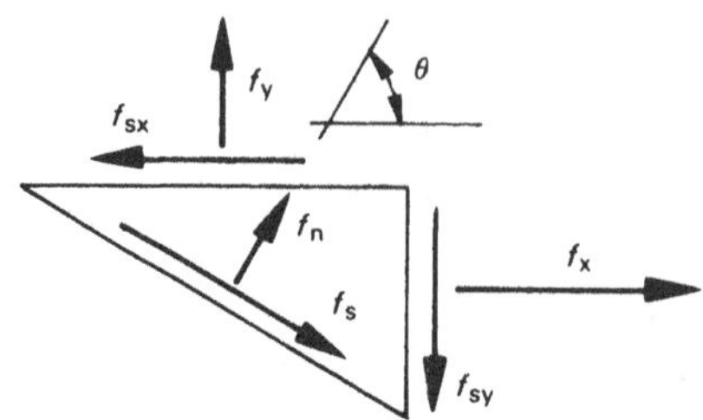

Fig. 16.16 *Combined stress on an inclined plane*

The combined effect of these stresses is a normal stress f_n and a shear stress f_s.

It was shown in section 16.4 that the f_x stress produces normal and shear stress components on the inclined plane given by:

$$f_n = f_x \cos^2 \theta, \quad f_s = f_x \sin \theta \cos \theta \qquad (1)$$

A similar analysis for the y direction stress f_y yields:

$$f_n = f_y \sin^2 \theta, \quad f_s = -f_y \sin \theta \cos \theta \qquad (2)$$

Also the shear stress f_{sy} produces stress components on the inclined plane given by:

$$f_n = -f_{sy} \sin \theta \cos \theta, \quad f_s = f_{sy} \cos^2 \theta \qquad (3)$$

Similarly the shear stress f_{sx} produces stress components:

$$f_n = -f_{sx} \sin \theta \cos \theta, \quad f_s = -f_{sx} \sin^2 \theta \qquad (4)$$

Adding equations (1), (2), (3) and (4) for the normal stress on the inclined plane yields:

$$f_n = f_x \cos^2 \theta + f_y \sin^2 \theta - f_{sy} \sin \theta \cos \theta - f_{sx} \sin \theta \cos \theta$$

But $f_{sx} = f_{sy}$, since horizontal and vertical shear stress are equal at any position.

$$\therefore \quad \boxed{f_n = f_x \cos^2 \theta + f_y \sin^2 \theta - 2 f_{sy} \sin \theta \cos \theta}$$

(16.5) combined normal stress on an inclined plane

Similarly, adding equations (1), (2), (3) and (4) for the shear stress on the inclined plane yields:

$$f_s = f_x \sin \theta \cos \theta - f_y \sin \theta \cos \theta + f_{sy} \cos^2 \theta - f_{sx} \sin^2 \theta$$

Since $f_{sx} = f_{sy}$, and using the trigonometric identities

$$\sin 2\theta = 2 \sin \theta \cos \theta, \ \cos 2\theta = \cos^2 \theta - \sin^2 \theta$$

this equation reduces to:

$$\boxed{f_s = \frac{f_x - f_y}{2} \sin 2\theta + f_{sy} \cos 2\theta}$$

(16.6) combined shear stress on an inclined plane

Example 16.8

A cylindrical pressure vessel diameter 500 mm and wall thickness 2.5 mm is welded along a helical seam that makes an angle of 55° with the longitudinal axis as shown in Figure 16.17.

If the internal pressure is 1 MPa, determine the normal and shear stress in the weld.

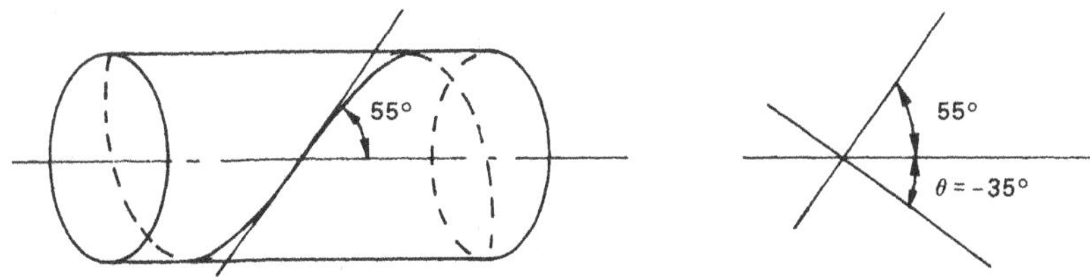

Fig. 16.17

Solution

$$f_x = \frac{pd}{4t} = \frac{1 \times 500}{4 \times 2.5} = 50 \text{ MPa}$$

$$f_y = \frac{pd}{2t} = \frac{1 \times 500}{2 \times 2.5} = 100 \text{ MPa}$$

$$f_{sy} = 0 \quad \text{(no applied shear stress)}$$

$$\theta = -35°$$

Using equation 16.5, the normal stress is:

$$f_n = 50 \cos^2(-35°) + 100 \sin^2(-35°)$$

$$= \mathbf{66.4\ MPa}$$

Using equation 16.6, the shear stress is:

$$f_s = \left(\frac{50 - 100}{2}\right) \sin(2 \times -35°)$$

$$= \mathbf{23.5\ MPa}$$

16.6 Maximum stress

When combined stresses act on an element, it is usually not important to calculate the normal and shear stress on a plane of given inclination but rather the *maximum* normal and shear stresses. The maximum (and minimum) normal stress is also known as principal stress.

Maximum normal stress and maximum shear stress may be obtained from equations 16.5 and 16.6 by differentiating and setting the derivative equal to zero.

When this is done, the result is:

$$f_s(\max) = \sqrt{f_{sy}^2 + \left(\frac{f_x - f_y}{2}\right)^2}$$

(16.7) maximum combined shear stress

$$\tan 2\theta = \frac{f_x - f_y}{2f_{sy}}$$

(16.8) angle of inclination of plane of maximum combined shear stress

$$f_n(\max) = \frac{f_x + f_y}{2} + f_s(\max)$$ **(16.9) maximum combined normal stress**

$$\tan 2\theta = \frac{-2f_{sy}}{f_x - f_y}$$ **(16.10) angle of inclination of plane of maximum combined normal stress**

Notes

1. Minimum normal stress may be derived from equation 16.9 by subtracting $f_s(\max)$ rather than adding. Minimum stress takes into account the sign of the stress and does not mean numerically smallest stress.
2. Minimum shear stress is numerically the same as maximum shear stress but negative rather than positive.
3. Minimum normal stress may be positive or negative. Since tension stress is considered positive, a negative normal stress indicates compression.
4. Maximum and minimum normal stresses (principal stresses) always occur on planes where the shear stress is zero. Planes of zero shear stress are inclined at 45° to the planes of maximum and minimum shear stress.
5. Planes of maximum normal stress and minimum normal stress are 90° apart from one another.
6. Planes of maximum shear stress and minimum shear stress are 90° apart from one another.
7. When the applied shear stress f_{sy} is zero:

$$f_s(\max) = \sqrt{0 + \left(\frac{f_x - f_y}{2}\right)^2} = \frac{f_x - f_y}{2}$$

$$\text{and } f_n(\max) = \frac{f_x + f_y}{2} + \frac{f_x - f_y}{2} = f_x$$

$$\text{and } f_n(\min) = \frac{f_x + f_y}{2} - \frac{f_x - f_y}{2} = f_y$$

   This shows that in cases where normal stress occurs in perpendicular directions without shear stress (e.g. in pressure vessels), the maximum combined normal stress is no greater than the larger of the normal stresses considered individually. This is, of course, very different to the combination of two forces in perpendicular directions which will produce a resultant greater than either of them acting alone. The reason for this difference is that in the case of stress, the area also increases on an inclined plane.
8. If the applied shear stress f_{sy} is zero and the applied perpendicular stress f_y is also zero, then:

$$f_s(\max) = \frac{f_x}{2}$$

   This is the same result as derived in section 16.4 where it was shown that an axial load causes a shear stress which has a maximum value equal to one-half the axial stress on a plane inclined at 45° to the axis.

Example 16.9

Show that equations 16.3 and 16.4 may be derived using the principal stress formulas.

Solution

For a solid shaft in bending:

$$f_x = f_b = \frac{My}{I} = \frac{32M}{\pi d^3}$$

Also for a shaft in torsion:

$$f_{sy} = f_s = \frac{\tau r}{J} = \frac{16\tau}{\pi d^3}$$

Using equation 16.7 with f_x and f_{sy} as above (note $f_y = 0$),

$$f_s(\max) = \sqrt{f_{sy}^2 + \left(\frac{f_x - f_y}{2}\right)^2}$$

$$= \sqrt{\left(\frac{16\tau}{\pi d^3}\right)^2 + \left(\frac{\frac{32M}{\pi d^3} - 0}{2}\right)^2}$$

$$= \frac{16}{\pi d^3}\sqrt{\tau^2 + M^2}$$

$$\text{or } f_s(\max) = \frac{16\tau_E}{\pi d^3} \quad \text{where} \quad \tau_E = \sqrt{\tau^2 + M^2}$$

Using equation 16.9 with f_x and f_{sy} as above ($f_y = 0$):

$$f_n(\max) = \frac{f_x + f_y}{2} + f_s(\max)$$

$$= \frac{\frac{32M}{\pi d^3}}{2} + \frac{16\tau_E}{\pi d^3}$$

$$= \frac{32}{\pi d^3}\left[\frac{M + \tau_E}{2}\right]$$

$$\text{or } f_n(\max) = \frac{32M_E}{\pi d^3} \quad \text{where} \quad M_E = \tfrac{1}{2}(M + \tau_E)$$

Example 16.10

A small section of a material under stress is loaded as follows:

Horizontal stress: $f_x = 100$ MPa (tension)
Vertical stress: $f_y = -50$ MPa (compression)
Vertical shear stress: $f_{sy} = 40$ MPa (clockwise)

Determine the maximum and minimum combined normal and shear stresses and the inclination of the planes on which they occur.

Solution

The stresses on the element are shown in Figure 16.18.

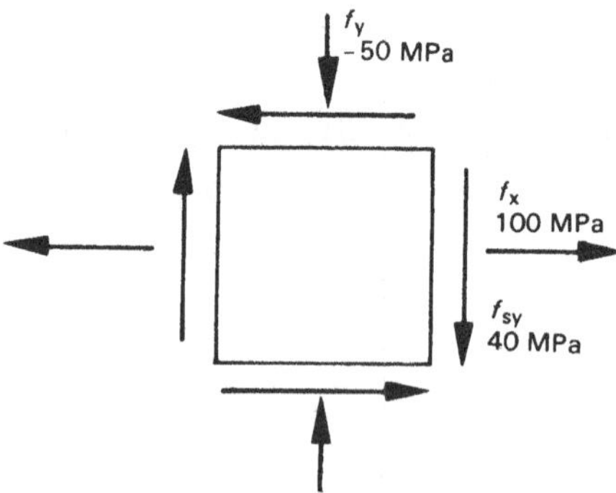

Fig. 16.18

Using equation 16.7:

$$f_s(\max) = \sqrt{f_{sy}^2 + \left(\frac{f_x - f_y}{2}\right)^2}$$

$$= \sqrt{40^2 + \left(\frac{100 - (-50)}{2}\right)^2}$$

$= \mathbf{85\ MPa}$ (which is more than twice the direct shear stress).

The inclination of the plane on which this stress occurs is given by equation 16.8:

$$\tan 2\theta = \frac{f_x - f_y}{2f_{sy}} = \frac{100 - (-50)}{2 \times 40} = \frac{150}{80} = 1.875$$

$\therefore\ 2\theta = 61.9°$ or $\theta = \mathbf{31°}$

(Remember that this is actually the angle made by the perpendicular axis.)

Using equation 16.9:

$$f_n(\max) = \frac{f_x + f_y}{2} + f_s(\max), \quad f_n(\min) = \frac{f_x + f_y}{2} - f_s(\max)$$

$$= \frac{100 - 50}{2} + 85 \qquad = \frac{100 - 50}{2} - 85$$

$= \mathbf{110\ MPa}$ $\qquad$ $= \mathbf{-60\ MPa}$

The angle of inclination of the plane of maximum normal stress may be determined from equation 16.10:

$$\tan 2\theta = \frac{-2f_{sy}}{f_x - f_y} = \frac{-2 \times 40}{100 - (-50)} = -0.5333$$

$\therefore\ 2\theta = -28°$ or $\theta = \mathbf{-14°}$

The angle of inclination of the plane of minimum normal stress is at 90° to this plane, i.e. at an angle of **76°**.

The plane inclination angles may be shown diagrammatically as in Figure 16.19.

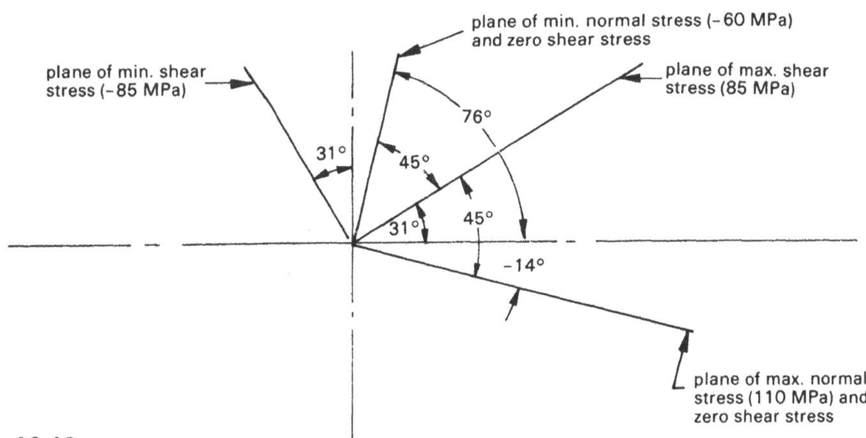

Fig. 16.19

16.7 Computer solution

The computer program PRINSTR was written to solve problems involving combined stress. The program determines maximum and minimum normal and shear stress and the plane angles at which they occur. The program will also determine the normal and shear stress on any inclined plane if so desired.

The program is listed in Appendix 3.18 with the solution to examples 16.8 and 16.10 and should be self-explanatory.

16.8 Graphical solution

A graphical solution for determining combined stresses is known as the **Mohr's circle**. Referring to Figure 16.20, the method is as follows:

1. Draw f_x to scale on a reference axis from O (point A). If f_x is positive (tension) it is drawn to the right of O and if negative (compression) to the left of O.
2. Draw f_y to scale on the same axis from O (point B). Use the same convention, if f_y is positive it is drawn to the right of O and if negative to the left.
3. Draw f_{sy} to scale perpendicular to f_x (point C). If f_{sy} is positive it is drawn above the axis and if negative it is drawn below the axis.
4. Draw $-f_{sy}$ to scale perpendicular to f_y (point D).
5. Join C and D and where this line crosses the x axis (at point E) draw a circle with centre E and radius CE to intersect the reference axis at points F and G.
6. Draw a perpendicular at E to the reference axis to intersect the circle at H and I.

From the diagram the following may be obtained:

Maximum normal stress = OF

Minimum normal stress = OG

Maximum shear stress = EH

Minimum shear stress = EI

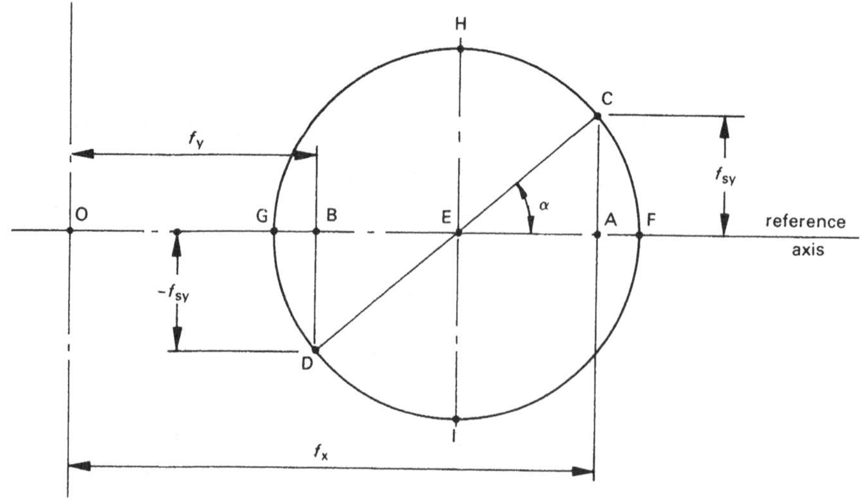

Fig. 16.20 *Mohr's circle for combined stress*

$$\text{Angle of inclination of plane of maximum normal stress} = \frac{-\alpha}{2}$$

Example 16.11

Solve example 16.10 using Mohr's circle.

Solution

The Mohr's circle is shown in Figure 16.21.

By scaling from the diagram:

f_n (max) = OF = **110 MPa**

f_n (min) = OG = **−60 MPa**

f_s (max) = EH = **85 MPa**

f_s (min) = EI = **−85 MPa**

$\alpha = 28° \therefore \theta = \mathbf{-14°}$

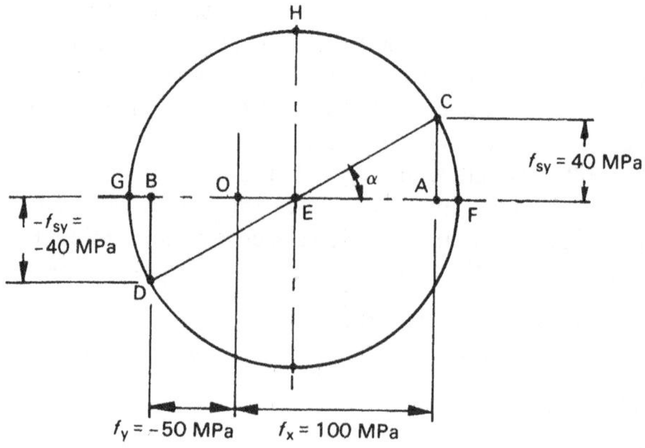

Fig. 16.21

16.9 Combined bending and shear in beams

An interesting example of combined stress occurs in beams where both bending and shear force exist at a section and hence both bending stress (axial stress) and shear stress occur. Typical distributions of these stresses for a rectangular section beam are shown in Figure 16.22.

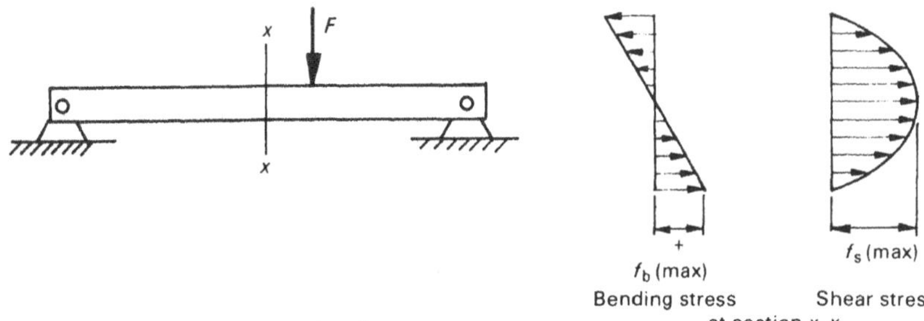

Fig. 16.22 *Bending and shear stress in a beam*

It will be noted that the bending stress is a maximum where the shear stress is zero (outer fibres) and the shear stress is a maximum where the bending stress is zero (neutral axis). The interesting problem is whether the combined effect of the bending and shear stress can produce a greater stress at some point than either of the maximum values of these stresses considered individually.

Analysis of this problem produces the following conclusions:

1. The combined stress may be greater than either of the maximum stresses considered separately but this only occurs in beams of very short span where the maximum bending and shear stresses calculated individually have the same order of magnitude.
2. Depending upon the relative magnitude of the shear stress and bending stress, there can be a shift in maximum combined stress position so that maximum combined shear stress can occur on the outer fibres and maximum combined normal stress can occur at the neutral axis (or very close to it).
3. For a flange beam section, maximum combined stress may occur in the web just inside the junction of the web and flange and good design practice is to check this position for combined stress.
4. Due to combined stress effect, tension can occur across the entire section of the beam (both above and below the neutral axis) on inclined planes. Similarly compression can occur across the entire section on inclined planes.

Rectangular or circular beams

For these beam sections it will be found that generally the maximum combined stress is only slightly larger than the individually calculated bending or shear stress.

However, the maximum combined shear stress may occur either at the neutral axis or the outer fibre. If the maximum combined shear stress occurs at the outer fibre, its value is equal to one-half the bending stress (which may be greater than the maximum shear stress at the neutral axis). Similarly, the maximum bending stress may occur at the neutral axis in which case it is equal to the maximum shear stress. Hence combined stresses should be checked at both the neutral axis and outer fibre positions.

Example 16.12

Determine the maximum normal and shear stresses for the cantilever pin shown in Figure 16.23. The load may be considered uniformly distributed over the length of the pin.

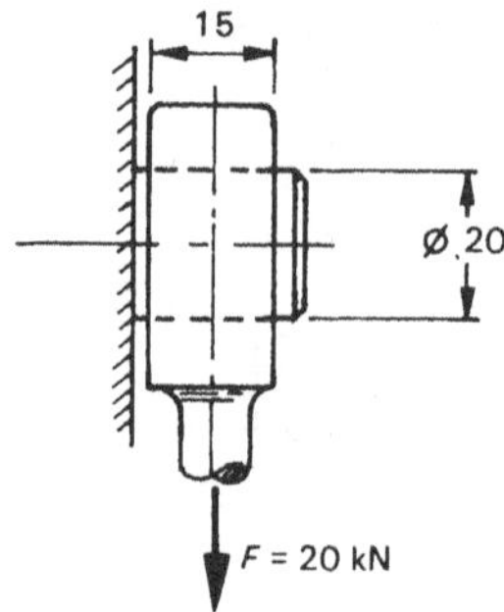

Fig. 16.23

Solution

For a circular section f_s (max) $= \frac{4}{3}\bar{f}_s$ and occurs at the centre.

$$f_s\,(\text{max}) = \frac{4}{3} \times \frac{F}{A} = \frac{4}{3} \times \frac{20 \times 10^3}{\frac{\pi \times 20^2}{4}} = 85 \text{ MPa}$$

Maximum bending moment at the support is:

$$M = 20 \times 10^3 \times \frac{15}{2} = 150 \times 10^3 \text{ Nmm}$$

Maximum bending stress for a circular section is:

$$f_b\,(\text{max}) = \frac{My}{I} = \frac{150 \times 10^3 \times 10}{\frac{\pi \times 20^4}{64}} = 191 \text{ MPa}$$

The combined stresses are:

f_n (max) = **191 MPa** (larger of 191 and 85) and occurs at the outer fibre.

f_s (max) = **95.5 MPa** $\left(\text{larger of 85 and } \frac{191}{2}\right)$ and occurs at the outer fibre.

Flange beams

Of particular interest is the flange beam. The following positions should be checked for maximum stress:

(a) neutral axis for shear
(b) outer fibre for bending
(c) in web just inside flange for combined bending and shear.

The method is illustrated in the following example.

Example 16.13

A cantilever flange beam (360 UB 57) is loaded as shown in Figure 16.24.

For the most highly stressed position (just outside the support), determine the following stresses in the section above the neutral axis:

(a) shear stress at the neutral axis
(b) bending stress at the outer fibre
(c) maximum combined shear stress in the web (just inside flange)
(d) maximum combined normal stress in the web (just inside flange)
(e) minimum combined normal stress in the web (just inside flange).
(f) Hence determine the maximum tension, compression and shear stresses which occur in the beam and their location.

The radius of the corner fillet between web and flange may be neglected.

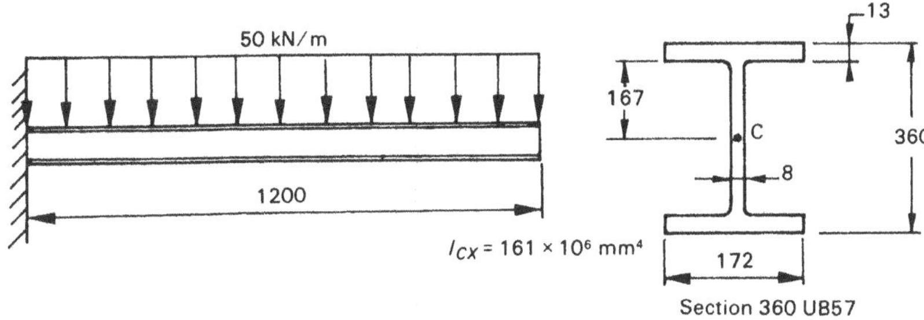

Fig. 16.24

Solution

(a) At the support, the shear force $V = 60$ kN.

At the neutral axis of the section, the first moment of the area above the neutral axis is:

$$Q = 172 \times 13 \left(167 + \frac{13}{2}\right) + 167 \times 8 \times \left(\frac{167}{2}\right)$$
$$= 388 \times 10^3 + 111.6 \times 10^3$$
$$= 499.6 \times 10^3 \text{ mm}^3$$

The shear stress at the neutral axis is:

$$f_s = \frac{QV}{Ib} = \frac{499.6 \times 10^3 \times 60 \times 10^3}{161 \times 10^6 \times 8} = \mathbf{23.3\ MPa}$$

(b) At the support, the bending moment $M = 60 \times \frac{1.2}{2} = 36$ kNm

The bending stress at the outer fibre is:

$$f_b = \frac{My}{I} = \frac{36 \times 10^6 \times 180}{161 \times 10^6} = \mathbf{40.2\ MPa}$$

(c) In the web, just inside the flange:

$Q = 388 \times 10^3$, hence the shear stress is:

$$f_s = \frac{QV}{Ib} = \frac{388 \times 10^3 \times 60 \times 10^3}{161 \times 10^6 \times 8} = 18.1 \text{ MPa}$$

and the bending stress is:

$$f_b = \frac{My}{I} = \frac{36 \times 10^6 \times 167}{161 \times 10^6} = 37.3 \text{ MPa}$$

Using equation 16.7, the maximum combined shear stress is:

$$f_s\,(\max) = \sqrt{f_{sy}^{\,2} + \left(\frac{f_x - f_y}{2}\right)^2}$$

where $f_x = 37.3$ MPa, $f_y = 0$, $f_{sy} = 18.1$ MPa

Substituting:

$$f_s\,(\max) = \sqrt{18.1^2 + \left(\frac{37.3 - 0}{2}\right)^2}$$

$$= \mathbf{26\ MPa}$$

(d) Using equation 16.9, the maximum combined normal stress is:

$$f_n\,(\max) = \frac{f_x + f_y}{2} + f_s\,(\max)$$

$$= \frac{37.3 + 0}{2} + 26$$

$$= \mathbf{44.6\ MPa}\ \text{(tension)}$$

(e) Using equation 16.9, the minimum combined normal stress (above the neutral axis) is:

$$f_n\,(\min) = \frac{f_x + f_y}{2} - f_s\,(\max)$$

$$= \frac{37.3 + 0}{2} - 26$$

$$= \mathbf{-7.35\ MPa}\ \text{(compression)}$$

(f) By comparison of the above results it is evident that:

Maximum tension stress = **44.6 MPa** and occurs above the neutral axis in the web (just inside the flange).

Maximum compression stress = **−44.6 MPa** and occurs below the neutral axis in the web (just inside the flange).

Maximum shear stress = **26 MPa** and occurs above and below the neutral axis in the web (just inside the flange).

Note: Even though the section above the neutral axis is in tension as a result of bending stress alone, there is actually compression on an inclined plane due to the combined effect of bending and shear. Similarly, tensile stress occurs on an inclined plane below the neutral axis.

Problems

Note: For these problems, neglect the load due to the weight of the member unless the mass of the member is given.

16.1 For the cantilever beam shown in Figure P16.1, determine:
(a) maximum combined tension stress
(b) maximum combined compression stress
(c) distance from the free end of the beam beyond which no compression stress occurs.

Also draw a diagram to scale showing the distribution of combined axial stress over the cross-section at the plane of maximum stress.

(a) 82.5 MPa (b) −67.5 MPa (c) 100 mm

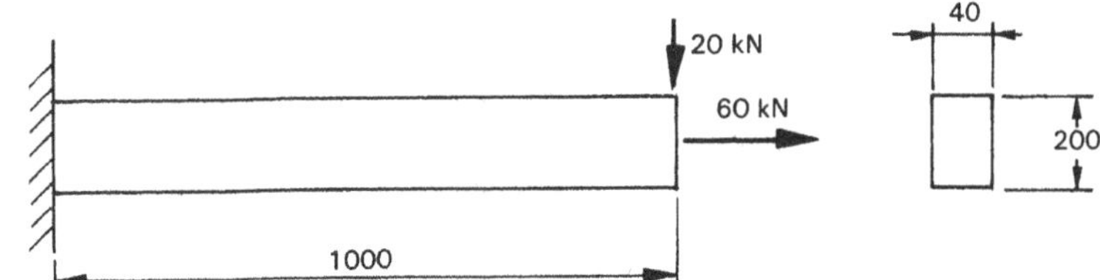

Fig. P16.1

16.2 A horizontal member of a frame is loaded as shown in Figure P16.2. Determine:

(a) maximum combined tension stress

(b) maximum combined compression stress.

Also draw a diagram to scale showing the distribution of combined axial stress at the plane of maximum stress.

(a) 93.3 MPa (b) −66.7 MPa

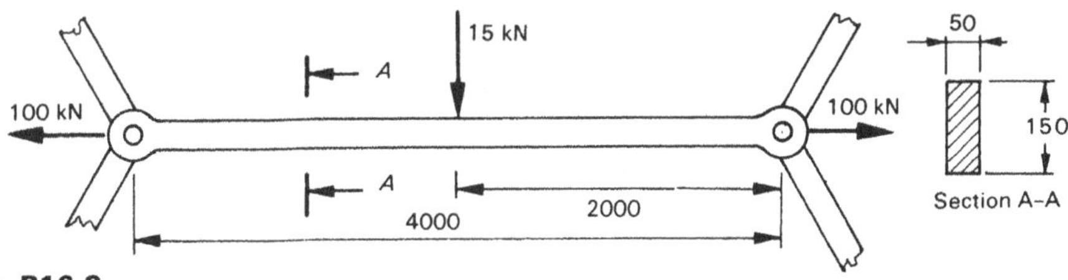

Fig. P16.2

16.3 Repeat problem 16.2 if a transverse distributed load of 10 kN/m acted on the member instead of the concentrated load, all other conditions being the same.

(a) 120 MPa (b)−93.3 MPa

16.4 A short column (150 UC 37) supports a load of 100 kN on a bracket as shown in Figure P16.4.

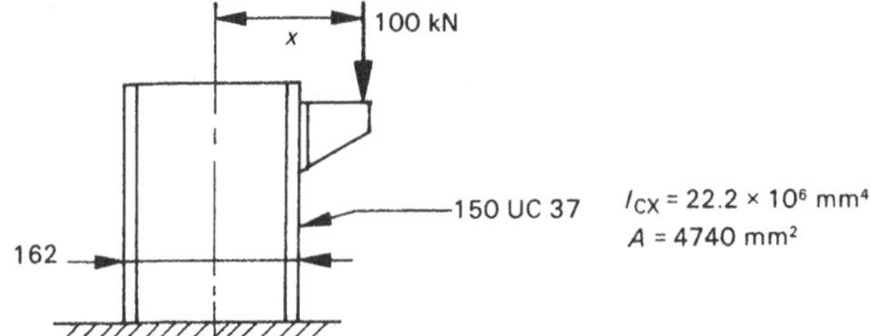

Fig. P16.4

Determine the distance x so that the maximum combined axial stress does not exceed 140 MPa.

326 mm

16.5 Determine the maximum load mass which may be held by the pulley system shown in Figure P16.5 so that the maximum combined axial stress in the bracket (just outside the support) does not exceed 100 MPa.

725 kg

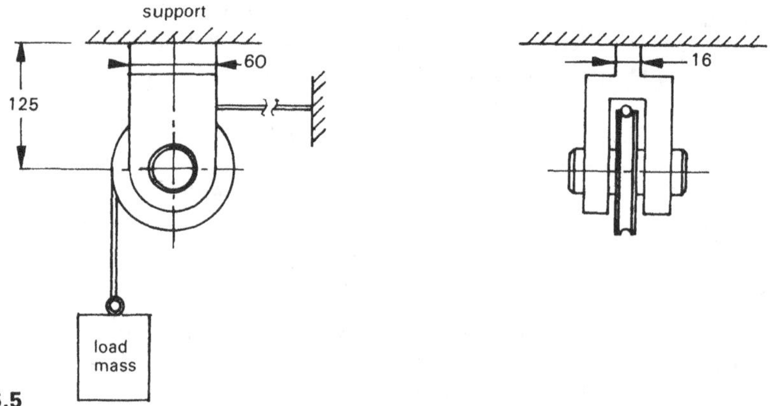

Fig. P16.5

16.6 Determine the maximum combined tension and compression stress which occur at the outer fibres in the flange beam shown in Figure P16.6.

119 MPa, −107 MPa

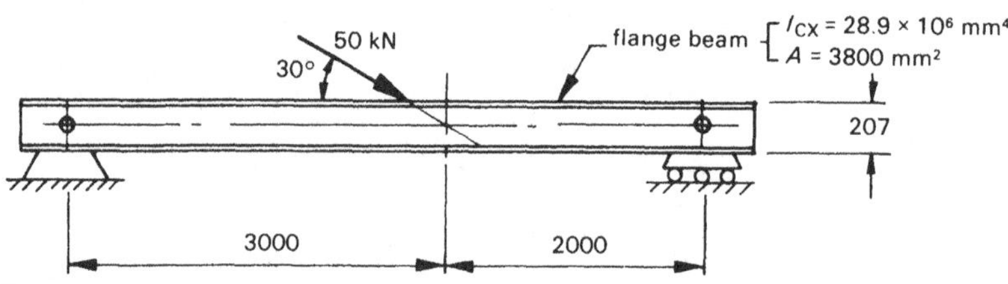

Fig. P16.6

16.7 Repeat problem 16.6 taking into account the weight of the beam which has a mass of 29.8 kg/m.

122 MPa, −111 MPa

16.8 The landing gear for an aircraft is shown in Figure P16.8.

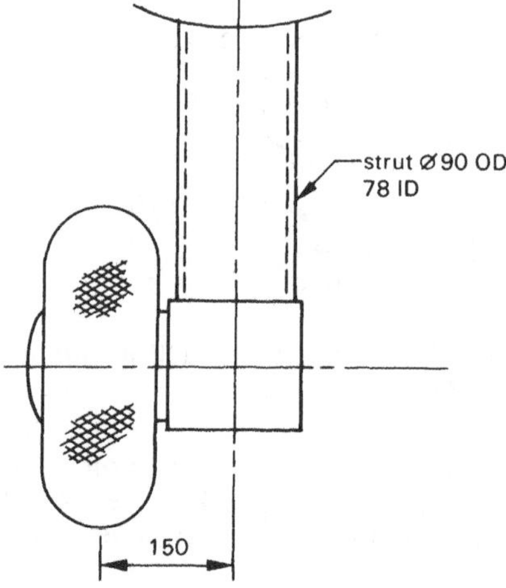

Fig. P16.8

Determine the maximum static weight which may be carried by the strut in order that the combined axial stress does not exceed 80 MPa.

1.5 t

16.9 The bolt shown in Figure P16.9 is used to support a load F and is made of steel which has a yield point of 250 MPa. The bolt is tightened so that the tensile stress in the shank is 40 per cent of the yield stress.

Determine the maximum value of the load F in order that the combined axial stress in the shank of the bolt does not exceed 80 per cent of the yield stress.
Assume the load and reactions are uniformly distributed on the bolt.

12.8 kN

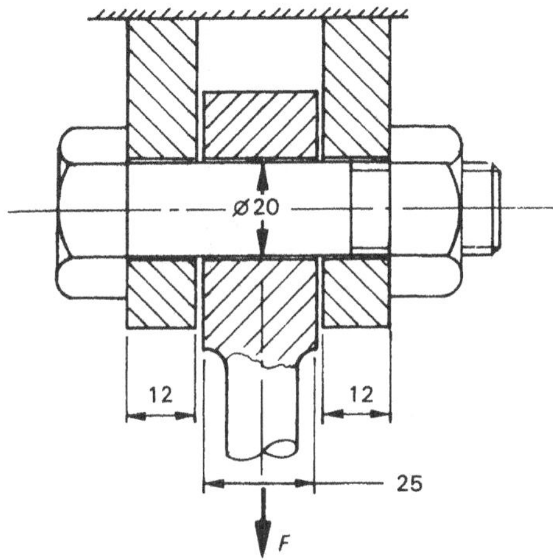

Fig. P16.9

16.10 An axial tensile load of 450 kN is applied to the ends of a cylindrical pressure vessel 1200 mm in diameter and containing gas at a pressure of 850 kPa (gauge).

Determine the required wall thickness so that:
(a) combined axial stress does not exceed 40 MPa
(b) hoop stress does not exceed 40 MPa.

(a) 9.36 mm (b) 12.75 mm

16.11 A cylindrical vessel, 800 mm in diameter and wall thickness 8 mm, stands with its axis vertical and supports a load mass placed on top.

Determine the combined mass of the load and vessel so that the combined axial stress and the hoop stress at the base of the vessel are equal when the pressure inside the vessel is 40 kPa (absolute).

3.14 t

16.12 A steel pressure vessel has a safety valve set to release when the pressure exceeds 600 kPa. The vessel dimensions are: diameter 600 mm, wall thickness 6 mm, length 2400 mm. When the temperature is 20°C the vessel contains gas at maximum pressure and just fits inside partially restrained end supports.

If the temperature rises to 42°C, determine the combined axial stress and the hoop stress if the supports allow 50 per cent of the thermal expansion in length.

Assume for steel: E = 200 GPa, α = 12 × 10^{-6}/°C.

− 11.4 MPa, 30 MPa

16.13 For problem 16.12, determine the safety valve setting so that at 42°C the combined axial stress is zero, all other conditions being the same.

1.056 MPa

16.14 A shaft 20 mm in diameter rotates at 240 rpm and transmits 1.5 kW at a pulley. In the plane of the pulley the bending moment in the shaft is 50 Nm.

Determine the maximum combined shear stress and tension stress in the shaft at the plane of the pulley.

49.6 MPa, 81.4 MPa

16.15 At the plane of maximum stress in a shaft, the bending moment is 100 Nm and the torque is 120 Nm.

If the combined shear stress in the shaft is not to exceed 80 MPa, determine the shaft diameter necessary.

What is the maximum combined tensile stress in the shaft with diameter so determined?

21.5 mm, 131 MPa

16.16 A machine shaft, diameter 30 mm, is supported between bearings 500 mm apart and is fitted with a chain sprocket at its midpoint. The sprocket has a pitch diameter of 230 mm and transmits 6 kW at 420 rpm.

Determine for the shaft:

(a) maximum combined shear stress

(b) maximum combined tension stress.

Neglect the weight of the sprocket and shaft and assume zero tension in the slack side chain.

(a) 38 MPa (b) 66 MPa

16.17 A 50 mm diameter horizontal shaft is supported between bearings 1200 mm apart and transmits torque to a flywheel located at the midpoint of the shaft. The torsional stress in the shaft is 70 MPa.

If the mass of the flywheel is 450 kg, determine the percentage reduction in torque necessary so that the combined shear stress in the shaft does not exceed 70 MPa.

36.3%

16.18 The shaft illustrated in Figure P16.18 is fixed rigidly at the support.

Determine the maximum permissible load F if the combined tensile stress in the shaft is not to exceed 130 MPa and the combined shear stress is not to exceed 90 MPa.

7.09 kN

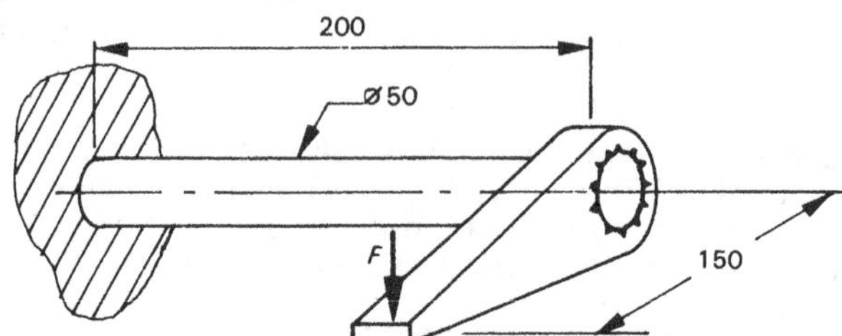

Fig. P16.18

16.19 Repeat problem 16.18 if the shaft were hollow, outside diameter 60 mm, inside diameter 50 mm, all other data being the same.

6.34 kN

16.20-16.24 For each of the stressed elements shown in Figures P16.20 to P16.24, determine:

(a) maximum combined shear stress and plane angle
(b) maximum combined normal stress and plane angle
(c) minimum combined normal stress and plane angle.

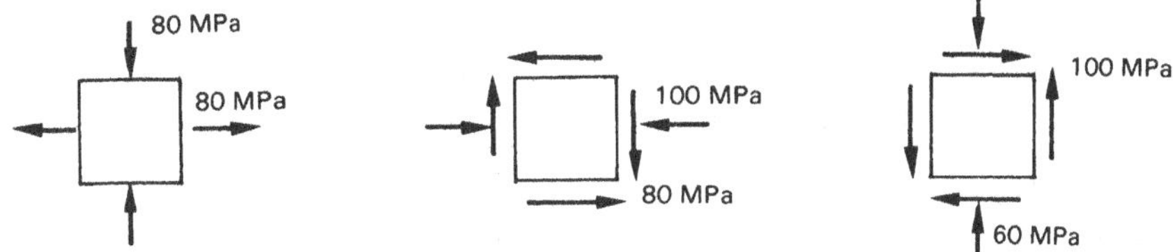

Fig. P16.20 **Fig. P16.21** **Fig. P16.22**

Fig. P16.23 **Fig. P16.24**

Solutions

	(a)		(b)		(c)	
	f_s (max) MPa	θ degrees	f_n (max) MPa	θ degrees	f_n (min) MPa	θ degrees
16.20	80	45	80	0	− 80	90
16.21	94.3	−16	44.3	29	−144.3	119
16.22	104.4	− 8.35	74.4	36.65	−134.4	126.65
16.23	64	19.3	144	−25.7	16	64.3
16.24	100	18.4	80	−26.6	−120	63.4

16.25-16.29 For each of the stressed elements shown in Figures P16.20 to P16.24 determine the combined normal and shear stress on planes inclined at angles of (a) 45° and (b) 120°.

Solutions

	(a)		(b)	
	f_n MPa	f_s MPa	f_n MPa	f_s MPa
16.25	0	80	− 40	−69.3
16.26	−130	−50	44.3	3.3
16.27	70	30	−131.6	24
16.28	30	40	103.3	−59.6
16.29	60	−60	− 59.3	92

16.30 An axial tensile load of 30 kN is applied to a horizontal bar which has a diameter of 15 mm.

Determine the combined normal and shear stress on planes whose perpendicular axis is inclined to the axis of the bar at angles of (a) 30°, (b) 45° and (c) 60°.

Solution

	(a)	(b)	(c)
f_n (MPa)	127	84.9	42.4
f_s (MPa)	73.5	84.9	73.5

16.31 A compressive load of 20 kN is applied to a cast-iron bar.

Determine the required diameter of the bar if the following combined stresses are not to be exceeded: tension 30 MPa, compression 120 MPa, shear 50 MPa.

16 mm

16.32 A tensile load of 80 kN is applied to a shaft 40 mm in diameter which is transmitting 30 kW at 440 rpm.

Determine:

(a) maximum torsional shear stress
(b) maximum combined shear stress
(c) maximum combined tension stress
(d) maximum combined compression stress.

(a) 51.8 MPa (b) 60.8 MPa (c) 92.6 MPa (d) −29 MPa

16.33 The propeller shaft of a vessel is 50 mm in diameter and transmits 200 kW at 1220 rpm.

Determine:

(a) shear stress due to torsion
(b) maximum propeller thrust which could be applied to the shaft so that the combined shear stress does not exceed 65 MPa
(c) maximum combined compressive stress in the shaft when maximum propeller thrust is applied.

(a) 63.8 MPa (b) 49.2 kN (c) 77.5 MPa

16.34 For the flange beam shown in Figure P16.34 determine for the plane of the load:

(a) shear stress at the neutral axis
(b) bending stress at the outer fibres.

Also determine for the web (just inside the flange) above the neutral axis, the following maximum combined stresses:

(c) shear
(d) tension
(e) compression.

(a) 31.3 MPa (b) 38.6 MPa (c) 29.3 MPa (d) 12.5 MPa (e) 46.0 MPa

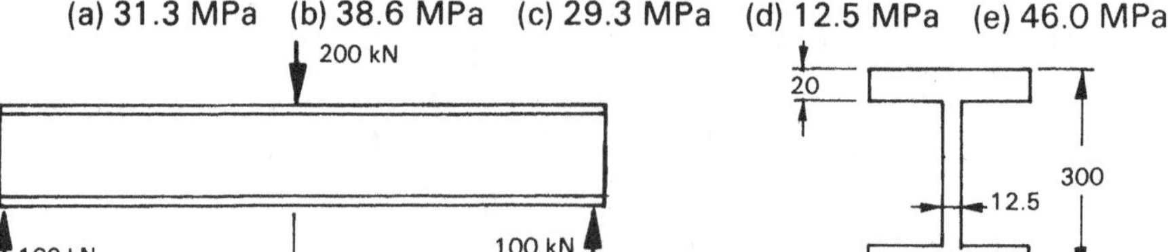

Fig. P16.34

17

Beam deflection

So far, the analysis of beams has been concerned with the stresses produced by the loads acting on the beams. Where the beam is of relatively short span, analysis of stresses may be sufficient consideration, but in other cases the deflection produced by the loads may be just as important a factor as the stresses. Indeed in some cases, such as in the design of machine tools or other equipment where accuracy and alignment are critical, analysis of deflection may be a more important factor than analysis of stresses. Similarly where beams are used in engineering construction, allowable maximum deflection is usually of prime importance and is stipulated by relevant building standards and codes.

Numerous methods are available for analysis of beam deflection such as:

1. double integration method (Macaulay's method)
2. area moment method
3. strain energy method
4. use of deflection formulas
5. formulas or charts prescribed by Standards or Codes
6. computer programs.

Although formulas and codes are widely used for particular applications in practical design, these will not be considered here because of their specific applicability and the fact that these methods do not involve basic analysis. The use of formulas and codes is usually treated as part of engineering design rather than strength of materials.

In this chapter one theoretical method, Macaulay's method, will be treated as this is a general method suitable for any combination of concentrated or distributed loads on simply supported or cantilever beams. Also a computer program to solve these problems is presented.

Notes

1. With all deflection methods, the principle of superposition may be used. This principle, applied thus far to stress analysis, may also be used for strain (or deflection) analysis. The principle is that the deflection at any point in a beam as a result of a number of loads is the algebraic sum of the deflections which would occur at this point by each load acting alone.
2. Only elastic deformation will be considered, that is the analysis is only valid provided the maximum stresses in the beam do not exceed the proportional limit (or elastic limit).
3. Only beams of uniform section will be treated.
4. It is more convenient to orient cantilever beams with the support to the right (free end to the left).
5. It is also more convenient to use units of kN for force and m for distance along the beam when using the double integration method.

17.1 Double integration method: Theory

The double integration method is a method based on the use of calculus to solve the differential equation for the elastic curve of a beam. For small deflections of a beam this equation is:

$$E\,I\frac{d^2y}{dx^2} = M \qquad \textbf{(17.1) elastic curve of a beam}$$

Since $\frac{dy}{dx}$ = slope of the beam, the equation for the slope may be obtained by integrating this equation. This results in:

$$E\,I\frac{dy}{dx} = \int (M\,dx) + A \qquad \textbf{(17.2) slope of a beam}$$

The deflection equation may now be obtained by integrating this equation. The result it:

$$E\,I\,y = \int\int (M\,dx) + Ax + B \qquad \textbf{(17.3) deflection of a beam}$$

In these formulas the meaning and units of each term are given below, units being based upon kN for force and m for displacement (or distance):

E = modulus of elasticity of the beam material (kN/m^2)

I = second moment of area of the beam section about the neutral axis (m^4)

x = distance along the beam from the origin of the x-y frame of reference (m)

y = deflection of the beam at position x (m)

M = bending moment at any position x along the beam (kNm). (The same sign convention is adopted as previously used, i.e. clockwise bending moment is positive when working from the left-hand end of the beam).

$\frac{dy}{dx}$ = slope of the beam at position x (no units)

A, B = constants of integration (A has units kNm^2 and B has units kNm^3).

The general configuration of the elastic curve for simply supported and cantilever beams is shown in Figure 17.1 (deflection greatly exaggerated). The elastic curve is shown by the dashed line and the origin for the x-y frame of reference is denoted by the point O.

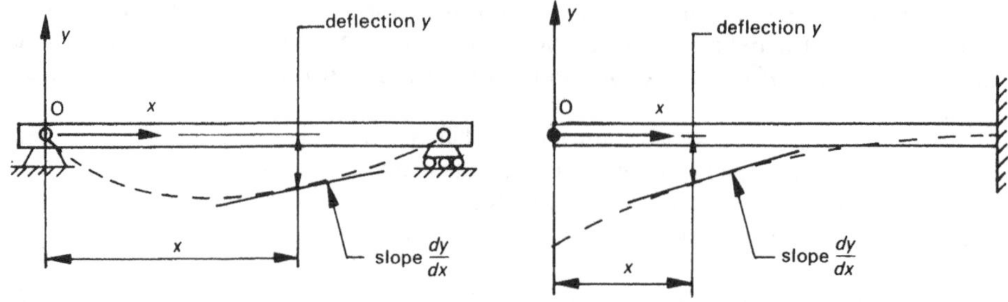

Fig. 17.1 *Elastic curves for simply supported and cantilever beams*

Note that in most cases the loads act downward on beams and, with the location of the frame of reference as given above, the deflection will be negative.

Example 17.1 Cantilever beam, single point load

Determine the maximum deflection and slope for the steel cantilever beam shown in Figure 17.2. The second moment of area of the beam section about the neutral axis is 300×10^6 mm^4 and the modulus of elasticity is 200 GPa.

Also determine the deflection and slope of the beam at the midpoint.

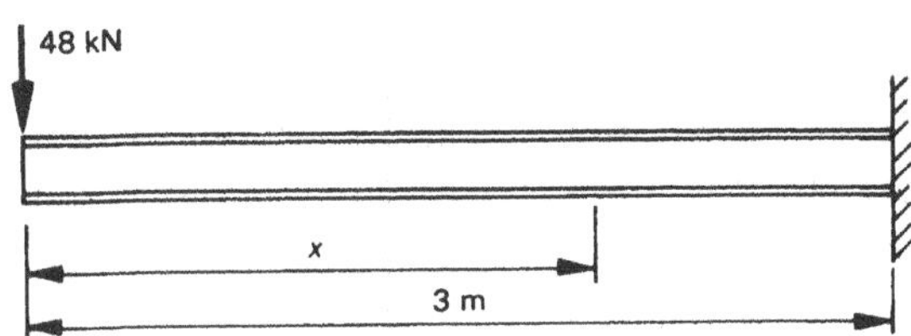

Fig. 17.2

Solution

Working from left to right and using units of kN for force and m for length, the moment equation is:

$$M = -48\,x$$

$$\therefore EI\frac{d^2y}{dx^2} = -48\,x$$

Integrating once yields:

$$EI\frac{dy}{dx} = -\frac{48}{2}x^2 + A \qquad (1)$$

Integrating again:

$$EIy = -\frac{48}{2 \times 3}x^3 + Ax + B$$

$$= -8\,x^3 + Ax + B \qquad (2)$$

In order to evaluate the constants A and B, it is necessary to consider the boundary conditions. It is evident that at the support, when $x = 3$ m, both the slope and the deflection are zero.

When $x = 3$ m, $\frac{dy}{dx} = 0$

Substituting in (1):

$$0 = -24 \times 3^2 + A$$

$$\therefore A = 216 \text{ kNm}^2$$

Similarly when $x = 3$ m, $y = 0$.

Substituting in (2) with $A = 216$ as determined above:

$$0 = -8 \times 3^3 + 216 \times 3 + B$$

$$B = -432 \text{ kNm}^3$$

Hence equation (2) is:

$$EIy = -8x^3 + 216x - 432$$

$$\text{or } y = \frac{-8x^3 + 216x - 432}{EI}$$

Before substituting in this equation it is necessary to adjust the units of E and I since E was given in GPa and I in mm^4, whereas the required units for E are kN/m^2 and for I are m^4. The conversion factors are:

$$\text{for } E, \left(\frac{N}{mm^{-2}}\right) \times \frac{10^{-3}}{10^{-6}} = 10^3$$

$$\text{for } I, (mm^4) \times 10^{-12}$$

Therefore EI must be multiplied by 10^{-9}.
The final deflection equation is:

$$y = \frac{-8x^3 + 216x - 432}{EI \times 10^{-9}}$$

$$= \frac{-8x^3 + 216x - 432}{200 \times 10^3 \times 300 \times 10^6 \times 10^{-9}}$$

$$= \frac{-8x^3 + 216x - 432}{60 \times 10^3}$$

Maximum deflection occurs when $x = 0$, hence the maximum deflection is:

$$y = \frac{0 + 0 - 432}{60 \times 10^3} \text{ (m)}$$

$$= \mathbf{-7.2\ mm}$$

The slope equation (1) with $A = 216 \text{ kNm}^2$ is:

$$EI\frac{dy}{dx} = -24x^2 + 216$$

or adjusting EI by the factor 10^{-9} (see above),

$$\frac{dy}{dx} = \frac{-24x^2 + 216}{EI \times 10^{-9}}$$

The maximum slope occurs when $x = 0$, that is:

$$\frac{dy}{dx} = \frac{216}{60 \times 10^3} = \mathbf{0.0036}$$

The deflection at the midpoint may be obtained by substitution in the deflection equation with $x = 1.5$.

$$y = \frac{-8 \times 1.5^3 + 216 \times 1.5 - 432}{60 \times 10^3}$$

$$= \mathbf{-2.25\ mm}$$

Similarly the slope at the midpoint is:

$$\frac{dy}{dx} = \frac{-24 \times 1.5^2 + 216}{60 \times 10^3}$$

$$= \mathbf{0.0027}$$

17.2 Macaulay's method (outline)

The double integration method as given in example 17.1 is applicable for any section of a beam for which a single moment equation is valid, for example a cantilever beam with a single, concentrated load at the end. However, if a beam is simply supported and carries a concentrated load, a single equation for the bending moment which will be valid for positions on the beam both to the left and to the right of the load, cannot be written. In this case, two equations would be necessary and constants A and B would have to be evaluated separately for each equation. This complication also arises if the load on the cantilever acts inside the span rather than at the end or whenever more than one load acts on a beam.

In order to resolve these difficulties and enable a single equation to be written, a mathematical procedure known as Macaulay's method (after its originator) may be used. The features of this method are as follow:

1. A single moment equation is written in terms of x, where x is a position on the beam between the load furthest to the right and the right-hand support.
2. Bracketed terms for the moment of loads to the left of this position are never simplified and must be retained at all times, that is during integration or when the constants of integration are evaluated. When integrating a bracketed term, it is treated as a single term. For example:

   $$\int a[x - b]\,dx = \frac{a}{2}[x - b]^2$$

   Note that square brackets are conventionally used.
3. When substituting in an equation in order to determine the moment, slope or deflection, if a quantity inside the bracket is negative, it is treated as zero.

17.3 Cantilever beam: Concentrated loads

Macaulay's method is suitable for cantilever or simply supported beams with concentrated or distributed loads. The use of the method for a cantilever beam with concentrated loads is illustrated in the following example.

Example 17.2

For the cantilevered steel rod shown in Figure 17.3,

(a) determine the deflection at the free end
(b) check the maximum bending stress
(c) draw the elastic curve of the beam to scale.

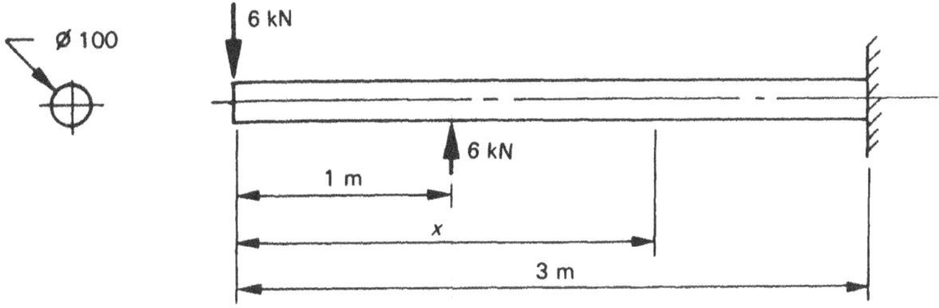

Fig. 17.3

Solution

This problem could be solved by writing separate equations for each of the loads and then using the principle of superposition. However, with Macaulay's method only one equation is required by choosing the position x to the right of the furthest load. Using units of kN and m, this equation is:

$$M = -6x + 6[x - 1]$$

Note that by using Macaulay's method this equation is also valid when x is a distance intermediate between the loads, that is for values of x between 0 and 1. At these positions, the term $6[x - 1]$ is negative and therefore treated as zero and the moment is simply: $M = -6x$.

Since $M = EI\dfrac{d^2y}{dx^2}$, the differential equation for the elastic curve of the entire beam is:

$$EI\frac{d^2y}{dx^2} = -6x + 6[x - 1]$$

Integrating:

$$EI\frac{dy}{dx} = -\frac{6}{2}x^2 + \frac{6}{2}[x - 1]^2 + A$$

$$= -3x^2 + 3[x - 1]^2 + A \qquad \textbf{(1)}$$

Integrating again:

$$EIy = \frac{-3}{3}x^3 + \frac{3}{3}[x - 1]^3 + Ax + B$$

$$= -x^3 + [x - 1]^3 + Ax + B \qquad \textbf{(2)}$$

The constant A may now be evaluated from the boundary condition when $x = 3, \dfrac{dy}{dx} = 0$.

Substituting in (1):

$$0 = -3 \times 3^2 + 3[3 - 1]^2 + A$$

$$\therefore A = \mathbf{15\ kNm^2}$$

Also when $x = 3, y = 0$.

Substituting in (2) (with $A = 15$):

$$0 = -3^3 + [3 - 1]^3 + 15 \times 3 + B$$

$$\therefore B = \mathbf{-26\ kNm^3}.$$

Hence the deflection equation is:

$$y = \frac{-x^3 + [x - 1]^3 + 15x - 26}{EI}$$

Now $E = 200$ GPa (steel) and for a circular section $I = \dfrac{\pi d^4}{64}$

$$\therefore I = \frac{\pi \times 100^4}{64} = 4.91 \times 10^6 \text{ mm}^4$$

Substituting these values in the deflection equation (remembering to multiply by the factor 10^{-9} as shown in example 17.1):

$$y = \frac{-x^3 + [x - 1]^3 + 15x + 26}{200 \times 10^3 \times 4.91 \times 10^6 \times 10^{-9}}$$

$$= \frac{-x^3 + [x - 1]^3 + 15x - 26}{982} \quad \text{(m)}$$

$$\therefore y = \frac{-x^3 + [x - 1]^3 + 15x - 26}{0.982} \quad \text{(mm)} \qquad \textbf{(3)}$$

(a) At the free end of the beam, $x = 0$.
Substituting in equation (3), remembering that any negative term inside the bracket is treated as zero:

$$y = \frac{0 + 0 + 0 - 26}{0.982} = \mathbf{-26.5\ mm}$$

(b) Maximum bending moment and therefore maximum bending stress occurs between the support and 2 metres out from the support where the bending moment is:

$M = -6 \text{ kNm} = -6 \times 10^6 \text{ Nmm}$

The maximum bending stress is:

$$f_b = \frac{My}{I} = \frac{6 \times 10^6 \times 50}{4.91 \times 10^6} = \mathbf{61.1\ MPa}$$

Note that the negative sign is ignored when determining bending stress because both positive and negative stress occur (above and below the neutral axis).

(c) In order to draw the elastic curve, several points along the beam may be chosen and the value of x at each point substituted in equation (3) in order to determine the deflection y at each point. The result of this calculation is tabulated below:

x (m)	0	0.5	1	1.5	2	2.5	3
y (mm)	−26.5	−19.0	−12.2	−6.87	−3.06	−0.764	0

These points may now be plotted in order to obtain the shape of the elastic curve as shown in Figure 17.4.

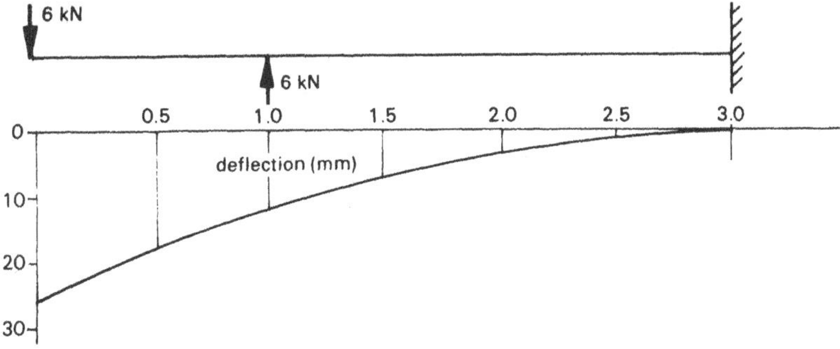

Fig. 17.4

17.4 Cantilever beam: Distributed load

When a distributed load acts over the whole span of a cantilever beam as shown in Figure 17.5, the bending moment at position x is:

$$M = -\frac{Wx^2}{2}$$

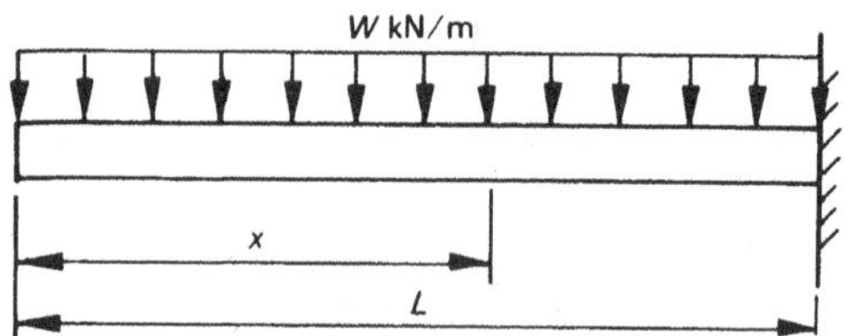

Fig. 17.5 *Distributed load over the whole span of a cantilever beam*

The use of Macaulay's method as outlined previously is straightforward. When the distributed load acts on part of the span only, as shown in Figure 17.6, a complication arises because a single moment equation cannot be written.

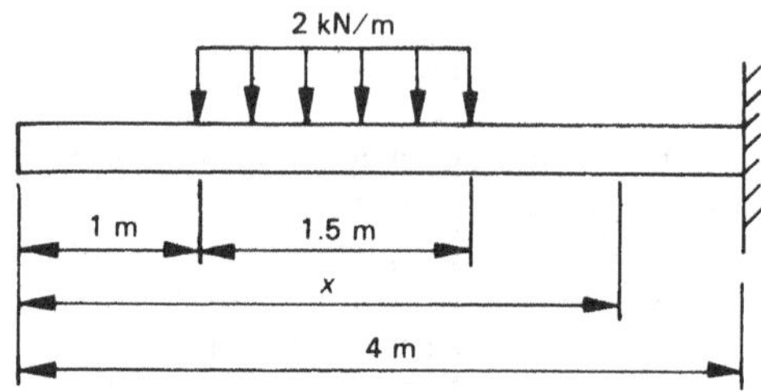

Fig. 17.6 *Distributed load over part of the span of a cantilever beam*

For example, if the bending moment equation is written with x chosen as shown, the equation is:

$$\begin{aligned} M &= -2 \times 1.5 \times [x - 1.75] \\ &= -3[x - 1.75] \end{aligned}$$

However if x lies between 1 m and 2.5 m, this equation is not valid. For example, if $x = 1.75$ m, this equation gives $M = 0$, yet it is quite evident that at the centre of the distributed load the moment is not zero.

This difficulty can be overcome by using a compensating distributed load as shown in Figure 17.7.

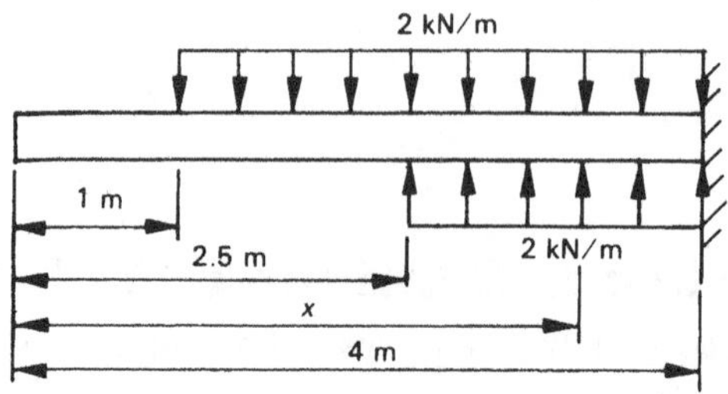

Fig. 17.7 *Compensating distributed load*

The moment equation now becomes:

$$M = \frac{-2[x - 1]^2}{2} + \frac{2[x - 2.5]^2}{2} = -[x - 1]^2 + [x - 2.5]^2$$

This equation is valid at any position because if $x < 1$, all the bracketed terms are negative and therefore are treated as being zero.

If $1 \leqslant x \leqslant 2.5$, the equation is also valid.

For example if $x = 1$, $M = 0$ (bracketed terms negative or zero)

$x = 1.75$, $M = -[0.75]^2 + 0 = -0.5625$ kNm

$x = 2.5$, $M = -[1.5]^2 + 0 = -2.25$ kNm

The equation is also valid if $x > 2.5$.

For example, if $x = 4$, $M = -[3]^2 + [1.5]^2 = -6.75$ kNm

Example 17.3

The cantilever beam shown in Figure 17.6 is of rectangular steel section of width 100 mm and height 60 mm.

Determine the deflection at positions when (a) $x = 0$, (b) $x = 1$ m, (c) $x = 2$ m, (d) $x = 3$ m.

(e) Also check the maximum bending stress in the beam.

Solution

The bending moment equation is:

$$M = EI\frac{d^2y}{dx^2} = -[x - 1]^2 + [x - 2.5]^2$$

Integrating:

$$EI\frac{dy}{dx} = -\tfrac{1}{3}[x - 1]^3 + \tfrac{1}{3}[x - 2.5]^3 + A \qquad \textbf{(1)}$$

Integrating again:

$$EIy = -\tfrac{1}{12}[x - 1]^4 + \tfrac{1}{12}[x - 2.5]^4 + Ax + B \qquad \textbf{(2)}$$

With boundary condition $x = 4$, $\dfrac{dy}{dx} = 0$, substitute in (1):

$$0 = -\tfrac{1}{3}[3]^3 + \tfrac{1}{3}[1.5]^3 + A$$

$$\therefore A = 7.875 \text{ kNm}^2$$

With boundary condition $x = 4$, $y = 0$, substitute in (2):

$$0 = -\tfrac{1}{12}[3]^4 + \tfrac{1}{12}[1.5]^4 + 7.875 \times 4 + B$$

$$\therefore B = -25.17 \text{ kNm}^3$$

Substituting the values of A and B in equation (2):

$$EIy = -\tfrac{1}{12}[x - 1]^4 + \tfrac{1}{12}[x - 2.5]^4 + 7.875x - 25.17$$

Now $E = 200$ GPa and $I = \dfrac{bh^3}{12} = \dfrac{100 \times 60^3}{12} = 1.8 \times 10^6 \text{ mm}^4$

Substituting and remembering to adjust the units of E and I by the factor 10^{-9} yields:

$$y = \frac{-\frac{1}{12}[x - 1]^4 + \frac{1}{12}[x - 2.5]^4 + 7.875x - 25.17}{200 \times 10^3 \times 1.8 \times 10^6 \times 10^{-9}} \quad \text{(m)}$$

That is, the deflection in mm is given by:

$$y = \frac{-\frac{1}{12}[x - 1]^4 + \frac{1}{12}[x - 2.5]^4 + 7.875x - 25.17}{0.36}$$

The deflection at the required positions may now be determined by substitution in this equation with the appropriate value of x.

(a) $x = 0, y = \dfrac{0 + 0 + 0 - 25.17}{0.36} = \mathbf{-70\ mm}$

(b) $x = 1, y = \dfrac{0 + 0 + 7.875 \times 1 - 25.17}{0.36} = \mathbf{-48\ mm}$

(c) $x = 2, y = \dfrac{-\frac{1}{12}[1]^4 + 0 + 7.875 \times 2 - 25.17}{0.36} = \mathbf{-26.4\ mm}$

(d) $x = 3, y = \dfrac{-\frac{1}{12}[2]^4 + \frac{1}{12}[0.5]^4 + 7.875 \times 3 - 25.17}{0.36} = \mathbf{-8\ mm}$

(e) The maximum bending stress occurs at the position of maximum bending moment, i.e. at the support where $x = 4$ m and the bending moment M is:

$$M = -[3]^2 + [1.5]^2 = -6.75 \text{ kNm}$$

$$\therefore f_b(\text{max}) = \frac{My}{I} = \frac{6.75 \times 10^6 \times 30}{1.8 \times 10^6} = \mathbf{112.5\ MPa}$$

17.5 Cantilever beam: Combined loads

When combined loads act on a cantilever beam, the method follows the same general principles as previously outlined. Notice that if any of the loads act at an angle to the beam, and therefore have a component in an axial direction, the axial component has no effect on the deflection.

Consider the beam loaded as shown in Figure 17.8.

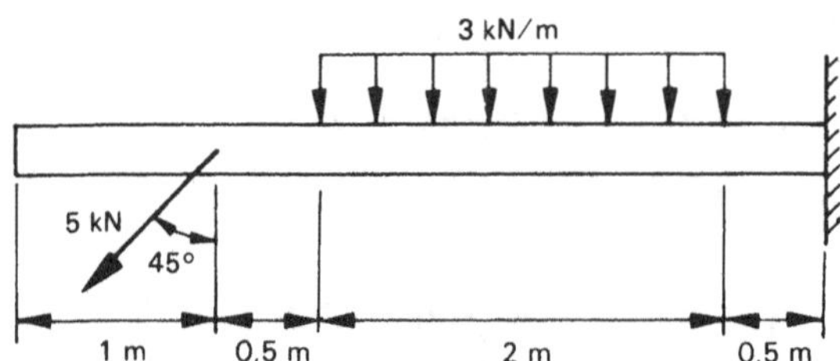

Fig. 17.8 *General case of combined loads on a cantilever beam*

For determining the deflection, equivalent loading as shown in Figure 17.9 may be used.

The bending moment in kNm at position x is:

$$M = -3.535[x - 1] - \frac{3[x - 1.5]^2}{2} + \frac{3[x - 3.5]^2}{2}$$

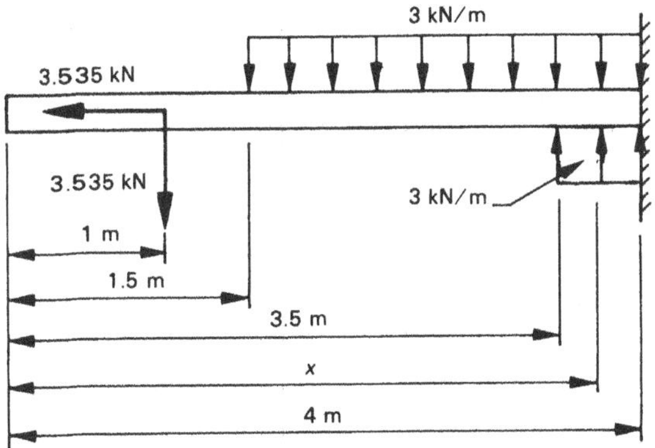

Fig. 17.9 *Equivalent loading*

Example 17.4

Determine the required second moment of area (and hence the dimensions) of a square section steel cantilever beam loaded as shown in Figure 17.8 if the maximum deflection is not to exceed 0.5 per cent of the length.

Also check the maximum axial stress in the beam with dimensions so determined in order to ensure that this stress does not exceed 100 MPa.

Solution

It was shown that the moment equation is:

$$M = EI\frac{d^2y}{dx^2} = -3.535[x - 1] - 1.5[x - 1.5]^2 + 1.5[x - 3.5]^2$$

Integrating:

$$EI\frac{dy}{dx} = \frac{-3.535}{2}[x - 1]^2 - \frac{1.5}{3}[x - 1.5]^3 + \frac{1.5}{3}[x - 3.5]^3 + A \qquad (1)$$

Integrating again:

$$EIy = \frac{-3.535}{6}[x - 1]^3 - \frac{1.5}{12}[x - 1.5]^4 + \frac{1.5}{12}[x - 3.5]^4 + Ax + B \qquad (2)$$

From (1) with boundary condition $x = 4$ m, $\frac{dy}{dx} = 0$:

$$0 = -\frac{3.535}{2}[3]^2 - \frac{1.5}{3}[2.5]^3 + \frac{1.5}{3}[0.5]^3 + A$$

$$\therefore A = 23.66 \text{ kNm}^2$$

From (2) with boundary condition $x = 4$ m, $y = 0$:

$$0 = -\frac{3.535}{6}[3]^3 - \frac{1.5}{12}[2.5]^4 + \frac{1.5}{12}[0.5]^4 + 23.66 \times 4 + B$$

$$\therefore B = -73.86 \text{ kNm}^3$$

The deflection equation with $E = 200$ GPa and I in mm^4 is:

$$y = \frac{\frac{-3.535}{6}[x - 1]^3 - \frac{1.5}{12}[x - 1.5]^4 + \frac{1.5}{12}[x - 3.5]^4 + 23.66x - 73.86}{200 \times 10^3 \times I \times 10^{-9}}$$

The maximum deflection is required to be 0.5 per cent of 4 m = −0.02 m (20 mm) which occurs when $x = 0$.

That is, $-0.02 = \dfrac{0 - 0 + 0 + 0 - 73.86}{200 \times 10^{-6} \times I}$

$\therefore I = 18.5 \times 10^6$ mm^4

If the beam is of square section width b (and height $h = b$):

$$I = \frac{bh^3}{12} = \frac{b^4}{12}$$

$$\therefore \frac{b^4}{12} = 18.5 \times 10^6 \quad \text{or} \quad b = \mathbf{122\ mm}$$

The beam is subject to combined bending and axial loading so that the combined axial stress is:

$$f = \frac{My}{I} + \frac{F}{A}$$

The maximum bending moment occurs at the support and is:

$$\begin{aligned} M(\max) &= -3.535[3] - 1.5[2.5]^2 + 1.5[0.5]^2 \\ &= -19.6 \text{ kNm} \end{aligned}$$

The maximum axial stress (tensile) may now be obtained:

$$\begin{aligned} f(\max) &= \frac{19.6 \times 10^6 \times 61}{18.5 \times 10^6} + \frac{3.535 \times 10^3}{122 \times 122} \\ &= \mathbf{64.9\ MPa} \end{aligned}$$

Note that in this case the axial stress due to the axial component of the load is very small compared to the bending stress.

17.6 Simply supported beam: Concentrated loads

When a beam is simply supported rather than a cantilever, the same general principles are followed. The moment equation is written for a position between the furthest load to the right and the right-hand support. This equation is integrated twice and the constants of integration evaluated from the boundary conditions. The differences in application of Macaulay's method are:

1. The reactions must be calculated first, particularly the vertical component of the reaction at the left-hand support must be known before the moment equation can be written. This was not necessary in a cantilever beam when the support was to the right and the moment equation could be written without knowing the support reaction.
2. The boundary conditions are different. Usually the position of zero slope is unknown but positions of zero deflection occur at both supports. If the beam is symmetrically

loaded about the centre of the span, then zero slope occurs at the centre and coincides with the position of maximum deflection.

3. The position of maximum deflection in a simply supported beam always occurs at the position where the slope is zero. However unless the beam is symmetrically loaded, this position cannot be determined by inspection as it does not occur at the centre of the beam or necessarily under one of the loads.

Example 17.5

Determine the deflection at 500 mm intervals along the simply supported steel flange beam shown in Figure 17.10 and hence draw the elastic curve of the beam to scale. From the elastic curve, determine the maximum deflection and the position where it occurs. Also check the maximum bending stress.

The second moment of the beam section about the neutral axis is 12.6×10^6 mm^4 and the depth of the section is 152 mm.

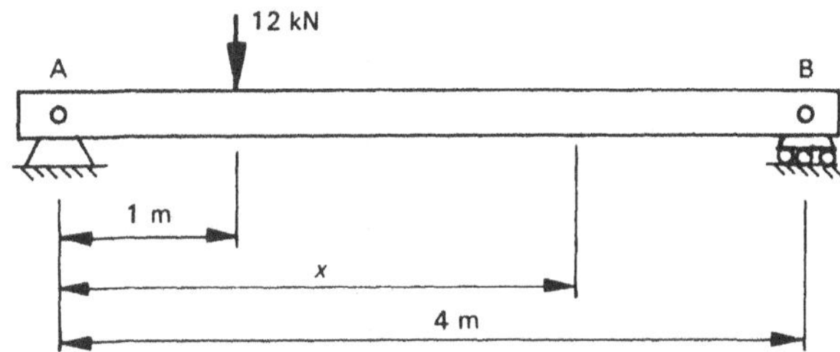

Fig. 17.10

Solution

First determine the reactions at A and B.

Moments about B: $R_A \times 4 - 3 \times 12 = 0$

$$\therefore R_A = 9 \text{ kN},$$
$$R_B = 3 \text{ kN}$$

The moment equation at position x may now be written:

$$M = 9x - 12[x - 1] = EI\frac{d^2y}{dx^2}$$

Integrating:

$$EI\frac{dy}{dx} = \frac{9}{2}x^2 - \frac{12}{2}[x - 1]^2 + A$$
$$= 4.5x^2 - 6[x - 1]^2 + A$$

Integrating again:

$$EIy = \frac{4.5}{3}x^3 - \frac{6}{3}[x - 1]^3 + Ax + B$$
$$= 1.5x^3 - 2[x - 1]^3 + Ax + B$$

From the boundary condition: $y = 0$ when $x = 0$ (support at A)

$$\therefore 0 = 0 - 0 + 0 + B$$
$$\therefore B = 0$$

Also $y = 0$ when $x = 4$ m (support at B)

$$\therefore 0 = 1.5 \times 4^3 - 2[3]^3 + A \times 4$$

$$\therefore A = -10.5 \text{ kNm}^2$$

$$\therefore EIy = 1.5x^3 - 2[x - 1]^3 - 10.5x$$

Now $E = 200$ GPa and $I = 12.6 \times 10^6 \text{ mm}^4$

$$\therefore EI = 200 \times 10^3 \times 12.6 \times 10^6 \times 10^{-9}$$ (note the 10^{-9} factor is required as before)

$$= 2520$$

Hence the deflection equation with x and y both in m is:

$$y = \frac{1.5x^3 - 2[x - 1]^3 - 10.5x}{2520}$$

Or if y is in mm:

$$y = \frac{1.5x^3 - 2[x - 1]^3 - 10.5x}{2.52}$$

Substituting in this equation, the deflection at 500 mm intervals along the span may be calculated and tabulated:

x (m)	0	0.5	1	1.5	2	2.5	3	3.5	4
y (mm)	0	−2.0	−3.57	−4.34	−4.365	−3.79	−2.78	−1.46	0

The elastic curve may now be drawn to scale (Fig. 17.11).

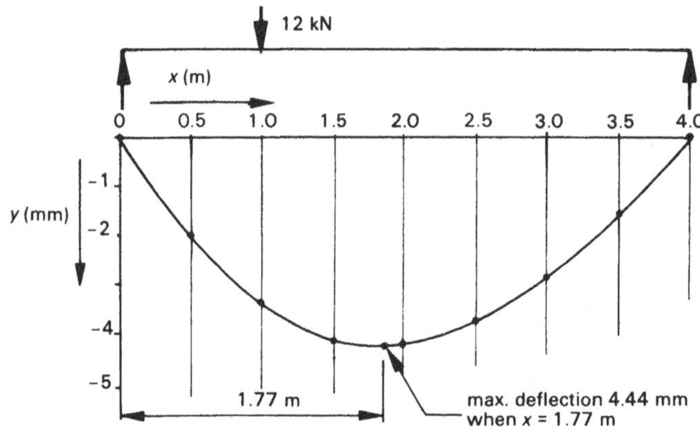

Fig. 17.11

From the curve it is clear that the maximum deflection occurs when x = **1.77 m** and is **−4.44 mm.**

Finally the maximum bending stress may be checked. Maximum bending moment occurs under the 12 kN load and is:

$$M = 9 \text{ kNm} = 9 \times 10^6 \text{ Nmm}$$

$$f_b = \frac{My}{I} = \frac{9 \times 10^6 \times \frac{152}{2}}{12.6 \times 10^6} = \mathbf{54.3\ MPa}$$

17.7 Simply supported beam: Distributed load

When a distributed load acts over the entire span of a simply supported beam as shown in Figure 17.12, the bending moment at position x is:

$$M = \frac{WL}{2}x - \frac{Wx^2}{2}$$

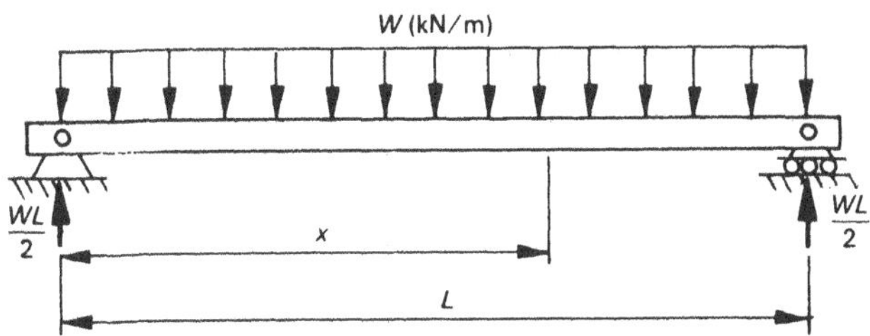

Fig. 17.12 *Distributed load on a simply supported beam*

The use of Macaulay's method is straightforward. However if the distributed load acts over only part of the span, a compensating distributed load must be used as was the case with this type of loading on a cantilever beam (as previously discussed). The method is illustrated in example 17.6.

Example 17.6

A timber beam has a rectangular section, width 250 mm and height 400 mm, and is loaded as shown in Figure 17.13.

Determine the deflection at the centre of the beam (maximum deflection). The modulus of elasticity of the timber is 12 GPa.

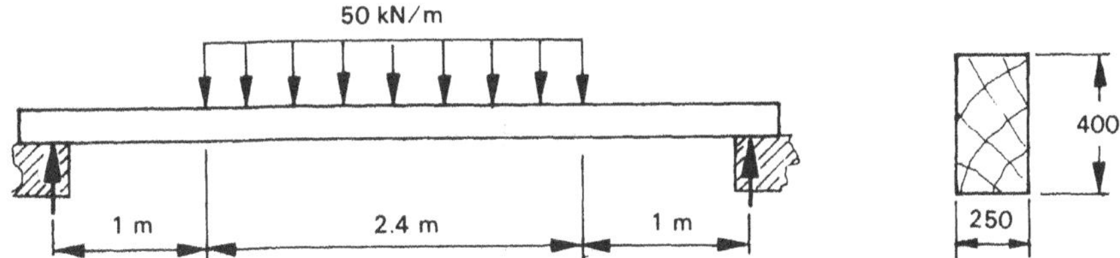

Fig. 17.13

Solution

The reactions are equal and are: $\dfrac{50 \times 2.4}{2} = 60$ kN

Using a compensating distributed load and choosing position x as shown in Figure 17.14, the bending moment equation is:

$$M = EI\frac{d^2y}{dx^2} = 60x - \frac{50[x - 1]^2}{2} + \frac{50[x - 3.4]^2}{2}$$

Integrating:

$$EI\frac{dy}{dx} = \frac{60}{2}x^2 - \frac{50}{6}[x - 1]^3 + \frac{50}{6}[x - 3.4]^3 + A \qquad (1)$$

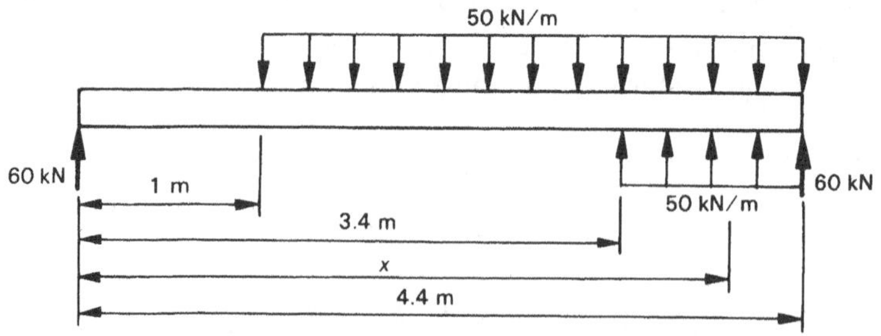

Fig. 17.14

Integrating again:

$$EIy = \frac{60}{6}x^3 - \frac{50}{24}[x - 1]^4 + \frac{50}{24}[x - 3.4]^4 + Ax + B \qquad \textbf{(2)}$$

With boundary condition $x = 0, y = 0 \quad \therefore B = 0$
and when $x = 4.4, y = 0$. Substituting in equation (2):

$$0 = \frac{60}{6} \times 4.4^3 - \frac{50}{24}[3.4]^4 + \frac{50}{24}[1]^4 + A \times 4.4$$

$$\therefore A = -130.8 \text{ kNm}^2$$

Note: Since the beam is symmetrically loaded, the slope is zero at the centre, that is $\frac{dy}{dx} = 0$ when $x = 2.2$ m. Substituting in (1):

$$0 = \frac{60}{2} \times 2.2^2 - \frac{50}{6}[1.2]^3 + A$$

$$\therefore A = -130.8 \text{ kNm}^2 \quad \text{(as above)}$$

Now $EI \times 10^{-9} = 12 \times 10^3 \times \frac{250 \times 400^3}{12} \times 10^{-9} = 16 \times 10^3$

Hence the deflection equation (y in mm) is:

$$y = \frac{10x^3 - \frac{50}{24}[x - 1]^4 + \frac{50}{24}[x - 3.4]^4 - 130.8x}{16}$$

The deflection at the centre (when $x = 2.2$ m) is:

$$y = \frac{10 \times 2.2^3 - \frac{50}{24}[1.2]^4 - 130.8 \times 2.2}{16}$$

$$= \mathbf{-11.6 \text{ mm}}$$

17.8 Simply supported beam: Combined loads

When a simply supported beam carries both distributed and concentrated loads, each load could be treated separately and the principle of superposition used but this is laborious. It is simpler to use Macaulay's method and write one equation for the bending moment for the entire beam taking into account all the loads. If the loading is not symmetrical, then the maximum deflection does not occur at the centre of the span. The maximum deflection and the position where it occurs may be determined by drawing the elastic curve of the beam to scale (by calculating the deflection at several positions along the span). Alternatively a numerical method (preferably using a programmable calculator or computer) may be used to determine the maximum value.

When the beam section is to be selected to ensure a certain deflection is not exceeded, then the easiest method is to choose an arbitrary value of I (say 1×10^6 mm^4) and determine the deflection accordingly. Given the required maximum deflection, the value of I required may now be determined by proportioning, since for a given loading on the beam, the deflection is inversely proportional to I. This procedure is illustrated in the following example.

Example 17.7

The steel beam shown in Figure 17.15 is loaded with a load of 20 kN (which may be considered concentrated) and a load of 30 kN distributed over 800 mm of the span.

Determine the required second moment of area of the beam about the neutral axis so that the maximum deflection does not exceed 0.5 per cent of the span.

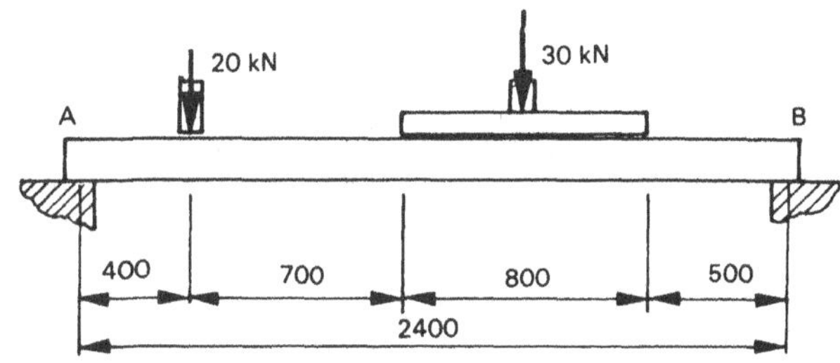

Fig. 17.15

Solution

Taking moments about B:

$$R_A \times 2.4 - 20 \times 2 - 30 \times 0.9 = 0$$

$$\therefore R_A = 27.92 \text{ kN}$$

$$R_B = 22.08 \text{ kN}$$

The equivalent loading on the beam is shown in Figure 17.16.

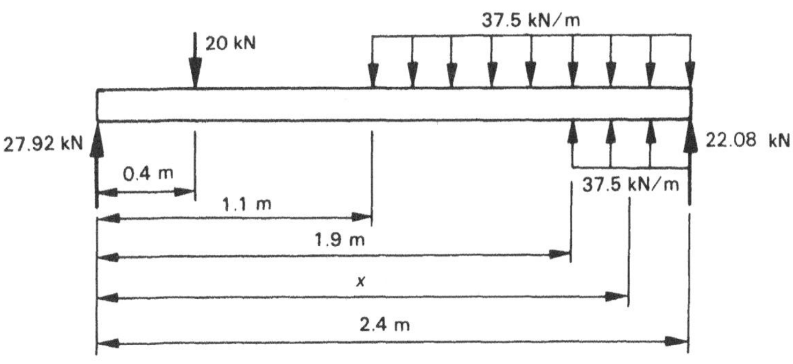

Fig. 17.16

Writing the moment equation at position x:

$$M = EI\frac{d^2y}{dx^2} = 27.92x - 20[x - 0.4] - \frac{37.5}{2}[x - 1.1]^2 + \frac{37.5}{2}[x - 1.9]^2$$

Integrating:

$$EI\frac{dy}{dx} = \frac{27.92}{2}x^2 - \frac{20}{2}[x - 0.4]^2 - \frac{37.5}{6}[x - 1.1]^3 + \frac{37.5}{6}[x - 1.9]^3 + A \qquad (1)$$

Integrating again:

$$EIy = \frac{27.92}{6}x^3 - \frac{20}{6}[x - 0.4]^3 - \frac{37.5}{24}[x - 1.1]^4 + \frac{37.5}{24}[x - 1.9]^4 + Ax + B \qquad (2)$$

With boundary condition $x = 0, y = 0 \therefore B = 0$
With boundary condition $x = 2.4$, $y = 0$, substituting in (2):

$$0 = \frac{27.92}{6} \times 2.4^3 - \frac{20}{6}[2]^3 - \frac{37.5}{24}[1.3]^4 + \frac{37.5}{24}[0.5]^4 + A \times 2.4$$

$$\therefore A = -13.87 \text{ kNm}^2$$

From equation (2) with $E = 200 \times 10^3$ MPa and giving I a nominal value of 1×10^6 mm^4, the deflection equation is (y in m):

$$y = \frac{\frac{27.92}{6}x^3 - \frac{20}{6}[x - 0.4]^3 - \frac{37.5}{24}[x - 1.1]^4 + \frac{37.5}{24}[x - 1.9]^4 - 13.87x}{200}$$

Using a numerical method with a computer, the maximum (absolute) value of this equation is found to be -0.0516 when $x = 1.21$ m. That is the maximum deflection is 51.6 mm (0.0516 m).

The required maximum deflection is 0.5 per cent of 2400 mm or 12 mm.

Hence the required value of I to give this deflection is:

$$I = 1 \times 10^6 \times \frac{51.6}{12} = 4.3 \times 10^6 \text{ mm}^4$$

17.9 Computer solution

The lengthy numerical computations involved in the determination of beam deflections are readily amenable to a computer program. A typical program BEAMDEF as listed in Appendix 3.19 may be used to determine the deflection of simply supported or cantilever beams with any combination of concentrated or distributed loads. The solution to examples 17.2, 17.3 and 17.7 using this program is also given in Appendix 3.19 and should be self-explanatory.

Notes

1. The following sign convention is used for loads and deflections:
   Loads acting up are positive and deflections upward are positive.
2. The program outputs the deflection at ten equally spaced positions along the beam (numbered 0–10) starting with position 0 at the left-hand end. Therefore the centre of the beam is position 5.
3. The program also calculates the deflection at any required position on the span. This

facility may be used to determine the maximum deflection by trial and error (which may be done rapidly by the computer).

4. Cantilever beams are oriented with the support to the right (free end to the left).
5. The sequence in which the loads are input is immaterial. However if both concentrated and distributed loads act on the beam, the concentrated loads are all input before any of the distributed loads.
6. If it is necessary to determine the second moment of area of the section to ensure that a maximum deflection value is not exceeded, the program may be run using a nominal value of I (say 1×10^6 mm^4) and the deflection determined. The required value of I may then be obtained by proportioning the deflection obtained to the maximum deflection required. This procedure was outlined in example 17.7.

Problems

For these problems assume E for steel = 200 GPa in all cases.

17.1 A steel flange beam for which the second moment of area about the centroid is 370×10^6 mm^4 is used as a cantilever of length 6 m and carries a load of 15 kN at its midpoint.

Determine:
(a) deflection at the free end
(b) deflection at the midpoint
(c) maximum bending stress if the depth of section is 460 mm.

(a) −4.56 mm (b) −1.83 mm (c) 28 MPa

17.2 Repeat problem 17.1 if the beam carries a uniformly distributed load of 3 kN/m over the entire span instead of the concentrated load.

(a) −6.57 mm (b) −2.33 mm (c) 33.6 MPa

17.3 If both the loadings given in problems 17.1 and 17.2 occur together on the beam, repeat the calculations.

(a) −11.1 mm (b) −4.15 mm (c) 61.6 MPa

17.4 Determine the slope at the free end and midpoint of the cantilever beam given in problem 17.1 and explain the result.

0.912×10^{-3}

17.5 Determine the slope at the free end and midpoint of the cantilever beam given in problem 17.2.

1.46×10^{-3}, 1.28×10^{-3}

17.6 A steel rod, diameter 70 mm, is loaded as shown in Figure P17.6.

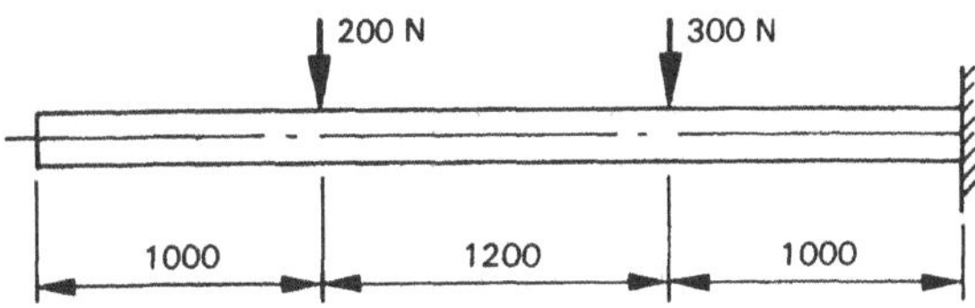

Fig. P17.6

Determine:
(a) deflection at the free end
(b) deflection at the centre
(c) maximum bending stress.

(a) −6.89 mm (b) −2.62 mm (c) 22 MPa

17.7 Repeat problem 17.6 including the self-weight loading of the rod which may be taken as 300 N/m.

(a) −23.6 mm (b) −8.53 mm (c) 67.6 MPa

17.8 Determine the deflection at the free end of the steel cantilever beam shown in Figure P17.8 if the second moment of area of the section about the neutral axis is 100×10^6 mm^4.

−4.71 mm

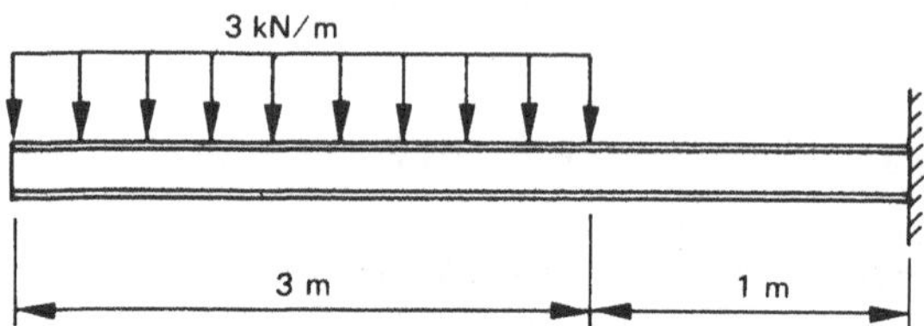

Fig. P17.8

17.9 Determine the deflection at points a, b and c for the timber cantilever beam shown in Figure P17.9. The modulus of elasticity of the timber is 12 GPa.

(a) −93.4 mm (b) −64.5 mm (c) −11.1 mm

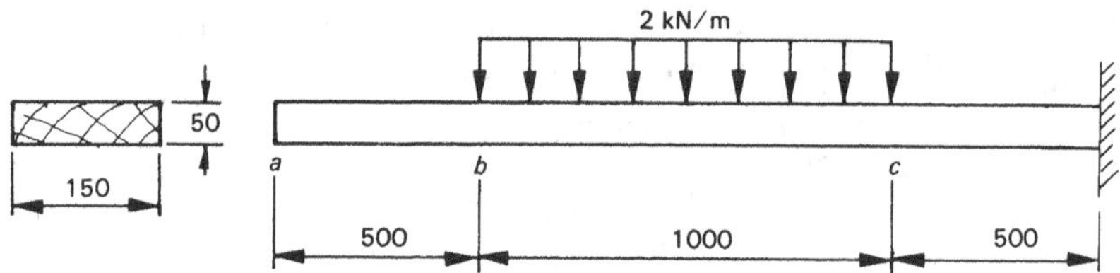

Fig. P17.9

17.10 Determine the deflection at points a, b and c for the steel cantilever beam loaded as shown in Figure P17.10. The second moment of area about the neutral axis is 2.5×10^6 mm^4.

(a) −26.7 mm (b) −23.3 mm (c) −6 mm

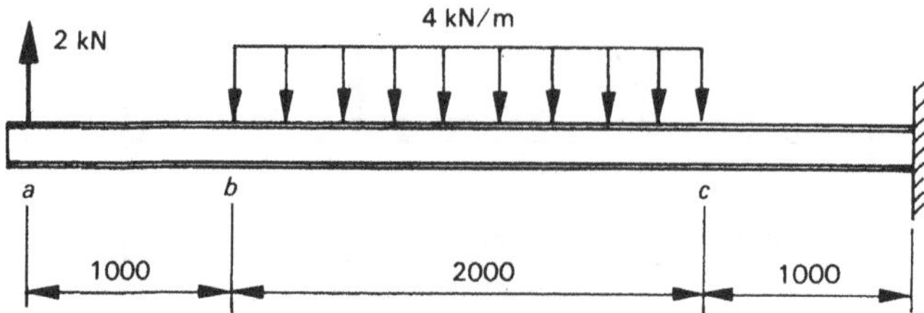

Fig. P17.10

17.11 The steel shaft shown in Figure P17.11 is rigidly supported at A.

Determine the shaft diameter necessary so that the vertical deflection at E does not exceed 3 mm.

31.4 mm

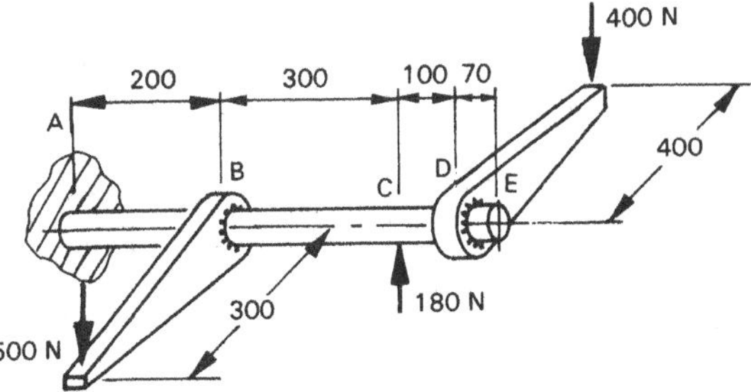

Fig. P17.11

17.12 A mild steel shaft is supported by bearings 500 mm apart and is fitted with a pulley at its midpoint. The sum of the belt tensions at the pulley is 1.2 kN.

(a) Determine the shaft diameter required so that the maximum deflection in the shaft does not exceed 2 mm.

(b) Check the maximum bending stress for the shaft with diameter calculated in (a) and hence determine if this diameter is feasible.

(a) 20 mm (b) 192 MPa (this stress is too high)

17.13 A simply supported steel flange beam, for which the second moment of area about the neutral axis is 215×10^6 mm^4, carries a concentrated load of 30 kN at the midpoint of the 5 m span.

Determine:

(a) maximum deflection

(b) maximum bending stress if the depth of section is 406 mm.

(a) −1.82 mm (b) 35.4 MPa

17.14 Repeat problem 17.13 if the beam carries a uniformly distributed load of 5 kN/m over the entire span rather than the concentrated load.

(a) −0.95 mm (b) 14.75 MPa

17.15 If both the loadings given in problems 17.13 and 17.14 occur together on the beam, repeat the calculations.

(a) −2.77 mm (b) 50.15 MPa

17.16 Determine the slope at the right-hand end of the beam given in problem 17.13.

1.09×10^{-3}

17.17 Determine the slope at the right-hand end of the beam given in problem 17.14.

0.606×10^{-3}

17.18 A simply supported steel beam, 5 m span, carries a load of 10 kN at a distance of 2 m from the left-hand end support. The second moment of area of the beam section about the neutral axis is 23.6×10^6 mm^4.

Determine:

(a) deflection under the load

(b) deflection at the midpoint

(c) maximum bending stress if the depth of section is 203 mm.

(a) −5.08 mm (b) −5.21 mm (c) 51.6 MPa

17.19 For the beam given in problem 17.18, determine the maximum deflection and position at which it occurs.

−5.23 mm, 2.35 m from the left-hand support

17.20 Two 100 × 50 mm timber beams are used to support a mass of 1 tonne distributed uniformly over a length of 1500 mm as shown in Figure P17.20.

Determine the maximum deflection if the modulus of elasticity of the timber is 12 GPa.

−27.1 mm

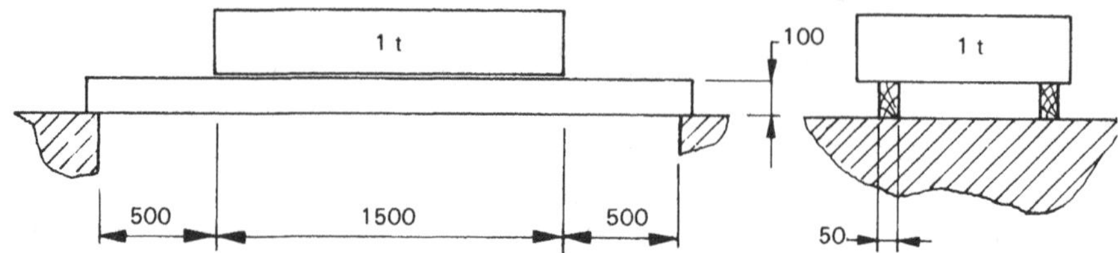

Fig. P17.20

17.21 The steel beam shown in Figure P17.21 has a second moment of area about the neutral axis of 28.9×10^6 mm^4.

Use Macaulay's method to determine the deflection at the centre of the beam when loads P and Q are as given below:

(a) $P = 30$ kN, $Q = 0$
(b) $P = 0$, $Q = 20$ kN
(c) $P = 30$ kN, $Q = 20$ kN

Hence show that the principle of superposition is valid for beam deflection.

(a) −19.9 mm (b) −13.3 mm (c) −33.2 mm

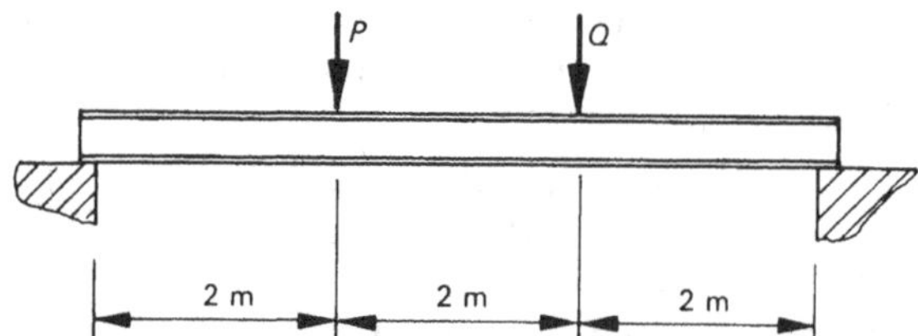

Fig. P17.21

17.22 A steel shaft, diameter 30 mm is loaded as shown in Figure P17.22.

Determine the deflection at ten positions equally spaced at 100 mm intervals along the shaft. Hence draw the elastic curve of the shaft to scale and determine the maximum deflection and position at which it occurs.

Solution (deflections in mm)

0	1	2	3	4	5	6	7	8	9	10
0	0.07	0.11	0.07	-0.01	-0.1	-0.16	-0.17	-0.14	-0.08	0

Maximum deflection = −0.175 mm at 0.67 m from the left-hand end.

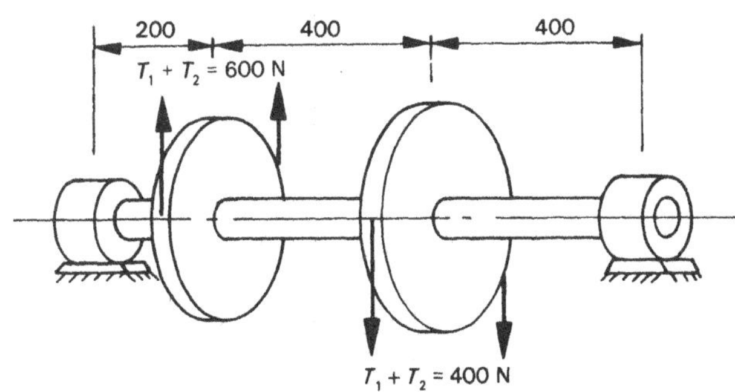

Fig. P17.22

17.23 Repeat problem 17.22 if the belt tensions both acted upward, all other data being the same.

Solution (deflections in mm)

0	1	2	3	4	5	6	7	8	9	10
0	0.63	1.18	1.58	1.82	1.88	1.77	1.49	1.07	0.56	0

Maximum deflection = 1.88 mm at 0.49 m from the left-hand end.

17.24 The composite bonded beam shown in Figure P17.24 is manufactured from material with a modulus of elasticity of 12 GPa.

Determine:

(a) second moment of area about the neutral axis
(b) maximum shear stress in the bonded joints
(c) maximum bending stress
(d) maximum deflection.

(a) 16.2×10^6 mm^4 (b) 0.111 MPa (c) 1.73 MPa (d) −5.53 mm

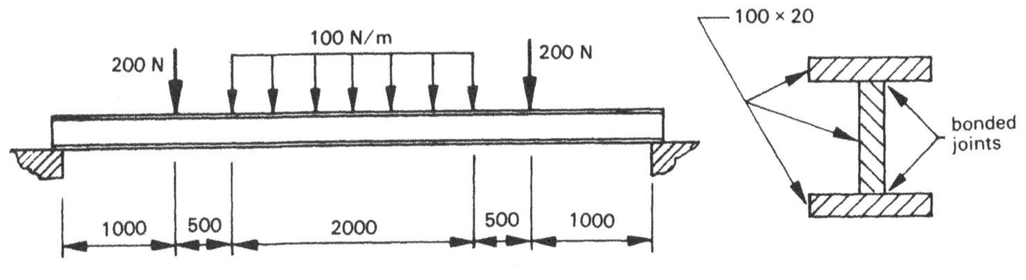

Fig. P17.24

18

Columns

Frequently, members loaded in compression will fail by buckling rather than by excessive compression stress or shear stress on an inclined plane. The possible modes of failure of a compression member are shown in Figure 18.1.

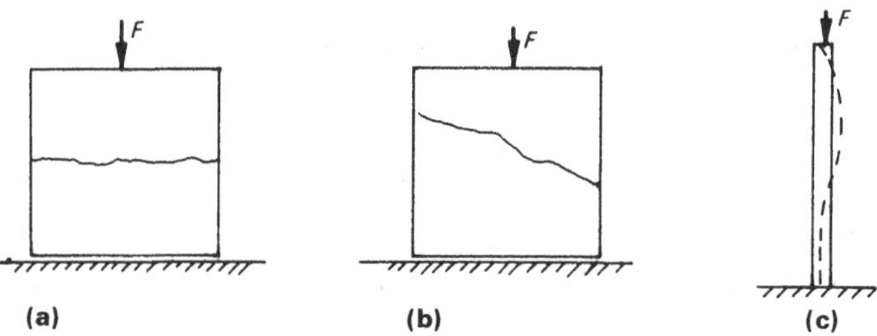

Fig. 18.1 *Possible modes of failure of a compression member: (a) compression, (b) shear on an inclined plane, (c) buckling*

For a compression member with given cross-section, buckling failure becomes increasingly more probable as the length of the member increases, that is, as the member becomes more slender. Compression members where buckling is the most likely mode of failure are known as **columns**.

Buckling failure in columns (or struts) is virtually the same as bending failure except that the load is applied axially and not transversely. In theory, if the load were exactly in line with the neutral axis of the compression member and the member itself were perfectly straight and of uniform properties, then buckling failure would not occur even with a long compression member. However, in practice these idealistic conditions are impossible to achieve and hence buckling failure is the most likely mode of failure in long compression members (columns).

18.1 Effective length of a column (L_e)

The buckling load for a column of given cross-section depends upon the length of the column and the end conditions or end restraints. Obviously, a column of given length with both ends firmly fixed can carry a greater load than the same column with the ends pinned (strut) or when one end is fixed and the other end free (cantilever column).

One method of dealing with the end restraints is to use the actual length of the column in column formulas and apply a factor which takes into account the different degrees of fixation associated with the various end restraints.

Another method is to write the column formulas in terms of the effective length of the column rather than the actual length, where the effective length takes into account the fixation of the end restraints. In this case no additional factor is necessary in the formulas to take into account the effect of the different possible end restraints.

The approach of using effective length rather than actual length is felt to be more logical and easier to apply and hence is the method used in this book. The effective length depends upon the shape of the deformed elastic curve of the column and for a column pin jointed at both ends, the effective length is the same as the actual length. For other types of end restraints the effective length relative to the actual length of the column is shown in Figure 18.2.

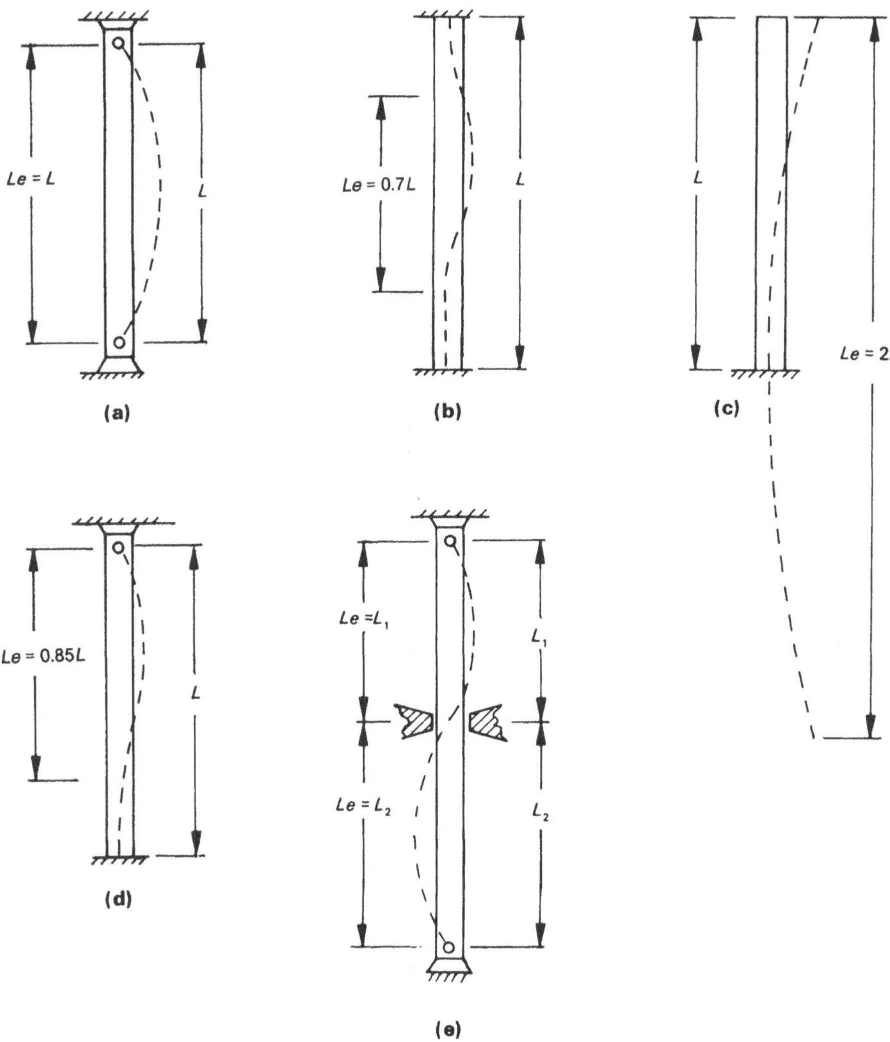

Fig. 18.2 *Effective length of columns with various end restraints: (a) ends pinned, (b) ends fixed, (c) cantilever, (d) one end fixed, other end pinned, (e) pinned ends, intermediate lateral restraint*

Notes

1. For restraints intermediate between those shown in Figure 18.2, effective length values may be estimated or obtained from codes or handbooks. For example, if the ends are fixed but not fully rigid, an effective length value between $0.5L$ and L is appropriate.
2. In Figure 18.2(e), the column acts effectively as two columns in series, each one with the ends pinned and the larger effective length should be used to check buckling.
3. In practice, a pinned column is closely approximated by one in which the ends are restrained from lateral movement but not against flexure. Typical examples are illustrated in Figure 18.3.

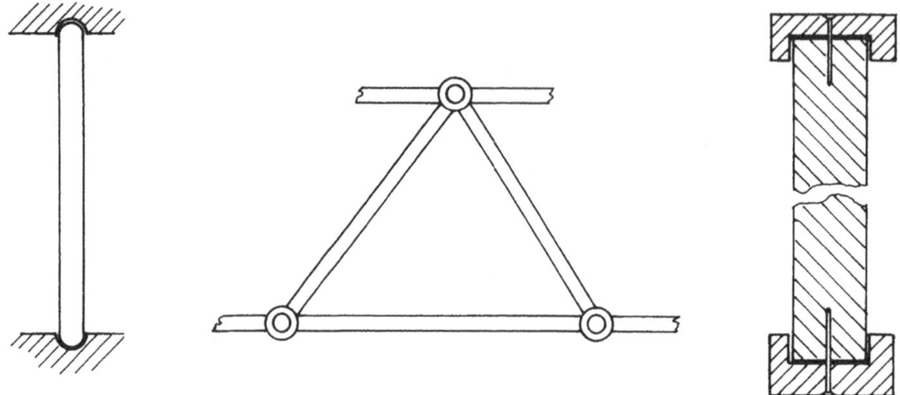

Fig. 18.3 *Examples of end restraints which may be considered as pinned*

18.2 Slenderness ratio

The most useful property of the cross-section of a column is the radius of gyration (based on area) as defined in Chapter 14 and given by equation 14.7:

$$k = \sqrt{\frac{I}{A}}$$

A measure of the slenderness of the column is obtained by dividing the effective length of the column by the radius of gyration. This ratio is called the slenderness ratio and is given by:

$$\text{slenderness ratio} = \frac{L_e}{k}$$

where L_e = effective length of the column (mm)
k = radius of gyration (mm) measured from the centroid (neutral axis of the section)

Notes

1. The larger the slenderness ratio, the greater the tendency for buckling, that is the smaller the load the column can carry. As the slenderness ratio decreases, buckling failure becomes less likely and, for slenderness ratios below about 30, the member is so stiff that buckling failure will not occur. In such cases, the member need not be considered as a column at all and may be treated purely as a compression member.

2. Both the effective length and the radius of gyration must be relevant to the plane of buckling, if these are different for different planes around the section of the column. Buckling failure will occur on that plane which has the largest slenderness ratio.
3. Flange beam sections are not very efficient as columns because for these sections I_{cy} is considerably lower than I_{cx} and hence k_y is considerably less than k_x (k_y varying from about 20–30 per cent of k_x). However, rolled steel column sections are available in a range of sizes and these have a larger flange width and therefore are more efficient as columns (k_y being about 50–60 per cent of k_x).

Example 18.1

A tubular section is used as a column as shown in Figure 18.4. In (a) the ends are both firmly fixed, whereas in (b) only one end is fixed (cantilever column).

Determine the slenderness ratio in each case.

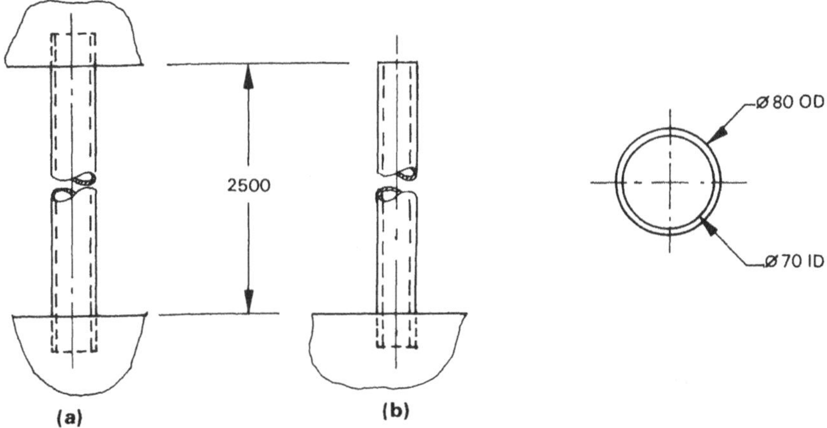

Fig. 18.4

Solution

Since the properties of the section are the same about any plane, buckling failure is equally likely about any plane and the radius of gyration is given by:

$$k = \sqrt{\frac{I}{A}} = \sqrt{\frac{\pi(D^4 - d^4)}{64 \times \frac{\pi}{4}(D^2 - d^2)}} = \sqrt{\frac{(D^4 - d^4)}{16(D^2 - d^2)}} = \frac{1}{4}\sqrt{D^2 + d^2}$$

$$\therefore k = \frac{1}{4}\sqrt{80^2 + 70^2} = 26.575 \text{ mm}$$

(a) Since both ends are fixed, $L_e = 0.7$ and the slenderness ratio is:

$$\frac{L_e}{k} = \frac{0.7 \times 2500}{26.575} = \mathbf{65.85}$$

(b) Since one end is fixed and the other end free, $L_e = 2L$ and the slenderness ratio is:

$$\frac{L_e}{k} = \frac{2 \times 2500}{26.575} = \mathbf{188}$$

Example 18.2

The 32 × 32 angle with properties as shown in Figure 18.5 is used as a strut in a frame.

Determine the maximum length of the strut so that the slenderness ratio does not exceed 200. The end restraints are such that the effective length is 90 per cent of the actual length (about any axis).

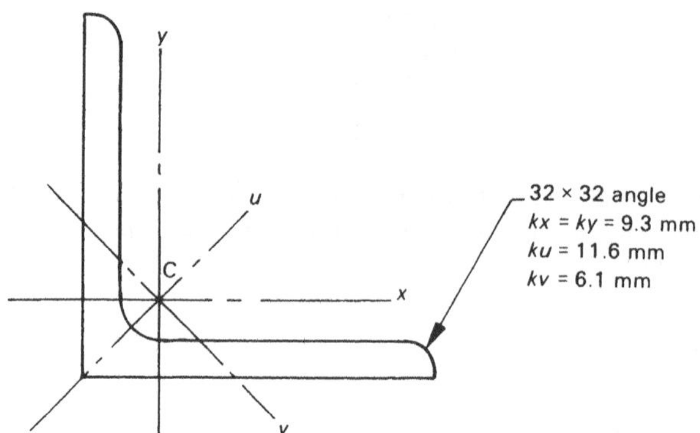

Fig. 18.5

Solution

Buckling failure is most likely about the plane for which the slenderness ratio is greatest. In this case, since the effective length is the same about any axis, maximum slenderness ratio occurs about the axis for which the radius of gyration is the smallest. For the angle given, the minimum radius of gyration occurs about the v axis where $k = 6.1$ mm.

$$\text{Slenderness ratio} = \frac{L_e}{k}$$

$$\therefore 200 = \frac{L_e}{6.1}$$

$$\therefore L_e = 1220 \text{ mm}$$

Since $L_e = 0.9L$

$$L = \frac{1220}{0.9} = \mathbf{1356\ mm}$$

Example 18.3

A connecting rod for an engine has a rectangular section as illustrated in Figure 18.6.

Determine the slenderness ratio for buckling about both the x and y axes planes and hence determine about which plane buckling is most likely to occur.

Assume neat clearance fits at both ends of the rod so that about the x axis plane $L_e = L$ (pinned restraints) and about the y axis plane $L_e = 0.5L$ (fixed restraints).

Solution

For a rectangular section $I = \dfrac{bh^3}{12}$ and $A = bh$

$$k = \sqrt{\frac{I}{A}} = \sqrt{\frac{bh^3}{12bh}} = \sqrt{\frac{h^2}{12}} = \frac{h}{\sqrt{12}}$$

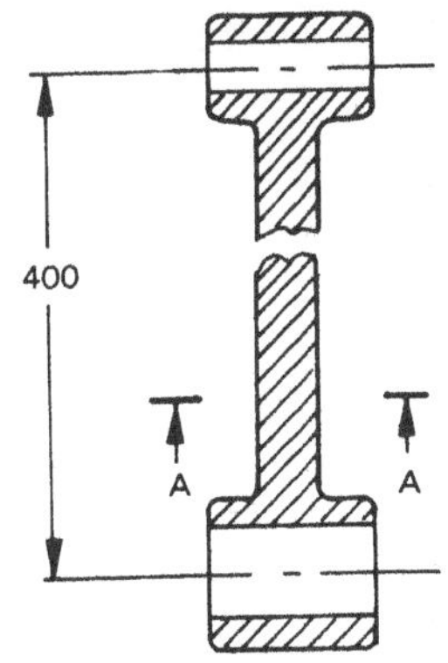

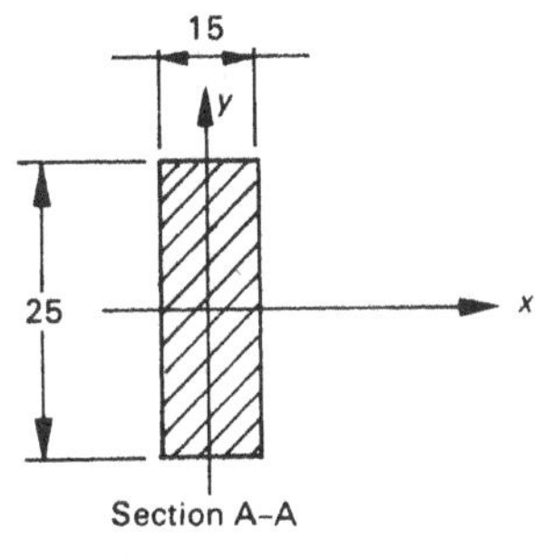

Fig. 18.6

Buckling about the x axis plane:

$$\text{slenderness ratio} = \frac{L_e}{k} = \frac{400}{25/\sqrt{12}} = \mathbf{55.4}$$

Buckling about the y axis plane:

$$\text{slenderness ratio} = \frac{L_e}{k} = \frac{0.7 \times 400}{15/\sqrt{12}} = \mathbf{64.7}$$

Therefore the rod is more likely to buckle about the y axis plane than the x axis plane because the slenderness ratio is greater about this plane.

18.3 Slender column buckling

One of the earliest column buckling formulas was derived theoretically in 1759 by Leonard Euler. This formula is still applicable today for slender columns (that is columns with high slenderness ratios) under elastic deformation conditions.

Consider the slender pinned column shown in Figure 18.7(a) in which an axial load F is applied and causes the column to deform elastically.

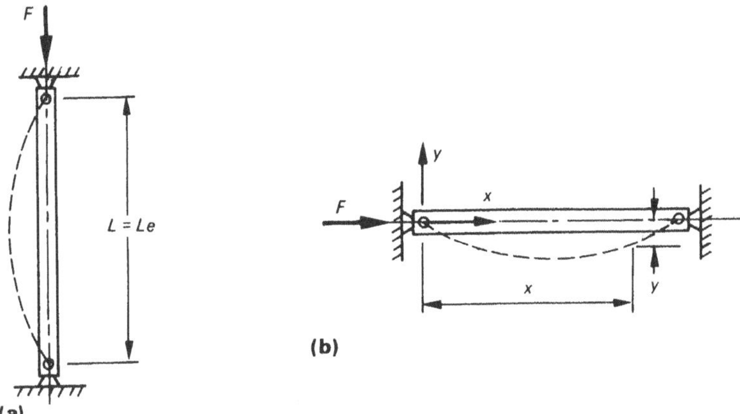

Fig. 18.7 *Slender column deformation*

If the column is rotated through 90°, the familiar elastic curve for a beam is obtained (see Fig. 18.7(b)).

The equation of the elastic curve (as used in Chap. 17) is:

$$EI\frac{d^2y}{dx^2} = M$$

Now the moment M (at any position x) is caused by the axial load F acting over the deflected distance y.

Hence the equation for the elastic curve may be written:

$$EI\frac{d^2y}{dx^2} = Fy$$

This is a differential equation and the mathematics of the solution are beyond the scope of this book. In fact there are several solutions, but the smallest value of F which causes buckling is known as the **critical load**. In practice, therefore, the critical load is taken as the maximum load which may be applied under elastic deformation conditions. The value of the critical load derived by Euler is:

$$F_{CR} = \frac{\pi^2 EA}{(L_e/k)^2}$$

(18.1) critical load, slender columns

Notes

1. Since the slenderness ratio L_e/k is dimensionless, and π is dimensionless, F_{CR} has units N when E is in MPa and A is in mm^2.
2. The critical load as calculated by the Euler formula does not include a safety allowance. Therefore, for design purposes a safety factor should be applied to F_{CR} to obtain the allowable or safe load.
3. The Euler formula was derived for pinned columns but is applicable for any conditions of end restraint since the formula uses the effective length rather than the actual length.

Example 18.4

Determine the critical buckling load for the steel angle strut given in example 18.2 if the cross-sectional area of the angle is 361 mm^2. The slenderness ratio is 200 and E = 200 GPa.

Solution

$$\text{Now } \frac{L_e}{k} = 200$$

$$A = 361 \text{ mm}^2$$

$$E = 200 \times 10^3 \text{ MPa}$$

The column is a slender one and the Euler formula may be applied:

$$F_{CR} = \frac{\pi^2 EA}{(L_e/k)^2}$$

$$= \frac{\pi^2 \times 200 \times 10^3 \times 361}{200^2} \quad \text{(N)}$$

$$= \mathbf{17.8\ kN}$$

18.4 Thick column buckling

It has been found experimentally that the Euler formula does not apply to thick columns (that is columns with relatively low slenderness ratios) because the yield point may be reached before sideways bowing occurs. Under these conditions there may be plastic deformation at lower loads than predicted by the Euler formula.

The theory of thick columns is complex and most formulas have been derived experimentally. There are many in use and design codes and standards usually specify the formulas applicable for the particular materials of construction.

A formula of general applicability which may be used for thick columns is known as the **Johnson formula** (derived from the work of J. B. Johnson).

The formula is:

$$F_{CR} = Af_y\left[1 - \frac{f_y(L_e/k)^2}{4\,\pi^2\,E}\right]$$

(18.2) critical load, thick columns

The symbols used in this formula have the same meaning and units as in the Euler formula but there is an additional variable, f_y, which is the yield point of the material (MPa).

Note that the quantity inside the square bracket has a maximum possible value of 1, and occurs when the slenderness ratio is zero. At this point the critical stress in the material $= f_y$, the yield stress, and the column behaves purely as a compression member. As the slenderness ratio increases, the quantity inside the square bracket becomes less than 1 and the critical stress is reduced compared to the yield stress.

Example 18.5

For the circular steel column shown in Figure 18.8, where both ends are firmly fixed, determine the critical load and stress using the Johnson formula. For the steel used, the yield point is 260 MPa and the modulus of elasticity is 200 GPa.

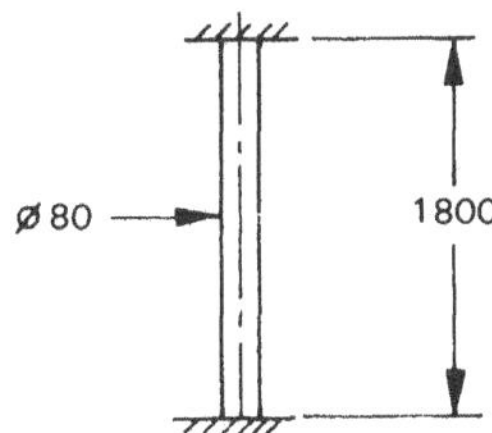

Fig. 18.8

Solution

For a circular section, $I = \dfrac{\pi d^4}{64}$ and $A = \dfrac{\pi d^2}{4}$

The radius of gyration $k = \sqrt{\dfrac{I}{A}} = \sqrt{\dfrac{\pi\,d^4}{64 \times \dfrac{\pi d^2}{4}}} = \dfrac{d}{4}$

$\therefore\ k = \dfrac{80}{4} = 20$ mm

Since both ends are fixed, the effective length $L_e = 0.7\ L$

$\therefore\ L_e = 1260$ mm

The slenderness ratio $\dfrac{L_e}{k} = \dfrac{1260}{20} = 63$

The column is therefore a stiff one and the Johnson formula is applicable:

$$F_{CR} = Af_y\left[1 - \frac{f_y(L_e/k)^2}{4\pi^2 E}\right]$$

$$= \frac{\pi \times 80^2}{4} \times 260\left[1 - \frac{260 \times 63^2}{4 \times \pi^2 \times 200 \times 10^3}\right] \quad \text{(N)}$$

$$= \mathbf{1.14\ MN}$$

The critical stress is:

$$f_{CR} = \frac{F_{CR}}{A} = \frac{1.14 \times 10^6}{\pi \times \dfrac{80^2}{4}} = \mathbf{226\ MPa}$$

Note that this stress is not very much less than the yield stress (93.3 per cent f_y) because the slenderness ratio is comparatively low.

18.5 Limiting slenderness ratio

It has been noted that the Euler formula is valid for slender columns and the Johnson formula is valid for thick columns. It is now necessary to define the range of slenderness ratios for which each formula is valid, that is the maximum slenderness ratio for which the Johnson formula is valid and the minimum slenderness ratio for which the Euler formula is valid. The limiting slenderness ratio for the validity of each formula is determined by the point at which both formulas give the same critical load.

That is, it is necessary to solve the following equation:

$$\frac{\pi^2 EA}{(L_e/k)^2} = Af_y\left[1 - \frac{f_y(L_e/k)^2}{4\pi^2 E}\right]$$

Cancelling the common A from both sides and using the symbol B for $(L_e/k)^2$ and the symbol C for $\pi^2 E$:

$$\frac{C}{B} = f_y\left[1 - \frac{f_y B}{4C}\right]$$

$$= f_y - \frac{f_y^2 B}{4C}$$

$$\therefore C = f_y B - \frac{f_y^2 B^2}{4C}$$

$$\text{or} \quad \frac{f_y^2}{4C}B^2 - f_y B + C = 0$$

This is a quadratic equation in B which may be solved using the general quadratic equation solution formula:

$$x = \frac{-b \pm \sqrt{b^2 - 4ac}}{2a}$$

$$\therefore B = \frac{f_y \pm \sqrt{f_y^2 - f_y^2}}{\dfrac{f_y^2}{2C}} \qquad = \frac{2C}{f_y}$$

Substituting for B and C:

$$\left(\frac{L_e}{k}\right)^2 = \frac{2\pi^2 E}{f_y}$$

$$\therefore \quad \boxed{\left(\frac{L_e}{k}\right)_{\text{lim}} = \sqrt{\frac{2\pi^2 E}{f_y}}}$$

(18.3) limiting slenderness ratio

Hence if the actual slenderness ratio for the column is less than the limiting value, the Johnson formula should be used and if the actual slenderness ratio is greater than the limiting value, the Euler formula should be used.

The region of validity of each formula may also be related to the compressive stress. It may readily be shown that, at the limiting slenderness ratio, the compressive stress in the column is 50 per cent of the yield stress. Hence the region of validity of each formula as related to the compressive stress is:

Euler $\quad 0 \leqslant f_{CR} \leqslant 0.5 f_y$

Johnson $\quad 0.5 f_y \leqslant f_{CR} \leqslant f_y$

Example 18.6

Determine the limiting slenderness ratio for a column made of standard grade structural steel for which the yield point is 250 MPa and the modulus of elasticity 200 GPa.

Solution

Using equation 18.3:

$$\left(\frac{L_e}{k}\right)_{\text{lim}} = \sqrt{\frac{2\pi^2 E}{f_y}}$$

$$= \sqrt{\frac{2 \times \pi^2 \times 200 \times 10^3}{250}}$$

$$= 126$$

Example 18.7

Use both the Johnson and Euler formulas to draw a graph of critical load against slenderness ratio for a compression member made of low carbon steel (soft condition) for which the yield point is 210 MPa and the modulus of elasticity 200 GPa. Use a nominal area of 100 mm^2 for the column and draw the graph for slenderness ratios in the range 60–260 (in steps of 20).

Show that both formulas give the same result at the limiting slenderness ratio.

Solution

Using the Johnson formula:

$$F_{CR} = A f_y \left[1 - \frac{f_y (L_e/k)^2}{4\pi^2 E}\right]$$

Substituting with $f_y = 210$ MPa, $A = 100\ \text{mm}^2$, $E = 200 \times 10^3$ MPa

$$F_{CR} = 100 \times 210 \left[1 - \frac{210 \times (L_e/k)^2}{4 \times \pi^2 \times 200 \times 10^3}\right] \quad \text{(N)}$$

$$= 21\left[1 - \frac{(L_e/k)^2}{37\,598}\right] \text{ kN} \qquad \textbf{(1)}$$

Using the Euler formula:

$$F_{CR} = \frac{\pi^2 EA}{(L_e/k)^2}$$

Substituting with the values of A and E as before:

$$F_{CR} = \frac{\pi^2 \times 200 \times 10^3 \times 100}{(L_e/k)^2} \quad \text{(N)}$$

$$= \frac{197\,392}{(L_e/k)^2} \text{ kN} \qquad \textbf{(2)}$$

Values of L_e/k may now be substituted in equations (1) and (2) and the following table derived:

L_e/k	60	80	100	120	140	160	180	200	220	240	260
F_{CR} (kN) Johnson	19.0	17.4	15.4	13.0	10.0	6.7	2.9	−1.34	−6.0	−11.2	−16.8
F_{CR} (kN) Euler	54.8	30.8	19.7	13.7	10.1	7.7	6.1	4.93	4.08	3.43	2.92

These results may now be graphed as shown in Figure 18.9.

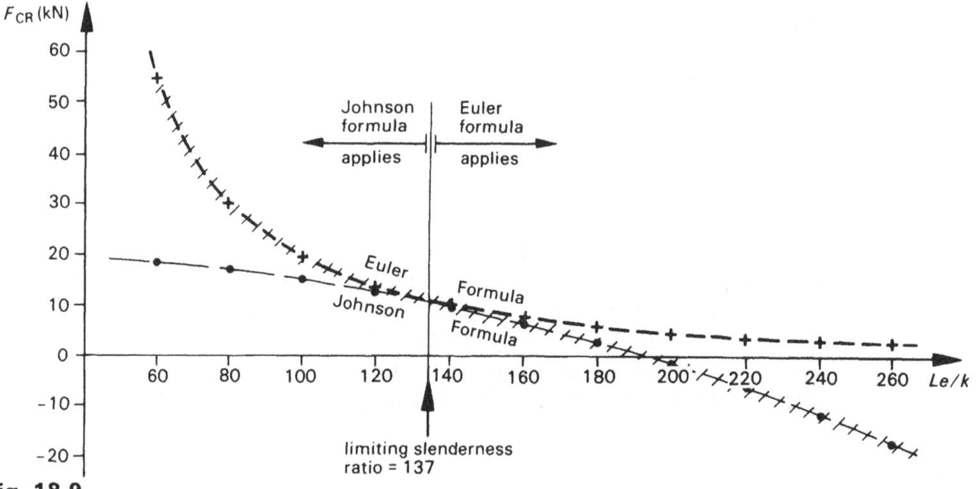

Fig. 18.9

The limiting slenderness ratio is:

$$\left(\frac{L_e}{k}\right)_{\text{lim}} = \sqrt{\frac{2\pi^2 E}{f_y}} = \sqrt{\frac{2 \times \pi^2 \times 200 \times 10^3}{210}} = 137$$

It is evident that both formulas give the same result at the limiting slenderness ratio of 137 (for this steel). The region where each formula is not valid is shown hatched on the diagram.

It is also evident from the diagram that at the limiting slenderness ratio of 137, the critical load is 10.5 kN.

Hence the compressive stress at this point is:

$$f_{CR} = \frac{10.5 \times 10^3}{100} = 105 \text{ MPa}$$

which is one-half the yield stress.

Example 18.8

A structural steel column is 5.5 m long and is of flange beam section with the following properties:

$k_x = 149$ mm, $k_y = 39.2$ mm, $A = 7220$ mm^2.

The yield stress of the steel is 250 MPa and the modulus of elasticity is 200 GPa.

If a factor of safety of 3 is used, determine the safe load on the column if:

(a) the ends of the column are located against movement but not firmly fixed
(b) the ends of the column are firmly fixed (e.g. welded to a cross beam or set in concrete).

Solution

The limiting slenderness ratio for structural steel with the properties given was found to be 126 (see example 18.6).

(a) The effective length of the column L_e is 5500 mm (end restraints equivalent to pinned) and the radius of gyration k is 39.2 mm (use the smallest value).

Hence the actual slenderness ratio is:

$$\frac{L_e}{k} = \frac{5500}{39.2} = 140.3$$

Since the actual slenderness ratio is greater than the limiting, the Euler formula should be used.

$$F_{CR} = \frac{\pi^2 EA}{(L_e/k)^2}$$

$$= \frac{\pi^2 \times 200 \times 10^3 \times 7220}{140.3^2} \quad \text{(N)}$$

$$= 724 \text{ kN}$$

Since a factor of safety of 3 is used, the safe load is:

$$F = \frac{724}{3} = \mathbf{241\ kN}$$

(b) Since both ends are firmly fixed the effective length is:

$$L_e = 0.7 \times 5500 = 3850 \text{ mm}$$

The actual slenderness ratio is:

$$\frac{L_e}{k} = \frac{3850}{39.2} = 98.2$$

Since the actual slenderness ratio is less than the limiting, the Johnson formula should be used.

$$F_{CR} = Af_y\left[1 - \frac{f_y(L_e/k)^2}{4\pi^2 E}\right]$$

$$= 7220 \times 250\left[1 - \frac{250 \times 98.2^2}{4 \times \pi^2 \times 200 \times 10^3}\right] \quad \text{(N)}$$

$$= 1254 \text{ kN}$$

Hence the safe load is $F = \frac{1254}{3} = \mathbf{418\ kN}$

18.6 Column selection: General procedure

If a circular, square or rectangular section is used then the section size of the column to withstand a given load may be determined directly (because there is a definite mathematical relationship between the size of the section and its properties of area and radius of gyration). However if any other type of section, such as a flange section or angle section, is used it is necessary to follow a trial and error procedure for the selection of the column section dimensions.

This procedure is outlined below:

1. Determine the limiting slenderness ratio for the material from which the column is to be manufactured.
2. Choose a section for the column and note the area and radius of gyration.
3. Determine the actual slenderness ratio for the column section chosen, taking into account the length of the column and the end restraints.
4. If the actual slenderness ratio (step 3) is less than the limiting slenderness ratio (step 1) use the Johnson formula, and if greater use the Euler formula. Hence determine the critical load.
5. Divide the critical load found in step 4 by the factor of safety to determine the safe load.
6. If the safe load (step 5) is greater than the applied load, the column is safe but it may be necessary to check if a smaller section could be satisfactory (and more economical). If the safe load is less than the applied load, the column is unsafe and a larger section is necessary. Repeat the procedure from step 2 onward.

Example 18.9

A copper rod is used as a compression member with an axial load of 1200 N. The ends are pinned and the distance between the pins is 500 mm.

Determine the required diameter of the rod if a factor of safety of 4 is to be used. For copper the following properties apply:

yield point 70 MPa, modulus of elasticity 110 GPa.

Solution

The limiting slenderness ratio for the material is:

$$\left(\frac{L_e}{k}\right)_{lim} = \sqrt{\frac{2\pi^2 E}{f_y}}$$

$$= \sqrt{\frac{2 \times \pi^2 \times 110 \times 10^3}{70}}$$

$$= 176$$

For a circular section the radius of gyration is:

$$k = \frac{d}{4} \quad \text{(see example 18.5)}$$

Choose a diameter of 10 mm as a first trial.

$$\text{Then } k = \frac{10}{4} = 2.5 \text{ mm}$$

$$\text{The area } A = \frac{\pi d^2}{4} = \frac{\pi \times 10^2}{4} = 78.5 \text{ mm}^2$$

The actual slenderness ratio is:

$$\frac{L_e}{k} = \frac{500}{2.5} = 200$$

Since this is greater than the limiting, the Euler formula should be used.

$$F_{CR} = \frac{\pi^2 EA}{(L_e/k)^2} = \frac{\pi^2 \times 110 \times 10^3 \times 78.5}{200^2} = 2.12 \text{ kN}$$

$$\text{The safe load is } F = \frac{2.12}{4} = 0.53 \text{ kN} = 530 \text{ N}$$

Since the safe load is less than the applied load, the section chosen is too small.
Try a 12.5 mm diameter rod.

$$k = \frac{12.5}{4} = 3.125 \text{ mm}, \; A = 122.7 \text{ mm}^2$$

$$\frac{L_e}{k} = \frac{500}{3.125} = 160$$

Since the actual slenderness ratio is now less than the limiting, the Johnson formula should be used.

$$F_{CR} = A f_y \left[1 - \frac{f_y (L_e/k)^2}{4\pi^2 E}\right]$$

$$= 122.7 \times 70 \left[1 - \frac{70 \times 160^2}{4 \times \pi^2 \times 110 \times 10^3}\right] \quad \text{(N)}$$

$$= 5.04 \text{ kN}$$

$$\text{Hence the safe load is } F = \frac{5.04}{4} = 1.26 \text{ kN} = 1260 \text{ N}$$

Since this is greater than the applied load, the **12.5 mm** diameter section is safe.

18.7 Computer solution

The computer program COLUMN was written to solve column problems using the Euler and Johnson formulas. If the properties of the section are known, the program determines the safe load on the column and also outputs the compressive stress and slenderness ratio. If the section properties are not known but a loading is given, the program determines the required circular or rectangular section for the given load and safety factor.

The program is listed in Appendix 3.20 with computer solution to examples 18.5 and 18.9 and should be self-explanatory.

Notes

1. The user has a choice of end restraints from the standard ones (a)–(d) shown in Figure 18.2 or else may use any end restraint conditions for which the effective length to actual length ratio is known.
2. If the section does not have the same radius of gyration in both the x and y directions, the radius of gyration input by the user must be applicable to the axis of buckling chosen. Of course the end restraints must also be applicable to this axis.
3. If the critical load and stress rather than the safe load and stress is required, this may be done by inputting a safety factor of 1.

Problems

For these problems assume E for steel is 200 GPa in all cases.

18.1 A rod, diameter 50 mm and length 1 m, supports a compressive load.

Determine the slenderness ratio if the ends of the rod are:

(a) both pinned
(b) both firmly fixed
(c) one end firmly fixed and the other end free
(d) one end firmly fixed and the other end pinned.

(a) 80 (b) 56 (c) 160 (d) 68

18.2 A flange column section as illustrated in Figure P18.2 is welded using steel plate.

Determine the slenderness ratio for buckling about (a) x axis plane and (b) y axis plane if the length of the column is 4 m and the ends are restrained so that the effective length is 85 per cent of the actual length.

(a) 37.8 (b) 101

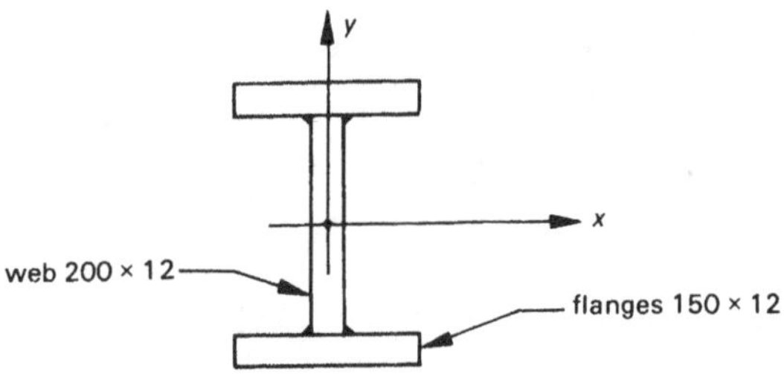

Fig. P18.2

18.3 The 127 × 76 unequal angle with properties shown in Figure P18.3 is used as a compression member. The length of the member is 3.5 m and the ends are restrained such that the effective length is 90 per cent of the actual length.

Determine the slenderness ratio about the following planes:

(a) x axis
(b) y axis
(c) u axis
(d) v axis.

(a) 78.4 (b) 151 (c) 74.7 (d) 193

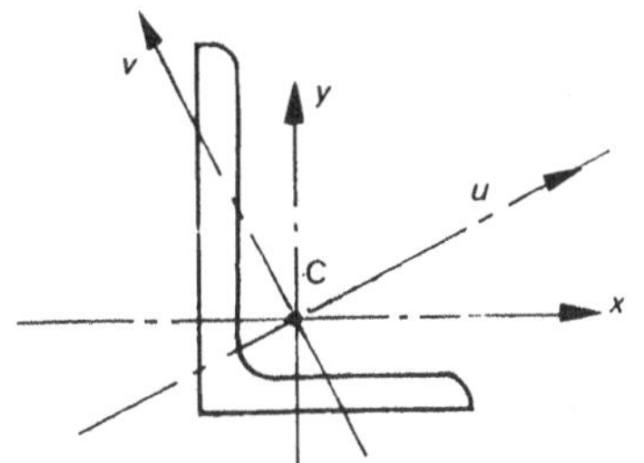

$A = 2410\ mm^2$
I ($\times 10^6\ mm^4$)
$x = 3.89$
$y = 1.05$
$u = 4.29$
$v = 0.643$

Fig. P18.3

18.4 The support joist illustrated in Figure P18.4 is restrained at the midpoint against sideways movement in the x direction but not in the y direction.

Determine the slenderness ratio for buckling about:

(a) x axis plane
(b) y axis plane.

(a) 104 (b) 104

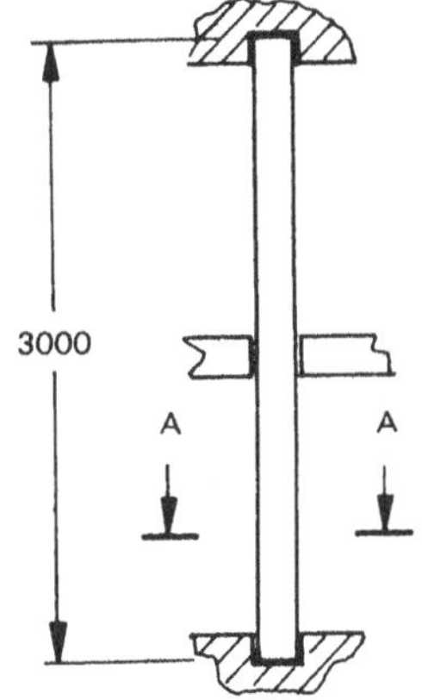

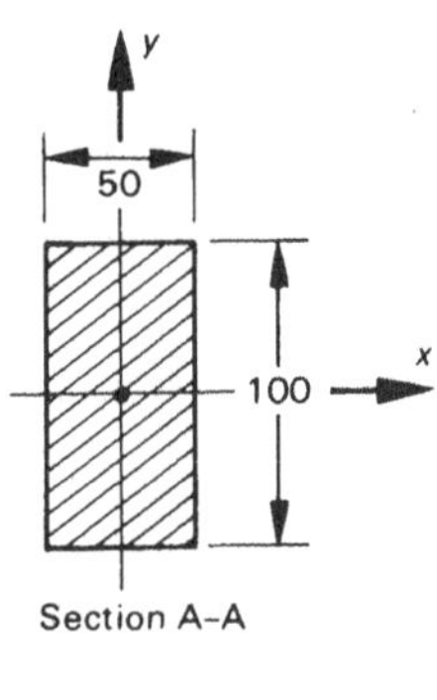

Fig. P18.4

18.5 Determine the limiting slenderness ratio for the following materials used as compression members:

(a) structural steel: $E = 200$ GPa, $f_y = 260$ MPa
(b) alloy steel: $E = 200$ GPa, $f_y = 440$ MPa
(c) aluminium alloy: $E = 70$ GPa, $f_y = 280$ MPa.

(a) 123 (b) 94.7 (c) 70.2

18.6 An alloy steel rod, diameter 45 mm and length 1100 mm, is used as a compression member in a machine. The yield stress of the steel is 480 MPa. The ends of the rod are pinned.

Determine:

(a) slenderness ratio
(b) limiting slenderness ratio
(c) which formula should be used for determining the critical load
(d) critical load
(e) maximum safe load (safety factor 4)
(f) compressive stress when the load is equal to the maximum safe load.

(a) 97.8 (b) 90.7 (c) Euler (d) 328 kN (e) 82.1 kN (f) 51.6 MPa

18.7 Repeat problem 18.6 if the rod is partially restrained at the ends instead of being pinned so that the effective length is 75 per cent of the actual length.

(a) 73.3 (b) 90.7 (c) Johnson (d) 514 kN (e) 128 kN (f) 80.8 MPa

18.8 A structural steel column with both ends firmly fixed has a length of 10 m and is of rolled flange section with the following properties:
$I_{cx} = 161 \times 10^6$ mm^4, $I_{cy} = 11.1 \times 10^6$ mm^4, $A = 7220$ mm^2, $f_y = 250$ MPa.

Determine the maximum safe axial load on the column if a safety factor of 3 is used.

149 kN

18.9 A structural steel member has a length of 1300 mm and the end restraints in all directions may be considered as equivalent to pinned. The member is made of unequal angle with properties as shown in Figure P18.9.

Determine:

(a) critical buckling load
(b) critical buckling stress.

For the grade of structural steel used, the yield stress is 240 MPa.

(a) 129.5 kN (b) 136.5 MPa

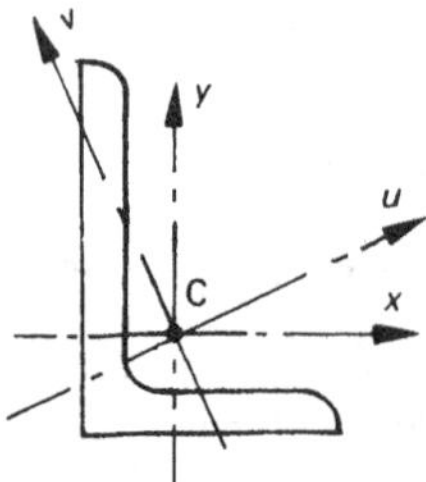

A = 949 mm^2
I ($\times 10^6$ mm^4)
x = 0.545
y = 0.194
u = 0.626
v = 0.113

Fig. P18.9

18.10 A tubular steel section is used as a compression member with one end firmly fixed as shown in Figure P18.10(a).

Determine:

(a) compressive stress in the rod at the point of buckling
(b) maximum safe axial load F which may be applied using a safety factor of 3.

The steel has a yield stress of 280 MPa.

(a) 71.1 MPa (b) 151 kN

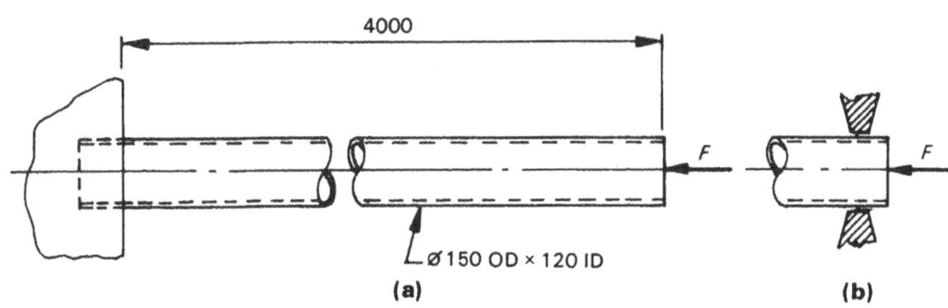

Fig. P18.10

18.11 Repeat problem 18.10 if the end of the member at which the load is applied is restrained all round from lateral movement as shown in Figure P18.10(b), all other conditions being the same.

(a) 230 MPa (b) 488 kN

18.12 A structural steel compression member as shown in Figure P18.12 is of rectangular section 100 × 40 mm. It is pinned at both ends and laterally restrained at the position shown against buckling on the *y* axis plane.

Determine:

(a) the plane about which buckling is most likely to occur

(b) maximum compressive stress which can safely be applied to the member using a safety factor of 3.

The steel has a yield point stress of 250 MPa.

(a) *x* axis (b) 34.3 MPa

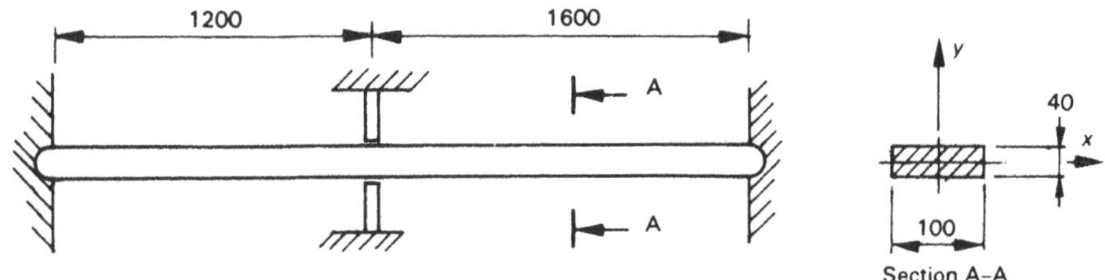

Fig. P18.12

18.13 Show that at the limiting slenderness ratio (that is when the Johnson and Euler formulas give the same result), the compressive stress at the point of buckling is one-half the yield stress.

(*Hint*: Substitute for limiting slenderness ratio in both formulas.)

Appendix 1 Principal symbols

Symbol	Quantity	Units/value
a	acceleration	m/s^2
a_c	centripetal acceleration	,,
a_{cor}	Coriolis acceleration	,,
A	area	m^2, mm^2
	amplitude of a vibration	mm
b	width of a section	,,
C	centroid of an area	—
CG	centre of mass (centre of gravity)	—
d	diameter of a circle	m, mm
	pitch diameter of a screw thread	mm
E	modulus of elasticity	GPa, MPa
E_R	modulus of resilience	,,
f	frequency of a vibration	Hz
	stress	MPa
f_A	axial pressure stress	,,
f_b	bending stress	,,
f_E	elastic limit stress	,,
f_H	hoop stress	,,
f_S	shear stress	,,
f_{st}	static stress	,,
f_n	normal stress	,,
f_x	stress on x axis	,,
f_y	stress on y axis	,,
	yield stress	,,
f_{sy}	vertical shear stress	,,
F	force	N, kN
	friction force	,,
F_C	centrifugal force	,,
F_{CR}	critical load on a column	,,
F_x, F_y	x and y components of a force	,,
$\hat{F}$	reduced force	kgm
FOS	factor of safety	—
g	acceleration due to gravity (gravitational constant)	9.81 m/s^2 (N/kg)
G	modulus of rigidity	GPa, MPa
h	height	m, mm
I	second moment of mass	kgm^2
	second moment of area	mm^4
J	polar moment of inertia	,,
k	radius of gyration	mm
	spring constant	N/mm

Symbol	Quantity	Units/value
KE	kinetic energy	J, kJ
l, L	length	m, mm
L_e	effective length of a column	mm
m	mass	kg
M	total mass of a vibrating system	kg
	moment of a force	Nm, Nmm
M_E	equivalent moment	,,
$\hat{M}$	reduced moment	kgm^2
MA	mechanical advantage	—
n	ratio of elastic moduli in parallel bars	—
N	normal reaction force	N, kN
	rotational speed	rpm
p	pressure	kPa, MPa
	thread pitch	mm
P	force	N, kN
	power	W, kW
PE	potential energy	J, kJ
Q	first moment of area	mm^3
r	radius	m, mm
R	reaction force (combined normal force and friction force)	N, kN
	radius of curvature of a beam	m, mm
s	displacement	m
S	horizontal shear force in a beam	N, kN
S'	horizontal shear force per unit length	N/m, kN/m
t	time	s
	thickness	mm
T	temperature	°C
	period of a vibration	s
	belt or rope tension	N, kN
T_R	transmissibility	—
U	strain energy	J, kJ
v	velocity	m/s
V	vertical shear force	N, kN
VR	velocity ratio	—
W	work	J, kJ
	weight	N, kN
	axial force on a screw thread	N, kN
x	distance or displacement	m, mm
	deflection under load	mm
x_{st}	static deflection	,,
$\bar{x}$	horizontal centroidal distance	,,
y	distance or displacement	,,
	distance from neutral axis to outer fibre	,,
$\bar{y}$	vertical centroidal distance	,,

Symbol	*Quantity*	*Units/value*
Z	section modulus	mm^3
θ	angle	deg
	angular displacement	rad
α	angle	deg
	angular acceleration	rad/s^2
	coefficient of linear expansion	/°C
β	angle	deg
	included angle of a vee belt	,,
ϕ	angle of friction	,,
μ	coefficient of friction	—
ω	angular velocity	rad/s
	equivalent angular velocity of free vibration	,,
ω_F	equivalent angular velocity of forced vibration	,,
ρ	density	kgm^3
η	efficiency	—
ϵ	axial strain	—
ϵ_s	shear strain	—
τ	torque	Nm, kNm
τ_E	equivalent torque	,,
Σ	summation	,,

Appendix 2 Principal formulas

1.1 $F_x = F \cos \theta$ horizontal component of a force

1.2 $F_y = F \sin \theta$ vertical component of a force

1.3 $M = Fx$ moment of a force

6.1 $F = \mu N$ friction force

6.2 $\phi = \tan^{-1}(\mu)$ angle of friction

6.3 $P = \dfrac{W(\sin \theta + \mu \cos \theta)}{\cos \alpha + \mu \sin \alpha}$ motion up plane

6.4a $P = \dfrac{W(\mu \cos \theta - \sin \theta)}{\cos \alpha' + \mu \sin \alpha'}$ motion down plane $\theta < \phi$

6.4b $P = \dfrac{W(\sin \theta - \mu \cos \theta)}{\cos \alpha - \mu \sin \alpha}$ motion down plane $\theta > \phi$

6.5 $\tan \theta = \dfrac{p}{\pi d}$ helix angle

6.6 $\tau = W\dfrac{d}{2}\left(\dfrac{\tan \theta + \mu}{1 - \mu \tan \theta}\right)$ tightening a screw thread

6.7 $\tau = W\dfrac{d}{2}\left(\dfrac{\mu - \tan \theta}{1 + \mu \tan \theta}\right)$ loosening a screw thread

6.8 $\tau = W\dfrac{d}{2}\tan(\theta + \phi)$ tightening a screw thread

6.9 $\tau = W\dfrac{d}{2}\tan(\phi - \theta)$ loosening a screw thread

6.10 $\eta = \dfrac{\tan \theta}{\tan(\theta + \phi)}$ efficiency of a screw thread

6.11 $\mu' = \dfrac{\mu}{\cos\left(\dfrac{\alpha}{2}\right)}$, $\phi' = \tan^{-1}(\mu')$ vee thread

6.12 $\dfrac{T_2}{T_1} = e^{\mu\theta}$ band or belt friction

6.13 $\tau = (T_2 - T_1)\dfrac{d}{2}$ braking torque

6.14 $P = \tau\omega$ braking power

6.15 $\dfrac{T_2}{T_1} = e^{\frac{\mu\theta}{\sin(\beta/2)}}$ vee belt drive

6.16 $\tau = \mu W r_M$ disc or collar friction (uniform wear)

7.1 $v = \dfrac{s}{t}$ constant (uniform) velocity

7.2 $v = \dfrac{ds}{dt}$ actual (instantaneous) velocity

7.3 $s = \displaystyle\int_{t_1}^{t_2} v\,dt$ displacement

7.4 $a = \dfrac{v_2 - v_1}{t}$ constant (uniform) acceleration

7.5 $a = \dfrac{dv}{dt}$ actual (instantaneous) acceleration

7.6 $v_2 - v_1 = \displaystyle\int_{t_1}^{t_2} a\,dt$ change in velocity

7.7 $s = \frac{1}{2}(v_1 + v_2)t$ displacement

7.8 $s = v_1 t + \frac{1}{2}at^2$ displacement

7.9 $v_2^2 = v_1^2 + 2as$ velocity

7.10 $v = r\omega$ linear and rotational velocity

7.11 $a = r\alpha$ linear and rotational acceleration

7.12 $a_C = r\omega^2$ centripetal acceleration

7.13 $a_C = \dfrac{v^2}{r}$ centripetal acceleration

7.14 $a_{COR} = 2v_R\,\omega$ Coriolis acceleration

8.1 $F = ma$ accelerating force

8.2 $a = \dfrac{P}{m}(\cos\alpha + \mu\sin\alpha) - g(\sin\theta + \mu\cos\theta)$ accelerated motion up plane

8.3 $a = \dfrac{P}{m}(\mu\sin\alpha - \cos\alpha) + g(\sin\theta - \mu\cos\theta)$ accelerated motion down plane

8.4 $F_C = mr\omega^2$ centrifugal force

8.5 $F_C = m\dfrac{v^2}{r}$ centrifugal force

8.6 $v = \sqrt{rg\tan(\theta + \phi)}$ motion up a rotating inclined plane

8.7 $v = \sqrt{rg\tan(\theta - \phi)}$ motion down a rotating inclined plane

8.8 $v = \sqrt{rg\tan\theta}$ balanced forces on a rotating inclined plane

8.9 $\tau = mr^2\alpha$ torque applied to a point mass

8.10 $I = \Sigma\,\Delta mr^2$ second moment of mass or mass moment of inertia

8.11 $\tau = I\alpha$ torque applied to a non-point mass

8.12 $I = \frac{1}{2}mr^2$ mass moment of inertia of a solid disc

8.13 $I_{XX} = I_{CG} + mx^2$ parallel axis theorem (mass)

8.14 $k = \sqrt{\dfrac{I}{m}}$ radius of gyration (mass)

9.1 $PE = mgh$ potential energy

9.2 $KE = \frac{1}{2}mv^2$ kinetic energy

9.3 $KE = \frac{1}{2}I\omega^2$ rotational kinetic energy

9.4 $W = Fs$ linear work

9.5 $P = Fv$ linear power

9.6 $W = \tau\theta$ rotational work

9.7 $P = \tau\omega$ rotational power

9.8 $PE_1 + KE_1 \pm W = PE_2 + KE_2$ energy equation (mechanical systems)

9.9 $\eta = \dfrac{\text{useful power output}}{\text{power input}}$ efficiency

9.10 $\eta = \dfrac{MA}{VR}$ efficiency

9.11 $\eta = \dfrac{MA\ (\text{actual})}{MA\ (\text{ideal})}$ efficiency

10.1 $T = \dfrac{1}{f}$ period of a vibration

10.2 $x = A\sin(\omega t)$ displacement in SHM

10.3 $v = A\omega\cos(\omega t)$

10.4 $v = \omega\sqrt{A^2 - x^2}$ } velocity in SHM

10.5 $a = A\omega^2\sin(\omega t)$

10.6 $a = \omega^2 x$ } acceleration in SHM

10.7 $f = \dfrac{\omega}{2\pi}$

10.8 $f = \dfrac{1}{2\pi}\sqrt{\dfrac{a}{x}}$ } frequency in SHM

10.9 $\omega = \sqrt{\dfrac{k}{m}}$ equivalent angular velocity

10.10 $a = \dfrac{g}{l}x$ acceleration of a simple pendulum

10.11 $f = \dfrac{1}{2\pi}\sqrt{\dfrac{g}{l}}$ frequency of a simple pendulum

10.12 $v = \sqrt{\dfrac{g}{l}(A^2 - x^2)}$ velocity of a simple pendulum

10.13 $A = \dfrac{F}{M(\omega_F^2 - \omega^2)}$ amplitude of a forced vibration

10.14 $T_R = \dfrac{1}{\left(\dfrac{\omega_F}{\omega}\right)^2 - 1}$ transmissibility

12.1 $f = \dfrac{F}{A}$ axial stress

12.2 $f_s = \dfrac{F}{A}$ shear stress

12.3 $\epsilon = \dfrac{x}{L}$ axial strain

12.4 $\epsilon_s = \dfrac{x}{L} = \theta$ shear strain

12.5 $E = \dfrac{f}{\epsilon} = \dfrac{FL}{xA}$ modulus of elasticity

12.6 $G = \dfrac{f_s}{\epsilon_s} = \dfrac{FL}{xA}$ modulus of rigidity

12.7 $f_A = \dfrac{pd}{4t}$ axial pressure stress

12.8 $f_H = \dfrac{pd}{2t}$ hoop stress due to pressure

12.9 $x = \alpha L \Delta T$ thermal expansion

12.10 $f_H = \rho v^2$ hoop stress in a rotating ring

12.11 $f_1 = \dfrac{P}{A_1 + \dfrac{A_2}{n}}$ stress in parallel bars

13.1 $U = \frac{1}{2}Fx$

13.2 $U = \dfrac{f^2 AL}{2E}$

13.3 $U = \dfrac{x^2 EA}{2L}$

(13.1–13.3) axial strain energy

13.4 $E_R = \dfrac{f_E^2}{2E}$ modulus of resilience

13.5 $U_1 + PE_1 + KE_1 = U_2 + PE_2 + KE_2$ energy equation (elastic systems)

13.6 $\dfrac{f}{f_{st}} = \dfrac{x}{x_{st}} = 1 + \sqrt{1 + \dfrac{2h}{x_{st}}}$ dynamic stress and extension (axial impact loads)

13.7 $\dfrac{f}{f_{st}} = \dfrac{x}{x_{st}} = 1 + \dfrac{v}{\sqrt{gx_{st}}}$ dynamic stress and extension (suddenly stopped load)

14.1 $Q = \Sigma y\, \Delta A = \int y\, dA$ first moment of area

14.2 $\bar{y} = \dfrac{Q}{A}$ centroid location

14.3 $I = \Sigma y^2 \Delta A = \int y^2 dA$ second moment of area

14.4 $I_{CX} = \dfrac{bh^3}{12}$ second moment of area for a rectangle about centroid

14.5 $I_{CX} = \dfrac{\pi d^4}{64}$ second moment of area for a circle about the centre

14.6 $I_{XX} = I_{CX} + Ax^2$ parallel axis theorem (area)

14.7 $k = \sqrt{\dfrac{I}{A}}$ radius of gyration (area)

14.8 $J = \Sigma r^2 \Delta A = \int r^2 dA$ polar moment of inertia

14.9 $J = \dfrac{\pi d^4}{32}$ polar moment of inertia for a circular section

15.1 $f_b = \dfrac{My}{I}$ } bending stress

15.2 $f_b = \dfrac{M}{Z}$ }

15.3 $R = \dfrac{EI}{M}$ radius of curvature of a beam

15.4 $f_b = \dfrac{Ey}{R}$ bending stress

15.5 $S' = \dfrac{S}{x} = \dfrac{QV}{I}$ horizontal shear force per unit length in a beam

15.6 $f_s = \dfrac{QV}{Ib}$ shear stress in beams

15.7 $f_s = \dfrac{\tau r}{J}$ torsional shear stress

15.8 $\theta = \dfrac{\tau L}{GJ}$ angle of twist in torsion

16.1 $\tau_E = \sqrt{\tau^2 + M^2}$ equivalent torque

16.2 $M_E = \frac{1}{2}[M + \tau_E]$ equivalent moment

16.3 $f_s = \dfrac{16\,\tau_E}{\pi d^3}$ combined shear stress (circular shaft)

16.4 $f = \dfrac{32\,M_E}{\pi d^3}$ combined tension/compression stress (circular shaft)

16.5 $f_n = f_x \cos^2\theta + f_y \sin^2\theta - 2f_{sy}\sin\theta\cos\theta$ combined normal stress on an inclined plane

16.6 $f_s = \dfrac{f_x - f_y}{2}\sin 2\theta + f_{sy}\cos 2\theta$ combined shear stress on an inclined plane

16.7 $f_s(\text{max}) = \sqrt{f_{sy}^2 + \left(\dfrac{f_x - f_y}{2}\right)^2}$ maximum combined shear stress

16.8 $\tan 2\theta = \dfrac{f_x - f_y}{2f_{sy}}$ angle of inclination of plane of maximum combined shear stress

16.9 $f_n(\text{max}) = \dfrac{f_x + f_y}{2} + f_s(\text{max})$ maximum combined normal stress

16.10 $\tan 2\theta = \dfrac{-2f_{sy}}{f_x - f_y}$ angle of inclination of plane of maximum combined normal stress

17.1 $EI\dfrac{d^2y}{dx^2} = M$ elastic curve of a beam

17.2 $EI\dfrac{dy}{dx} = \int(M\,dx) + A$ slope of a beam

17.3 $EIy = \iint(M\,dx) + Ax + B$ deflection of a beam

18.1 $F_{CR} = \dfrac{\pi^2 EA}{(L_e/k)^2}$ critical load, slender columns

18.2 $F_{CR} = Af_y\left[1 - \frac{f_y(L_e/k)^2}{4\pi^2 E}\right]$ critical load, thick columns

18.3 $\left(\frac{L_e}{k}\right)_{lim} = \sqrt{\frac{2\pi^2 E}{f_y}}$ limiting slenderness ratio

Appendix 3 Computer programs

Number	*Name*	*Description*	*Examples*
3.1	FORCE 1	Resultant of any number of concurrent forces	1.2
3.2	FORCE 2	Resultant of any number of non-concurrent forces	1.8
3.3	CANTRE	Reactions—cantilever beam	3.2
3.4	REACT	Reactions—simply supported beam	3.3
3.5	FORCE 3	Method of joints—two unknown forces	4.1
3.6	INPLANE 1	Equilibrium on the inclined plane	6.1
3.7	WEDGE	Wedges	6.3(b)
3.8	MOTION 1	Motion with constant velocity or constant acceleration	7.4, 7.10
3.9	MOTION 2	Relative motion	7.5, 7.6, 7.7, 7.8
3.10	INPLANE 2	Motion on the inclined plane	8.1, 8.5
3.11	FLYINT	Mass moment of inertia of a flywheel	8.12, 8.13
3.12	COPBAL	Balance of coplanar masses	11.2
3.13	COPRE	Reaction of rotating coplanar masses	11.3
3.14	NCBAL	Balance of non-coplanar masses	11.4
3.15	NCREACT	Reactions for rotating non-coplanar masses	11.5
3.16	DYNLOAD	Dynamic axial loads on circular sections	13.5, 13.7, 13.8
3.17	CENTINT	Centroid and second moment of area	14.1, 14.2, 14.3, 14.4, 14.6
3.18	PRINSTR	Combined stresses	16.8, 16.10
3.19	BEAMDEF	Beam deflections	17.2, 17.3, 17.7
3.20	COLUMN	Columns	18.5, 18.9

APPENDIX 3.1

```
5  PRINT
10  PRINT "FILE NAME FORCE 1"
15  PRINT "RESULTANT OF CONCURRENT FORCES"
20  REM  BY ROGER KINSKY 29/11/84
30  PRINT "NOTE THAT FORCES POINTING AWAY FROM THE ORIGIN ARE POSITIVE"
40  PRINT "ANGLES MEASURED ANTI-CLOCKWISE FROM THE X AXIS ARE POSITIVE"
50  PRINT
55 F3 = 0:F4 = 0:W = 0
60  PRINT "HOW MANY FORCES ARE THERE ?"
65  INPUT V
70  PRINT "NOW INPUT THE MAGNITUDE OF EACH FORCE AND ITS ANGLE IN TURN "
80  INPUT F,X
90  PRINT
100 F1 = F *  SIN (X * .0174533)
110 F2 = F *  COS (X * .0174533)
120 F3 = F3 + F1
130 F4 = F4 + F2
140 W = W + 1
150  IF  ABS (V - W) > .2 THEN 80
160 F5 = (F3 ^ 2 + F4 ^ 2) ^ .5
170 F5 =  INT (F5 * 100 + .5) / 100
180  IF F4 = 0 THEN F4 = 1E - 6
190 Y =  ATN ( ABS (F3 / F4)) * 57.296
200 Y =  INT (Y * 100 + .5) / 100
210  IF F3 < 0 THEN 300
220  IF F4 < 0 THEN Y = 180 - Y
230  GOTO 400
300  IF F4 < 0 THEN Y = 180 + Y
310  IF F4 > 0 THEN Y = 360 - Y
400  PRINT : PRINT
410  PRINT "THE RESULTANT FORCE IS   ";F5
420  PRINT "THE ANGLE MADE BY THE RESULTANT IN DEGREES IS  ";Y
430  PRINT
440 F3 =  INT (F3 * 100 + .5) / 100
450  PRINT "THE SUM OF THE VERTICAL COMPONENTS IS  ";F3
460 F4 =  INT (F4 * 100 + .5) / 100
470  PRINT "THE SUM OF THE HORIZONTAL COMPONENTS IS  ";F4
999  END
```

EXAMPLE 1.2

```
]RUN

FILE NAME FORCE 1
RESULTANT OF CONCURRENT FORCES
NOTE THAT FORCES POINTING AWAY FROM THE ORIGIN ARE POSITIVE
ANGLES MEASURED ANTI-CLOCKWISE FROM THE X AXIS ARE POSITIVE

HOW MANY FORCES ARE THERE ?
?3
NOW INPUT THE MAGNITUDE OF EACH FORCE AND ITS ANGLE IN TURN
?-3,45

?5,135

?6,-110
```

```
THE RESULTANT FORCE IS   8.79
THE ANGLE MADE BY THE RESULTANT IN DEGREES IS  208.72

THE SUM OF THE VERTICAL COMPONENTS IS  -4.22
THE SUM OF THE HORIZONTAL COMPONENTS IS  -7.71
```

APPENDIX 3.2

```
5  PRINT
10  PRINT "FILE NAME FORCE 2"
15  PRINT "RESULTANT FORCE AND MOMENT OF NON-CONCURRENT FORCES"
20  REM  BY ROGER KINSKY 29/11/84
30  PRINT "NOTE THAT FORCES POINTING AWAY FROM THE ORIGIN ARE POSITIVE"
40  PRINT "ANGLES MEASURED ANTI-CLOCKWISE WITH THE X AXIS ARE POSITIVE"
50  PRINT
55 F3 = 0:F4 = 0:W = 0
60  PRINT "HOW MANY FORCES ARE THERE ?"
65  INPUT V
70  PRINT "FOR EACH FORCE INPUT THE MAGNITUDE, ANGLE, X & Y CO-ORDINATES"
80  INPUT F,X,A,B
90  PRINT
100 F1 = F *  SIN (X * .0174533)
110 F2 = F *  COS (X * .0174533)
115 M = F2 * B - F1 * A
120 F3 = F3 + F1
130 F4 = F4 + F2
135 M1 = M1 + M
140 W = W + 1
150  IF  ABS (V - W) > .2 THEN 80
160 F5 = (F3 ^ 2 + F4 ^ 2) ^ .5
170 F5 =  INT (F5 * 100 + .5) / 100
180  IF F4 = 0 THEN F4 = 1E - 6
190 Y =  ATN ( ABS (F3 / F4)) * 57.296
200 Y =  INT (Y * 100 + .5) / 100
210  IF F3 < 0 THEN 300
220  IF F4 < 0 THEN Y = 180 - Y
230  GOTO 400
300  IF F4 < 0 THEN Y = 180 + Y
310  IF F4 > 0 THEN Y = 360 - Y
400  PRINT : PRINT
410  PRINT "THE RESULTANT FORCE IS   ";F5
420  PRINT "THE ANGLE MADE BY THE RESULTANT IN DEGREES IS  ";Y
430  PRINT
440 F7 =  INT (F3 * 100 + .5) / 100
450  PRINT "THE SUM OF THE VERTICAL COMPONENTS IS  ";F7
460 F8 =  INT (F4 * 100 + .5) / 100
470  PRINT "THE SUM OF THE HORIZONTAL COMPONENTS IS  ";F8
480  PRINT
490  IF  ABS (F3) < 1E - 5 THEN 600
500 C = (M1 *  - 1) / F3
```

```
510 C =  INT (C * 100 + .5) / 100
520 M1 =  INT (M1 * 100 + .5) / 100
530  PRINT "THE RESULTANT MOMENT IS  ";M1
540  PRINT "THE RESULTANT FORCE CROSSES THE X AXIS AT  ";C
550  GOTO 999
600  IF  ABS (F4) < 1E - 5 THEN 620
620 C =  INT (C * 100 + .5) / 100
630 M1 =  INT (M1 * 100 + .5) / 100
640  PRINT "THE RESULTANT MOMENT IS  ";M1
650  PRINT "THE RESULTANT FORCE CROSSES THE Y AXIS AT  ";C
999  END
```

EXAMPLE 1.8

```
]RUN

FILE NAME FORCE 2
RESULTANT FORCE AND MOMENT OF NON-CONCURRENT FORCES
NOTE THAT FORCES POINTING AWAY FROM THE ORIGIN ARE POSITIVE
ANGLES MEASURED ANTI-CLOCKWISE WITH THE X AXIS ARE POSITIVE

HOW MANY FORCES ARE THERE ?
?4
FOR EACH FORCE INPUT THE MAGNITUDE, ANGLE, X & Y CO-ORDINATES
?120,45,-450,200

?-60,90,0,0

?-140,30,600,300

?80,-20,600,150

THE RESULTANT FORCE IS   82.23
THE ANGLE MADE BY THE RESULTANT IN DEGREES IS  298.14

THE SUM OF THE VERTICAL COMPONENTS IS  -72.51
THE SUM OF THE HORIZONTAL COMPONENTS IS  38.78

THE RESULTANT MOMENT IS  88474.57
THE RESULTANT FORCE CROSSES THE X AXIS AT  1220.19
```

APPENDIX 3.3

```
5  PRINT
10  PRINT "FILE NAME CANTRE"
15  PRINT "REACTIVE FORCE AND MOMENT FOR A CANTILEVER BEAM"
20  REM  BY ROGER KINSKY 31/10/84
```

```
30  PRINT "NOTE THAT FORCES POINTING AWAY FROM THE ORIGIN (SUPPORT) ARE
     POSITIVE "
40  PRINT "ANGLES MEASURED ANTI-CLOCKWISE FROM THE X AXIS ARE POSITIVE"
50  PRINT
55 F3 = 0:F4 = 0:W = 0
56 M1 = 0
60  PRINT "HOW MANY FORCES ACT ON THE BEAM ?"
65  INPUT V
70  PRINT "NOW INPUT THE MAGNITUDE OF EACH FORCE, ITS ANGLE, ITS X AND Y
     COORDINATES IN TURN "
80  INPUT F,X,A,B
90  PRINT
100 F1 = F *  SIN (X * .0174533)
110 F2 = F *  COS (X * .0174533)
115 M = F2 * B - F1 * A
120 F3 = F3 + F1
130 F4 = F4 + F2
135 M1 = M1 + M
140 W = W + 1
150  IF  ABS (V - W) > .2 THEN 80
160 F5 = (F3 ^ 2 + F4 ^ 2) ^ .5
170 F5 =  INT (F5 * 100 + .5) / 100
180  IF F4 = 0 THEN F4 = 1E - 6
190 Y =  ATN ( ABS (F3 / F4)) * 57.296
200 Y =  INT (Y * 100 + .5) / 100
210  IF F3 < 0 THEN 300
220  IF F4 < 0 THEN Y = 180 - Y
230  GOTO 320
300  IF F4 < 0 THEN Y = 180 + Y
310  IF F4 > 0 THEN Y = 360 - Y
320  IF Y = 0 THEN Y = 180: GOTO 400
330  IF Y = 180 THEN Y = 0: GOTO 400
340  IF Y < 180 THEN Y = Y + 180: GOTO 400
350 Y = Y - 180
400  PRINT : PRINT
410  PRINT "THE REACTIVE FORCE IS   ";F5
420  PRINT "THE ANGLE MADE BY THE REACTIVE FORCE IN DEGREES IS   ";Y
430  PRINT
440 F7 =  INT (F3 * 100 + .5) / 100
445 F7 = F7 *  - 1
450  PRINT "VERTICAL COMPONENT OF THE REACTIVE FORCE IS   ";F7
460 F8 =  INT (F4 * 100 + .5) / 100
465 F8 = F8 *  - 1
470  PRINT "HORIZONTAL COMPONENT OF THE REACTIVE FORCE IS  ";F8
480  PRINT
520 M1 =  INT (M1 * 100 + .5) / 100
525 M1 = M1 *  - 1
530  PRINT "THE REACTIVE MOMENT IS  ";M1
999  END
```

EXAMPLE 3.2

```
]RUN

FILE NAME CANTRE
REACTIVE FORCE AND MOMENT FOR A CANTILEVER BEAM
NOTE THAT FORCES POINTING AWAY FROM THE ORIGIN (SUPPORT) ARE POSITIVE
ANGLES MEASURED ANTI-CLOCKWISE FROM THE X AXIS ARE POSITIVE
```

```
HOW MANY FORCES ACT ON THE BEAM ?
?4
NOW INPUT THE MAGNITUDE OF EACH FORCE, ITS ANGLE, ITS X AND Y COORDINATES
IN TURN
?-1.413,90,0,0

?-2.825,90,0.7,0

?3,0,1.4,3.15

?3,-66,1.4,3.15


THE REACTIVE FORCE IS   8.16
THE ANGLE MADE BY THE REACTIVE FORCE IN DEGREES IS   121.16

VERTICAL COMPONENT OF THE REACTIVE FORCE IS   6.98
HORIZONTAL COMPONENT OF THE REACTIVE FORCE IS  -4.22

THE REACTIVE MOMENT IS  -19.11
```

APPENDIX 3.4

```
5  PRINT
10  PRINT "FILE NAME   REACT"
15  PRINT "RESULTANT FORCE AND MOMENT OF NON-CONCURRENT FORCES"
17  PRINT "AND SUPPORT REACTIONS FOR A SIMPLY SUPPORTED BEAM"
20  REM  BY ROGER KINSKY 29/11/84
30  PRINT "NOTE THAT FORCES POINTING AWAY FROM THE ORIGIN ARE POSITIVE"
40  PRINT "AND ANGLES MEASURED ANTICLOCKWISE FROM THE X AXIS ARE POSITIVE"
42  PRINT
44  PRINT "THE PINNED JOINT A MUST BE LOCATED AT THE ORIGIN "
50  PRINT
55 F3 = 0:F4 = 0:W = 0
56 M1 = 0:J = 0
60  PRINT "HOW MANY APPLIED FORCES ACT ON THE BEAM ?"
65  INPUT V
70  PRINT "FOR EACH FORCE INPUT THE MAGNITUDE, ANGLE, X AND Y CO-ORDINATES"
80  INPUT F,X,A,B
90  PRINT
100 F1 = F *  SIN (X * .0174533)
110 F2 = F *  COS (X * .0174533)
115 M = F2 * B - F1 * A
120 F3 = F3 + F1
130 F4 = F4 + F2
135 M1 = M1 + M
140 W = W + 1
150  IF  ABS (V - W) > .2 THEN 80
160 F5 = (F3 ^ 2 + F4 ^ 2) ^ .5
```

```
170 F5 =  INT (F5 * 100 + .5) / 100
180  IF F4 = 0 THEN F4 = 1E - 6
190 Y =  ATN ( ABS (F3 / F4)) * 57.296
200 Y =  INT (Y * 100 + .5) / 100
210  IF F3 < 0 THEN 300
220  IF F4 < 0 THEN Y = 180 - Y
230  GOTO 320
300  IF F4 < 0 THEN Y = 180 + Y
310  IF F4 > 0 THEN Y = 360 - Y
320  PRINT
330  REM  NOW CALCULATE THE REACTIONS
340  GOTO 730
400  PRINT : PRINT
410  PRINT "THE RESULTANT FORCE IS   ";F5
420  PRINT "THE ANGLE MADE BY THE RESULTANT IN DEGREES IS  ";Y
430  PRINT
440 F7 =  INT (F3 * 100 + .5) / 100
450  PRINT "THE SUM OF THE VERTICAL COMPONENTS IS   ";F7
460 F8 =  INT (F4 * 100 + .5) / 100
470  PRINT "THE SUM OF THE HORIZONTAL COMPONENTS IS   ";F8
480  PRINT
485  IF F3 = 0 THEN J = 10
490  IF J = 10 THEN F3 = F4 *  - 1
494  IF F3 = 0 THEN C = 0
496  IF F3 = 0 THEN 520
500 C = (M1 *  - 1) / F3
510 C =  INT (C * 100 + .5) / 100
520 M1 =  INT (M1 * 100 + .5) / 100
530  PRINT "THE RESULTANT MOMENT IS  ";M1
535  IF J = 10 THEN 560
540  PRINT "THE RESULTANT FORCE CROSSES THE X AXIS AT  ";C
550  GOTO 570
560  PRINT "THE RESULTANT FORCE CROSSES THE Y AXIS AT  ";C
570  PRINT
580  PRINT "THE REACTION AT B IS AS FOLLOWS"
590 R2 =  INT (R2 * 100 + .5) / 100
600 R3 =  INT (R3 * 100 + .5) / 100
610  PRINT "VERTICAL COMPONENT  ";R2;"  HORIZONTAL COMPONENT  ";R3
620 R1 =  INT (R1 * 100 + .5) / 100
630  PRINT "RESULTANT  ";R1;"  AT  ";D;"  DEGREES"
640  PRINT
650  PRINT "THE REACTION AT A IS AS FOLLOWS"
660 R4 =  INT (R4 * 100 + .5) / 100
670 R5 =  INT (R5 * 100 + .5) / 100
680  PRINT "VERTICAL COMPONENT  ";R4;"   HORIZONTAL COMPONENT  ";R5
690 R6 =  INT (R6 * 100 + .5) / 100
700 E =  INT (E * 100 + .5) / 100
710  PRINT "RESULTANT  ";R6"  AT  ";E;"  DEGREES"
720  GOTO 999
730  PRINT "WHAT IS THE KNOWN DIRECTION OF REACTION B IN DEGREES ?"
735  INPUT D
740  PRINT "WHAT ARE THE X AND Y CO-ORDINATES OF REACTION B ?"
745  INPUT A,B
750 R1 = M1 / (B *  COS (D * .0174533) - A *  SIN (D * .0174533))
760 R1 = R1 *  - 1
770 R2 = R1 *  SIN (D * .0174533)
780 R3 = R1 *  COS (D * .0174533)
790 R4 = F3 *  - 1 - R2
800 R5 = F4 *  - 1 - R3
810  IF R5 = 0 THEN E = 90
820  IF R5 = 0 THEN 840
830 E =  ATN ( ABS (R4 / R5)) * 57.296
```

```
840 R6 =  SQR (R4 ^ 2 + R5 ^ 2)
850  IF R4 < 0 THEN 880
860  IF R5 < 0 THEN E = 180 - E
870  GOTO 400
880  IF R5 < 0 THEN E = 180 + E
890  IF R5 > 0 THEN E = 360 - E
900  GOTO 400
999  END
```

<u>EXAMPLE 3.3</u>

```
]RUN

FILE NAME   REACT
RESULTANT FORCE AND MOMENT OF NON-CONCURRENT FORCES
AND SUPPORT REACTIONS FOR A SIMPLY SUPPORTED BEAM
NOTE THAT FORCES POINTING AWAY FROM THE ORIGIN ARE POSITIVE
AND ANGLES MEASURED ANTICLOCKWISE FROM THE X AXIS ARE POSITIVE

THE PINNED JOINT A MUST BE LOCATED AT THE ORIGIN

HOW MANY APPLIED FORCES ACT ON THE BEAM ?
?2
FOR EACH FORCE INPUT THE MAGNITUDE, ANGLE, X AND Y CO-ORDINATES
?-12,120,1.732,1

?-11,90,6.214,2


WHAT IS THE KNOWN DIRECTION OF REACTION B IN DEGREES ?
?90
WHAT ARE THE X AND Y CO-ORDINATES OF REACTION B ?
?7.464,2


THE RESULTANT FORCE IS   22.22
THE ANGLE MADE BY THE RESULTANT IN DEGREES IS  285.67

THE SUM OF THE VERTICAL COMPONENTS IS   -21.39
THE SUM OF THE HORIZONTAL COMPONENTS IS   6

THE RESULTANT MOMENT IS  92.35
THE RESULTANT FORCE CROSSES THE X AXIS AT  4.32

THE REACTION AT B IS AS FOLLOWS
VERTICAL COMPONENT  12.37  HORIZONTAL COMPONENT  0
RESULTANT  12.37  AT  90  DEGREES

THE REACTION AT A IS AS FOLLOWS
VERTICAL COMPONENT  9.02   HORIZONTAL COMPONENT  -6
RESULTANT  10.83  AT  123.63  DEGREES
```

APPENDIX 3.5

```
5  PRINT
10  PRINT "FILE NAME FORCE 3"
20  REM  BY ROGER KINSKY 15/12/84
30  PRINT "FOR CONCURRENT FORCE SYSTEMS WITH 2 UNKNOWNS"
40  PRINT "NOTE THAT FORCES POINTING AWAY FROM THE ORIGIN ARE POSITIVE"
50  PRINT "AND ANGLES MEASURED ANTI-CLOCKWISE FROM THE X AXIS ARE POSITIVE"
60  PRINT
70 F3 = 0:F4 = 0:N = 0
80 P = .0174533
100  PRINT "WHAT IS THE JOINT NUMBER"
110  INPUT J
120  PRINT "INPUT THE NUMBER OF KNOWN FORCES ACTING AT JOINT  ";J
130  INPUT M
140  PRINT "NOW INPUT EACH KNOWN FORCE AND ITS ANGLE ACTING AT JOINT  ";J
     ;"  IN TURN"
150  INPUT F,X
160 N = N + 1
170 F1 = F *  SIN (X * P)
180 F2 = F *  COS (X * P)
190 F3 = F3 + F1
200 F4 = F4 + F2
210  IF  ABS (M - N) > .5 THEN 150
220  PRINT
230  PRINT "NOW INPUT THE DIRECTIONS OF THE 2 UNKNOWN FORCES AT JOINT  ";J
240  INPUT A,B
250 A = A * P
260 B = B * P
300 F6 = (F4 *  TAN (B) - F3) / ( SIN (A) -  COS (A) *  TAN (B))
310 F7 =  - 1 * F4 /  COS (B) - F6 *  COS (A) /  COS (B)
320  PRINT
400  PRINT "THE UNKNOWN FORCES AT JOINT  ";J;"  ARE"
410 A = A / P
420 B = B / P
430 F6 =  INT (F6 * 100 + .5) / 100
440 F7 =  INT (F7 * 100 + .5) / 100
500  PRINT "  FORCE   ";F6;"  AT  ";A;"  DEGREES"
510  PRINT
520  PRINT "  FORCE   ";F7;"  AT  ";B;"  DEGREES"
530  PRINT
540  PRINT "DO YOU WISH TO SOLVE ANY MORE JOINTS ? ANSWER Y OR N"
550  INPUT A$
560  IF A$ = "Y" THEN 70
999  END
```

EXAMPLE 4.1

```
]LOAD REACT

]RUN

FILE NAME   REACT
RESULTANT FORCE AND MOMENT OF NON-CONCURRENT FORCES
AND SUPPORT REACTIONS FOR A SIMPLY SUPPORTED BEAM
NOTE THAT FORCES POINTING AWAY FROM THE ORIGIN ARE POSITIVE
```

```
AND ANGLES MEASURED ANTICLOCKWISE FROM THE X AXIS ARE POSITIVE

THE PINNED JOINT A MUST BE LOCATED AT THE ORIGIN

HOW MANY APPLIED FORCES ACT ON THE BEAM ?
?2
FOR EACH FORCE INPUT THE MAGNITUDE, ANGLE, X AND Y CO-ORDINATES
?-5,60,2,0

?6,270,4,0

WHAT IS THE KNOWN DIRECTION OF REACTION B IN DEGREES ?
?-30
WHAT ARE THE X AND Y CO-ORDINATES OF REACTION B ?
?0,2.3094


THE RESULTANT FORCE IS   10.63
THE ANGLE MADE BY THE RESULTANT IN DEGREES IS  256.4

THE SUM OF THE VERTICAL COMPONENTS IS   -10.33
THE SUM OF THE HORIZONTAL COMPONENTS IS   -2.5

THE RESULTANT MOMENT IS  32.66
THE RESULTANT FORCE CROSSES THE X AXIS AT  3.16

THE REACTION AT B IS AS FOLLOWS
VERTICAL COMPONENT  8.17  HORIZONTAL COMPONENT  -14.14
RESULTANT  -16.33  AT  -30  DEGREES

THE REACTION AT A IS AS FOLLOWS
VERTICAL COMPONENT  2.17   HORIZONTAL COMPONENT  16.64
RESULTANT  16.78  AT  7.41  DEGREES


]LOAD FORCE 3

]RUN


FILE NAME FORCE 3
FOR CONCURRENT FORCE SYSTEMS WITH 2 UNKNOWNS
NOTE THAT FORCES POINTING AWAY FROM THE ORIGIN ARE POSITIVE
AND ANGLES MEASURED ANTI-CLOCKWISE FROM THE X AXIS ARE POSITIVE

WHAT IS THE JOINT NUMBER
?6
INPUT THE NUMBER OF KNOWN FORCES ACTING AT JOINT  6
?1
NOW INPUT EACH KNOWN FORCE AND ITS ANGLE ACTING AT JOINT  6  IN TURN
?6,270

NOW INPUT THE DIRECTIONS OF THE 2 UNKNOWN FORCES AT JOINT  6
?150,180


THE UNKNOWN FORCES AT JOINT  6  ARE
  FORCE   12  AT  150  DEGREES

  FORCE   -10.39  AT  180  DEGREES

DO YOU WISH TO SOLVE ANY MORE JOINTS ? ANSWER Y OR N
?Y
```

```
WHAT IS THE JOINT NUMBER
?5
INPUT THE NUMBER OF KNOWN FORCES ACTING AT JOINT  5
?2
NOW INPUT EACH KNOWN FORCE AND ITS ANGLE ACTING AT JOINT  5  IN TURN
?-5,90
?12,0

NOW INPUT THE DIRECTIONS OF THE 2 UNKNOWN FORCES AT JOINT  5
?180,270


THE UNKNOWN FORCES AT JOINT  5  ARE
  FORCE   12  AT  180  DEGREES

  FORCE   -5  AT  270  DEGREES

DO YOU WISH TO SOLVE ANY MORE JOINTS ? ANSWER Y OR N
?Y
WHAT IS THE JOINT NUMBER
?4
INPUT THE NUMBER OF KNOWN FORCES ACTING AT JOINT  4
?2
NOW INPUT EACH KNOWN FORCE AND ITS ANGLE ACTING AT JOINT  4  IN TURN
?-10.39,0
?-5,60

NOW INPUT THE DIRECTIONS OF THE 2 UNKNOWN FORCES AT JOINT  4
?120,180

THE UNKNOWN FORCES AT JOINT  4  ARE
  FORCE   5  AT  120  DEGREES

  FORCE   -15.39  AT  180  DEGREES

DO YOU WISH TO SOLVE ANY MORE JOINTS ? ANSWER Y OR N
?Y
WHAT IS THE JOINT NUMBER
?3
INPUT THE NUMBER OF KNOWN FORCES ACTING AT JOINT  3
?2

NOW INPUT EACH KNOWN FORCE AND ITS ANGLE ACTING AT JOINT  3  IN TURN
?12,0
?5,-30


NOW INPUT THE DIRECTIONS OF THE 2 UNKNOWN FORCES AT JOINT  3
?180,270

THE UNKNOWN FORCES AT JOINT  3  ARE
  FORCE   16.33  AT  180  DEGREES

  FORCE   -2.5  AT  270  DEGREES

DO YOU WISH TO SOLVE ANY MORE JOINTS ? ANSWER Y OR N
?Y
WHAT IS THE JOINT NUMBER
?1
INPUT THE NUMBER OF KNOWN FORCES ACTING AT JOINT  1
?1
NOW INPUT EACH KNOWN FORCE AND ITS ANGLE ACTING AT JOINT  1  IN TURN
?16.78,7.41
```

```
NOW INPUT THE DIRECTIONS OF THE 2 UNKNOWN FORCES AT JOINT  1
?0,60

THE UNKNOWN FORCES AT JOINT  1  ARE
  FORCE   -15.39  AT   0  DEGREES

  FORCE   -2.5  AT  60  DEGREES

DO YOU WISH TO SOLVE ANY MORE JOINTS ? ANSWER Y OR N
?N
```

APPENDIX 3.6

```
10  PRINT "PROGRAM NAME INPLANE 1"
20  REM  BY ROGER KINSKY 19/9/84
25  PRINT " FOR NON-ACCELERATING MASSES ON THE INCLINED PLANE"
30  PRINT
40  PRINT "ENTER U FOR MOTION UP THE PLANE AND D FOR MOTION DOWN THE PLANE "
50  INPUT Z$
60  PRINT
70  PRINT "WHAT IS THE INCLINED PLANE ANGLE IN DEGREES TO THE HORIZONTAL "
80  INPUT B
90 F = 0.017453
100 B = B * F
110  PRINT
120  PRINT "WHAT IS THE ANGLE OF FORCE P WITH THE INCLINED PLANE"
125  PRINT "IN DEGREES ANTI-CLOCKWISE WITH P POINTING AWAY FROM THE BLOCK
      ?"
130  INPUT A
140 A = A * F
150  PRINT
160  PRINT "WHAT IS THE COEFFICIENT OF FRICTION   "
170  INPUT U
180  PRINT
190  PRINT "WHAT IS THE MASS OF THE BLOCK IN kg"
200  INPUT M
210 W = M * 9.81
220  PRINT
230  IF Z$ = "D" THEN U =  - 1 * U
250 P = W * ( SIN (B) + U *  COS (B)) / ( COS (A) + U *  SIN (A))
255 P =  INT (P * 100 + .5) / 100
256 P =  ABS (P)
257  IF Z$ = "D" THEN 300
260  PRINT "THE FORCE P IN N ACTING UP THE PLANE IS    ";P
270  GOTO 999
300  IF  TAN (B) <  ABS (U) THEN 320
310  GOTO 260
320  PRINT "THE FORCE P ACTING DOWN THE PLANE IN N IS  ";P
999  END
```

EXAMPLE 6.1 (a)

```
]RUN

PROGRAM NAME INPLANE 1
 FOR NON-ACCELERATING MASSES ON THE INCLINED PLANE

ENTER U FOR MOTION UP THE PLANE AND D FOR MOTION DOWN THE PLANE
?U

WHAT IS THE INCLINED PLANE ANGLE IN DEGREES TO THE HORIZONTAL
?15

WHAT IS THE ANGLE OF FORCE P WITH THE INCLINED PLANE
IN DEGREES ANTI-CLOCKWISE WITH P POINTING AWAY FROM THE BLOCK ?
?-20

WHAT IS THE COEFFICIENT OF FRICTION
?.4

WHAT IS THE MASS OF THE BLOCK IN kg
?5

THE FORCE P IN N ACTING UP THE PLANE IS   39.42
```

EXAMPLE 6.1 (b)

```
]RUN

PROGRAM NAME INPLANE 1
 FOR NON-ACCELERATING MASSES ON THE INCLINED PLANE

ENTER U FOR MOTION UP THE PLANE AND D FOR MOTION DOWN THE PLANE
?D

WHAT IS THE INCLINED PLANE ANGLE IN DEGREES TO THE HORIZONTAL
?15

WHAT IS THE ANGLE OF FORCE P WITH THE INCLINED PLANE
IN DEGREES ANTI-CLOCKWISE WITH P POINTING AWAY FROM THE BLOCK ?
?-20

WHAT IS THE COEFFICIENT OF FRICTION
?.4

WHAT IS THE MASS OF THE BLOCK IN kg
?5

THE FORCE P ACTING DOWN THE PLANE IN N IS  5.81
```

APPENDIX 3.7

```
10  PRINT "PROGRAM NAME WEDGE"
20  PRINT "BY ROGER KINSKY 17/9/84"
30  PRINT "01 IS THE WEDGE ANGLE BELOW HORIZONTAL"
40  PRINT "02 IS THE WEDGE ANGLE ABOVE THE HORIZONTAL"
50  PRINT "03 IS THE BLOCK ANGLE MEASURED ANTI-CLOCKWISE FROM THE VERTICAL"
60  PRINT
100  PRINT "PLEASE ENTER ANGLES 01,02,03 IN DEGREES"
110  INPUT 01,02,03
115  PRINT
120  PRINT "NOW ENTER MASS OF BLOCK AND WEDGE IN kg "
130  INPUT M,M1
140 W = M * 9.81
142 W1 = M1 * 9.81
145  PRINT
150  PRINT "PLEASE ENTER THE WEDGE FORCE P IN N"
160  INPUT P
165  PRINT
170  PRINT "NOW ENTER THE COEFFICIENT OF FRICTION"
180  INPUT U
200 0 =  ATN (U)
210 PI = 3.141592
220 0 = 0 * 180 / PI
300 A = 90 - 01 - 0
310 C = 90 - A + 02 + 0
320 E = 90 - 03 + 0
330 F = 180 - E - 02 - 0
340 PI = 3.141592
350 A = A * PI / 180
360 C = C * PI / 180
370 E = E * PI / 180
380 F = F * PI / 180
390 P = P - W1 *  TAN (PI / 2 - A)
400 Q = P * ( SIN (A) /  SIN (C)) * ( SIN (F) /  SIN (E)) - W
410  PRINT : PRINT
420 Q =  INT (Q * 100 + .5) / 100
450  PRINT "THE SIDE THRUST FORCE Q IN N IS   ";Q
999  END
```

EXAMPLE 6.3(b)

```
]RUN

PROGRAM NAME WEDGE
BY ROGER KINSKY 17/9/84
01 IS THE WEDGE ANGLE BELOW HORIZONTAL
02 IS THE WEDGE ANGLE ABOVE THE HORIZONTAL
03 IS THE BLOCK ANGLE MEASURED ANTI-CLOCKWISE FROM THE VERTICAL

PLEASE ENTER ANGLES 01,02,03 IN DEGREES
?0,10,0

NOW ENTER MASS OF BLOCK AND WEDGE IN kg
?0,0

PLEASE ENTER THE WEDGE FORCE P IN N
?1000
```

```
NOW ENTER THE COEFFICIENT OF FRICTION
?.25


THE SIDE THRUST FORCE Q IN N IS   1276.61
```

APPENDIX 3.8

```
5  PRINT "FILE NAME   MOTION 1"
10  PRINT "PROGRAM FOR CONST ACCEL OR CONST VEL MOTION"
15  REM  BY ROGER KINSKY 5/12/84
20  PRINT "DO YOU HAVE LINEAR OR ROTATIONAL MOTION ,ANSWER L OR R"
30  INPUT Z$
40  PRINT "NOW INPUT THE VARIABLE WHEN PROMPTED"
50  PRINT "IF ANY VARIABLE IS UNKNOWN RETURN X WHEN PROMPTED"
55  PRINT "IF MOTION IS CONST VEL, INPUT 0 FOR ACCEL"
57  IF Z$ = "L" THEN 64
60  IF Z$ = "R" THEN 94
62  GOTO 20
64  PRINT
66  PRINT "TO CONVERT KM/H TO M/S, DIVIDE BY 3.6"
68  PRINT
70  PRINT "INITIAL VELOCITY IN M/S ?"
72  INPUT U$
74  PRINT "FINAL VELOCITY IN M/S ?"
76  INPUT V$
78  PRINT "ACCELERATION IN M/S^2 ?"
80  INPUT A$
82  PRINT "ELAPSED TIME IN S ?"
84  INPUT T$
86  PRINT "DISPLACEMENT IN M ?"
88  INPUT S$
90  GOTO 130
94  PRINT
96  PRINT "TO CONVERT RPM TO RAD/S DIVIDE BY 9.5493"
98  PRINT "TO CONVERT REVS TO RADS MULTIPLY BY 6.2832"
99  PRINT
100  PRINT "INITIAL ROTATIONAL SPEED IN RAD/S ?"
102  INPUT U$
104  PRINT "FINAL ROTATIONAL SPEED IN RAD/S ?"
106  INPUT V$
108  PRINT "ANGULAR ACCELERATION IN RAD/S^2 ?"
109  INPUT A$
110  PRINT "ELAPSED TIME IN S ?"
112  INPUT T$
114  PRINT "DISPLACEMENT IN RADIANS ?"
116  INPUT S$
130  IF U$ <  > "X" THEN U =  VAL (U$)
135  IF V$ <  > "X" THEN V =  VAL (V$)
140  IF A$ <  > "X" THEN A =  VAL (A$)
```

```
145  IF T$ <  > "X" THEN T =  VAL (T$)
150  IF S$ <  > "X" THEN S =  VAL (S$)
160  IF U$ = "X" THEN 200
165  IF V$ = "X" THEN 250
170  IF A$ = "X" THEN 300
175  IF T$ = "X" THEN 350
200  IF V$ = "X" THEN 400
210  IF A$ = "X" THEN 450
220  IF T$ = "X" THEN 500
230  IF S$ = "X" THEN 550
250  IF A$ = "X" THEN 600
260  IF T$ = "X" THEN 650
270  IF S$ = "X" THEN 700
300  IF T$ = "X" THEN 750
310  IF S$ = "X" THEN 800
330  REM
340  REM
350 T = (V - U) / A
360 S = .5 * T * (U + V)
370  GOTO 900
400 U = (S - .5 * A * T ^ 2) / T
410 V = U + A * T
420  GOTO 900
450 U = 2 * S / T - V
460 A = (V - U) / T
470  GOTO 900
500 U =  SQR (V ^ 2 - 2 * A * S)
510 T = (V - U) / A
520  GOTO 900
550 U = V - A * T
560  GOTO 360
600 V = 2 * S / T - U
610  GOTO 460
650 V =  SQR (U ^ 2 + 2 * A * S)
660  GOTO 510
700 V = U + A * T
710  GOTO 360
750 A = (V ^ 2 - U ^ 2) / (2 * S)
760  GOTO 510
800 A = (V - U) / T
810  GOTO 360
900 U1 =  INT (U * 954.93 + .5) / 100
901 U =  INT (U * 100 + .5) / 100
902 V1 =  INT (V * 954.93 + .5) / 100
903 V =  INT (V * 100 + .5) / 100
904 A =  INT (A * 100 + .5) / 100
906 T =  INT (T * 100 + .5) / 100
907 S1 =  INT (S * 15.9155 + .5) / 100
908 S =  INT (S * 100 + .5) / 100
910  IF Z$ = "R" THEN 950
920  PRINT
925  PRINT "INITIAL VELOCITY   ";U;"  m/s"
930  PRINT "FINAL VELOCTIY   ";V;"  m/s"
935  PRINT "ACCELERATION   ";A;"  m/s^2"
940  PRINT "ELAPSED TIME   ";T;"  s"
945  PRINT "DISPLACEMENT   ";S;"  m"
947  GOTO 999
950  PRINT
955  PRINT "INITIAL VELOCITY   ";U1;" rpm   ";U;" rad/s"
960  PRINT "FINAL VELOCITY   ";V1;" rpm   ";V;" rad/s"
```

```
965  PRINT "ACCELERATION    ";A;" rad/s^2"
970  PRINT "ELAPSED TIME    ";T;" s"
975  PRINT "DISPLACEMENT    ";S1;" revs    ";S;" rad"
999  END
```

EXAMPLE 7.4(a)

```
]RUN

FILE NAME   MOTION 1
PROGRAM FOR CONST ACCEL OR CONST VEL MOTION
DO YOU HAVE LINEAR OR ROTATIONAL MOTION ,ANSWER L OR R
?L
NOW INPUT THE VARIABLE WHEN PROMPTED
IF ANY VARIABLE IS UNKNOWN RETURN X WHEN PROMPTED
IF MOTION IS CONST VEL, INPUT 0 FOR ACCEL

TO CONVERT KM/H TO M/S, DIVIDE BY 3.6

INITIAL VELOCITY IN M/S ?
?0
FINAL VELOCITY IN M/S ?
?X
ACCELERATION IN M/S^2 ?
?2
ELAPSED TIME IN S ?
?5
DISPLACEMENT IN M ?
?X

INITIAL VELOCITY   0  m/s
FINAL VELOCTIY   10  m/s
ACCELERATION   2  m/s^2
ELAPSED TIME   5  s
DISPLACEMENT   25  m
```

EXAMPLE 7.4(b)

```
]RUN

FILE NAME   MOTION 1
PROGRAM FOR CONST ACCEL OR CONST VEL MOTION
DO YOU HAVE LINEAR OR ROTATIONAL MOTION ,ANSWER L OR R
?L
NOW INPUT THE VARIABLE WHEN PROMPTED
IF ANY VARIABLE IS UNKNOWN RETURN X WHEN PROMPTED
IF MOTION IS CONST VEL, INPUT 0 FOR ACCEL


TO CONVERT KM/H TO M/S, DIVIDE BY 3.6

INITIAL VELOCITY IN M/S ?
?10
FINAL VELOCITY IN M/S ?
?10
```

```
ACCELERATION IN M/S^2 ?
?0
ELAPSED TIME IN S ?
?10
DISPLACEMENT IN M ?
?X

INITIAL VELOCITY   10  m/s
FINAL VELOCTIY   10  m/s
ACCELERATION   0  m/s^2
ELAPSED TIME   10  s
DISPLACEMENT   100  m
```

EXAMPLE 7.10

```
]RUN

FILE NAME   MOTION 1
PROGRAM FOR CONST ACCEL OR CONST VEL MOTION
DO YOU HAVE LINEAR OR ROTATIONAL MOTION ,ANSWER L OR R
?R
NOW INPUT THE VARIABLE WHEN PROMPTED
IF ANY VARIABLE IS UNKNOWN RETURN X WHEN PROMPTED
IF MOTION IS CONST VEL, INPUT 0 FOR ACCEL

TO CONVERT RPM TO RAD/S DIVIDE BY 9.5493
TO CONVERT REVS TO RADS MULTIPLY BY 6.2832

INITIAL ROTATIONAL SPEED IN RAD/S ?
?544.54
FINAL ROTATIONAL SPEED IN RAD/S ?
?0
ANGULAR ACCELERATION IN RAD/S^2 ?
?X
ELAPSED TIME IN S ?
?30
DISPLACEMENT IN RADIANS ?
?X

INITIAL VELOCITY   5199.98 rpm   544.54 rad/s
FINAL VELOCITY   0 rpm   0 rad/s
ACCELERATION   -18.15 rad/s^2
ELAPSED TIME   30 s
DISPLACEMENT   1299.99 revs    8168.1 rad
```

APPENDIX 3.9

```
5  PRINT
10  PRINT "FILE NAME MOTION 2"
15  PRINT "RESULTANT OR RELATIVE MOTION"
20  REM  BY ROGER KINSKY 14/12/84
30  PRINT "NOTE THAT VECTORS POINTING AWAY FROM THE ORIGIN ARE POSITIVE"
40  PRINT "AND ANGLES MEASURED ANTI-CLOCKWISE FROM THE X AXIS ARE POSITIVE"
41 F3 = 0:F4 = 0:W = 0:V = 2: PRINT
42  PRINT "DO YOU HAVE COMBINED (RESULTANT) OR RELATIVE MOTION ?"
43  PRINT "ANSWER C (COMBINED) OR R (RELATIVE)"
44  INPUT Z$
45  IF Z$ = "R" THEN 500
46  IF Z$ = "C" THEN 50
47  GOTO 42
50  PRINT
70  PRINT "NOW INPUT THE MAGNITUDE OF EACH VECTOR AND ITS ANGLE IN TURN"
80  INPUT F,X
90  PRINT
100 F1 = F *  SIN (X * .0174533)
110 F2 = F *  COS (X * .0174533)
120 F3 = F3 + F1
130 F4 = F4 + F2
140 W = W + 1
150  IF  ABS (V - W) > .2 THEN 80
160 F5 = (F3 ^ 2 + F4 ^ 2) ^ .5
170 F5 =  INT (F5 * 100 + .5) / 100
180  IF F4 = 0 THEN F4 = 1E - 6
190 Y =  ATN ( ABS (F3 / F4)) * 57.296
200 Y =  INT (Y * 100 + .5) / 100
210  IF F3 < 0 THEN 300
220  IF F4 < 0 THEN Y = 180 - Y
230  GOTO 400
300  IF F4 < 0 THEN Y = 180 + Y
310  IF F4 > 0 THEN Y = 360 - Y
400  PRINT : PRINT
405  IF Z$ = "R" THEN 600
410  PRINT "THE RESULTANT MOTION HAS MAGNITUDE   ";F5
420  PRINT "THE ANGLE MADE BY THE RESULTANT IN DEGREES IS  ";Y
430  PRINT
440  GOTO 620
500  PRINT "NOW INPUT THE MAGNITUDE AND ANGLE OF THE VECTOR THE MOTION IS
      TO BE RELATIVE TO"
510  INPUT F,X
520 F = F *  - 1
530 F1 = F *  SIN (X * .0174533)
540 F2 = F *  COS (X * .0174533)
550 F3 = F3 + F1
560 F4 = F4 + F2
570 W = 1
580  PRINT "NOW INPUT THE MAGNITUDE AND ANGLE MADE BY THE OTHER VECTOR"
590  GOTO 80
600  PRINT "THE RELATIVE MOTION VECTOR HAS MAGNITUDE   ";F5
610  PRINT "THE ANGLE MADE BY THE RELATIVE MOTION VECTOR IN DEGREES IS";Y
620  PRINT "DO YOU WISH TO GO AGAIN, ANSWER Y OR N"
630  INPUT Y$
640  IF Y$ = "Y" THEN 41
999  END
```

EXAMPLE 7.5

```
]RUN

FILE NAME MOTION 2
RESULTANT OR RELATIVE MOTION
NOTE THAT VECTORS POINTING AWAY FROM THE ORIGIN ARE POSITIVE
AND ANGLES MEASURED ANTI-CLOCKWISE FROM THE X AXIS ARE POSITIVE

DO YOU HAVE COMBINED (RESULTANT) OR RELATIVE MOTION ?
ANSWER C (COMBINED) OR R (RELATIVE)
?C

NOW INPUT THE MAGNITUDE OF EACH VECTOR AND ITS ANGLE IN TURN
?15,45

?5,180


THE RESULTANT MOTION HAS MAGNITUDE   12
THE ANGLE MADE BY THE RESULTANT IN DEGREES IS  62.14

DO YOU WISH TO GO AGAIN, ANSWER Y OR N
?Y
```

EXAMPLE 7.6

```
DO YOU HAVE COMBINED (RESULTANT) OR RELATIVE MOTION ?
ANSWER C (COMBINED) OR R (RELATIVE)
?C

NOW INPUT THE MAGNITUDE OF EACH VECTOR AND ITS ANGLE IN TURN
?120,0

?-49,90


THE RESULTANT MOTION HAS MAGNITUDE   129.62
THE ANGLE MADE BY THE RESULTANT IN DEGREES IS  337.79

DO YOU WISH TO GO AGAIN, ANSWER Y OR N
?Y
```

EXAMPLE 7.7(a)

```
DO YOU HAVE COMBINED (RESULTANT) OR RELATIVE MOTION ?
ANSWER C (COMBINED) OR R (RELATIVE)
?R
NOW INPUT THE MAGNITUDE AND ANGLE OF THE VECTOR THE MOTION IS TO BE
 RELATIVE TO
?40,0
NOW INPUT THE MAGNITUDE AND ANGLE MADE BY THE OTHER VECTOR
?-50,90
```

```
THE RELATIVE MOTION VECTOR HAS MAGNITUDE   64.03
THE ANGLE MADE BY THE RELATIVE MOTION VECTOR IN DEGREES IS   231.34
DO YOU WISH TO GO AGAIN, ANSWER Y OR N
?Y
```

EXAMPLE 7.7(b)

```
DO YOU HAVE COMBINED (RESULTANT) OR RELATIVE MOTION ?
ANSWER C (COMBINED) OR R (RELATIVE)
?R
NOW INPUT THE MAGNITUDE AND ANGLE OF THE VECTOR THE MOTION IS TO BE
 RELATIVE TO
?80,0
NOW INPUT THE MAGNITUDE AND ANGLE MADE BY THE OTHER VECTOR
?-50,90


THE RELATIVE MOTION VECTOR HAS MAGNITUDE   94.34
THE ANGLE MADE BY THE RELATIVE MOTION VECTOR IN DEGREES IS   212.01
DO YOU WISH TO GO AGAIN, ANSWER Y OR N
?Y
```

EXAMPLE 7.8

```
DO YOU HAVE COMBINED (RESULTANT) OR RELATIVE MOTION ?
ANSWER C (COMBINED) OR R (RELATIVE)
?R
NOW INPUT THE MAGNITUDE AND ANGLE OF THE VECTOR THE MOTION IS TO BE
 RELATIVE TO
?10,135
NOW INPUT THE MAGNITUDE AND ANGLE MADE BY THE OTHER VECTOR
?25,180


THE RELATIVE MOTION VECTOR HAS MAGNITUDE   19.27
THE ANGLE MADE BY THE RELATIVE MOTION VECTOR IN DEGREES IS   201.52
DO YOU WISH TO GO AGAIN, ANSWER Y OR N
?N
```

APPENDIX 3.10

```
10  PRINT "PROGRAM NAME INPLANE 2"
20  REM  BY ROGER KINSKY 15/10/84
25  PRINT "FOR BOTH ACCELERATING AND NON-ACCELERATING MASSES ON THE
     INCLINED PLANE"
```

```
30   PRINT
40   PRINT "ENTER U FOR MOTION UP THE PLANE AND D FOR MOTION DOWN THE PLANE"
50   INPUT Z$
52   PRINT
55   PRINT "ENTER A FOR ACCELERATED MOTION AND N FOR NON-ACCELERATED MOTION"
57   INPUT W$
60   PRINT
70   PRINT "WHAT IS THE INCLINED PLANE ANGLE IN DEGREES TO THE HORIZONTAL"
80   INPUT B
90 F = 0.017453
100 B = B * F
110  PRINT
120  PRINT "WHAT IS THE ANGLE OF FORCE P WITH THE INCLINED PLANE"
125  PRINT "IN DEGREES ANTI-CLOCKWISE WITH P POINTING AWAY FROM THE BLOCK?"
130  INPUT A
140 A = A * F
150  PRINT
160  PRINT "WHAT IS THE COEFFICIENT OF FRICTION  "
170  INPUT U
180  PRINT
190  PRINT "WHAT IS THE MASS OF THE BLOCK IN kg"
200  INPUT M
210 W = M * 9.81
220  PRINT
225  IF W$ = "A" THEN 500
230  IF Z$ = "D" THEN U =  - 1 * U
250 P = W * ( SIN (B) + U *  COS (B)) / ( COS (A) + U *  SIN (A))
255 P =  INT (P * 100 + .5) / 100
256 P =  ABS (P)
257  IF Z$ = "D" THEN 300
260  PRINT "THE FORCE P IN N ACTING UP THE PLANE IS   ";P
270  GOTO 999
300  IF  TAN (B) <  ABS (U) THEN 320
310  GOTO 260
320  PRINT "THE FORCE P ACTING DOWN THE PLANE IN N IS  ";P
330  GOTO 999
500  PRINT
510  PRINT "WHAT IS THE MAGNITUDE OF THE APPLIED FORCE IN N "
520  INPUT P
525  IF Z$ = "D" THEN 720
530 C = P / M * ( COS (A) + U *  SIN (A)) - ( SIN (B) + U *  COS (B)) * 9
    .81
540 C =  INT (C * 1000 + .5) / 1000
550  PRINT "THE ACCELERATION OF THE MASS UP THE PLANE IN m/s^2 IS  ";C
560  GOTO 999
640 C =  INT (C * 1000 + .5) / 1000
650  PRINT "THE ACCELERATION OF THE MASS DOWN THE PLANE IN m/s^2 IS  ";C
660  GOTO 999
720 C = P / M * (U *  SIN (A) -  COS (A)) + ( SIN (B) - U *  COS (B)) * 9
    .81
730  GOTO 640
999  END
```

EXAMPLE 8.1

```
]RUN

PROGRAM NAME INPLANE 2
FOR BOTH ACCELERATING AND NON-ACCELERATING MASSES ON THE INCLINED PLANE
```

```
ENTER U FOR MOTION UP THE PLANE AND D FOR MOTION DOWN THE PLANE
?U

ENTER A FOR ACCELERATED MOTION AND N FOR NON-ACCELERATED MOTION
?A

WHAT IS THE INCLINED PLANE ANGLE IN DEGREES TO THE HORIZONTAL
?0

WHAT IS THE ANGLE OF FORCE P WITH THE INCLINED PLANE
IN DEGREES ANTI-CLOCKWISE WITH P POINTING AWAY FROM THE BLOCK ?
?-30

WHAT IS THE COEFFICIENT OF FRICTION
?.3

WHAT IS THE MASS OF THE BLOCK IN kg
?5


WHAT IS THE MAGNITUDE OF THE APPLIED FORCE IN N
?50
THE ACCELERATION OF THE MASS UP THE PLANE IN m/s^2 IS  4.217
```

EXAMPLE 8.5(a)

```
]RUN

PROGRAM NAME INPLANE 2
FOR BOTH ACCELERATING AND NON-ACCELERATING MASSES ON THE INCLINED PLANE

ENTER U FOR MOTION UP THE PLANE AND D FOR MOTION DOWN THE PLANE
?U

ENTER A FOR ACCELERATED MOTION AND N FOR NON-ACCELERATED MOTION
?A

WHAT IS THE INCLINED PLANE ANGLE IN DEGREES TO THE HORIZONTAL
?45

WHAT IS THE ANGLE OF FORCE P WITH THE INCLINED PLANE
IN DEGREES ANTI-CLOCKWISE WITH P POINTING AWAY FROM THE BLOCK ?
?-15

WHAT IS THE COEFFICIENT OF FRICTION
?.25

WHAT IS THE MASS OF THE BLOCK IN kg
?50


WHAT IS THE MAGNITUDE OF THE APPLIED FORCE IN N
?180
THE ACCELERATION OF THE MASS UP THE PLANE IN m/s^2 IS  -5.426
```

Note negative answer, therefore motion is down plane

```
]RUN

PROGRAM NAME INPLANE 2
FOR BOTH ACCELERATING AND NON-ACCELERATING MASSES ON THE INCLINED PLANE

ENTER U FOR MOTION UP THE PLANE AND D FOR MOTION DOWN THE PLANE
?D

ENTER A FOR ACCELERATED MOTION AND N FOR NON-ACCELERATED MOTION
?A

WHAT IS THE INCLINED PLANE ANGLE IN DEGREES TO THE HORIZONTAL
?45

WHAT IS THE ANGLE OF FORCE P WITH THE INCLINED PLANE
IN DEGREES ANTI-CLOCKWISE WITH P POINTING AWAY FROM THE BLOCK ?
?-15

WHAT IS THE COEFFICIENT OF FRICTION
?.25

WHAT IS THE MASS OF THE BLOCK IN kg
?50


WHAT IS THE MAGNITUDE OF THE APPLIED FORCE IN N
?180
THE ACCELERATION OF THE MASS DOWN THE PLANE IN m/s^2 IS  1.492
```

EXAMPLE 8.5(b)

```
]RUN

PROGRAM NAME INPLANE 2
FOR BOTH ACCELERATING AND NON-ACCELERATING MASSES ON THE INCLINED PLANE

ENTER U FOR MOTION UP THE PLANE AND D FOR MOTION DOWN THE PLANE
?D

ENTER A FOR ACCELERATED MOTION AND N FOR NON-ACCELERATED MOTION
?A

WHAT IS THE INCLINED PLANE ANGLE IN DEGREES TO THE HORIZONTAL
?45

WHAT IS THE ANGLE OF FORCE P WITH THE INCLINED PLANE
IN DEGREES ANTI-CLOCKWISE WITH P POINTING AWAY FROM THE BLOCK ?
?-15

WHAT IS THE COEFFICIENT OF FRICTION
?.25

WHAT IS THE MASS OF THE BLOCK IN kg
?50


WHAT IS THE MAGNITUDE OF THE APPLIED FORCE IN N
?0
THE ACCELERATION OF THE MASS DOWN THE PLANE IN m/s^2 IS  5.202
```

APPENDIX 3.11

```
2  REM  PROGRAM NAME FLYINT
4  REM  BY ROGER KINSKY 30/12/84
5  PRINT "THIS PROGRAM CALCULATES THE MASS MOMENT OF INERTIA OF A FLYWHEEL
    WHICH MAY HAVE HOLES IN THE WEB"
7  PRINT "NOTE THE INERTIA OF THE BOSS IS NEGLECTED"
10  REM  W=DENSITY (CI 7200,STEEL 7800)
20  REM  D=OUTSIDE DIA OF RIM(MM)
30  REM  D1=INSIDE DIA OF RIM(mm)
40  REM  T=THICKNESS OF RIM(mm)
50  REM  T1= THICKNESS OF WEB (mm)
60  REM  D2=DIA. OF HOLES (MM)
70  REM  N=NUMBER OF HOLES
80  REM  D3=PCD OF HOLES (mm)
90  REM  M=MASS OF FLYWHEEL (kg)
92  REM  MASS IN KG M1 (RIM),M2 (HOLES),M3 (SOLID WEB),M4 (EACH HOLE)
94  REM  I=MASS MOMENT OF INERTIA (kgm^2)96 REM INERTIAS I1 (RIM),I2 (WEB)
     ,I3 (SOLID WEB), I4 (EACH HOLE)
96  REM  I1=INERTIA OF FLANGE, I2=INERTIA OF WEB
110  REM  K=RADIUS OF GYRATION (mm)
120  PRINT "WHAT MATERIAL IS THE FLYWHEEL MADE OF ?"
130  PRINT "ANSWER 1 STEEL, 2 CAST IRON,  3 OTHER"
140  INPUT J
150  IF J = 1 THEN W = 7800
160  IF J = 2 THEN W = 7200
170  IF J = 3 THEN 500
190 PI = 3.141593
210  PRINT "WHAT IS THE OUTSIDE DIAMETER OF THE RIM IN mm ?"
220  INPUT D
230 D = D / 1000
240  PRINT "WHAT IS THE INSIDE DIAMETER OF THE RIM IN mm ?"
250  INPUT D1
260 D1 = D1 / 1000
270  PRINT "WHAT IS THE WIDTH OF THE RIM IN mm ?"
280  INPUT T
290 T = T / 1000
300 M1 = PI / 4 * (D ^ 2 - D1 ^ 2) * W * T
310 I1 = M1 / 8 * (D ^ 2 + D1 ^ 2)
320  PRINT "WHAT IS THE THICKNESS OF THE WEB IN mm ?"
330  INPUT T1
335 T1 = T1 / 1000
340  PRINT "HOW MANY HOLES IN THE WEB ?"
342  INPUT N
344  IF N = 0 THEN 360
346  PRINT "NOW INPUT THE HOLE DIA. AND PCD IN mm "
350  INPUT D2,D3
354 D2 = D2 / 1000
356 D3 = D3 / 1000
360 M3 = PI / 4 * D1 ^ 2 * T1 * W
365 M4 = PI / 4 * D2 ^ 2 * T1 * W
370 M2 = M3 - N * M4
375 I3 = .5 * M3 * (D1 / 2) ^ 2
380 I4 = M4 * (.5 * (D2 / 2) ^ 2 + (D3 / 2) ^ 2)
382 I2 = I3 - N * I4
384 M1 =  INT (M1 * 10 + .5) / 10
386 M2 =  INT (M2 * 10 + .5) / 10
388 M = M1 + M2
390 I1 =  INT (I1 * 10 + .5) / 10
392 I2 =  INT (I2 * 10 + .5) / 10
```

```
394 I = I1 + I2
396  PRINT : PRINT
400  PRINT "THE MASS OF THE RIM IN kg IS  ";M1
405  PRINT "THE MOMENT OF INERTIA OF THE RIM IN kgm^2 is  ";I1
410  PRINT : PRINT
420  PRINT "THE MASS OF THE WEB IN kg IS  ";M2
425  PRINT " MOMENT OF INERTIA OF THE WEB IN kgm^2 IS  ";I2
430  PRINT : PRINT
435  PRINT "MASS OF THE FLYWHEEL IN kg IS  ";M
440  PRINT "THE MASS MOMENT OF INERTIA OF THE FLYWHEEL IN kg m^2 IS   ";I
450  PRINT : PRINT
460 K =  SQR (I / M) * 1000
465 K =  INT (K + .5)
470  PRINT "RADIUS OF GYRATION OF THE FLYWHEEL IN mm IS   ";K
490  GOTO 999
500  PRINT "WHAT IS THE DENSITY OF THE MATERIAL IN kg/m^3 ?"
510  INPUT W
520  GOTO 190
999  END
```

EXAMPLE 8.12

```
]RUN

THIS PROGRAM CALCULATES THE MASS MOMENT OF INERTIA OF A FLYWHEEL WHICH
 MAY HAVE HOLES IN THE WEB
NOTE THE INERTIA OF THE BOSS IS NEGLECTED
WHAT MATERIAL IS THE FLYWHEEL MADE OF ?

ANSWER 1 STEEL, 2 CAST IRON,  3 OTHER
?1

WHAT IS THE OUTSIDE DIAMETER OF THE RIM IN mm ?
?600

WHAT IS THE INSIDE DIAMETER OF THE RIM IN mm ?
?400

WHAT IS THE WIDTH OF THE RIM IN mm ?
?350

WHAT IS THE THICKNESS OF THE WEB IN mm ?
?0

HOW MANY HOLES IN THE WEB ?
?0


THE MASS OF THE RIM IN kg IS  428.8
THE MOMENT OF INERTIA OF THE RIM IN kgm^2 is  27.9


THE MASS OF THE WEB IN kg IS  0
 MOMENT OF INERTIA OF THE WEB IN kgm^2 IS  0


MASS OF THE FLYWHEEL IN kg IS  428.8
THE MASS MOMENT OF INERTIA OF THE FLYWHEEL IN kg m^2 IS   27.9


RADIUS OF GYRATION OF THE FLYWHEEL IN mm IS   255
```

EXAMPLE 8.13

```
]RUN

THIS PROGRAM CALCULATES THE MASS MOMENT OF INERTIA OF A FLYWHEEL WHICH
 MAY HAVE HOLES IN THE WEB
NOTE THE INERTIA OF THE BOSS IS NEGLECTED
WHAT MATERIAL IS THE FLYWHEEL MADE OF ?
ANSWER 1 STEEL, 2 CAST IRON,  3 OTHER
?2
WHAT IS THE OUTSIDE DIAMETER OF THE RIM IN mm ?
?2000
WHAT IS THE INSIDE DIAMETER OF THE RIM IN mm ?
?1840
WHAT IS THE WIDTH OF THE RIM IN mm ?
?200
WHAT IS THE THICKNESS OF THE WEB IN mm ?
?60
HOW MANY HOLES IN THE WEB ?
?4
NOW INPUT THE HOLE DIA. AND PCD IN mm
?400,1000


THE MASS OF THE RIM IN kg IS  694.9
THE MOMENT OF INERTIA OF THE RIM IN kgm^2 is  641.5


THE MASS OF THE WEB IN kg IS  931.6
 MOMENT OF INERTIA OF THE WEB IN kgm^2 IS  427.5


MASS OF THE FLYWHEEL IN kg IS  1626.5
THE MASS MOMENT OF INERTIA OF THE FLYWHEEL IN kg m^2 IS   1069


RADIUS OF GYRATION OF THE FLYWHEEL IN mm IS   811
```

APPENDIX 3.12

```
5  PRINT
10  PRINT "FILE NAME COPBAL"
15  PRINT "BALANCING COPLANAR MASSES"
20  REM BY ROGER KINSKY 12/1/85
40  PRINT "ANGLES ARE MEASURED IN DEGREES ANTI-CLOCKWISE FROM THE X AXIS"
50  PRINT
55 F3 = 0:F4 = 0:W = 0
60  PRINT "HOW MANY OUT-OF-BALANCE MASSES ARE THERE ?"
65  INPUT V
70  PRINT "NOW INPUT FOR EACH MASS IN TURN MAGNITUDE, RADIUS AND ANGLE"
```

```
80  INPUT M,R,X
90  PRINT
95 F = M * R
100 F1 = F *  SIN (X * .0174533)
110 F2 = F *  COS (X * .0174533)
120 F3 = F3 + F1
130 F4 = F4 + F2
140 W = W + 1
150  IF  ABS (V - W)  > .2 THEN 80
160 F5 = (F3 ^ 2 + F4 ^ 2) ^ .5
180  IF F4 = 0 THEN F4 = 1E - 6
190 Y =  ATN ( ABS (F3 / F4)) * 57.296
200 Y =  INT (Y * 100 + .5) / 100
210  IF F3 < 0 THEN 300
220  IF F4 < 0 THEN Y = 180 - Y
230  GOTO 400
300  IF F4 < 0 THEN Y = 180 + Y
310  IF F4 > 0 THEN Y = 360 - Y
400  PRINT : PRINT
410  PRINT "DO YOU WISH TO SPECIFY THE MAGNITUDE OR RADIUS OF THE BALANCE
      MASS ?"
420  PRINT "ANSWER M OR R"
430  INPUT A$
440  IF A$ = "M" THEN 550
450  IF A$ = "R" THEN 470
460  GOTO 410
470  PRINT "WHAT IS THE DESIRED RADIUS OF THE BALANCE MASS ?"
480  INPUT R
490 M = F5 / R
500 M =  INT (M * 100 + .5) / 100
510  GOTO 600
550  PRINT "WHAT IS THE DESIRED MAGNITUDE OF THE BALANCE MASS ?"
560  INPUT M
570 R = F5 / M
580 R =  INT (R * 100 + .5) / 100
600 Y1 = 180 + Y
610  IF Y1 >  = 360 THEN Y1 = Y1 - 360
615  PRINT
620  PRINT "THE REQUIRED BALANCE MASS IS AS FOLLOWS :"
625  PRINT
630  PRINT "MASS","RADIUS","ANGLE"
635  PRINT
640  PRINT M,R,Y1
999  END
```

EXAMPLE 11.2

```
]RUN

FILE NAME COPBAL
BALANCING COPLANAR MASSES
ANGLES ARE MEASURED IN DEGREES ANTI-CLOCKWISE FROM THE X AXIS

HOW MANY OUT-OF-BALANCE MASSES ARE THERE ?
?2
NOW INPUT FOR EACH MASS IN TURN MAGNITUDE, RADIUS AND ANGLE
?1,300,30

?1.5,400,135
```

```
DO YOU WISH TO SPECIFY THE MAGNITUDE OR RADIUS OF THE BALANCE MASS ?
ANSWER M OR R
?M
WHAT IS THE DESIRED MAGNITUDE OF THE BALANCE MASS ?
?2

THE REQUIRED BALANCE MASS IS AS FOLLOWS :

MASS                RADIUS              ANGLE

2                   298.67              285.98
```

APPENDIX 3.13

```
5  PRINT
10  PRINT "FILE NAME COPRE"
15  PRINT "REACTIONS FOR ROTATING COPLANAR MASSES"
20  REM  BY ROGER KINSKY 16/1/85
40  PRINT "ANGLES ARE MEASURED IN DEGREES ANTI-CLOCKWISE FROM THE X AXIS"
50  PRINT
55 F3 = 0:F4 = 0:W = 0
57 W2 = 0
60  PRINT "HOW MANY OUT-OF-BALANCE MASSES ARE THERE ?"
62  INPUT V
65  PRINT "WHAT IS THE SHAFT SPEED IN RPM ?"
67  INPUT N
70  PRINT "NOW INPUT FOR EACH MASS IN TURN MAGNITUDE, RADIUS AND ANGLE"
75  PRINT "NOTE MASSES IN kg AND RADIUS IN m"
80  INPUT M,R,X
90  PRINT
95 F = M * R * (.10472 * N) ^ 2
97 W1 = M * 9.81
100 F1 = F *  SIN (X * .0174533)
110 F2 = F *  COS (X * .0174533)
115 W2 = W2 + W1
120 F3 = F3 + F1
130 F4 = F4 + F2
135  IF W = 0 THEN X1 = X
140 W = W + 1
150  IF  ABS (V - W) > .2 THEN 80
160 F5 = (F3 ^ 2 + F4 ^ 2) ^ .5
180  IF F4 = 0 THEN F4 = 1E - 6
190 Y =  ATN ( ABS (F3 / F4)) * 57.296
210  IF F3 < 0 THEN 300
220  IF F4 < 0 THEN Y = 180 - Y
230  GOTO 400
300  IF F4 < 0 THEN Y = 180 + Y
310  IF F4 > 0 THEN Y = 360 - Y
400  PRINT : PRINT
410 R1 = F3 - W2
```

```
420 R2 =  SQR (R1 ^ 2 + F4 ^ 2)
430 Y2 =  ATN ( ABS (R1 / F4)) * 57.296
450  IF R1 < 0 THEN 480
460  IF F4 < 0 THEN Y2 = 180 - Y2
470  GOTO 500
480  IF F4 < 0 THEN Y2 = 180 + Y2
490  IF F4 > 0 THEN Y2 = 360 - Y2
500  PRINT : PRINT
510 Y3 = 180 + Y2
520  IF Y3 >  = 360 THEN Y3 = Y3 - 360
525 R2 =  INT (R2 * 100 + .5) / 100
527 Y3 =  INT (Y3 * 100 + .5) / 100
530  PRINT "BEARING REACTION IS  ";R2;" N @ ";Y3;" DEGREES"
540  PRINT
545 R3 =  INT ((F5 + W2) * 100 + .5) / 100
547 Y4 = 270 - Y + X1
548 Y4 =  INT (Y4 * 100 + .5) / 100
549  IF  SGN (Y4) =  - 1 THEN Y4 = 360 + Y4
550  PRINT "MAX REACTION IS ";R3;" N DOWN WHEN ANGLE MADE BY MASS ONE IS"
      ;Y4;" DEGREES"
560  PRINT
565 R4 =  INT ((F5 - W2) * 100 + .5) / 100
567 Y5 = 90 - Y + X1
568 Y5 =  INT (Y5 * 100 + .5) / 100
569  IF  SGN (Y5) =  - 1 THEN Y5 = 360 + Y5
570  PRINT "MIN REACTION IS ";R4;" N UP WHEN ANGLE MADE BY MASS ONE IS ";
     Y5;" DEGREES"
999  END
```

EXAMPLE 11.3

```
]RUN

FILE NAME COPRE
REACTIONS FOR ROTATING COPLANAR MASSES
ANGLES ARE MEASURED IN DEGREES ANTI-CLOCKWISE FROM THE X AXIS

HOW MANY OUT-OF-BALANCE MASSES ARE THERE ?
?2
WHAT IS THE SHAFT SPEED IN RPM ?
?90
NOW INPUT FOR EACH MASS IN TURN MAGNITUDE, RADIUS AND ANGLE
NOTE MASSES IN kg AND RADIUS IN m
?2,.2,45

?1.2,.16,105

BEARING REACTION IS  23.09 N @ 206.23 DEGREES

MAX REACTION IS 77.86 N DOWN WHEN ANGLE MADE BY MASS ONE IS 251.47 DEGREES

MIN REACTION IS 15.08 N UP WHEN ANGLE MADE BY MASS ONE IS 71.47 DEGREES
```

APPENDIX 3.14

```
10  PRINT
20  PRINT "FILE NAME NCBAL"
30  PRINT "BALANCING NON-COPLANAR MASSES"
40  REM  BY ROGER KINSKY 18/1/85
50  PRINT "ANGLES MEASURED IN DEGREES ANTI-CLOCKWISE ARE POSITIVE"
60  PRINT "LENGTHS IN FRONT OF THE REFERENCE PLANE ARE POSITIVE"
80 F3 = 0:F4 = 0:A = 0
90 P = .0174533
100  PRINT "HOW MANY KNOWN MASSES ARE TO BE BALANCED"
110  INPUT V
120  PRINT "NOW INPUT THE INFORMATION BELOW FOR EACH KNOWN MASS"
125  PRINT
130  PRINT "MASS,RADIUS,ANGLE,AXIAL DIST"
135  PRINT
140  FOR I = 1 TO V
150  INPUT M(I),R(I),X(I),L(I)
155 A = A + M(I)
160  NEXT I
162 A = A / (V * 1000)
164 B = A * 1E6
165  PRINT
170  PRINT "NOW INPUT THE RADIUS AND AXIAL DISTANCE OF THE BALANCE MASS NOT
      IN THE REFERENCE PLANE"
180  INPUT R(V + 1),L(V + 1)
200  FOR I = 1 TO V
210 F(I) = M(I) * R(I) * L(I)
220  GOSUB 700
250  NEXT I
260  GOSUB 750
270 M = F5 / (R(V + 1) * L(V + 1))
271  IF  SGN (M) =  - 1 THEN Y1 = Y1 + 180
272  IF Y1 >  = 360 THEN Y1 = Y1 - 360
273 M =  ABS (M)
274 M(V + 1) = M
275  IF A >  = M THEN 900
277  IF M > B THEN 920
280  PRINT
285  PRINT "THE REQUIRED BALANCE MASS NOT IN THE REFERENCE PLANE IS AS
     FOLLOWS"
290 M =  INT (M * 1000 + .5) / 1000
295  PRINT
300  PRINT "MASS             RADIUS          ANGLE          AXIAL DIST"
302  PRINT
303  PRINT M;"            ";R(V + 1);"           ";Y1;"            ";L(V
     + 1)
305  PRINT
307 X(V + 1) = Y1
309 F3 = 0:F4 = 0
310  PRINT "NOW INPUT THE RADIUS OF THE BALANCE MASS IN THE REFERENCE PLANE"
320  INPUT R(V + 2)
330  FOR I = 1 TO V + 1
340 F(I) = M(I) * R(I)
350  GOSUB 700
360  NEXT I
370  GOSUB 750
375 M = F5 / R(V + 2)
380  IF A >  = M THEN 940
385  IF M > B THEN 960
```

```
390  PRINT "THE REQUIRED BALANCE MASS IN THE REFERENCE PLANE IS AS FOLLOWS"
395  PRINT
400 M =  INT (M * 1000 + .5) / 1000
410  PRINT "MASS","RADIUS","ANGLE"
412  PRINT
415  PRINT M,R(V + 2),Y1
420  GOTO 999
700 F1 = F(I) *  SIN (X(I) * P)
710 F2 = F(I) *  COS (X(I) * P)
720 F3 = F3 + F1
730 F4 = F4 + F2
740  RETURN
750 F5 =  SQR (F3 ^ 2 + F4 ^ 2)
760  IF F4 = 0 THEN F4 = 1E - 6
770 Y =  ATN ( ABS (F3 / F4)) / P
780  IF F3 < 0 THEN 800
790  IF F4 < 0 THEN Y = 180 - Y
795  GOTO 820
800  IF F4 < 0 THEN Y = 180 + Y
810  IF F4 > 0 THEN Y = 360 - Y
820 Y1 = 180 + Y
830  IF Y1 > = 360 THEN Y1 = Y1 - 360
840 Y1 =  INT (Y1 * 100 + .5) / 100
850  RETURN
900  PRINT "NO BALANCE MASS REQUIRED IN THIS PLANE "
910  GOTO 305
920  PRINT "BALANCE MASS TOO LARGE, TRY AGAIN"
930  GOTO 80
940  PRINT "NO BALANCE MASS REQUIRED IN REFERENCE PLANE"
950  GOTO 999
960  PRINT "BALANCE MASS IN REFERENCE PLANE IS TOO LARGE, TRY AGAIN"
970  GOTO 305
999  END
```

EXAMPLE 11.4

(a) Reference plane in plane of m_1

```
]RUN

FILE NAME NCBAL
BALANCING NON-COPLANAR MASSES
ANGLES MEASURED IN DEGREES ANTI-CLOCKWISE ARE POSITIVE
LENGTHS IN FRONT OF THE REFERENCE PLANE ARE POSITIVE

HOW MANY KNOWN MASSES ARE TO BE BALANCED
?3

NOW INPUT THE INFORMATION BELOW FOR EACH KNOWN MASS

MASS,RADIUS,ANGLE,AXIAL DIST

?.8,102,90,0
?.6,102,0,187
?.4,102,300,576

NOW INPUT THE RADIUS AND AXIAL DISTANCE OF THE BALANCE MASS NOT IN THE
 REFERENCE PLANE
?127,341
```

```
THE REQUIRED BALANCE MASS NOT IN THE REFERENCE PLANE IS AS FOLLOWS

MASS              RADIUS            ANGLE             AXIAL DIST

.713              127               138.73             341

NOW INPUT THE RADIUS OF THE BALANCE MASS IN THE REFERENCE PLANE
?127
THE REQUIRED BALANCE MASS IN THE REFERENCE PLANE IS AS FOLLOWS

MASS              RADIUS            ANGLE

.841              127               262.69
```

EXAMPLE 11.4

(b) Reference plane in plane of m_4

```
]RUN

FILE NAME NCBAL
BALANCING NON-COPLANAR MASSES
ANGLES MEASURED IN DEGREES ANTI-CLOCKWISE ARE POSITIVE
LENGTHS IN FRONT OF THE REFERENCE PLANE ARE POSITIVE
HOW MANY KNOWN MASSES ARE TO BE BALANCED
?3
NOW INPUT THE INFORMATION BELOW FOR EACH KNOWN MASS

MASS,RADIUS,ANGLE,AXIAL DIST

?.4,102,300,235
?.6,102,0,-154
?.8,102,90,-341

NOW INPUT THE RADIUS AND AXIAL DISTANCE OF THE BALANCE MASS NOT IN THE
 REFERENCE PLANE
?127,-341

THE REQUIRED BALANCE MASS NOT IN THE REFERENCE PLANE IS AS FOLLOWS

MASS              RADIUS            ANGLE             AXIAL DIST

.841              127               262.7              -341

NOW INPUT THE RADIUS OF THE BALANCE MASS IN THE REFERENCE PLANE
?127
THE REQUIRED BALANCE MASS IN THE REFERENCE PLANE IS AS FOLLOWS

MASS              RADIUS            ANGLE

.713              127               138.74
```

APPENDIX 3.15

```
5  PRINT "FILE NAME NCREACT"
10  REM  BY ROGER KINSKY 24/1/85
20  PRINT "BEARING REACTIONS FOR NON-COPLANAR ROTATING MASSES"
30  PRINT "CALL THE LH BRG A, THE RH BRG B"
60 Q1 = 0:Q2 = 0:Q3 = 0:Q4 = 0
65 P = .0174533
70  PRINT "WHAT IS THE DISTANCE BETWEEN THE BEARINGS IN m"
75  INPUT L1
80  PRINT "WHAT IS THE SHAFT SPEED IN rpm"
85  INPUT N
90  PRINT "HOW MANY ROTATING MASSES"
95  INPUT V
100  PRINT "NOW INPUT THE MASS (kg),RADIUS (m),ANGLE (DEG.ANTI-CLOCK),AND
     AXIAL DIST (m) FOR EACH MASS IN TURN"
110  PRINT "MASS , RADIUS ,  ANGLE , AXIAL DIST"
130  FOR I = 1 TO V
140  INPUT M,R,X,L
142 X(I) = X
144 L(I) = L
146 Y(I) = X(I) - X(1)
148 Y(1) = 0
150 F = M * R * (.10472 * N) ^ 2
155 F(I) = F
160 W = 9.81 * M
165 W(I) = W
170 Q1 = Q1 + (W - F *  SIN (X * P)) * L
180 Q2 = Q2 + (F *  COS (X * P)) * L
190 Q3 = Q3 + (F *  SIN (X * P) - W) * (L1 - L)
200 Q4 = Q4 + ( - 1 * F *  COS (X * P)) * (L1 - L)
210  NEXT I
220 R1 = Q1 / L1
230 R2 =  - Q2 / L1
240 R3 =  - Q3 / L1
250 R4 = Q4 / L1
270 R5 =  SQR (R1 ^ 2 + R2 ^ 2)
280  IF R2 = 0 THEN R2 = 1E - 6
290 Y1 =  ATN ( ABS (R1 / R2)) / P
300  IF R1 < 0 THEN 330
310  IF R2 < 0 THEN Y1 = 180 - Y1
320  GOTO 350
330  IF R2 < 0 THEN Y1 = 180 + Y1
340  IF R2 > 0 THEN Y1 = 360 - Y1
350  PRINT
370 R6 =  SQR (R3 ^ 2 + R4 ^ 2)
380  IF R4 = 0 THEN R4 = 1E - 6
390 Y2 =  ATN ( ABS (R3 / R4)) / P
400  IF R3 < 0 THEN 430
410  IF R4 < 0 THEN Y2 = 180 - Y2
420  GOTO 450
430  IF R4 < 0 THEN Y2 = 180 + Y2
440  IF R4 > 0 THEN Y2 = 360 - Y2
450  PRINT
460 R5 =  INT (R5 * 100 + .5) / 100
470 R6 =  INT (R6 * 100 + .5) / 100
480 Y1 =  INT (Y1 * 100 + .5) / 100
490 Y2 =  INT (Y2 * 100 + .5) / 100
500  PRINT "THE REACTION AT B IS  ";R5;" N @  ";Y1;" DEGREES"
510  PRINT
```

```
520  PRINT "THE REACTION AT A IS  ";R6;" N @  ";Y2;" DEGREES"
530  PRINT
540  PRINT "DO YOU WANT THE MAX AND MIN REACTIONS AS THE SHAFT ROTATES ?"
545  PRINT "ANSWER Y OR N"
550  INPUT A$
560  IF A$ = "N" THEN 999
570  PRINT
580  PRINT "PLEASE WAIT AND BE PATIENT, THIS PART OF THE RUN TAKES SEVERAL
      MINUTES"
585  PRINT : PRINT
590 E = R5:B = R5:C = R6:D = R6
600  FOR J = 0 TO 360
605 Q1 = 0:Q2 = 0:Q3 = 0:Q4 = 0
610  FOR I = 1 TO V
620 E(I) = (J + Y(I)) * P
630 Q1 = Q1 + (W(I) - F(I) *  SIN (E(I))) * L(I)
650 Q2 = Q2 + (F(I) *  COS (E(I))) * L(I)
670 Q3 = Q3 + (F(I) *  SIN (E(I)) - W(I)) * (L1 - L(I))
690 Q4 = Q4 + ( - 1 * F(I) *  COS (E(I))) * (L1 - L(I))
710  NEXT I
720 R1 = Q1 / L1
725 R2 =  - Q2 / L1
728 R3 =  - Q3 / L1
730 R4 = Q4 / L1
732 G =  SQR (R1 ^ 2 + R2 ^ 2)
735 H =  SQR (R3 ^ 2 + R4 ^ 2)
740  IF G > E THEN Z1 = J
750  IF G > E THEN E = G
760  IF G < B THEN Z2 = J
770  IF G < B THEN B = G
780  IF H > C THEN Z3 = J
790  IF H > C THEN C = H
800  IF H < D THEN Z4 = J
810  IF H < D THEN D = H
820  NEXT J
830  PRINT
840 E =  INT (E * 100 + .5) / 100
850 B =  INT (B * 100 + .5) / 100
860 C =  INT (C * 100 + .5) / 100
870 D =  INT (D * 100 + .5) / 100
880  PRINT "MAX AND MIN REACTIONS AND ANGLE MADE BY MASS 1 WHEN THEY OCCUR
      ARE :"
890  PRINT
900  PRINT "MAX           ANGLE         MIN           ANGLE"
910  PRINT "BEARING B"
920  PRINT E;"          ";Z1;"          ";B;"          ";Z2
930  PRINT
940  PRINT "BEARING A"
950  PRINT C;"          ";Z3;"          ";D;"          ";Z4
999  END
```

EXAMPLE 11.5

```
]RUN

FILE NAME NCREACT
BEARING REACTIONS FOR NON-COPLANAR ROTATING MASSES
CALL THE LH BRG A, THE RH BRG B
WHAT IS THE DISTANCE BETWEEN THE BEARINGS IN m
?.95
```

```
WHAT IS THE SHAFT SPEED IN rpm
?90
HOW MANY ROTATING MASSES
?2
NOW INPUT THE MASS (kg),RADIUS (m),ANGLE (DEG.ANTI-CLOCK),AND AXIAL DIST
 (m) FOR EACH MASS IN TURN
MASS , RADIUS ,  ANGLE , AXIAL DIST
?1.5,.3,90,.35
?1.2,.25,0,.6


THE REACTION AT B IS  16.93 N @  186.34 DEGREES

THE REACTION AT A IS  15.21 N @  229.79 DEGREES

DO YOU WANT THE MAX AND MIN REACTIONS AS THE SHAFT ROTATES ?
ANSWER Y OR N
?Y

PLEASE WAIT AND BE PATIENT, THIS PART OF THE RUN TAKES SEVERAL MINUTES


MAX AND MIN REACTIONS AND ANGLE MADE BY MASS 1 WHEN THEY OCCUR ARE :

MAX          ANGLE        MIN          ANGLE
BEARING B
35.22         319         9.51         139

BEARING A
40.72        291         13.46         111
```

APPENDIX 3.16

```
5  PRINT "PROGRAM NAME DYNLOAD "
10  REM  BY ROGER KINSKY 3/2/85
15  PRINT "FOR DETERMINING STRESS AND DEFLECTION FOR DYNAMIC AXIAL LOADS ON
     CIRCULAR SECTIONS"
20  PRINT
25  PRINT "INPUT THE LENGTH AND DIAMETER OF THE ROD IN mm"
30  INPUT L,D
35  PRINT
40  PRINT "INPUT THE LOAD MASS IN kg "
45  INPUT M
55  PRINT
60  PRINT "INPUT THE MODULUS OF ELASTICITY IN GPa "
65  INPUT E
70  PRINT
90  PRINT "DOES THE PROBLEM INVOLVE AN IMPACTING MASS FALLING FROM HEIGHT H"
100  PRINT "OR A SUDDENLY STOPPED MASS MOVING WITH VELOCITY V ?"
```

```
110  PRINT "ANSWER H FOR FORMER AND V FOR LATTER"
120  INPUT A$
122  PRINT "
123  PRINT "IS THE MAXIMUM STRESS GIVEN OR REQUIRED ?"
124  PRINT "ANSWER G FOR FORMER AND R FOR LATTER "
126  INPUT B$
128  IF B$ = "G" THEN 400
130  IF A$ = "H" THEN 185
140  IF A$ = "V" THEN 150
145  GOTO 100
150  PRINT "WHAT IS THE VELOCITY OF THE LOAD IN m/s "
160  INPUT V
170  GOTO 200
185  PRINT "WHAT IS THE HEIGHT OF THE LOAD IN mm "
190  INPUT H
200  PRINT
210 X = M * 9.81 * L / (E * 785.4 * D ^ 2)
220  IF A$ = "H" THEN Y = X * (1 +  SQR (1 + 2 * H / X))
225  IF A$ = "V" THEN Y = X * (1 + V /  SQR (.00981 * X))
230 F = E * 1000 * Y / L
240 F1 = F * X / Y
250 Y =  INT (Y * 100 + .5) / 100
260 F =  INT (F * 100 + .5) / 100
270 F1 =  INT (F1 * 100 + .5) / 100
275 X =  INT (X * 100 + .5) / 100
280  PRINT
290  PRINT "THE MAXIMUM STRESS IS  ";F;" MPa"
300  PRINT
310  PRINT "THE MAXIMUM DEFLECTION IS  ";Y;" mm"
320  PRINT : PRINT
330  PRINT "THE STATIC STRESS IS  ";F1;" MPa"
340  PRINT
350  PRINT "THE STATIC DEFLECTION IS  ";X;" mm"
360  GOTO 999
400 X = M * 9.81 * L / (E * 785.4 * D ^ 2)
410  PRINT "WHAT IS THE MAXIMUM ALLOWABLE STRESS IN MPa ?"
420  INPUT F
430 Y = F * L * .001 / E
440 H = ((Y / X - 1) ^ 2 - 1) * X / 2
450 V = (Y / X - 1) *  SQR (.00981 * X)
455 F1 = F * X / Y
460 H =  INT (H * 100 + .5) / 100
470 V =  INT (V * 100 + .5) / 100
480 Y =  INT (Y * 100 + .5) / 100
500  PRINT
510  IF A$ = "V" THEN 550
520  PRINT "THE MAXIMUM HEIGHT FROM WHICH THE LOAD MAY BE DROPPED WITHOUT
     THE STRESS EXCEEDING ";F;" MPa  IS ";H;" mm"
525  PRINT
530  PRINT "AND THE CORROSPONDING EXTENSION IS  ";Y;" mm"
540  GOTO 600
550  PRINT "THE MAXIMUM VELOCITY OF THE LOAD SO THAT THE STRESS DOES NOT
     EXCEED ";F;" MPa IS ";V;" m/s"
560  PRINT
570  PRINT "AND THE CORROSPONDING EXTENSION IS  ";Y"  mm"
600  PRINT
610 F1 =  INT (F1 * 100 + .5) / 100
620 X =  INT (X * 100 + .5) / 100
630  GOTO 320
999  END
```

EXAMPLE 13.5

```
]RUN

PROGRAM NAME DYNLOAD
FOR DETERMINING STRESS AND DEFLECTION FOR DYNAMIC AXIAL LOADS ON CIRCULAR
SECTIONS

INPUT THE LENGTH AND DIAMETER OF THE ROD IN mm
?1500,0.8

INPUT THE LOAD MASS IN kg
?5

INPUT THE MODULUS OF ELASTICITY IN GPa
?110

DOES THE PROBLEM INVOLVE AN IMPACTING MASS FALLING FROM HEIGHT H
OR A SUDDENLY STOPPED MASS MOVING WITH VELOCITY V ?
ANSWER H FOR FORMER AND V FOR LATTER
?H

IS THE MAXIMUM STRESS GIVEN OR REQUIRED ?
ANSWER G FOR FORMER AND R FOR LATTER
?R
WHAT IS THE HEIGHT OF THE LOAD IN mm
?0


THE MAXIMUM STRESS IS  195.16 MPa

THE MAXIMUM DEFLECTION IS  2.66 mm


THE STATIC STRESS IS  97.58 MPa

THE STATIC DEFLECTION IS  1.33 mm
```

EXAMPLE 13.7

```
]RUN

PROGRAM NAME DYNLOAD
FOR DETERMINING STRESS AND DEFLECTION FOR DYNAMIC AXIAL LOADS ON CIRCULAR
SECTIONS

INPUT THE LENGTH AND DIAMETER OF THE ROD IN mm
?3000,5

INPUT THE LOAD MASS IN kg
?100

INPUT THE MODULUS OF ELASTICITY IN GPa
?200

DOES THE PROBLEM INVOLVE AN IMPACTING MASS FALLING FROM HEIGHT H
OR A SUDDENLY STOPPED MASS MOVING WITH VELOCITY V ?
ANSWER H FOR FORMER AND V FOR LATTER
?H
```

```
IS THE MAXIMUM STRESS GIVEN OR REQUIRED ?
ANSWER G FOR FORMER AND R FOR LATTER
?R
WHAT IS THE HEIGHT OF THE LOAD IN mm
?50



THE MAXIMUM STRESS IS  629.25 MPa

THE MAXIMUM DEFLECTION IS  9.44 mm



THE STATIC STRESS IS  49.96 MPa

THE STATIC DEFLECTION IS  .75 mm
```

EXAMPLE 13.8

```
]RUN

PROGRAM NAME DYNLOAD
FOR DETERMINING STRESS AND DEFLECTION FOR DYNAMIC AXIAL LOADS ON CIRCULAR
SECTIONS

INPUT THE LENGTH AND DIAMETER OF THE ROD IN mm
?18000,45

INPUT THE LOAD MASS IN kg
?5000

INPUT THE MODULUS OF ELASTICITY IN GPa
?110

DOES THE PROBLEM INVOLVE AN IMPACTING MASS FALLING FROM HEIGHT H
OR A SUDDENLY STOPPED MASS MOVING WITH VELOCITY V ?
ANSWER H FOR FORMER AND V FOR LATTER
?V

IS THE MAXIMUM STRESS GIVEN OR REQUIRED ?
ANSWER G FOR FORMER AND R FOR LATTER
?R
WHAT IS THE VELOCITY OF THE LOAD IN m/s
?0.9



THE MAXIMUM STRESS IS  155.59 MPa

THE MAXIMUM DEFLECTION IS  25.46 mm



THE STATIC STRESS IS  30.84 MPa

THE STATIC DEFLECTION IS  5.05 mm
```

APPENDIX 3.17

```
10  PRINT "PROGRAM NAME CENTINT"
20  REM  BY ROGER KINSKY 27/2/85
30  PRINT "CALCULATES CENTROIDAL POSITION AND SECOND MOMENT OF AREA"
40  PRINT "OF COMPOSITE SECTIONS OF UP TO 10 INDIVIDUAL AREA SEGMENTS"
55  PRINT "NOTE THAT ALL DIMENSIONS ARE IN mm"
60  PRINT "THERE ARE 12 CHOICES OF AREA SEGMENTS, CONSULT THE TEXT FOR
      REFERENCE NUMBERS OF THESE"
70 I = 0:A = 0:M = 0:K = 0:L = 0
80  PRINT "INPUT THREE DETAILS (SEPARATED BY A COMMA) FOR EACH SEGMENT WHEN
      PROMPTED"
90  PRINT "THESE ARE - AREA TYPE (NUMBER), S FOR SOLID OR V FOR VOID,
      NUMBER OF IDENTICAL AREA SEGMENTS"
95  PRINT
96  PRINT "IS THE COMPOSITE SECTION SYMMETRICAL ABOUT THE REFERENCE AXIS ?"
97  PRINT "ANSWER Y OR N"
98  INPUT C$
100 I = I + 1:J = I
110  PRINT "NOW INPUT THE THREE DETAILS OF THE AREA SEGMENT ";I
120  INPUT T,A$,N
130  IF A$ = "V" THEN E =  - 1
140  IF A$ = "S" THEN E = 1
150  PRINT
200  IF T = 1 THEN 230
202  IF T = 2 THEN 250
204  IF T = 3 THEN 270
206  IF T = 4 THEN 290
207  IF T = 5 THEN 290
208  IF T = 6 THEN 310
210  IF T = 7 THEN 310
212  IF T = 8 THEN 330
214  IF T = 9 THEN 330
216  IF T = 10 THEN 350
218  IF T = 11 THEN 350
220  IF T = 12 THEN 370
230  PRINT "INPUT BASE AND HEIGHT OF RECTANGLE"
232  INPUT B,H
234  PRINT ":INPUT DISTANCE FROM BASE OF RECTANGLE TO REFERENCE AXIS"
236  INPUT Y
238 A(I) = B * H
240 I(I) = (B * H ^ 3) / 12
242 Y(I) = Y + H / 2
244  GOTO 500
250  PRINT "INPUT THE DIAMETER OF THE CIRCLE"
252  INPUT D
254  PRINT "INPUT DISTANCE FROM CENTER OF THE CIRCLE TO REFERENCE AXIS"
256  INPUT Y
258 A(I) = .7854 * D ^ 2
260 I(I) = .049087 * D ^ 4
262 Y(I) = Y
264  GOTO 500
270  PRINT "INPUT RADIUS OF SEMI-CIRCLE"
272  INPUT R
274  PRINT "INPUT DISTANCE FROM CENTER OF SEMI CIRCLE TO REFERENCE AXIS"
276  INPUT Y
278 A(I) = 1.5708 * R ^ 2
280 I(I) = .024544 * (2 * R) ^ 4
282 Y(I) = Y
284  GOTO 500
```

```
290  PRINT "INPUT RADIUS OF SEMI-CIRCLE"
292  INPUT R
294  PRINT "INPUT DISTANCE FROM CENTRE OF SEMI-CIRCLE TO REFERENCE AXIS"
296  INPUT Y
298 A(I) = 1.5708 * R ^ 2
300 I(I) = .10976 * R ^ 4
302  IF T = 5 THEN Y(I) = Y + .424413 * R
304  IF T = 4 THEN Y(I) = Y - .424413 * R
306  GOTO 500
310  PRINT "INPUT RADIUS OF THE QUADRANT"
312  INPUT R
314  PRINT "INPUT DISTANCE FROM THE CENTRE OF THE QUADRANT TO THE REFERENCE
      AXIS"
316  INPUT Y
318 A(I) = .7854 * R ^ 2
320 I(I) = .0549 * R ^ 4
322  IF T = 7 THEN Y(I) = Y + .424413 * R
324  IF T = 6 THEN Y(I) = Y - .4244413 * R
326  GOTO 500
330  PRINT "INPUT BASE AND HEIGHT OF TRIANGULAR FILLET"
332  INPUT B,H
334  PRINT "INPUT DISTANCE FROM BASE OF FILLET TO REFERENCE AXIS"
336  INPUT Y
338 A(I) = B * H / 2
340 I(I) = (B * H ^ 3) / 36
342  IF T = 8 THEN Y(I) = Y - H / 3
344  IF T = 9 THEN Y(I) = Y + H / 3
346  GOTO 500
350  PRINT "INPUT RADIUS OF CIRCULAR FILLET"
352  INPUT R
354  PRINT "INPUT DISTANCE FROM CENTRE OF FILLET TO REFERENCE AXIS"
356  INPUT Y
358 A(I) = .2146 * R ^ 2
360 I(I) = .007546 * R ^ 4
362  IF T = 10 THEN Y(I) = Y + .776632 * R
364  IF T = 11 THEN Y(I) = Y - .776632 * R
366  GOTO 500
370  PRINT "INPUT KNOWN AREA IN mm^2"
372  INPUT A(I)
374  PRINT "INPUT KNOWN CENTROIDAL DISTANCE FROM REFERENCE AXIS IN mm"
376  INPUT Y(I)
378  PRINT "INPUT KNOWN MOMENT OF INERTIA IN mm^4 *1E6 ABOUT CENTROIDAL
      AXIS"
380  INPUT I(I)
382 I(I) = I(I) * 1E6
384  GOTO 500
500 I(I) = I(I) * N * E
510 A(I) = A(I) * N * E
520 M(I) = A(I) * Y(I)
530 A = A + A(I)
540 M = M + M(I)
550  PRINT
560  PRINT "ANY MORE SEGMENTS ? ANSWER Y OR N"
570  INPUT B$
580  IF B$ = "Y" THEN 100
585  PRINT
590 C = M / A
595  IF C$ = "Y" THEN C = 0
600  FOR I = 1 TO J
610 K = K + I(I) + A(I) * (Y(I) - C) ^ 2
620 L = L + I(I) + A(I) * Y(I) ^ 2
630  NEXT I
```

```
640 G =  SQR (K / A)
650 H =  SQR (L / A)
660 G =  INT (G * 10 + .5) / 10
670 H =  INT (H * 10 + .5) / 10
700 A =  INT (A * 10 + .5) / 10
710 C =  INT (C * 10 + .5) / 10
720 K = K / 1E6
724  IF K < 1 THEN K =  INT (K * 10000 + .5) / 10000
726  IF K < 1 THEN 740
730 K =  INT (K * 100 + .5) / 100
740 L = L / 1E6
744  IF L < 1 THEN L =  INT (L * 10000 + .5) / 10000
746  IF L < 1 THEN 800
750 L =  INT (L * 100 + .5) / 100
800  PRINT "THE DETAILS OF THE COMPOSITE SECTION ARE AS FOLLOWS"
810  PRINT
820  PRINT "AREA IN mm^2  ";A
830  PRINT
840  PRINT "CENTROID DISTANCE FROM REF AXIS  ";C
850  PRINT
860  PRINT "I ABOUT CENTROIDAL AXIS IN mm^4  ";K;" E6"
870  PRINT
880  PRINT "I ABOUT REF AXIS IN mm^4  ";L;" E6"
890  PRINT
900  PRINT "K ABOUT CENTROIDAL AXIS IN mm  ";G
910  PRINT
920  PRINT "K ABOUT REFERENCE AXIS  ";H
999  END
```

EXAMPLE 14.1 (X AXIS)

```
PROGRAM NAME CENTINT
CALCULATES CENTROIDAL POSITION AND SECOND MOMENT OF AREA
OF COMPOSITE SECTIONS OF UP TO 10 INDIVIDUAL AREA SEGMENTS
NOTE THAT ALL DIMENSIONS ARE IN mm
THERE ARE 12 CHOICES OF AREA SEGMENTS, CONSULT THE TEXT FOR REFERENCE
 NUMBERS OF THESE
INPUT THREE DETAILS (SEPARATED BY A COMMA) FOR EACH SEGMENT WHEN PROMPTED
THESE ARE - AREA TYPE (NUMBER), S FOR SOLID OR V FOR VOID, NUMBER OF
 IDENTICAL AREA SEGMENTS

IS THE COMPOSITE SECTION SYMMETRICAL ABOUT THE REFERENCE AXIS ?
ANSWER Y OR N
?N
NOW INPUT THE THREE DETAILS OF THE AREA SEGMENT 1
?1,S,1

INPUT BASE AND HEIGHT OF RECTANGLE
?20,100
:INPUT DISTANCE FROM BASE OF RECTANGLE TO REFERENCE AXIS
?0

ANY MORE SEGMENTS ? ANSWER Y OR N
?Y
NOW INPUT THE THREE DETAILS OF THE AREA SEGMENT 2
?1,S,1

INPUT BASE AND HEIGHT OF RECTANGLE
?130,20
```

```
:INPUT DISTANCE FROM BASE OF RECTANGLE TO REFERENCE AXIS
?0

ANY MORE SEGMENTS ? ANSWER Y OR N
?N

THE DETAILS OF THE COMPOSITE SECTION ARE AS FOLLOWS

AREA IN mm^2  4600

CENTROID DISTANCE FROM REF AXIS  27.4

I ABOUT CENTROIDAL AXIS IN mm^4  3.56 E6

I ABOUT REF AXIS IN mm^4  7.01 E6

K ABOUT CENTROIDAL AXIS IN mm  27.8

K ABOUT REFERENCE AXIS  39
```

EXAMPLE 14.2

```
IS THE COMPOSITE SECTION SYMMETRICAL ABOUT THE REFERENCE AXIS ?
ANSWER Y OR N
?N
NOW INPUT THE THREE DETAILS OF THE AREA SEGMENT 1
?1,S,1

INPUT BASE AND HEIGHT OF RECTANGLE
?100,150
:INPUT DISTANCE FROM BASE OF RECTANGLE TO REFERENCE AXIS
?0

ANY MORE SEGMENTS ? ANSWER Y OR N
?Y
NOW INPUT THE THREE DETAILS OF THE AREA SEGMENT 2
?5,S,1

INPUT RADIUS OF SEMI-CIRCLE
?50
INPUT DISTANCE FROM CENTRE OF SEMI-CIRCLE TO REFERENCE AXIS
?150

ANY MORE SEGMENTS ? ANSWER Y OR N
?Y
NOW INPUT THE THREE DETAILS OF THE AREA SEGMENT 3
?2,V,1

INPUT THE DIAMETER OF THE CIRCLE
?50
INPUT DISTANCE FROM CENTER OF THE CIRCLE TO REFERENCE AXIS
?40

ANY MORE SEGMENTS ? ANSWER Y OR N
?N

THE DETAILS OF THE COMPOSITE SECTION ARE AS FOLLOWS

AREA IN mm^2  16963.5
```

```
CENTROID DISTANCE FROM REF AXIS  101.3

I ABOUT CENTROIDAL AXIS IN mm^4  50.7 E6

I ABOUT REF AXIS IN mm^4  224.86 E6

K ABOUT CENTROIDAL AXIS IN mm  54.7

K ABOUT REFERENCE AXIS  115.1
```

EXAMPLE 14.3 (X AXIS)

```
IS THE COMPOSITE SECTION SYMMETRICAL ABOUT THE REFERENCE AXIS ?
ANSWER Y OR N
?N
NOW INPUT THE THREE DETAILS OF THE AREA SEGMENT 1
?1,S,1

INPUT BASE AND HEIGHT OF RECTANGLE
?150,250
:INPUT DISTANCE FROM BASE OF RECTANGLE TO REFERENCE AXIS
?0

ANY MORE SEGMENTS ? ANSWER Y OR N
?Y
NOW INPUT THE THREE DETAILS OF THE AREA SEGMENT 2
?1,V,1

INPUT BASE AND HEIGHT OF RECTANGLE
?150,40
:INPUT DISTANCE FROM BASE OF RECTANGLE TO REFERENCE AXIS
?105

ANY MORE SEGMENTS ? ANSWER Y OR N
?N

THE DETAILS OF THE COMPOSITE SECTION ARE AS FOLLOWS

AREA IN mm^2  31500

CENTROID DISTANCE FROM REF AXIS  125

I ABOUT CENTROIDAL AXIS IN mm^4  194.51 E6

I ABOUT REF AXIS IN mm^4  686.7 E6

K ABOUT CENTROIDAL AXIS IN mm  78.6

K ABOUT REFERENCE AXIS  147.6
```

EXAMPLE 14.3 (Y AXIS)

```
IS THE COMPOSITE SECTION SYMMETRICAL ABOUT THE REFERENCE AXIS ?
ANSWER Y OR N
?N
NOW INPUT THE THREE DETAILS OF THE AREA SEGMENT 1
?1,S,1
```

```
INPUT BASE AND HEIGHT OF RECTANGLE
?250,150
:INPUT DISTANCE FROM BASE OF RECTANGLE TO REFERENCE AXIS
?0

ANY MORE SEGMENTS ? ANSWER Y OR N
?Y
NOW INPUT THE THREE DETAILS OF THE AREA SEGMENT 2
?1,V,1

INPUT BASE AND HEIGHT OF RECTANGLE
?40,150
:INPUT DISTANCE FROM BASE OF RECTANGLE TO REFERENCE AXIS
?0

ANY MORE SEGMENTS ? ANSWER Y OR N
?N

THE DETAILS OF THE COMPOSITE SECTION ARE AS FOLLOWS

AREA IN mm^2  31500

CENTROID DISTANCE FROM REF AXIS  75

I ABOUT CENTROIDAL AXIS IN mm^4  59.06 E6

I ABOUT REF AXIS IN mm^4  236.25 E6

K ABOUT CENTROIDAL AXIS IN mm  43.3

K ABOUT REFERENCE AXIS  86.6
```

EXAMPLE 14.4 (X AXIS)

```
IS THE COMPOSITE SECTION SYMMETRICAL ABOUT THE REFERENCE AXIS ?
ANSWER Y OR N
?Y
NOW INPUT THE THREE DETAILS OF THE AREA SEGMENT 1
?1,S,1

INPUT BASE AND HEIGHT OF RECTANGLE
?7.8,380.4
:INPUT DISTANCE FROM BASE OF RECTANGLE TO REFERENCE AXIS
?-190.2

ANY MORE SEGMENTS ? ANSWER Y OR N
?Y
NOW INPUT THE THREE DETAILS OF THE AREA SEGMENT 2
?1,S,2

INPUT BASE AND HEIGHT OF RECTANGLE
?178,12.8
:INPUT DISTANCE FROM BASE OF RECTANGLE TO REFERENCE AXIS
?190.2

ANY MORE SEGMENTS ? ANSWER Y OR N
?N

THE DETAILS OF THE COMPOSITE SECTION ARE AS FOLLOWS
```

```
AREA IN mm^2  7523.9

CENTROID DISTANCE FROM REF AXIS  0

I ABOUT CENTROIDAL AXIS IN mm^4  211.97 E6

I ABOUT REF AXIS IN mm^4  211.97 E6

K ABOUT CENTROIDAL AXIS IN mm  167.8

K ABOUT REFERENCE AXIS  167.8
```

EXAMPLE 14.4 (Y AXIS)

```
IS THE COMPOSITE SECTION SYMMETRICAL ABOUT THE REFERENCE AXIS ?
ANSWER Y OR N
?N
NOW INPUT THE THREE DETAILS OF THE AREA SEGMENT 1
?1,S,2

INPUT BASE AND HEIGHT OF RECTANGLE
?12.8,178
:INPUT DISTANCE FROM BASE OF RECTANGLE TO REFERENCE AXIS
?0

ANY MORE SEGMENTS ? ANSWER Y OR N
?Y
NOW INPUT THE THREE DETAILS OF THE AREA SEGMENT 2
?1,S,1

INPUT BASE AND HEIGHT OF RECTANGLE
?380.4,7.8
:INPUT DISTANCE FROM BASE OF RECTANGLE TO REFERENCE AXIS
?85.1

ANY MORE SEGMENTS ? ANSWER Y OR N
?N

THE DETAILS OF THE COMPOSITE SECTION ARE AS FOLLOWS

AREA IN mm^2  7523.9

CENTROID DISTANCE FROM REF AXIS  89

I ABOUT CENTROIDAL AXIS IN mm^4  12.05 E6

I ABOUT REF AXIS IN mm^4  71.64 E6

K ABOUT CENTROIDAL AXIS IN mm  40

K ABOUT REFERENCE AXIS  97.6
```

EXAMPLE 14.6

```
IS THE COMPOSITE SECTION SYMMETRICAL ABOUT THE REFERENCE AXIS ?
ANSWER Y OR N
?N
```

```
NOW INPUT THE THREE DETAILS OF THE AREA SEGMENT 1
?1,S,1

INPUT BASE AND HEIGHT OF RECTANGLE
?14,200
:INPUT DISTANCE FROM BASE OF RECTANGLE TO REFERENCE AXIS
?0

ANY MORE SEGMENTS ? ANSWER Y OR N
?Y
NOW INPUT THE THREE DETAILS OF THE AREA SEGMENT 2
?12,S,2

INPUT KNOWN AREA IN mm^2
?1120
INPUT KNOWN CENTROIDAL DISTANCE FROM REFERENCE AXIS IN mm
?185.8
INPUT KNOWN MOMENT OF INERTIA IN mm^4 *1E6 ABOUT CENTROIDAL AXIS
?.405

ANY MORE SEGMENTS ? ANSWER Y OR N
?N

THE DETAILS OF THE COMPOSITE SECTION ARE AS FOLLOWS

AREA IN mm^2  5040

CENTROID DISTANCE FROM REF AXIS  138.1

I ABOUT CENTROIDAL AXIS IN mm^4  19.3 E6

I ABOUT REF AXIS IN mm^4  115.47 E6

K ABOUT CENTROIDAL AXIS IN mm  61.9

K ABOUT REFERENCE AXIS  151.4
```

APPENDIX 3.18

```
5  REM PROGRAM BY ROGER KINSKY 3/4/85
10  PRINT "FILE NAME PRINSTR"
20  PRINT "CALCULATES COMBINED STRESS"
30  PRINT
40  PRINT "PLEASE INPUT fx,fy,fs IN THAT ORDER"
50  INPUT X,Y,S
55  IF X = Y THEN X = Y + 1E - 6
60 S1 =  SQR (S ^ 2 + ((X - Y) / 2) ^ 2)
70 F1 = (X + Y) / 2 + S1
80 F2 = (X + Y) / 2 - S1
90  IF S = 0 THEN S = 1E - 6
```

```
100 01 =  ATN ((X - Y) / (2 * S))
110 01 = 01 * 28.648
120 02 =  ATN ((S *  - 2) / (X - Y))
130 02 = 02 * 28.648
200 V = S1: GOSUB 600
205 S1 = V
210 V = F1: GOSUB 600
215 F1 = V
220 V = 01: GOSUB 600
225 01 = V
230 V = F2: GOSUB 600
235 F2 = V
240 V = 02: GOSUB 600
245 02 = V
250  PRINT
300  PRINT "MAX.SHEAR STRESS ";S1;"  AT ANGLE ";01;" DEGREES"
310  PRINT
320  PRINT "MAX.NORMAL STRESS ";F1;"  AT ANGLE ";02" DEGREES"
330  PRINT
340  PRINT "MIN.NORMAL STRESS ";F2;"  AT ANGLE ";02 + 90;" DEGREES"
400  PRINT
410  PRINT "DO YOU WISH TO KNOW THE STRESS AT ANY OTHER ANGLE, ANSWER Y OR
      N"
420  INPUT A$
430  IF A$ = "N" THEN 999
440  PRINT
450  PRINT "WHAT IS THE ANGLE OF THE PLANE IN DEG. ANTI-CLOCKWISE ?"
460  INPUT 03
470 0 = 03 / 57.296
480 F3 =  SIN (2 * 0) * (X - Y) / 2 + S *  COS (2 * 0)
490 F4 = X *  COS (0) ^ 2 + Y *  SIN (0) ^ 2 - 2 * S *  SIN (0) *  COS (0)
500 V = F3: GOSUB 600
505 F3 = V
510 V = F4: GOSUB 600
515 F4 = V
520  PRINT
530  PRINT "SHEAR STRESS ON PLANE INCLINED AT  ";03;" DEGREES IS ";F3
540  PRINT
550  PRINT "NORMAL STRESS ON THIS PLANE IS  ";F4
560  PRINT
570  PRINT
580  GOTO 400
600 V =  INT (V * 100 + .5) / 100
610  RETURN
999  END
```

<u>EXAMPLE 16.8</u>

```
]RUN

FILE NAME PRINSTR
CALCULATES COMBINED STRESS

PLEASE INPUT fx,fy,fs IN THAT ORDER
?50,100,0

MAX.SHEAR STRESS 25  AT ANGLE -45 DEGREES

MAX.NORMAL STRESS 100  AT ANGLE 0 DEGREES
```

```
MIN.NORMAL STRESS 50  AT ANGLE 90 DEGREES

DO YOU WISH TO KNOW THE STRESS AT ANY OTHER ANGLE, ANSWER Y OR N
?Y

WHAT IS THE ANGLE OF THE PLANE IN DEG. ANTI-CLOCKWISE ?
?-35

SHEAR STRESS ON PLANE INCLINED AT  -35 DEGREES IS 23.49

NORMAL STRESS ON THIS PLANE IS  66.45

DO YOU WISH TO KNOW THE STRESS AT ANY OTHER ANGLE, ANSWER Y OR N
?N
```

EXAMPLE 16.10

```
]RUN
FILE NAME PRINSTR
CALCULATES COMBINED STRESS

PLEASE INPUT fx,fy,fs IN THAT ORDER
?100,-50,40

MAX.SHEAR STRESS 85  AT ANGLE 30.96 DEGREES

MAX.NORMAL STRESS 110  AT ANGLE -14.04 DEGREES

MIN.NORMAL STRESS -60  AT ANGLE 75.96 DEGREES

DO YOU WISH TO KNOW THE STRESS AT ANY OTHER ANGLE, ANSWER Y OR N
?N
```

APPENDIX 3.19

```
5  REM  PROGRAM BY ROGER KINSKY 4/5/85
7  PRINT "FILE NAME BEAMDEF"
10  PRINT "PROGRAM FOR DEFLECTION SIMPLY SUPPORTED OR CANTILEVER BEAMS"
15  PRINT
20  PRINT "NOTE - LOADS AND DEFLNS ARE POSITIVE UPWARD"
22  PRINT
25  DIM Y(20)
30  PRINT "IS THE BEAM SIMPLY SUPPORTED OR CANTILEVER ? ANSWER S OR C"
35  INPUT E$
40  PRINT "IS THE BEAM MADE OF STEEL ? ANSWER Y OR N "
45  INPUT D$
50  IF D$ = "N" THEN 65
55 E = .2
```

```
60  GOTO 75
65  PRINT "WHAT IS THE MODULUS OF ELASTICITY IN GPa ?"
70  INPUT E
72 E = E / 1000
75  PRINT "WHAT IS THE SECOND MOMENT OF AREA IN mm^4 *1E6 ?"
80  INPUT S
86  PRINT "WHAT IS THE LENGTH OF THE BEAM IN m ?"
90  INPUT L
95 I = 0:K = 0:Z = 0
100  PRINT "HOW MANY CONCENTRATED LOADS ?"
110  INPUT K
120  IF K = 0 THEN 210
130 I = I + 1
140  PRINT "INPUT SIGN AND MAGNITUDE IN KN OF CONC.LOAD ";I
150  INPUT P(I)
160  PRINT "AND DISTANCE IN m FROM LH END OF BEAM ?"
170  INPUT L(I)
172 L1 = L(I)
175  IF E$ = "S" THEN 190
180 A(I) =  - P(I) * .5 * (L - L1) ^ 2
185 B(I) =  - P(I) / 6 * (L - L1) ^ 3 - A(I) * L
187  GOTO 200
190 R(I) =  - P(I) * (L - L1) / L
195 A(I) = ( - R(I) * L ^ 3 / 6 - P(I) * (L - L1) ^ 3 / 6) / L
200  IF I < K THEN 130
210  PRINT "HOW MANY DISTRIBUTED LOADS ?"
220  INPUT M
230  IF M = 0 THEN 360
250 I = I + 1
255  PRINT "INPUT SIGN AND MAGNITUDE IN kN/m OF DIST. LOAD ";I
260  INPUT P(I)
270  PRINT "AND DISTANCE IN m (START OF DIST. LOAD) FROM LH END ?"
280  INPUT L(I)
282 L1 = L(I)
290  PRINT "AND DISTANCE IN m (END OF DIST. LOAD) FROM LH END ?"
300  INPUT H(I)
302 L2 = H(I)
305  IF E$ = "S" THEN 320
310 A(I) =  - P(I) / 6 * (L - L1) ^ 3 + P(I) / 6 * (L - L2) ^ 3
315 B(I) =  - P(I) / 24 * (L - L1) ^ 4 + P(I) / 24 * (L - L2) ^ 4 - A(I) *
    L
317  GOTO 330
320 R(I) =  - P(I) * (L2 - L1) * (L - (L1 + L2) / 2) / L
325 A(I) = ( - R(I) * L ^ 3 / 6 - P(I) * (L - L1) ^ 4 / 24 + P(I) * (L -
    L2) ^ 4 / 24) / L
330  IF I < K + M THEN 250
360  PRINT
370  PRINT "NUMBER      POSITION m      DEFLN mm"
380 Y = 0
400 X =  - L / 10
410  FOR J = 0 TO 10
415 Y(J) = 0
420 X = X + L / 10
430 I = 0
434 I = I + 1
435 L1 = L(I):L2 = H(I)
436  IF I <  = K THEN 500
450  IF I <  = K + M THEN 550
470  PRINT J, INT (X * 100 + .5) / 100, INT (Y(J) * 100 + .5) / 100
480  NEXT J
490  GOTO 600
500  IF E$ = "S" THEN 800
```

```
510  IF X <  = L1 THEN Y = (A(I) * X + B(I)) / (E * S)
515  IF X <  = L1 THEN 525
520 Y = (P(I) / 6 * (X - L1) ^ 3 + A(I) * X + B(I)) / (E * S)
525 Y(J) = Y(J) + Y
527  IF J = 11 THEN 654
530  GOTO 434
550  IF E$ = "S" THEN 850
560  IF X <  = L1 THEN Y = (A(I) * X + B(I)) / (E * S)
565  IF X <  = L1 THEN 585
570  IF X > L2 THEN Y = (P(I) / 24 * (X - L1) ^ 4 - P(I) / 24 * (X - L2) ^
     4 + A(I) * X + B(I)) / (E * S)
575  IF X > L2 THEN 585
580 Y = (P(I) / 24 * (X - L1) ^ 4 + A(I) * X + B(I)) / (E * S)
585 Y(J) = Y(J) + Y
587  IF J = 11 THEN 654
590  GOTO 434
600  PRINT
605  PRINT "DO YOU WANT THE DEFLN AT A SELECTED POSITION ? ANSWER Y OR N"
610  INPUT F$
620  IF F$ = "N" THEN 999
625 Y(11) = 0
630  PRINT "AT WHAT POSITION IN m DO YOU WANT THE DEFLN ?"
640  INPUT X
645 J = 11
650 I = 0
654 I = I + 1
655 L1 = L(I):L2 = H(I)
656  IF I <  = K THEN 500
662  IF I <  = K + M THEN 550
670 Y(J) =  INT (Y(J) * 100 + .5) / 100
700  PRINT
710  PRINT "DEFLN AT POSITION ";X;" m  IS ";Y(J);" mm"
720  GOTO 600
800  IF X <  = L1 THEN Y = (R(I) * X ^ 3 / 6 + A(I) * X) / (E * S)
805  IF X <  = L1 THEN 815
810 Y = (R(I) / 6 * X ^ 3 + P(I) * (X - L1) ^ 3 / 6 + A(I) * X) / (E * S)
815 Y(J) = Y(J) + Y
817  IF J = 11 THEN 654
820  GOTO 434
850  IF X <  = L1 THEN Y = (R(I) * X ^ 3 / 6 + A(I) * X) / (E * S)
855  IF X <  = L1 THEN 875
860  IF X > L2 THEN Y = (R(I) * X ^ 3 / 6 + P(I) * (X - L1) ^ 4 / 24 - P(
     I) * (X - L2) ^ 4 / 24 + A(I) * X) / (E * S)
865  IF X > L2 THEN 875
870 Y = (R(I) * X ^ 3 / 6 + P(I) * (X - L1) ^ 4 / 24 + A(I) * X) / (E * S)
875 Y(J) = Y(J) + Y
877  IF J = 11 THEN 654
880  GOTO 434
999  END
```

EXAMPLE 17.2

```
]RUN

FILE NAME BEAMDEF
PROGRAM FOR DEFLECTION SIMPLY SUPPORTED OR CANTILEVER BEAMS

NOTE - LOADS AND DEFLNS ARE POSITIVE UPWARD
```

```
IS THE BEAM SIMPLY SUPPORTED OR CANTILEVER ? ANSWER S OR C
?C
IS THE BEAM MADE OF STEEL ? ANSWER Y OR N
?Y
WHAT IS THE SECOND MOMENT OF AREA IN mm^4 *1E6 ?
?4.91
WHAT IS THE LENGTH OF THE BEAM IN m ?
?3
HOW MANY CONCENTRATED LOADS ?
?2
INPUT SIGN AND MAGNITUDE IN KN OF CONC.LOAD 1
?-6
AND DISTANCE IN m FROM LH END OF BEAM ?
?0
INPUT SIGN AND MAGNITUDE IN KN OF CONC.LOAD 2
?6

AND DISTANCE IN m FROM LH END OF BEAM ?
?1
HOW MANY DISTRIBUTED LOADS ?
?0

NUMBER      POSITION m      DEFLN mm
0              0                -26.48
1              .3               -21.92
2              .6               -17.53
3              .9               -13.47
4              1.2              -9.9
5              1.5              -6.87
6              1.8              -4.4
7              2.1              -2.47
8              2.4              -1.1
9              2.7              -.27
10             3                0

DO YOU WANT THE DEFLN AT A SELECTED POSITION ? ANSWER Y OR N
?N
```

EXAMPLE 17.3

```
]RUN

FILE NAME BEAMDEF
PROGRAM FOR DEFLECTION SIMPLY SUPPORTED OR CANTILEVER BEAMS

NOTE - LOADS AND DEFLNS ARE POSITIVE UPWARD

IS THE BEAM SIMPLY SUPPORTED OR CANTILEVER ? ANSWER S OR C
?C
IS THE BEAM MADE OF STEEL ? ANSWER Y OR N
?Y
WHAT IS THE SECOND MOMENT OF AREA IN mm^4 *1E6 ?
?1.8
WHAT IS THE LENGTH OF THE BEAM IN m ?
?4
HOW MANY CONCENTRATED LOADS ?
?0
HOW MANY DISTRIBUTED LOADS ?
?1
```

```
INPUT SIGN AND MAGNITUDE IN kN/m OF DIST. LOAD 1
?-2
AND DISTANCE IN m (START OF DIST. LOAD) FROM LH END ?
?1
AND DISTANCE IN m (END OF DIST. LOAD) FROM LH END ?
?2.5

NUMBER       POSITION m       DEFLN mm
0                0               -69.92
1                .4              -61.17
2                .8              -52.42
3                1.2             -43.67
4                1.6             -34.95
5                2               -26.4
6                2.4             -18.31
7                2.8             -11.1
8                3.2             -5.29
9                3.6             -1.41
10               4               0

DO YOU WANT THE DEFLN AT A SELECTED POSITION ? ANSWER Y OR N
?N
```

EXAMPLE 17.7

```
]RUN

FILE NAME BEAMDEF
PROGRAM FOR DEFLECTION SIMPLY SUPPORTED OR CANTILEVER BEAMS

NOTE - LOADS AND DEFLNS ARE POSITIVE UPWARD


IS THE BEAM SIMPLY SUPPORTED OR CANTILEVER ? ANSWER S OR C
?S
IS THE BEAM MADE OF STEEL ? ANSWER Y OR N
?Y
WHAT IS THE SECOND MOMENT OF AREA IN mm^4 *1E6 ?
?1
WHAT IS THE LENGTH OF THE BEAM IN m ?
?2.4
HOW MANY CONCENTRATED LOADS ?
?1
INPUT SIGN AND MAGNITUDE IN KN OF CONC.LOAD 1
?-20
AND DISTANCE IN m FROM LH END OF BEAM ?
?.4
HOW MANY DISTRIBUTED LOADS ?
?1
INPUT SIGN AND MAGNITUDE IN kN/m OF DIST. LOAD 2
?-37.5
AND DISTANCE IN m (START OF DIST. LOAD) FROM LH END ?
?1.1
AND DISTANCE IN m (END OF DIST. LOAD) FROM LH END ?
?1.9
```

```
NUMBER        POSITION m        DEFLN mm
0                0                 0
1                .24               -16.32
2                .48               -30.72
3                .72               -41.8
4                .96               -48.92
5                1.2               -51.55
6                1.44              -49.25
7                1.68              -42.04
8                1.92              -30.56
9                2.16              -16.04
10               2.4               0

DO YOU WANT THE DEFLN AT A SELECTED POSITION ? ANSWER Y OR N
?Y
AT WHAT POSITION IN m DO YOU WANT THE DEFLN ?
?1.21

DEFLN AT POSITION 1.21 m  IS -51.56 mm

DO YOU WANT THE DEFLN AT A SELECTED POSITION ? ANSWER Y OR N
?Y
AT WHAT POSITION IN m DO YOU WANT THE DEFLN ?
?1.22

DEFLN AT POSITION 1.22 m  IS -51.56 mm

DO YOU WANT THE DEFLN AT A SELECTED POSITION ? ANSWER Y OR N
?Y
AT WHAT POSITION IN m DO YOU WANT THE DEFLN ?
?1.23

DEFLN AT POSITION 1.23 m  IS -51.54 mm

DO YOU WANT THE DEFLN AT A SELECTED POSITION ? ANSWER Y OR N
?N
```

APPENDIX 3.20

```
5  REM  PROGRAM BY ROGER KINSKY 18/5/85
10  PRINT "FILE NAME COLUMN"
15  PRINT "IS THE COLUMN MADE OF STRUCTURAL STEEL  YIELD 250MPa ? ANSWER
     Y OR N"
20  INPUT A$
25  IF A$ = "N" THEN 40
30 E = 200:F = 250
35  GOTO 50
40  PRINT "PLEASE INPUT PROPERTIES OF MATERIAL , E IN GPa , fy IN MPa "
45  INPUT E,F
50  PRINT "WHAT SAFETY FACTOR IS REQUIRED ?"
55  INPUT S
60  PRINT "WHAT IS THE LENGTH OF THE COLUMN IN mm ?"
65  INPUT L
70  PRINT "CHOOSE FROM THE FOLLOWING END RESTRAINTS"
75  PRINT "1 PINNED, 2 CANTILEVER, 3 BOTH ENDS FIXED, 4 ONE END PINNED
     OTHER END FIXED 5 OTHER"
80  INPUT R
85  IF R = 2 THEN L = 2 * L
```

```
90  IF R = 3 THEN L = .7 * L
95  IF R = 4 THEN L = .85 * L
100  IF R <  = 4 THEN 120
105  PRINT "WHAT IS THE EFFECTIVE LENGTH TO ACTUAL LENGTH RATIO ?"
110  INPUT T
115 L = L * T
120 L2 =  SQR (19739 * E / F)
125  PRINT "ARE THE DIMENSIONS OF THE SECTION KNOWN ? ANSWER Y OR N "
130  INPUT B$
135  IF B$ = "N" THEN 400
137  GOTO 340
140  PRINT "NOW INPUT  K (mm), A (mm^2)
145  INPUT K,A
150 L1 = L / K
155  IF L1 < L2 THEN  GOSUB 200
160  IF L1 >  = L2 THEN  GOSUB 300
170  PRINT
172 P1 = P / S:P2 =  INT (P1 * 100 + .5) / 100
175  PRINT "SAFE LOAD ON THE COLUMN IN kN IS  ";P2
177  PRINT
180  PRINT "SLENDERNESS RATIO IS "; INT (L1 * 100 + .5) / 100
181  PRINT
182 P3 = P1 * 1000 / A:P3 =  INT (P3 * 100 + .5) / 100
183  PRINT "THE COMPRESSIVE STRESS IN MPa AT MAX SAFE LOAD IS ";P3
184  PRINT
185  PRINT "DO YOU WANT TO TRY ANOTHER SECTION ? ANSWER Y OR N"
190  INPUT C$
195  IF C$ = "Y" THEN 340
197  GOTO 999
200  REM JOHNSON FORMULA
210 P = .001 * A * F * (1 - F * L1 ^ 2 / (39478.4 * E))
230  RETURN
300  REM EULER FORMULA
310 P = 9.87 * E * A / (L1 ^ 2)
330  RETURN
340  PRINT "CHOOSE FROM THE FOLLOWING SECTIONS "
342  PRINT "1 CIRCLE , 2 RECTANGLE , 3 TUBE , 4 OTHER"
344  INPUT J
346  IF J = 1 THEN 700
348  IF J = 2 THEN 725
350  IF J = 3 THEN 750
352  IF J = 4 THEN 140
400  PRINT "WHAT IS THE LOAD ON THE COLUMN IN kN ?"
410  INPUT P1
420 P1 = P1 * S
430  PRINT "IS THE COLUMN CIRCULAR OR RECTANGULAR ? ANSWER C OR R"
440  INPUT D$
460  IF D$ = "R" THEN 600
480 L1 = 200
490 K = L / L1
500 D = 4 * K
510 A = .7854 * D ^ 2
520  IF L1 < L2 THEN  GOSUB 200
530  IF L1 >  = L2 THEN  GOSUB 300
540  GOSUB 800
550  GOTO 490
600  PRINT "INPUT HEIGHT/BREADTH RATIO REQUIRED"
610  INPUT R
620 L1 = 200
630 K = L / L1
640 H = 3.4641 * K
650 A = H ^ 2 / R
```

```
660  IF L1 < L2 THEN  GOSUB 200
670  IF L1 >  = L2 THEN  GOSUB 300
680  GOSUB 800
690  GOTO 630
692  PRINT
693 B = H / R
694  PRINT "BREADTH OF SECTION IN mm IS "; INT (B * 100 + .5) / 100
696  PRINT "HEIGHT OF SECTION IN mm IS "; INT (H * 100 + .5) / 100
698  GOTO 999
700  PRINT "INPUT DIAMETER OF CIRCULAR SECTION IN mm "
705  INPUT D
710 A = .7854 * D ^ 2
715 K = D / 4
720  GOTO 150
725  PRINT "INPUT HEIGHT AND BREADTH OF RECTANGULAR SECTION IN mm"
730  INPUT H,B
735 K = H / 3.4641
740 A = B * H
745  GOTO 150
750  PRINT "INPUT OD AND ID OF THE TUBE IN mm"
755  INPUT D2,D1
760 K = .25 *  SQR (D2 ^ 2 + D1 ^ 2)
765 A = .7854 * (D2 ^ 2 - D1 ^ 2)
770  GOTO 150
800 Q = (P1 - P) / P1
810  IF Q <  = .01 THEN 900
820  IF  SGN (Q) = 1 THEN 850
830 L1 = L1 + 1
840  RETURN
850 L1 = L1 - 1
860  RETURN
900  IF D$ = "R" THEN 692
910  PRINT
915  PRINT "DIAMETER OF CIRCULAR COLUMN IN mm IS "; INT (D * 100 + .5) /
     100
920  PRINT
925  PRINT "SLENDERNESS RATIO IS  ";L1
999  END
```

EXAMPLE 18.5

```
]RUN

FILE NAME COLUMN
IS THE COLUMN MADE OF STRUCTURAL STEEL  YIELD 250MPa ? ANSWER Y OR N
?N
PLEASE INPUT PROPERTIES OF MATERIAL , E IN GPa , fy IN MPa
?200,260
WHAT SAFETY FACTOR IS REQUIRED ?
?1
WHAT IS THE LENGTH OF THE COLUMN IN mm ?
?1800
CHOOSE FROM THE FOLLOWING END RESTRAINTS
1 PINNED, 2 CANTILEVER, 3 BOTH ENDS FIXED, 4 ONE END PINNED OTHER END FIXED
5 OTHER
?3
ARE THE DIMENSIONS OF THE SECTION KNOWN ? ANSWER Y OR N
?Y
```

```
CHOOSE FROM THE FOLLOWING SECTIONS
1 CIRCLE , 2 RECTANGLE , 3 TUBE , 4 OTHER
?1
INPUT DIAMETER OF CIRCULAR SECTION IN mm
?80

SAFE LOAD ON THE COLUMN IN kN IS  1136.1

SLENDERNESS RATIO IS 63

THE COMPRESSIVE STRESS IN MPa AT MAX SAFE LOAD IS 226.02

DO YOU WANT TO TRY ANOTHER SECTION ? ANSWER Y OR N
?N
```

EXAMPLE 18.9

```
]RUN

FILE NAME COLUMN
IS THE COLUMN MADE OF STRUCTURAL STEEL  YIELD 250MPa ? ANSWER Y OR N
?N
PLEASE INPUT PROPERTIES OF MATERIAL , E IN GPa , fy IN MPa
?110,70
WHAT SAFETY FACTOR IS REQUIRED ?
?4
WHAT IS THE LENGTH OF THE COLUMN IN mm ?
?500
CHOOSE FROM THE FOLLOWING END RESTRAINTS
1 PINNED, 2 CANTILEVER, 3 BOTH ENDS FIXED, 4 ONE END PINNED OTHER END FIXED
5 OTHER
?1
ARE THE DIMENSIONS OF THE SECTION KNOWN ? ANSWER Y OR N
?N
WHAT IS THE LOAD ON THE COLUMN IN kN ?
?1.2
IS THE COLUMN CIRCULAR OR RECTANGULAR ? ANSWER C OR R
?C

DIAMETER OF CIRCULAR COLUMN IN mm IS 12.35

SLENDERNESS RATIO IS  162
```

Index

F

G

H

I

J

K

L

M

N

P

R

S

T

U

V

W

Y